Approaches and Trends in Plant Disease Management

Dr. S.K. Gupta
Dr. Monica Sharma

Published by:

Scientific Publishers (India)
5-A, New Pali Road, P.O. Box 91,
Jodhpur – 342 001 (India)

E-mail: info@scientificpub.com
Website: www.scientificpubonline.com

Print: 2021

ISBN: 978-81-7233-887-9

Printed in India

Dr. Y.S. Parmar University of Horticulture and Forestry
Solan, (H.P.) 173 230

V.S. Thakur
Vice Chancellor

FOREWORD

There have been significant development in agricultural research and trade in the recent past focussing more on bio-security, crop health and food safety. These developments have had a dramatic impact on the roles and responsibilities of plant pathologists. Developing effective and viable plant disease management strategies has been the real challenge over the recent years. Conventional plant disease control based on chemical pesticides alone may not be acceptable in the present day WTO regime where protection of the environment and human health is given a serious consideration. New tools and techniques are being developed to improve crop productivity and health. Crop intensification and changing climate have already led to the increased frequency and magnitude of disease outbreaks and losses. Changing climate has also necessitated to look into the new management options. Disease cause significant yield reductions (20%) in principle food and cash crops. To tackle the emerging disease problems, there is a need to blend the new discoveries in genomics, information technologies and molecular science and other age old disease management technologies to keep the ravages by devastating diseases below economic threshold level.

I am happy to know that a book entitled ***"Approaches and Trends in Plant Disease Management"*** is being published by the scientists of Department of Plant Pathology and serious efforts have been made to compile available information on the latest developments in disease management approaches. Several option such as host resistance, cultural practices, biological agents, biotechnological and chemical options have been discussed in detail to develop better, sustainable and environmentally safe disease management strategies. I am sure that this book will not only serve the needs of researchers, students and extension workers but will also be helpful to the elite growers. I appreciate the efforts made by the authors in publishing this book.

8/8/13

(V.S. Thakur)

PREFACE

Ever-increasing population warrants for sustainable production of food materials. The predicted growth of world's population from 7 billion at present (1.2 billion in India alone) to 8.3 billion by 2030 presents a major global challenge to meet necessary food demand. At the same time, available land for agriculture is decreasing day by day due to increasing industrialization and urbanization. India has produced 255MT of food grains during 2012-13 and with the increase in rate of population growth; it would require about 400 MT by 2050. Diseases have become serious constraints in the cultivation of agricultural, horticultural, medicinal and fiber crops. Plant diseases are estimated to cause yield reductions of almost 20 per cent in the principal food and cash crops worldwide.. Besides fungi, many bacterial and viral and nematode diseases are also affecting crop plants which are difficult to manage. Knowledge of various factors leading to epidemic build-up of the diseases is necessary for timely application of various management methods. Earlier emphasis was on the chemicals to combat the ravages of various diseases but due to some ill effects like residues in food grains and pollution hazards, these are now taking back seat. Moreover, without proper knowledge of diseases, farmers are indiscriminately using agrochemicals to control insect-pests and diseases, which may lead to the development of resistant strains of the pathogen. In the WTO regime, for the export of food material to other countries, care has to be taken to keep the pesticide residues below the permissible limits. Since long, several approaches like host resistance, cultural, biological, chemical methods and regulatory measures have been developed for their control. Depending on the crop, the disease and availability of control methods, a different set of approaches is employed.

During the past decade, a lot of information has been generated by plant pathologists on different aspects of plant diseases and their management options. The present book on "***Approaches and Trends in Plant Disease Management***" offers a wide choice of topics for the readers to look into recent developments in diverse aspects of plant disease management. There are 20 articles covering disease management approaches in molecular plant pathology, biological control, cultural control, disease resistance, induced resistance and recent development in fungicides and their rational use and integrated disease management. Articles on protected cultivation, nematode problems and their management and climate variables and their impact on plant diseases: retrospect and prospect have also been included. The authors have done a commendable job by providing the latest information available in these nicely written articles. We are grateful to all the authors for their vivid contributions.

The help rendered by Dr. T.S. Thind, former Additional Director of Research and Head Department of Plant Pathology, Punjab Agricultural University, Ludhiana, in finalizing the chapters is gratefully acknowledged. We take this opportunity to thank Dr. V.S. Thakur, Hon'ble Vice Chancellor, Dr. R.C. Sharma, Director of Research, and Dr. B.C. Suman, Professor and Head, Department of Plant Pathology,

Dr. Y.S. Parmar University of Horticulture and Forestry, Nauni, Solan for encouraging and providing necessary facilities in preparing this book. We are also thankful to Dr. Y.S. Parmar University of Horticulture and Forestry, Nauni, Solan for the permission to write this book. We are indebted to our family members for allowing us to spare time for this arduous task.

S.K. Gupta
Monica Sharma

CONTENTS

1

Role of Cultural Practices in Plant Disease Management

S. K. Gupta and Sapna Devi

Department of Plant Pathology, Dr YS Parmar University of Horticulture and Forestry, Nauni, Solan - 173 230 (HP)

The burgeoning human population exerts pressure for more food production, which necessitates that crop losses due to insects, pests, diseases and weeds should be minimized. Varshney (2009) reported up to 29 per cent yield losses due to plant diseases. The disease management practices must have only limited detrimental impact on the environment. The use of chemicals in the management of diseases have residual effect and causes environmental pollution. Now, it is a time to encourage and use environmentally safe technologies for the management of diseases that includes use of cultural practices, which have immense value in the present time due to the pollution and health hazards resulting from the deposits of chemical sprays. The plant pathogens must be suppressed with some physical manipulation of plant, soil and environment. In the ancient civilization of India, there was a recommendation to expose plant material to bright sunlight for curing-apparently by solarization.

The rise of modern plant pathology in the 19th century provided the scientific basis for the systemic development of cultural practices for disease management and the elucidation of the mechanism involved. However, the concomitant in last few decades development of powerful and effective chemicals for disease management e.g. fumigants, resulted in declining the interest in the use of many time tested cultural practices for disease management. Moreover, the increased intensification of many crops (which also led to monoculture) made the use of cultural practices often quite difficult. Cultural practices are classical, non-chemical method(s) of plant disease control and have the potential to minimize or replace pesticides to some extent.

"Cultural control" implies control through modification of cultural practices. R.B. Stevens (Horsfall and Dimond 1960) defined "it includes measures like agricultural cropping practices, harvesting and storage methods, tillage, crop rotation, soil management, resistant varieties, land use planning and all like in nature". Management of soil borne diseases (soil being a major source of inoculum) should be implemented before, at and after planting (Katan 2000). Soil borne diseases are especially severe in intensive cropping systems due to frequent cultivation of susceptible crops. Therefore, sanitation of the soil from surviving inoculum between crops is essential. Cultural practices can play a pivotal role in the management of soil borne diseases, particularly because of the recent ban on the use of methyl bromide. Cultural practices can be used alone or as one of the component of pest-management programs because these practices if selected properly and followed, changes the microclimate in favour of host which not only reduce the disease but also enhance the efficacy of other control measures applied.

Types of cultural practices

Cultural practices can be categorized in to three types:

1. Used for regular agricultural purposes, which can also be effective for disease control, e.g., minerals, fertilization and irrigation.
2. Used solely or mainly for pest control, e.g. sanitation, solarization and flooding
3. Used for both agricultural purposes and pest control, e.g. crop rotation, grafting and composting.

Principles

1. Any potential control method may be considered, providing that method is environmentally, technologically and economically feasible.
2. Combining with other chemical or non-chemical methods, in minimizing pesticide use, should be the goal.

Priority should be given to diseases that are difficult to control, or those which involve harmful pesticides, e.g. methyl bromide. Economic, social and legal considerations should be adopted.

Cultural practices for disease management

The main cultural practices are crop rotation, flooding, soil solarization, adjustment in date of sowing, fertilization, composting, sanitation, etc. Physical methods like heating the soil or propagation material and UV - irradiation of contaminated water should also be considered when applicable. The emphasis should be placed on practices, which can be easily adopted for the management of diseases in different crops. Cultural practices viz., planting disease-free seeds, mulching, adjustment in sowing date and crop rotation has been found effective to keep the root rot and web blight of French bean under check (Sartorato et al 1988). Practices such as selection of healthy seed (Leonian 1922), rouging of diseased plants/parts (Beattie et al 1944), mulching with wheat straw (Osnitzakaya 1949), transplanting of plants on the top of furrow increased to 30 cm (Fereyra et al 1984), soil solarization with transparent polyethylene sheet along with the application of fumigants (Cartia et al 1989), amendment of soil with high C/N ratio material (Kim et al 1990), and cultivation of *Capsicum* with non-host plants like groundnut, sesame, onion, garlic, pea and spinach (Lee et al 1993) have been reported effective in reducing the leaf blight and fruit rot (*Phytophthora nicotianae* var. *nicotianae*) of bell pepper. The general cultural practices like use of disease-free seeds, three- four years crop rotation, destruction of diseased plant material, deep ploughing, wider spacing and removal of infected leaves was found effective for the control of angular leaf spot of French bean (Navarro et al 1981, Trutman and Kayitare 1991). Cultural practices reduce both initial inoculum (X_0) as well as infection rate (r). Different practices are described below:

1. Practices adopted to reduce the initial inoculum (X_0)

I. Crop rotation	VI. Soil solarization
II. Soil amendment	VII. Flooding
III. Compost	VIII. Fire and flaming
IV. Tillage	IX. Harrowing and mowing
V. Sanitation	X. Rougueing

2. Practices adopted to reduce the infection rate (r)

I. Mulching	V. Irrigation and water management
II. Multiple cropping	VI. Grafting
III. Cover crops	VII. Herbicides and Weed control
IV. Sowing and Spacing of crops	VIII. Mineral Nutrition

1. Practices to reduce the initial inoculum

I. Crop rotation

The principle underlying disease control by rotation has long been considered as starvation of the pathogen or its surviving propagules through the absence of suitable host. Soil borne pathogens that infect one or few plants of particular species can be possibly reduced by planting, for 3- 4 years, crops belonging to species or families not attacked by that particular pathogen. Crop rotation deals with plant health in general rather than with controlling a specific disease, although a certain crop sequence might be more effective in reducing the incidence of a specific disease than others. It can also be regarded as insurance against the rise of unknown pathogens (Curl 1963, Bruehl 1987). In long run, continuous or frequent cropping of the same crop in the same soil, and consequently the frequent incorporation of its tissues in the soil, results in a shift and a build- up of harmful biological and chemical and physical agents in the soil (Katan 1984). This can be due to exhaustion of a particular nutrient and its excessive use by same type of crop; accumulation of organic acids and toxic substances released by the crop and easy survival of the pathogen due to the regular presence of susceptible host (Singh 1983). It was suggested that disease might be controlled by crop rotation if clean cultivation is practised and all the susceptible carriers are destroyed. A group of fungi that cause mucilage disease in edible fungi can be effectively managed by following clean cultivation along with crop rotation.

The spread of the disease, the period of development and conditions which favour its development, were considered while discussing the merits of crop rotation. It should be kept in mind that complete eradication of a parasite is not attainable by this method nor it is expected or necessary. Even where careful rotation is practiced, weed host are nearly always present or the parasite may survive for longer periods in subsoil below cultivation depth where microbial activity and competition are reduced.

Crop rotation will continue to be used as a supplementary control measure along with proper seed selection and seedbed sanitation. Reduction in inoculum of *Plasmodiophora brassicae,* the cabbage club root fungus, requires a host-free period exceeding four years. When resistant varieties of peas have not been developed, crop rotation has been used with success to obtain profitable yields in spite of fungi that cause pea root rots. The adverse effect of continuous cultivation of same crop in same piece of land can be minimized by following sound crop rotation e.g. the pathogen having limited survival ability in soil, restricted only to their host residues are starved out due to the continuous absence of their living host or their inoculum level is reduced drastically. As number of disease-producing organisms live over in the soil on plant residues left in the field after harvest, it is advisable to rotate with non-susceptible crops to reduce inoculum.

Tomatoes should not be planted immediately after beans, beets, cabbage, celery, lettuce, peppers, potatoes, etc. where *Sclerotium rolfsii* is the major soil borne pathogen (Doocitile et al 1961). Crop rotation, especially as a tool for disease control, has been thoroughly discussed in many reviews (Bruehl 1987, Curl 1963, Garret 1970, Palti 1981, Schippers et al 1987). Crop rotation of cauliflower with rice proved highly effective in reducing *Sclerotinia sclerotiorum* infection up to 60 per cent and in increasing seed yield by more than 161 per cent (Sharma et al 1983, Gupta et al 1987). Club root is caused by *Plasmodiophora brassicae* and affects members of the *Brassicaceae* family such as cauliflowers and turnips, ornamentals and weeds. *P. brassicae*, once introduced in soil remains there indefinitely for decades. *P. brassicae* is encouraged by wet, poorly drained, acidic soils and liming is a key control measure. Boron and nitrate applications promote populations of soil bacteria antagonistic to *P. brassicae*. Rotational cropping, autumnal digging, winter fallowing, early control of *Brassicaceae* weeds, all contribute towards club root control (Dixon et al 2003). Crop rotation of cauliflower with summer savoy cabbage, peppermint and thyme for 2-3 consecutive years resulted in significant decrease in club root incidence and complete disease control was achieved when the forecrops were grown for 3-6 weeks in spring (Robak 1994). Wilt severity in tomato was significantly ($P<0.05$) reduced by rotation with maize, okra, cowpea or resistant tomato lines (CL1131-0-0-43-4 (CL1131), compared with continuous tomato (Adhikari and Basnyat 1998).

Different crops in rotation have many physical, chemical and biotic effects on the soil environment. Some of these may be unfavourable for the survival and growth of pathogens. Antagonistic microorganisms may play a significant role in changed environment. The use of rotation of broccoli and lettuce to control infection in lettuce caused by *Sclerotinia minor* is also practised. Continuous cultivation of lettuce crop in the same field increased sclerotial population in soil and incidence of lettuce drops. Although in low inoculum density fields with random inoculum distribution, continuous planting of lettuce did not result in higher disease incidence for at least 3 years. In high inoculum density fields with both uniform and random distribution of sclerotia, rotations of lettuce with broccoli resulted in significant reductions in sclerotia and lettuce drop incidence. Broccoli residue reduced sclerotia in soil by reducing the viability of the sclerotia soon after incorporation into the soil, and by reducing incidence of lettuce drop and the consequent number of sclerotia added to soil after this crop. Alternate crops of broccoli and lettuce provided greater reductions in the soil borne sclerotia than consecutive crops of broccoli and lettuce. The use of crop rotation with broccoli and irrigation management has also been found effective for control of microsclerotia of *Verticillium dahliae* in soil and wilt infection in cauliflower (Xiao et al 1998).

Crop residues used in rotation have different chemicals, which are released on decomposition that helps in modifying the soil environment and reduce the available inoculum. Effect of crop rotation on the incidence of bacterial wilt (*Ralstonia solanacearum*) and potato tuber yield was studied (Lemaga et al 2001). Pulses, cereals, root crops and vegetables commonly grown in the area were included in the different sets of rotation. The highest wilt incidence under rotation was below 13 per cent as compared to 62 per cent in the control. The highest wilt reduction was obtained with finger millet and sweet potatoes, while the least reduction was recorded in carrots and onions. The highest wilt reduction was obtained with potato-beans-maize-potato treatment, while the least wilt reduction was observed in the potato-maize-maize-potato treatment. Sweet potato was found to be a good rotation crop for reducing wilt, while beans can be a good rotation crop if used before or after cereals in a potato rotation scheme. Control of cavity spot (*Pythium sulcatum*) disease of carrots by rotation with a non-host, broccoli, was attempted on a badly infested site (Davison and McKay 2003). The incidence and severity of seedling infection by *P. sulcatum* was significantly reduced when carrots followed one, two or three broccoli crops. There was a decrease in the incidence and severity of cavity spot in two of the three plantings where carrots followed broccoli. Oospores of *P. sulcatum* are able to survive for at least 21 months in the absence of a host.

The host range of *P. sulcatum*, the cause of many diseases in most carrot-growing areas was determined by growing seedlings in infested soil in the field (Davison and McKay 2003). One-year crop rotation of cabbage with carrot and potato has been found effective to control club root disease (*Plasmodiophora brassicae*) in cabbage (Yamada et al 2003). The rotations comprised: continuous cropping (C); cabbage-carrot-potato rotation (R); cabbage-potato-carrot rotation (R_2); R rotation inserted with fallow cropping every two seasons (RF); R rotation but inserted with maize after every vegetable (RC); R_2 rotation inserted with maize after every vegetable (RC_2); R rotation with mixed vegetables (RM); and cabbage plants were not cultivated for more than two years (new). The C plot yields showed large fluctuations, however, from the second to the sixth season, all the values were lower than those from the R plots, indicating considerable yield reduction will occur when cabbage is continuously grown. Many cabbage plants in the C plots died from club root disease during the third season, and thereafter, severe club root damage was observed in C plots. The effectiveness of the crop rotations was confirmed in the fourth season. The R plots showed yields ranging from 7.07 to 7.39 kg/m^2 compared to 2.57 kg/m^2 in the C plots indicating the importance of crop rotation but no importance of crop sequence. In the second season, early growth stage and the yield in plots kept in fallow during the last season were significantly better than those of the C plots. In the sixth season, these were again compared and club root damage scores were 7 and 93, respectively, suggesting the importance of removing the host plants for brief period of time under moderate temperature.

In fields where proper rotation is followed, better growth of plants grown in rotation helps in the disease avoidance. The crops grown in rotation get proper nutrient and their root system is not liable to adverse effects of accumulated toxic substances. Crop rotation of sweet potato, carrot, beet root and cabbage with oats and macuna for two to three years, not only reduces the incidence of diseases but also increases the yield and reduced the weed growth. The level of phosphorus (P) in the soil significantly increased after the third year, whereas the soil organic matter gradually increased under all rotation systems (Silva and Althoff 2003).

II. Soil Amendments

The term 'Soil amendment' refers to any material mixed into the soil. This is defined as "any material added to a soil to improve its physical properties, such as water retention, permeability, water infiltration, drainage, aeration and structure". Amending the soil with a mixture of organic soil amendments is a promising, environmentally sound method for the management of soil borne diseases (Palti 1981, Gupta et al 1986). Organic amendments hold potential to suppress plant diseases, which may be due to changes in soil biology and other soil properties. The suppression of root rot pathogens (*Phytophthora* spp., *Pythium* spp., *Fusarium* spp., and *Rhizoctonia solani*) has been well documented (Lewis and Papavizas 1971, Singh 1983).

In world, a considerable proportion of agricultural production is lost every year due to diseases and insect pests. In India, the total losses due to insect – pest, diseases and weeds is approximately - 18 per cent of our total production. Synthetic pesticides are widely accepted as a solution to this but has adverse effects on environment. Besides these are expensive and become ineffective as pathogen and pests develop resistance. Numerous species of plants are known to have fungicidal and insecticidal properties. Natural chemicals of about 2000 species of plants play an important role in the pest management. The management of disease should be based on a sound understanding of the biology of the pathogen, including its modes of survival and dissemination, host range and the role of environmental factors in disease (Guest 2002).

One of the cheapest and effective methods of alteration of soil –environment is amendment of soil with decomposable organic matter. Organic amendment of soil has been considered as one of the methods of biological control of plant diseases. Organic amendments have potential to suppress plant diseases, which may be due to changes in soil biology and other soil properties.

Types of soil amendments

There are two types of soil amendments:

1. **Organic amendment :** Organic amendment may be defined as "material which comes from something that is or was alive". It includes sphagnum peat, wood chips, grass clippings, straw, compost, manure, bio-solids, sawdust and wood ash.
2. **Inorganic amendment:** Inorganic amendments are either mined or manmade. Inorganic amendments include vermiculite, perlite, tire chunks, pea gravel and sand.

Soil amendments are further classified into following types

1. Plant and animal byproducts: Plant byproducts are excellent fertilizer as these after decomposition modify the soil environment and improve the nutrient status of the soil (Table 1). Similarly, animal by-products are also rich in nutrients like N, P, and K (Table 1).
2. Manure and compost based products: Among all above mentioned soil amendments most commonly used amendments include compost, manure, dry or green crop residue (including green manure crops), oilcakes, meals, sawdust, wood shavings, straw and peat moss.
3. Seaweed products: Soil applications of seaweed extract (*Ascophyllum nodusum*) @2.5 g/m^2 increased the shoot length, number of shoots per plant, bulb weight, yield and reduced the downy mildew disease incidence in onion by 77.6 per cent (Dogra and Mandradia 2009).

Table 1. Per cent nitrogen, phosphorus and potash present in different plant and animal by-products

Product(s)	Nitrogen (%)	Potash (%)	Phosphorus (%)
Plant byproducts			
Alfa alfa meal	3%	-	-
Cottonseed meal	7%	-	-
Fruit pomaces	-	-	-
Leaf compost	-	-	-
Soybean meal	7%	-	-
Wood ash		6%	2%
Animal byproducts			
Blood meal	7-10%	-	-
Feather meal	10%	-	6%
Fish meal	-	-	-
Fish emulsions	10%	-	-
Leather meal	-	-	-

Factors to consider while selecting an amendment

There are at least four factors to be considered while choosing a soil amendment

(1) Longevity of the amendment

(2) Soil texture

(3) Soil salinity and plant sensitivities to salts, and

(4) Salt content and pH of the amendment

(1) Longevity of the amendment

The choice of the amendment depends upon the goals:

- If you want to improve soil physical properties quickly, then choose an amendment that decompose rapidly.
- Choose an amendment that decompose slowly for a long lasting improvement to soil.
- For a quick improvement that lasts a long time –choose a combination of amendments.

Different amended materials have different decomposition rates. Some decomposes at very fast rate (grass, manures), at slow rate (wood chips, hardwood bark, peat), whereas compost decomposes at moderate rate (Table 2).

Table 2. Decomposition rate of various amendments

Amendment	Decomposition rate
Grass clippings, manures	Rapid decomposition
Compost	Moderate decomposition
Wood chips, hardwood bark, peat	Slow decomposition

(2) Soil texture

Soil texture reflects the size of the soil particles. Sandy soils have large soil particles and feel gritty while, clay soils have small particles and feel sticky. When amending sandy soils, the goal is to increase the soil ability to hold moisture and store nutrients. To achieve this, organic amendments can be successfully used that are well decomposed, like composts and aged manure. With clay soils, the goal is to improve soil aggregation, increase porosity and permeability and improve aeration and drainage, fibrous amendments like peat, wood chips, tree bark and straw are most effective. Bettiol et al (1997) gave the effectiveness of several media (soil-sand-perlite, soil-sand-perlite-manure, soil-sand-perlite-compost, soil-sand-perlite-peat, soil-sand-perlite-wheat straw and sand-perlite-peat-compost-manure-wheat straw) as alternatives for chemical control of soil infection by *Pythium ultimum* on cucumber.

Soil amendments may be fibrous, humus and inorganic type (Table 3). Peat is a type of fibrous amendment and its water retention is very high, and permeability ranged from low to medium (Table 3). Vermiculite has very high permeability as well as water retention.

Table 3. Permeability and water retention of various soil amendments

	Amendment(s)	Permeability	Water retention
Fibrous	Peat	Low- medium	Very high
	Wood chips	High	Low -medium
	Hard wood bark	High	Low-medium
Humus	Compost	Low- medium	Medium-high
	Aged manure	Low-medium	Medium
Inorganic	Vermiculite	High	High
	Perlite	High	Low

(3) Soil salinity and plant sensitivity to salts

Avoid those amendments in soils that are already high in salts particularly when growing plants are also sensitive to salts. Bean, carrot, onion, is salt sensitive plants, choose sphagnum peat or ground leaves instead of compost or manures.

(4) Salt content and pH

Choose a low salt amendment for soils testing high in salts. Avoid amendments that are high in salts or that have a high pH. Amendments high in salts and /or pH are wood ash, composted manure and mountain peat.

These amendments are allowed to decompose in the field itself where the pathogen is supposed to be present. The decomposition of the organic mater helps in alternation of the physical, chemical and biotic conditions of the soil. The altered conditions reduce the inoculum potential of the pathogen. In addition, it improves soil structure which promotes root growth of the host. Various biochemical's released during decomposition of lignin containing materials includes antibiotics and phenols which in turn induce disease resistance on the root surface as well as in the tissue when absorbed.

The principle underlying organic amendment is that decomposable organic matter should be allowed to decompose in the field in such a way and for such a period that the process of decomposition and its associated activities suppress or destroy the pathogen, there should be no interference with normal cultural practices and there should be no harmful effects on the plants raised in amended field. Simultaneously the disease proneness of the plants is also reduced.

Mechanism of action

The mechanism of action of organic amendment in disease management is not yet fully understood due to complex nature of soil environments. It appears that disease control in amended soil is the result of not one specific mechanism but different mechanisms operate at different stages during decomposition of organic matter and under different soil environmental conditions.

There are following three pathways of action of organic amendments in controlling soil borne pathogens e.g., control of root knot of tomato, okra and black scurf of potato:-

(i) Disease avoidance through modification of physico-chemical environment of soil in favour of the host

(ii) Eradication of the pathogen and control of disease through effects of the decomposition products and microbial metabolites on the pathogen and the host.

(iii) Eradication of the pathogen through direct antagonism

(i) Disease avoidance

The physico-chemical changes in soil affect the host on one hand and affect the pathogens on the other hand. The effect of decomposition of organic matter on soil microorganisms determines the speed of decomposition and the nature of antagonistic microorganisms. The beneficial effects on the host enable it to absorb more nutrients. The water holding capacity of the soil is improved and root growth is promoted. These conditions make the plant vigorous and develop the capacity to avoid/ tolerate the disease; ultimately better yields mark the decrease in effects of disease.

(ii) Pathogen eradication

Some decomposition products or metabolites are directly toxic to pathogens. The organic sulphur compounds released during decomposition of leaves of cruciferous plants are directly toxic to many fungi and nematodes. Certain substances such as aldehydes are stimulatory for pathogens by stimulating germination of resting fungal structures or hatching of the eggs of nematodes. In the absence of the host, the germ tubes from fungal spores or sclerotia and larvae from eggs of nematodes are destroyed by strong antagonistic action (lysis, parasitism, predation etc.) of the enhanced microbial activity taking place during decomposition of organic matter.

Some substances can be absorbed by roots and act as precursors for synthesis of phenol like compounds in the tissues which induced biochemical resistance in the plant. By affecting the host physiology these substances can bring about such changes in the root exudates and rhizosphere effect which may be harmful for the pathogens.

It is well established fact that antibiotics are produced in soil on the surface of decomposing organic matter because maximum microbial activity occurs on or in close vicinity of particles of organic matter in soil. There is evidence in literature to suggest that organic amendment stimulate the activity of these organisms i.e. *Trichoderma* spp., *Penicillium* spp., *Streptomyces* spp., *Bacillus subtilis* etc.

During decomposition of organic matter, number and activity of microorganisms increased which puts heavy demand on limited supply of oxygen, nutrients, space etc. The pathogens with weak competitive saprophytic survival ability are unable to compete with other soil saprophytes for these requirements and either perish or become dormant by forming special structures.

iii) Pathogen eradication through direct antagonism

In a detailed investigation of the mechanism of disease suppression, a very close correlation was found between disease suppression and the activity of antagonistic endospore - forming bacteria and actinomycetes (Aryantha et al 2000) based on the hypothesis that composted manures (chicken, cow, sheep, horse) improve soil drainage and nutrition and stimulated soil biological activity in general (thus the saprophytic competition faced by the pathogen) but the suppression of *Phytophthora* is due to direct antagonism that is specifically stimulated by chicken manure compost. Thus composts and mulches work in a number of ways, by stimulating root growth and providing a source of plant nutrients, by promoting disease suppressive soils and suppressing competition from weeds (Guest 2002).

Soil amendments leading to reduced inoculum density and causing disease suppression have been found effective in various plant diseases as reviewed by Singh (1983). radish, watermelon, mustard, cabbage (Sun and Huang, 1985) and of *Plasmodiophora brasssicae* on Chinese cabbage, of *Phytophthora melonis* on cucumber, of *Pseudomonas solanocrarum* on tomato, of *Rhizoctonia solani* on rice and beans , of *Sclerotium rolfsii* on pepper and *Sclerotium rolfsii* causing foot rot of wheat.

For understanding the modes of action, amendments can be formulated to improve efficacy and reduce viability of inoculum. This includes basal application rates upon soil properties, manipulating amendment or soil pH, altering the ratio of nitrification and appropriate chemistries, improving storage conditions, altering their microbiology and improving the timing and mode of application.

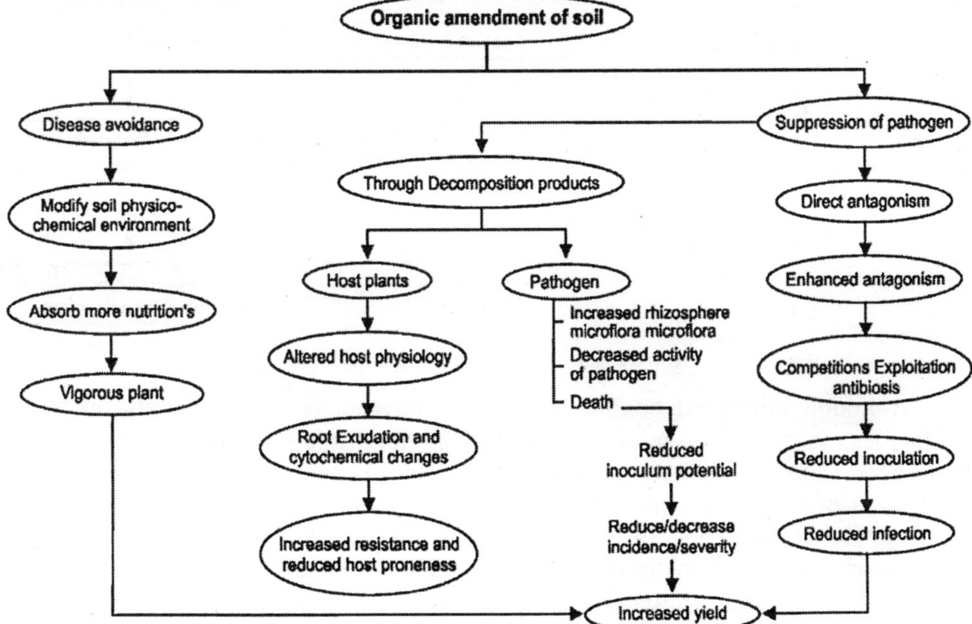

Fig. 1. Possible pathways of action of organic amendments against pathogen (Source: Chaube and Pundir, 2005)

But perhaps the most promising approach will rely on the use of cultural practices such as crop rotation and organic soil amendment to stimulate beneficial resident soil microbial populations and the development of disease suppressive soils.

The organic amendments help in the avoidance of disease by modifying the physico-chemical environment of the soil (Fig 1). These organic amendments suppress the pathogen by direct antagonism (competition, antibiosis and exploitation) and through the decomposition of amended products. These products are toxic to the pathogen and increased the resistance in host plants against the pathogen (Fig 1). The organic amendment releases substances like isothiocynate, methyl sulphide and mercaptan etc. which are toxic to pathogen. These decomposed substances also absorbed by roots of the plants where they act as precursor for synthesis of phenols and leads to the induction of resistance in host plants against pathogen (Fig 2).

Different soil amendments like saw dust, neem cake, mustard oil cake and peat moss etc. releases toxic substances and these substances are toxic to the *Ralstonia solanacearum* and controlled the bacterial wilt in tomato (Table 5). Similarly nematodes like *Meloidogyne incognita* and *Pratylenchus brachycerus* are also suppressed by these amendments.

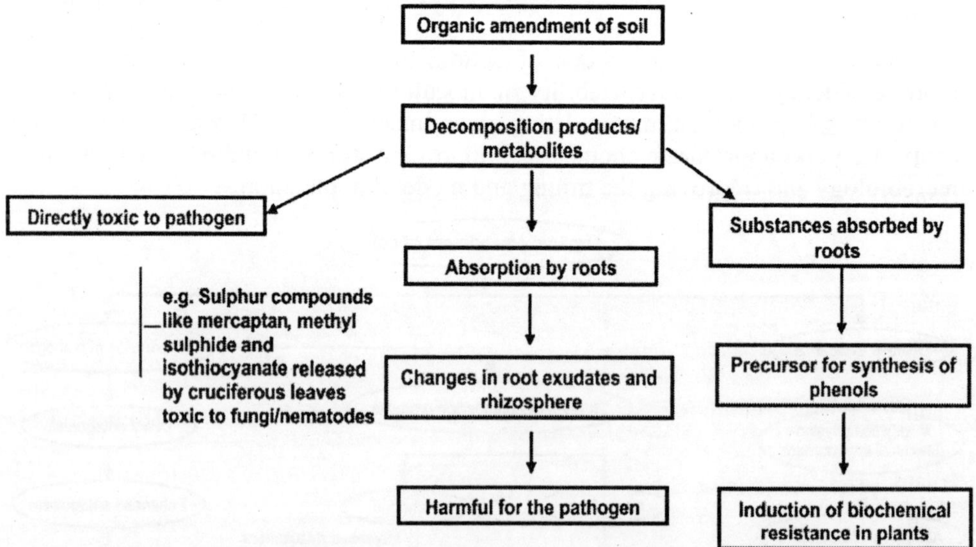

Fig. 2: Mode of action of different soil amendments in plant disease management

Although biological control has several advantages over synthetic chemical fungicides, its efficacy varies depending on the timing and the environmental conditions. Though synthetic chemicals can control many fungal diseases, but the rising cost of fungicides and their undesirable effects on the environment have lead to the exploitation of host plant resistance. Due to the residual effect of pesticides, peoples are interested in those practices which are environmentally suitable, and control the pathogen without harming the host plant. The soil amendments have positive effect on the growth of host and suppress the fungus like *Fusarium oxysporum*, *Sclerotium rolfsii*, *Phytophthora infestans*, *Aphanomyces euteiches* and *Verticillium dahliae* (Table 6).

Table 5. Suppression of soil borne bacterial pathogens and nematodes by different soil amendments

Crop	Disease	Amendment
Bacterial pathogens		
Brinjal	Bacterial wilt (*Ralstonia solanacearun*)	Karanj cake
Tomato	Bacterial wilt (*Ralstonia solanacearun*)	Saw dust, peat moss, Neem cake, muatrad oil cake, rice husk and Dhaincha
Nematode pathogen		
Okra	Root Knot (*Meloidogyne incognita*)	Neem leaves, straw, manure, castor oil, neem oil, saw dust, groundnut, lettuce and various oil cakes
Tomato	-do-	Linseed, sesame, cotton, castor, mustard oil cakes, wheat straw, biosolids, raw and composted sewage sludge
Okra	Lesion nematode (*Pratylenchus brachycerus*)	Manure, sheep manure, cow dung, raw sewage sludge and cassava peeling
Egg plant	Root knot (*Meliodogyne incognita*)	Animal organic waste (cow, horse, sheep dung)

Table 6. Suppression of soil borne fungal pathogens by soil amendments

Crop	Disease/ Causal organism	Amendment
Tomato	Fusarium wilt (*F. oxysporum* f. sp. *lycopersici*	Crab shell powder, chicken manure
Potato	Rhizoctonia stem canker (*R. solani*)	Oil cake
Potato	Soil borne pathogens	Soymeal
Tomato	Blight (*Sclerotium rolfsii*)	Composted cotton gin, trash, green manure, rye vetch green manure
Cauliflower	Verticillium wilt (*V. dahliae* var. *longisporum*)	Crop residues
Potato	Pink rot (*Phytophthora erythroseptica*)	Mixture of varius amendments
Bean	White mould (*Sclerotinia sclerotiorum*	Sweet clover straw, Fish meal
Cucumber	Fusarium wilt (*F. oxysporum* f. sp. *cucumerinum*)	Cow dung
Snapbean	Common root rot (*Aphanomyces euteiches*)	Fresh and composted paper mill residual byproducts

The effect of six organic amendments namely, neem cake, pongamia cake, pongamia fresh and dry leaves, eucalyptus dry leaves and farm yard manure (FYM) on seedling diseases in tomato (cv. Pusa Ruby) caused by *F. oxysporum* f.sp. *lycopersici* was studied (Padmodaya and Reddy 1999). In general, soil covering for a certain period of time reduced the use of herbicides by 75 per cent. The use of old straw for soil covering reduced the complete use of herbicide. Mulching with black polyethylene or paper or old straw increased the average head weight, yield and percentage of class I products of vegetables (Poll 1996). Soil covering resulted in less *Phytophthora porri* incidence in leek than no soil cover.

Soil amendments do not always decrease the incidence of diseases but sometimes in favourable environment for the development of disease may results in increase in disease incidence. The effect of different soil amendments chitin, cabbage, vetch and rye on stem canker and black scurf (*Rhizoctonia solani*) of potato was compared and found that chitin, cabbage, vetch and rye amendments reduced the severity of black scurf, while seaweed and manure increased its severity. These treatments have no effect on the length of stem canker lesions.

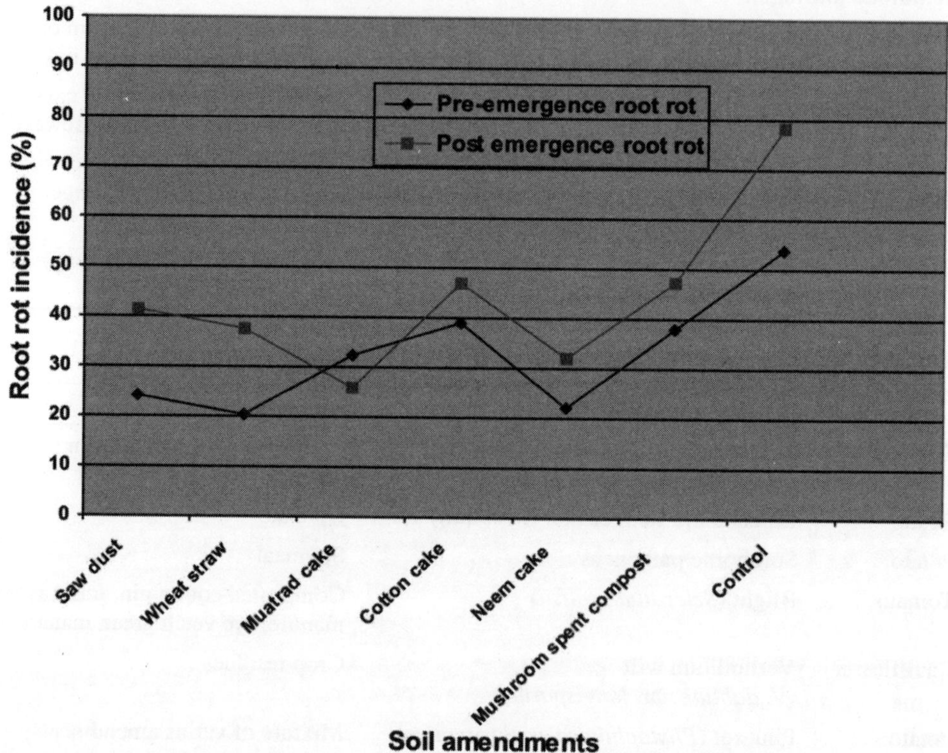

Fig. 4. Effect of different soil amendments on pre and post emergence of root rot of french bean (Source: Upmanyu et al 2002)

Soil borne diseases like white root rot, seedling blight, collar rot and crown gall are major problem in apple production and causes great economic losses in terms of money and labour. Soil amendments play an important role in reducing the incidence of soil borne diseases. The different organic soil amendments (cotton cake, mustard cake and neem cake) under *in vitro* and *in vivo* conditions reduced the incidence of soilborne diseases of apple, i.e. white root rot (*Dematophora necatrix* [*Rosellinia necatrix*]), collar rot (*Phytophthora cactorum*), and seedling blight (*Sclerotium rolfsii* [*Corticium rolfsii*]). Disease incidence was lowest with neem cake (60.88%) followed by mustard cake (70.07%) under pot conditions (Sharma and Sharma 2005). Total microbial population in the soil also increased as a result of amendment with oil cakes as compared to untreated control. The root rot disease incidence in

French bean has also been reduced by the use of soil amendments i.e. wheat straw, saw dust, mustrad cake, cotton cake, neem cake and spent mushroom compost (Upmanyu et al 2002). Minimum disease incidence of pre-emergence root rot was observed in soil amended with wheat straw (20.37%) whereas minimum incidence of post-emergence root rot was observed in mustard cake (25.93%) soil amendment (Fig 4). Soil amendment with neem cake reduced the rice sheath blight incidence from 77.5 to 27.5 per cent and disease severity from 23.75 to 2.55 per cent (Daisy, 2010). The amendment of soil with broccoli residues amendments also reduced the populations of *Verticillium dahliae* on cauliflower and *Sclerotinia minor* on lettuce in California, USA (Subbarao 2004).

Advantages of soil amendments

There are various advantages of soil amendments which are as follows :-

1. Soil amendments improve soil structure.
2. It provides the better aeration and temperature for the growth.
3. It supports the growth of other soil microorganisms, few of which are antagonistic to the pathogen.
4. It improves the water holding capacity of soil.
5. It acts as a buffer.
6. It helps in dissolving the minerals.
7. It plays an important role in recycling of minerals and water.
8. It also adds nutrients in to the soil.

III. Compost

Compost type and maturity level, as well as the composting method, affect the suppression of disease. Control of diseases through composting includes killing of plant pathogens present in waste material during and after the high-temperature process. Specific physical, chemical, and biological properties of compost may have a major effect on the suppressiveness of pathogens. Activity of anagonists involved in biological control is affected by nutrition present in compost.

The first report on control of Phytophthora disease in strawberry with the compost of tree bark appeared in 1950s (Vaughn et al 1954). The composted municipal sludge (cms) has long term effects on *Sclerotinia minor* in lettuce crop (Lumsden et al 1984). However, differences in efficacy among composts to control the club root of Chinese cabbage have been observed (Tamura and Taketani 1977).

Compost suppresses the major soil borne plant pathogens, thus reducing crop losses (Chef et al 1983, Daft et al 1979, Hoitink and Fahy 1986, Van Assche and Uytterbroeck 1981). Voland and Epstein (1994) reported the efficacy of fresh manure over the composted manure in reducing the bean hypocotyl lesions. The effectiveness of these treatments may be due to the enhanced populations of non parasitic microorganisms exhibiting the antagonistic effects on the pathogens thereby inhibiting the mycelial growth and sclerotia production of the pathogen. Various reports refer to the control of *Phytophthora* diseases with organic amendments

(Gilpatrick 1969, Hoitink et al 1977, Nesbitt et al 1979, Spencer and Benson 1982, Hardy and Sivasithamparam 1991, Tsao and Oster 1981). Organic materials which can be used for this purpose include hardwood and pine bark, municipal sludge, licorice roots, grape marc, cattle manure, poultry litter, cotton waste and other crop residues (Zentmyer 1963, Huber and Watson 1974, Daft et al 1979, Lumsden et al 1983, Nelson et al 1983, Hadar and Mandelbaum 1986, Hoitink and Fahy 1986, Chen et al 1988, Kuter et al 1988, Mandelbaum et al 1988, Mandelbaum and Hadar 1990, Gorodecki and Hadar 1991). However, when applying green manures for controlling soilborne pathogens, results are often inconsistent (Zentmyer 1963, Kuter 1983, Ramirez and Munnecke 1988) because effects of age and quality of the substance have been overlooked (Nelson et al 1983). Furthermore, many green manures result in side-effects due to phytotoxicity, nitrogen immobilisation, high salt content and structural incompatibility (Verdonck 1988). It is thus feasible to convert the raw material to a more stable product by composting.

IV. Tillage

Tillage may be defined as a cultural practice which changes the conditions of soil for crop production (Annonymous 1979). There are various tillage practices, including deep ploughing, reduced cultivation, sub-soiling, chiseling, conservation tillage, minimum tillage and zero or no tillage (Palti and Katan 1997, Summer et al 1981). A typical example of the effect of tillage on a soil borne disease has been advocated by Burke et al (1972). Loss of soil by sub-soiling to reduce soil compaction after seedbed preparation increased yield of bean plants whose roots were infected by *F. solani* f.sp. *phaseoli*. Yield of the crop increased not due to reduced infection but from counteraction of the effects of disease through greater regeneration of root as the season progressed. These results are further supported by other studies showing considerable suppression of bean growth under the stress of low oxygen and soil compaction in a soil infested with *F. solani* f. sp. *phaseoli* (Miller and Burke 1975, 1985). Tillage practices also directly influence chemical properties of soil, root growth and nutritive uptake and population of vectors of plant pathogens and the susceptibility or resistance of host (Summer et al 1981). In a long - term study, the intensity of common root rot disease caused by *Cochliobolus sativus* was generally lower under zero than conventional tillage (Tinline and Spurr 1991). Although some plant pathogens can be controlled by reduced tillage, many others are favoured if the soil is not tilled adequately. Deep tillage (30 cm sub soiling) can reduce the soil compaction and incidence of root rot in French bean in addition to increased yield (Tan and Tu 1995).

No tillage practice gave variable results with vegetables. Leaving plant residue on the soil often leads to increased stunting, root rot and damping off in vegetables but research on the influence of specific tillage practices on the plant disease of different vegetables is generally lacking (Cochrane 1949, Janes et al 1955). Soil quality, crop growth and the incidence of pests were compared for snapbean (*Phaseolus vulgaris* cv. Matador) planted in conventional tillage, in rye (*Secale cereale*) mulch without strips and in strip-tilled rye mulch (Bottenberg et al 1999). Soil bulk density was higher in the rye mulch treatments than in the conventionally tilled plots and may have reduced plant growth. Leaf nitrogen content was lower in the rye mulch treatments three weeks after planting, this may be related to nitrogen tie-up during rye decomposition or to the negative impact of soil compaction on the soil nitrogen cycle. Insect damage to snapbean pods and leaves was not affected by rye mulching. Potato leafhopper (*Empoasca fabae*) populations were significantly higher for conventional tillage than for rye treatments. The incidence of white mold (*Sclerotinia sclerotiorum*) was reduced by the rye treatments.

V. Sanitation

Sanitation reduces the rate (r) of disease epidemic. It generally reduces the initial inoculum (X_0). It includes the activities aimed at eliminating or reducing the amount of inoculum present in plant, field or a warehouse and also at preventing the spread of the pathogen to other healthy plants.

The inoculum destruction of *Fusarium* up to 90 per cent resulted in reduction in wilt incidence in tomato (Vander Plank 1963). The main aim of sanitation is to prevent the introduction of inoculum in to field, farm or community, and to reduce or eliminate the inoculum from diseased field (Palti 1981), thus reduces the availability of initial inoculum of the pathogen. This requires the identification of all inoculum sources and means of its dispersal, within a field or between fields or regions. Sanitation aims to destroy the inoculum at the source rather than in transit or in the infection court (Zentmyer and Bald 1977). Sanitation is particularly applicable to pathogens that do not spread from plant to plant in the same field. Studies revealed that *Fusarium* pathogen can produce conidia on stems which can be aerially dispersed (Rekah et al 2000, Gamliel et al 1996). Crop sanitation is a control measure designed to prevent the dispersal of pathogens through propagated material, crop residue, moving water etc (Garrett 1970). Although the effects of sanitation is not dramatic, nor is it immediate, in the long run it is very valuable and should constitute an integral part of disease management.

VI. Soil Solarization

Soil solarization is a technique to treat the field soil by trapping solar heat during hot summer months, by using transparent polyethylene sheets. Soil solarization involves the use of transparent polyethylene sheet to capture the energy from sun to heat the soil for controlling various soil pathogens (Katan 1981). The soil solarization technique for targeting soil pathogens was used for the first time by Jones et al (1966). Polyethylene (more commonly known as polythene) is most popular material used in agriculture, was first introduced on commercial scale in 1939 (Byrdson 1970). Among different polyethylene mulches, transparent mulch was able to transmit most of the solar radiations that heat the soil (Macfayden 1967). It was suggested that mulching should be carried out during the period of high temperature. Thinnest polyethylene mulch (25 to 30 μm) have been found to be more effective in heating due to better radiation transmittance than the other one (Horowitz and Regev 1980). Heating is achieved by mulching the soil with transparent plastic sheeting during the appropriate seasons, thereby controlling many pathogens and weed (Katan 1981). Mulching with polythene sheets increased soil temperature and resulted in pronouced reduction in the population of *Verticillium dahliae* and

Fusarium oxysporum f. sp. *lycopersici* as well as in disease caused by *V. dahliae* in eggplant and tomatoes.

Israel has made a landmark in soil solarization technology. Their experience over a period of time has revealed that mulching of the soil with transparent polyethylene sheet in hot season proved effective against soil borne pathogens and weeds (Katan et al 1976, 1980). Studies on soil solarization have also been carried out in several other countries like Jordon (Al-Raddad 1979), England (White and Buczacki 1979), Greece (Tjamos and Faridis 1980), California, USA (Pullman and DeVay 1977; Stapleton and DeVay 1982), Australia and India (Chauhan et al 1988). There was a decrease in population densities of *Agrobacterium* at lower depth due to cumulative effects of sub-lethal temperatures on the pathogens (Tjamos and Paplomatas 1988). Since then a lot of work has been carried out on this aspect alone as well as in combination with other management practices to combat the ravages of soil borne pathogens. The decrease in population densities of *Agrobacterium* at lower depth due to the cumulative effect of sub-lethal temperatures on the pathogen was much more in irrigated than in non-irrigated solarized plots (Raio et al 1997). Longer duration of soil solarizatin (90 days) was much more conducive in achieving desired thermal doses and its effectiveness was also correlated to soil type. Pinkerton et al (2000) also observed that incidence of crown gall on "Mazzard" cherry rootstock planted in solarized beds having silty clay soil was reduced significantly. Arora (2001) studied the effect of soil solarization on black scurf, russet scab and yield of potato and found that incidence of black scurf (6.5%) and russet scab (29.8%) was reduced to minimum as compared to control (11.5% and 92.3%) whereas yield was maximum in soil solarized plots (486 q/ha) as compared unsolarized plots (410 q/ha). Gupta and Khosla (2007) observed that integration of soil solarization for 90 days under irrigated soil conditions with *Bacillus subtilis* soil drench significantly reduced the incidence of crown gall in cherry to 11.04% and 11.93% as compared to non-solarized plots.

There are some important points which should be taken in to consideration before doing soil solarization

1. Soil solarization should be done during periods of high temperature and intense solar radiations. It is generally done from mid May to mid June for four weeks.
2. Light irrigation (around 50 mm) should be done in the field 48 hr before soil solarization. Cultivate and level the field thoroughly before laying the polyethylene sheet.
3. Polyethylene sheet should be clear, transparent, 25-100 μm thick and of low density.

VII. Flooding

Flooding is a pre-planting practice which somewhat resembles disinfection. The harmful effect of flooding on the soil-borne pathogens may be related to lack of oxygen, increased CO_2 or various microbial interactions, e.g. production of toxic products to the pathogens upon anaerobic processes (Bruehl 1987). Flooding also apparently destroy *Pseudomonas solanacearum*, and the nematode *Radopholus simili*

(Stover 1962). Flooding was also used for controlling other soil-borne pathogens like *Sclerotinia sclerotiorum* on a smaller scale (Steadman 1979, Newhall 1955). Flooding can be used as a cultural practice for disease control only in those countries where large resources of water are available or where fishes are grown in ponds (Palti 1981).

VIII. Fire and Flaming

Hardison (1976) defined this approach as thermo sanitation and described many examples of controlling disease(s) by fire and flaming. Application of heat to treat the fields with flames produced by oil or gas burning devices, as an alternative to open fires, was developed for various reasons: public opposition to smoke, low effectiveness of open fire, or climate conditions (Hardison 1976).

IX. Harrowing and mowing

Blind harrowing consists of waiting until the crop has been planted, the seed is sprouted and the shoot is not yet emerged. At this crucial time, the field is harrowed to kill the small weed seedlings that have already sprouted. The crop will emerge shortly afterwards, having gained a head start on the next flush of weeds. When properly carried out, this method can be most effective in controlling annual weeds in large seeded crops. In cereals, this blind harrowing will often provide adequate control of annual broadleaf weeds. In row crops, it can be used to keep weeds down until the crop plants are large enough to withstand other mechanical control methods. Any type of light harrow can be used. Blind harrowing must not be done on crops previously under seeded with grass or legume seeds. Mowing or cutting weeds may control weeds in orchards, roadsides, lawns, etc. Harvesting hay or cereal crops also helps to control weeds. Mowing cereal stubble in August can reduce weed growth and weed seed production, especially if the cereal had been under seeded to another crop such as red clover to compete with the new weed growth.

Loginov (2003) reported early spring harrowing for control of snow mould on the main cereal crop winter rye as well as for root rots in perennial grasses. Successful control is reported for the main pest of perennial leguminous grasses, clover seed-eater (*Cecidomyia leguminicola*), by cultivation of fallow land. Crop losses caused by the dangerous pest turnip moth (*Agrotis segetum*) on vegetables and winter rye are controlled by fallow soil cultivation during egg laying stage of the insect. Decrease is reported in the spread of smut (*Claviceps purpurea*) on winter rye due to annual mowing of diseased plants. Use is outlined of harrowing, mowing, combing and ploughing of fallow land for control of weeds, particularly wild oats (*Avena fatua*), couch grass (*Elymus trachycaulus*) and common winter grass (*Barbarea vulgaris*).

X. Roguing

Roguing can be defined as removal of diseased plant or plant part. The spread of tomato yellow leaf curl virus can be prevented by roguing of old, overwintered tomato crops (Ioannou 1987). The use of healthy transplants delayed subsequent disease development in the field but had little effect on final disease incidence. Growers generally use hand or herbicide roguing of virus infected plants on about 80 per cent of the lily acreage, 69 per cent of the tulip acreage and less than 1 per cent of the daffodil acreage to reduce the incidence of viral diseases (Chastagner et al 2002).

2. Practices to reduce the infection rate

I. Mulching

Mulching is an important component of integrated crop production system which offers potential benefits by maintaining high quality soil environment (Neilson et al 2003, Moor et al 2004). Mulch is a material placed on or spread over the soil surface to protect soil from soil erosion, conserve soil moisture and suppress weed growth besides creating a barrier in between the soil borne inoculum and host, thus reducing the splashing of soil borne pathogens to host through spattering rain splashes. Materials commonly used for mulch include polyethylene sheet, banana leaf, compost, grass clippings, wood chips or bark, pine needles and nut sheets. Mulching with rice straw has positive effect on the growth of hot pepper (Vos et al 1995). White and silvery plastic mulch reduced thrips injury and delayed virus epidemics. The overall effect of mulch on crop health contributed to improved crop production. Soil mulch with pine needles or sunflower inflorescence residues reduced the disease incidence by >50 per cent and increased the seed yield by >50 per cent in cauliflower seed crop (Sharma et al 1983).

The use of plastic materials for mulching is a very common practice for vegetable crops. Black polyethylene is the most widely used due to its excellent properties and low cost. However, the massive use of these materials imposes an environmental risk. In the last few years, the use of starch-based biodegradable films has been introduced as an alternative to conventional mulches. These materials can be incorporated into the soil at the end of the crop season and undergo biodegradation by soil microorganisms. Biodegradable films underwent early decomposition, but in general remained functional during use and did not affect yield and the fruit quality attributes (total soluble solids, firmness, dry weight, juice content and shape). The temperature under polyethylene films were always higher than under biodegradable films, which could be a disadvantage in certain circumstances, especially in hot climates, although may be advantageous in cool conditions. The use of polyethylene films resulted in the lowest values of soil microbial biomass and soil organic matter mineralization, probably due to the increase in temperature under mulches (Moreno 2008). Polythene mulch was as effective as a fungicide for controlling buckeye rot disease in tomato (Dodan et al 1994) where *Phytophthora nicotianae* var *parasitica* inoculum is splashed to the foliage through rain splashes.

Mulching and staking reduced the incidence of early and late leaf blight of tomato cv. Moneymaker (5-20%) caused by *Alternaria solani* and *Phytophthora infestans*, respectively (Lyimo et al 1998). Mulching and staking also increased fruit yield by more than two fold as compared to unmulched and unstaked controls. The apparent infection rate of the two pathogens was also significantly lower in mulched and staked tomato than in the check. Mulching alone was more effective than staking in suppressing early and late blight diseases in tomato. The effect of six different mulches i.e. polythene sheet, sawdust, wheat straw, eucalyptus leaves, pine needles and grasses, and five different plant heights for defoliation in the management of septoria leaf spot (*Septoria obesa*) of chrysanthemum were studied and found that polythene mulch gave the maximum (64.28%) disease control with minimum disease severity (18.33%) as compared to control (51.33%) whereas leaves plucked up to a height of 25 cm gave lowest (15.0%) disease severity and maximum disease control (Chandel et al 2005). Strawberry can be protected from grey mould, the most common and serious fruit rot disease, by covering the crops with plastic tunnels from the beginning of blooming onwards (Genggotti et al 2005). Mulching has also been reported to reduce hazards of phytototoxity occurring at high temperatures (Dawson et al 1965). Temperature achieved at the upper soil layers by mulching in the range of those found to be lethal to the pathogens (Nelson and Wilhelm 1958). Fruit rot (*Phytophthora nicotianae* var *nicotianae*) is an important disease of bell paper in mid hills of Himachal Pradesh which causes huge losses to the farmers by premature defoliation. The incidence of this disease can be kept in check if the field floor is mulched with pine needles before the onset of monsoon rains as the infection of the disease takes place with rain splashed zoospores (Gupta and Jarial, 2008).

II. Multiple cropping systems

Multiple cropping is the practice of planting several different crops on the same plot of land at the same time. It is common among small-scale farmers in Africa. Crops, livestock and trees can all be integrated into a small farm, making it much more productive. Such integrated farming is particularly relevant for small farms or fields. The integration of many farm enterprises gives farm families several advantages. More crops can be planted in a small space. For example, intercropping and relay cropping can allow a farmer to plant two crops i.e., maize and beans in the field at the same time. The production of crops is usually spread over a longer period of the year, allowing for better vegetative cover to protect the soil, but also spreading out the harvest throughout the year. The appropriate crops, crop combinations, planting times and planting patterns will vary from place to place depending on the local climate, soils, topography, water availability, pests and diseases, socio-economic conditions, and other factors. Pests and disease incidence is low in a series of 3-tier cropping systems; with coconut, banana, *Leucaena* and *Eucalyptus* as the first tier; cassava as the middle tier, and groundnuts, French beans (*Phaseolus vulgaris*) or vegetable cowpeas (*Vigna ungui*culata) as the ground tier. The incidence of brown leaf spot disease, transmitted by *Bemisia tabaci*, was lowest on cassava grown with *Eucalyptus* and *Leucaena* (Ghosh et al 1986)

Advantages

The advantages of multiple cropping systems are as follows:

1. Reduces the risk of total loss from drought, pests and diseases.
2. Optimizes production from small plots so can help farmers to cope with land shortages.

3. Including legumes in the cropping pattern helps to maintain soil fertility by fixing nitrogen in the soil.
4. It suppresses weeds. As the planting density is high, weeds cannot compete with the crops.
5. Different types of crops can be planted to take advantage of different seasons. For example, crops that require a lot of water can be grown in the wet season, intercropped with drought-resistant crops that can be harvested in the following dry season.

Disadvantages

The disadvantages of multiple cropping systems are as follows:

1. The presence of crops in the field throughout the year allows crop pests to survive more easily. Some pests can shift from one crop to another e.g., aphids can move to cotton plants during dry season.
2. The large number of different crops in the field allows the weeds to grow because soil is irrigated continuously.

III. Cover crops

Cover crops have never been as popular as they are today, and the information on cover crops is readily available. There is a cover crop available for every need. However, cover crops in vegetable production are used mainly as green manuring crops. When used for green manuring, cover crops are ploughed under. Benefits derived from ploughed under cover crops are temporary. However, when cover crops are used along with conservation tillage, this not only improved the quality of soil but also produced healthy crops which are resistant to pests, diseases and nematodes. Lewis et al (1997) established why a total system approach is more sustainable for pest, disease and nematode management and that the "treat the symptoms" mentality should be the last line of defense rather than the first. With the integration of cover crops and conservation tillage in vegetable production, it is possible to make pesticide intervention as a last line of defense. In most situations insecticides, fungicides, and nematicides are not needed or used. Weed control strategies are altered. No-till delays harvest of vegetables and thus, strip-tilling is essential to raise vegetables to match harvest with market window (Phatak et al 2002). Cover crops are effective in improving biomass, yield as well as carbon and nitrogen accumulation, enhancing soil physico-chemical properties, water qualities, and controlling weeds, pests and diseases of vegetables (Sainju 2005). The incidence of potyviruses and silver leaf can be greatly reduced or might be delayed by using light weight floating row cover (Webb and Linda 1992). The non host barrier crops pearl millet, sesame, wheat and barley were used in management of viral diseases of chilli (*Capsicum annuum*). Sesame and wheat were the most effective barrier crops in summer and winter crops (Dhawan and Rishi 1999).

IV. Sowing and Spacing of crops

The choice of growing season and adjustment of planting date to avoid the risk of the disease is an important to prevent the disease spread by air borne inoculum.

Increased spacing between plants in a row or between rows of plants or between commercial crops susceptible to the same pathogen will reduce disease spread from a seed borne sources. The disease can also be minimized by alteration in the date of sowing. This strategy can be very effective where pathogens are splash-borne (e.g. bacteria and pycnidial fungi) and are spread over relatively short distance. Spacing prevents splash spread of the halo blight bacterium from the newly introduced tropical pasture legume siratro bean to commercial bean.

Stemphylium vesicarium was widespread on onion and its management through cultural practices was studied by Jakhar et al (1994) and found that disease incidence was influenced by the sowing date and spacing: it was highest (52.2%) when the crop was sown on 30 September with spacing 45 x 30 cm and lowest when sown on 30 October at 60 x 45 cm. Similarly early sowing of broad bean reduced the incidence of subterranean clover red leaf virus infection (Johnstone and Repley 1979). Phyllody disease transmitted by *Orosius albicinctus* is one of the most important limiting factors in the cultivation of toria (*Brassica campestris* var. *toria*). The disease incidence was significantly low as the sowing was done late in the season, probably because of the lower vector incidence late in the growing season (Kaushik et al 1978). Low downy mildew (*Peronospora trigonella*) intensity in fenugreek was recorded when the crop was sown in the month of November as compared to October sown crop (Goyal and Sharma 2012).

The activity of the noctuid *Spodoptera eridania*, agromyzid *Liriomyza trifolii*, the severity of bacterial spot caused by *Xanthomonas campestris* pv. *vesicatoria* and yield of staked-tomatoes was affected by the plant row spacing (Stanley et al 1988). The incidence and severity of African cassava mosaic (CMD) increased with delayed planting and reached a peak in cassava planted in August. CMD incidence also increased with wider spacing and decreased significantly with increases in the number of cassava plants (Otim and Ingoot 1987). The early sowing (1st October) and removal of 50% lower leaves at 60 days of crucifer crop age resulted in disease escape from white rust (*Albugo candida*) and downy mildew (*Peronospora parasitica*) staghead formation. The incidence of Alternaria blight, white rust and downy mildew diseases was also reduced (Kaul and Singh 1999). However, both cultural management practices like date of sowing and removal of lower leaves have independent effects on seed yield. Sherbakoff (1917) advocated the staking of plants in the field for the control of buckeye rot of tomato and this was found effective by Wager (1935) but Wilson (1956) found the disease rather frequently on staked plants. Wider spacing of tomato plants 4-6 x 1-1.5 feet should be adopted in unstaked plants to control the buckeye rot (Loest 1948). Foliage and fruits upto a height of 30 cm should be removed to avoid the moist and stagnant air conditions (Sharma et al 1976). The progressive increase in row spacing of tomato (cv. Roma) resulted in decrease in the severity of Septoria leaf spot (*Septoria lycopersici*) in tomato. Row spacings of 45 and 60 cm significantly reduced the incidence to 57.6 and 57.8%, respectively. The disease developed slowly in widely spaced (90 cm) tomato resulting in 28.2% disease incidence after 3 weeks. Tomato plants which were staked with the lower leaves defoliated developed less disease as compared to staked but undefoliated plants. Similarly, defoliated but unstaked tomato also showed less disease development (Kumar and Sugha 2000).

A thorough understanding of the ecology and plant husbandry of a particular crop is absolutely essential for root disease control, appropriate modifications or innovation in crop husbandry practice have furnished the formations for success in all types of crop cultivation (Garrett 1970). The combinations of cultural practice as, deep sub-soiling, narrow row spacing and the use of resistant cultivars increased seed yields of bush snap beans (*Phaseolus vulgaris*) grown in fields infested with *Fusarium solani* f.sp. *phaseoli*, the principal causal agent of root rot (Silbernagel and Mills 1990). The medium row-to-row and plant-to-plant spacing (30-60 x 15 cm) and late sowing (first fortnight of August) reduced the disease incidence/severity of root rot and web blight of French bean and increased the yield (Gupta and Shyam 2001). Sclerotinia rot of mustard can be kept under check if the crop is sown around first week of October (Prasad et al 2009). Similarly closely spaced and densely planted fenugreek crop had more downy mildew intensity than widely spaced crop (Goyal and Sharma 2012).

V. Irrigation and water management

Irrigation provides free moisture in the soil and reduces temperature in the plant canopy. Therefore, optimizing irrigation schedule check the disease epidemic. Time, frequency, amount and mode of irrigation offer the farmers a wide choice of water-management option to reduce both foliar and soil borne diseases when applicable (Rotem and Patil 1969, Ristaino 1991). The understanding of moisture and aeration requirement of each of the biotic factors involved in the diseases enables the farmers to create an environment which is less favourable to disease development (Frank 1967). Irrigation may also reduce plant disease(s) through its effect on predisposition to the disease(s) (Duniway 1977, Blaker and MacDonald 1981). Probably no other single agricultural practice has a greater influence on the plant-microorganisms environment than does irrigation. Zoosporic fungi such as *Pythium* and *Phytophthora* depend on soil water for zoospore release and motility. Diseases caused by these fungi were frequently reported to be favoured by heavy rain or frequent irrigation. Thus the incidence of Phytophthora root rot of chilli peppers (*P. capsici*) in plants receiving alternate row irrigation was significantly less than in plants with irrigation of every row (Blies et al 1992). The decrease in the number of irrigations reduced the incidence of root rot caused by *Sclerotinia sclerotiorum* in lettuce.

Alternaria brassicicola conidia on plug trays or applied on to the soil rarely caused black sooty spot disease on cabbage plug seedlings. Sowing of cabbage seeds contaminated with the conidia of the pathogen hindered seed germination and increased disease severity. Ebb and flow irrigation did not differ from overhead irrigation in terms of the frequency of germination or the number of infected plugs from contaminated seeds. Ebb and flow irrigation decreased the frequency of infection and the degree of the symptoms after inoculation of the seedlings with the conidia of the pathogen. This type of irrigation also suppressed secondary infection. The effect of managing irrigation and fertilizer schedule on disease control was studied by Ganaieny et al (1998) in garlic for purple blotch and downy mildew. They found that purple blotch or downy mildew incidence was low and the best yields were obtained when plants were treated with potassium sulfate at 100 kg/feddan and irrigated after every 30 days. Sometimes overhead sprinkler irrigation helps in the spread of disease(s) particularly where the inoculum is splashed on the surface of host with spattering water splashes. The infection of web blight (*Rhizoctonia solani*) in French bean, buckeye rot in tomato, leaf blight and fruit rot of bell pepper (Shyam and Gupta 1998, Gupta et al 2002, Gupta et al 2002) takes place through splashing of inoculum from soil, in such crops irrigation with overhead sprinkler should be avoided.

VI. Grafting

Grafting comprises the union of two living plant parts so that they grow as a single plant. Grafting of vegetable plants is a common practice to control soil-borne diseases and nematodes (Edelstein 2004). In addition, grafted plants may have higher yields, improved tolerance to environmental stresses such as high boron, soil salinity and low soil temperatures. Grafting of vegetable crops is an old practice; grafting of cucurbits was briefly described in a seventeenth century book in Korea. Grafting was first used commercially in 20th century in vegetable production in Asia. Grafting of eggplants started in the 1950s followed by grafting of cucumber and tomato around 1960 and 1970, respectively. In 2000, a total of 700 million grafted vegetable crop plants were used in Japan and Korea. There are various manual grafting methods that suit each vegetable, and recently, grafting machines have been developed to produce the huge amount of grafted plants required.

In spite of its advantages, there are some problems associated with grafting. These include the additional cost, graft incompatibility that commonly appears to cause physiological disorders and reductions in yield, fruit quality and flower formation. Initiating or increasing the use of grafted plants should be done only after the benefits and risks of grafted seedlings have been fully understood. The grafting melon (*Cucumis melo* L.) seedlings on to the *Fusarium oxysporum* f. sp. *melonis* (Fom) commercial resistant squash rootstocks 'Mamouth' and 'Nun 9075 RT' was compared with soil sterilization with calcium cyanamide. The grafted plants on 'Mamouth' and 'Nun 9075 RT' and plants in the Perlka treatment developed mild symptoms, as indicated by the significantly lower leaf symptom index and disease index, respectively, as compared to controls (Bletsos 2005).

VII. Herbicides and Weed control

Herbicides may have positive or negative side effects on non target organisms like pathogens, the host or soil microorganisms, which may lead to an increase or decrease in disease incidence and severity (Katan and Eshel 1973; Altman and Campbell 1977). Weeds harbour many pathogen(s) and act as source of inoculum for many diseases particulary virus(s). Timely destruction of weeds may help in reducing the primary inoculum. Certain herbicides such as dinitroaniline (Grinstein et al 1979) and chloracetamides (Cohen et al 1992) decreased the incidence of Fusarium wilt in various plants at relatively low dosages. They can serve, in a controlled manner, as a potential agricultural approach for reducing disease incidence, as a part of integrated pest-management programme. Many weeds serve as hosts of pathogens and virus vectors. Hence, their elimination is important in terms of sanitation. Weed control is in itself an agricultural practice for reducing the incidence of soil borne and foliar disease (s).

VIII. Disease management with Mineral Nutrition

Mineral Nutrition of the plant (with both macro and trace elements) may also affect disease incidence through its effect on the biotic and abiotic components of the soil. There are many methods for the soil and foliar application of mineral nutrients, which can be highly accurate and methods for assessing the mineral status in plants and soil have become progressively more sophisticated.

FUTURE THURST

Cultural practices usually cannot provide complete control of diseases. Therefore, cultural practices cannot be used alone in expensive crops. They should be regarded as complementary methods or a component in an integrated disease management program. These can prevent the introduction of inoculum in a new area and reduce the inoculum level and its build up, resulting in improved control and reduced pesticides dosage. Different practices should be standardized in integrated disease management programmes for different crops, so that minimum use of chemicals is recommended.

References

Adhikari TB and Basnyat RC 1998 Effect of crop rotation and cultivar resistance on bacterial wilt of tomato in Nepal, *Canadian Journal of Plant Pathology* **20** 283-287

Al-Raddad AMM 1979 Soil disinfection by plastic tarping, M.Sc. Thesis, University of Jordon

Altman J and Campbell CL 1977 Effect of herbicides on plant diseases, *Annual Review of Phytopathology* **15** 361-385

Anonymous 1979 Terminology and definition for soil tillage and soil tool relationship, in *Agriculture Engineers Year Book 263-265*, St. Joprephs, Mich. Amerilatn Society Agriculture Engineering, 740 page

Anonymous 2006 Abstracts of the ASHS Southern Region 66th annual meeting, Orlando, Florida, USA, 2-4 February 2006, *Horticultural Science* **41** 497-520

Arora RK 2001 Annual Scientific Report, 2000-01, Central Potato Research Institute, Shimla

Aryantha IP, Cross R, and Guest DI 2000 Suppression of *Phytophthora cinnamoni* in potting mixture amended with uncomposted and composted animal manure, *Phytopathology* **90** 775-782

Beattie JH, Doolittle SP, Beattie WR, and Webster RE 1944 Production of pepper, *Leaflet, USDA* 140, 7 pages

Bettiol W, Migheli Q, and Garibaldi A 1997 Control, with organic matter, of cucumber damping-off caused by *Pythium ultimum, Trow Pesquisa-Agropecuaria-Brasileira* **32** 57-61

Blaker NS and MacDonald JD 1981 Predisposing effects of soil moisture extremes on the susceptibility of Rhododendron to Phytophthora root and crown rot, *Phytopathology* **71** 831-834

Bletsos FA 2005 Use of grafting and calcium cyanamide as alternatives to methyl bromide soil fumigation and their effects on growth, yield, quality and Fusarium wilt control in melon, *Journal of Phytopathology* **153** 155-161

Blies CL, Udsey DL, and Liddell CM 1992 Control of Phytophthora root rot of chilli peppers by irrigation practices and fungicides, *Crop Protection* **11** 225-228

Bottenberg H, Masiunas J and Eastman C 1999 Strip tillage reduces yield loss of snapbean planted in rye mulch, *Horticulture Technology* **9** 235-240

Bruehl GW 1987 *Soil borne plant pathogens,* New York: MacMillan Publishing Company

Burke DW, Holmes LD, and Barker AW 1972 Distribution of *Fusarium solani* f. sp. *phaseoli* and bean roots in relation to tillage and soil compaction, *Phyopathology* **62** 550-554

Byrdson JA 1970 *Plastic Materials,* London: Iliffe Books

Cartia G, Cipriano T and Greco N 1989 Effect of solarization and fumigation on soil borne pathogens of pepper in green house, *Acta Horticulturae* **255** 111-116

Chandel S, Chandel V and Tomar M 2005 Management of Septoria leaf spot of *Chrysanthemum* through cultural practices, 231-234. *In Integrated Disease Management Challenging Problems in Horticultural and Forest Pathology,* Solan, India

Chastagner GA, Antonelli A and Riley KL 2002 Management of aphid transmitted virus diseases of ornamental bulb crops in pacific North west, *Acta Horticulturae* **570** 297-300

Chaube HS and Pundhir VS 2005 *Crop diseases and their management,* New Delhi: Prentice Hall of India Private Limited

Chauhan YS, Nene YL, Johonson C, Hawara MP, Sexsena NP, Singh S, Sharma B, Sahrawat KL, Burford JR, Rupela OP, Kumar JV, Rao DK, Sithanamthan S and Singh S 1988 Effect of soil solarization on pigeonpea and chickpea, *Research Bulletin* ICRISAT **11** 16

Cheah LH and Falloon RE 2005 Integrated disease management for clubroot of vegetable Brassicas, in *Vegetables: Growing Environment and Mineral Nutrition,* 126-137 pages

Chef DG, Hoitink HAJ and Madden LV 1983 Effects of organic components in container medicinal on suppression of *Fusarium wilt* of chrysanthemum and flax, *Phytopathology* **73** 279-281

Chen W, Hoitink HAJ, and Madden LV 1988 Microbial activity and biomass in container media for predicting suppressiveness to damping-off caused by *Pythium ultimum, Phytopathology* **78** 1447-1450

Cochrane VW 1949 Crop residue as causative agents of root rots of vegetables, *Connecticut Agricultural Experiment Station New Heaven Bulletin* 526-534

Cohen R, Blairer B, and Katan J 1992 Chloracetamide herbicides reduce incidence of Fusarium wilt in melons, *Crop Protection* **11** 181-187

Cook RJ and Baker KF 1987 The nature and practice of biological control of plant pathogens, Ammerican Phytopathology Society St. Paul M N 539 p.

Curl E A 1963 Control of plant disease by crop rotation. *Botanical Review* **29** 413-479

Daft GC, Poole HA and Hoitink HAJ 1979 Composted hardwood and fungicide drenches for control of poinsettia crown and root rot, *Horticultural Science* **14** 185-187

Daisy S 2010 Efficacy of soil amendments for the management of rice sheath blight, *Indian Phytopathology* **63(1)** 94-95

Davison EM and McKay AG 2003 Host range of *Pythium sulcatum* and the effects of rotation on Pythium diseases of carrots, *Australian Plant Pathology* **32** (3) 339-346

Dawson JR, Johnson RAH, Adams P and Last FT 1965 Influence of steam/air mixtures when used for heating soil or biological and chemical properties that affect seedling growth, *Annals of Applied Biological Control* **56** 243-250

Dhawan P and Rishi N 1999 Eco-friendly management of viral diseases of chilli by non-host barrier trap crops, *Capsicum and EggplantNewsletter* **18** 52-55

Dixon GR 2003 Clubroot, a case for integrated control, *Plantsman* **2**(3) 179-183

Dodan DS, Shyam KR, and Gupta SK 1994 Integrated management of buckeye rot of tomato, *Plant Disease Research* **9**(2) 141-144

Dogra BS and Mandradia RK 2009 Effect of sea weed (*Ascophyllum nodusum*) extract on plant growth parameters and downy mildew in onion. 124 *In National Symposium on Rational use of fungicides in management of Horticultural crop diseases*, UHF, Nauni, Solan, HP

Doocittle SP, Taylor AL and Danielson LL 1961 *Tomato diseases and their control.* Department of Agriculture, Hand Book 203 pages

Duniway JM 1977 Predisposing effect of water stress on the severity of Phytophthora root rot in safflower, *Phytopathology* **67** 884-889

Eahel Y and Katan J 1972 Effect of time of application of diphenamid on pepper, weeds and disease, *Weed Science* **20** 468-471

Edelstein M 2004 Grafting vegetable-crop plants: pros and cons, *Acta Horticulturae* **659**(1) 235-238

Ferreyra ER, Tosso TJ, and Fernanddez MC 1984 Effect of irrigation water management on *Phytophthora capsici* Leonian, causal agent of *Capsicum annum* blight, *Agricultura Technica* **44** 319-324

Frank ZR 1967 Effect of irrigation procedure in Pythium rot of groundnut pods, *Plant Disease Reporter* **51** 414-416

Gamliel A, Katan J, Yunis H and Katan J 1996 *Fusarium wilt* and crown rot or sweet basal: involvement of soil borne and air borne inoculums, *Phytopathology* **86** 52-62

Ganaieny ERMA, Osman AZ and Mohamed HY 1998 Effect of some agricultural practices on garlic purple leaf blotch and downy mildew diseases, *Assiut Journal of Agricultural Science* **29**(1) 45-58

Garibaldi A and Tamietti G 1983 Attempts to use soil solarization in closed glass houses in northern Italy, *Acta Horticulturae* **152** 237-243

Garrett SD 1970 *Pathogenic root infecting Fungi,* Cambridge: Cambridge University Press

Gengotti S, Lucchi C and Antoniacci L 2005 Peat and disease management in organic strawbwrry (Fragaria × Anassa) production in Emilia Romagna (Italy), *Informatore Fitopatologico* **55**(5) 6-11

Ghosh SP, Pillai KS and Thankappan M 1986 Cassava based multiple cropping systems II. Incidence of pests and diseases, *Journal of Root Crops* **12** (2) 83-89

Gilpatrick JD 1969 Effect of soil amendments upon inoculum survival and function in *Phytophthora* root rot of avocado, *Phytopathology* **59** 979 - 985

Gorodecki B and Hadar Y 1991 Suppression of *Rhizoctonia solani* and *Sclerotium rolfsii* in container media containing composted separated cattle manure and composted grape marc, *Crop Protection* **9** 271 – 274

Goyal SK and Sharma P 2012 Effect of different cultural practices on downy mildew (*Peronospora trigonella*) of fenugreek and estimation of avoidable losses with seed yield, *Plant Disease Research* **27(2)** 138-143

Grinstein A, Elad Y, Katan J and Chet I 1979 Control of *sclerotium rolfsii* by means of a herbicide and *Trichoderma harzianum, Plant Disease Reporter* **63** 823-826

Guest D 2002 Integrated management of Phytophthora diseases in tropical Horticulture, *Acta Horticulturae* **575**(2) 447-451

Gupta AK and Khosla K 2007 Integration of soil solarization and potential native antagonist for the management of crown gall on cherry rootstock colt, *Scientia Horticulturae* **112** 51-57

Gupta SK and Shyam KR 2001 Third half yearly progress report of ICAR Adhoc Research Project "*Epidemiology and management of root rot and web blight of French bean caused by Rhizoctoina solani.* Department of Mycology and Plant Pathology, UHF, Nauni, Solan.

Gupta SK, Agarwal RK, and Sharma SL 1986 Efficacy of soil amendments in managing stalk rot (*Sclerotinia sclerotiorum*) of cauliflower. *Indian Journal of Plant Pathology* **4**(2) 129-131

Gupta S K, Agarwal R K and Sharma S L 1987 Effect of crop rotation on the management of stalk rot of cauliflower, *Indian Journal of Plant Pathology* **5**(1) 79-80

Gupta SK, Shayam KR, and Bhardwaj SS 2002 *Leaf blight and fruit rot of Bell paper, Research Information Bulletin*, Department of Mycology and Plant Pathology, UHF, Nauni, Solan, 24 Pages

Gupta SK, Shyam KR, and Upmanyu S 2002 *Root rot and web blight of French bean and its management, Reasearch Information Bulletin*, Department of Mycology and Plant Pathology, UHF, Nauni, Solan. 26 Pages

Gupta SK and Jarial K 2008 Management of fruit rot (*Phytophthora nicotianae* var. *nicotianae*) of bell paper (*Capsicum annum*) with mulch and fungicide application, *Journal of Mycology and Plant Pathology* **38**(3) 689-690

Hadar Y and Mandelbaum R 1986 Suppression of *Pythium aphanidermatum* damping off in container media containing composted liquorice roots, *Crop Protection* **5** 88 - 92

Hardison JR 1976 Fire and flame for plant disease control, *Annual Review of Phytopathology* **14** 355-379

Hardy GE StJ and Sivasithamparam K 1991 Effects of sterile and non-sterile leachates extracted from composted eucalyptus bark and pine-bark container media on *Phytophthora* spp, *Soil Biology Biochemistry* **23** 25 - 30

Hoitink HAJ, Vandoren DM, and Schmitthenner AE 1977 Suppression of *Phytophthora cinnamomi* in composted hardwood bark potting medium, *Phytopathology* **67** 561 - 565

Hoitink HAJ and Fahy PC 1986 Basis for the control of soilborne plant pathogens with composts, *Annual Review of Phytopathology* **24** 93 - 114

Horowitz M and Regev Y 1980 Mulching with plastic sheets as a method of weed control, *Hassadeh* **60** 395-399

Horsfall JG and Diamond AE 1960 *Plant Pathology, an advance treatise* Vol I and III

Hou XL and Wu ZH 2004 The prevention and control of plant disease and elimination of pests and seeds in non-pollution vegetable production, *China Vegas* **1** 58-62

Huber DM and Watson RD 1974 Nitrogen form and plant disease, *Annual Review of Phytopathology* **12** 139-165

Ioannou N 1987 Cultural management of tomato yellow leaf curl disease in Cyprus, *Plant Pathology* **36**(3) 367-373

Janobson R, Greenberger A, Katan J, Levi M and Alon H 1980 Control of Egyptian broomrape (*Orobanche aegyptica*) and other weeds by means of solar heating of the soil by polyethylene mulching, *Weed Science* **28** 312-316

Jakhar SS, Suhag LS, and Duhan JC 1994 Prevalence and incidence of Stemphylium blight of onion (*Allium cepa* L.) and its management through cultural practices, *Crop Research Hisar* **8**(3) 562-564

Janes BE, Drinkwater MO, Bcall G, Scarchuk J, and Lent JM 1955 A 14 year study of vegetable crop rotations on a Merrinac fine sundy town soil, Connecticut (*Storrs) Agricultural Experimental Station Bulletin* 319-355 pages

Johnstone GR and Rapley PEL 1979 The effect of time of sowing on the incidence of subterranean clover red leaf virus infection in broad bean (*Vicia faba*), *Annals of Applied Biology* **91**(3) 345-351

Jones JP, Overman AG, and Geraldson CM 1966 Effect of fumigants and plastic films on the control of several soil borne pathogens of tomato, *Phytopathology* **56** 929-932

Kaewruang W, Sivastithamparam K and Hardy GE 1989 Use of soil solarization to control root rots in gerberas (*Gerbera jamesonii*), *Biology and* Fertility *of Soil* **8** 38-47

Katan J and Devay JE 1991 *Soil solarization*, Coca R A: CRC Press

Katan J and Eshel Y 1973 Interaction between herbicides and plant pathogens, *Residue Review* **45** 145-171

Katan J, Greenberger A, Alon H and Feinstein A 1976 Solar heating by polyethylene mulching for the control of disease caused by soil borne pathogens, *Phytopathology* **66** 683-688

Katan J, Rotem I, Finkel Y and Daniel J 1980 Solar heating of the soil for the control of pink root and other soil borne diseases in onions, *Phytoparasitica* **8** 39-50

Katan J 1981 Solar heating (solarization) of soil for control of soil borne pests, *Annual Review of Phytopathology* **19** 211-236

Katan J 1984 The role of soil disinfection in achieving high production in Horticultural Crops, *Crop Protection* **3** 1189-1196

Katan J 2000 Physical cultural methods for the management of soil borne pathogens, *Crop Protection* **19** 725-731

Katan J 1981 International Horticultural Congress: Managing Soil-borne Pathogens: A sound rhizosphere to improve productivity in intensive Horticultural system ISHS, *Acta Horticulturae* 635

Kaul TK and Singh SK 1999 Cultural management of major foliar diseases in *Brassica juncea*, *Journal of Mycology and Plant Pathology* **29**(1) 87-90

Kaushik CD, Tripathi NN, and Satyavir 1978 Effect of date of sowing of toria on phyllody incidence and estimation of losses, *Haryana Agricultural University Journal of Research* **8**(1) 28-30

Kim KY, Park SK, Shin YA, and Lee EJ 1990 Survey on the actual condition of cultural methods and continuous cultivation injury of red pepper in field, *Research Report Rural Development Administration Horticulture* **32**(1) 1-10

Kumar S and Sugha SK 2000 Role of cultural practices in the management of Septoria leaf spot of tomato, *Indian Phytopathology* **53**(1) 105-106

Kuter GA, Hoitink HAJ and Chen W 1988 Effects of municipal sludge compost curing time on suppression of *Pythium* and *Rhizoctonia* diseases of ornamental plants, *Plant Disease* **72** 731-756

Kuter GA, Nelson EB, Hoitink HAJ, and Madden LV 1983 Fungal populations in container media amended with composted hardwood bark suppressive and conducive to *Rhizoctonia* damping-off, *Phytopathology* **73** 1450 -1456

Lee HU, Kim CD, Lee YS, and Cho CT 1993 Effects of soil chemical properties and soil micro flora by antiplants on Phytophthora blight incidence of red pepper, *RDA Journal Agricultural Science Soil Fertility* **35**(2) 305-310

Lemaga B, Kanzikwera R, Kakuhenzire R, Hakiza JJ and Manzi G 2001 The effect of crop rotation on bacterial wilt incidence and potato tuber yield, *African Crop Science Journal* **9**(1) 257-266

Leonian LH 1922 Stem and fruit blight of pepper caused by *Phytophthora capsici* sp. nov. *Phytopathology* **12** 401-408

Lewis JA and Papavizas GC 1971Contribution a I'etude de I' activite systemique du benzimidazole 2 methylcarbamate (B.M.C.) du benomyl et du Methylthiophonate, *Phytiatrie Phytopharmacie* **21** 237-254

Lewis WJ, Vanlentern JC, Phatak SC, and Tumlinson JH 1997 A total system appraoach to sustainable peat management, *Proceedings of National Academy of Science USA* **94** 12243- 12248

Lodha S 2006 Compost as an integral component of organic farming for disease suppression, 12. In National Symposium on Recent Trends in Diagnosis and Management of chronic and Emerging Plant Diseases, Nagpur

Loest FC 1948 Plant diseases and cultural operations, *Fmg South Africa* **23** 304-307

Loginov NG 2003 The basis is cultural control, *Zashchit I Karantin Rastenii* **7** 6-7

Lorbeer JW 1997 Management of diseases in alliums, *Acta Horticulturae* **433** 585-591

Lumsden RD, Lewis JA, and Millner PD 1983 Effects of composted sewage sludge on several soilborne pathogens and diseases, *Phytopathology* **73** 1543 -1548

Lyimo HFJ, Tiluhongelwa TDM, Maerere AP, and Njau P 1998 The effect of mulching and staking on the development of early and late blights of tomato, *Tanzania Journal Agricultural Science* **1** (2) 167-172

Macfayden A 1967 Thermal energy as a factor in the biology of soils, in *Thermology* (ed) Rose AH, New York: Academic Press, 653 pages

Maggoner PE, Miller PN and De Roo H 1960 Plastic mulching principles and benefits, *Canadian Agicultural Experiment Station Bullettin*

Mandelbaum R, Hadar Y, and Chen Y 1988 Compost of agricultural wastes for their use as container media. II. Effect of heat treatment on suppression of *Pythium aphanidermatum* and microbial activities in substrates containing compost, *Biological Wastes* **26** 261 - 274

Mandelbaum R and Hadar Y 1990 Effects of available carbon source on microbial activity of *Pythium aphanidermatum* in compost and peat container media, *Phytopathology* **80** 794-804

Mc Call T M 1933 Depth of planting experiments for the control of *Rhizoctonia* on potatoes, 413-414. *In Proceedings of American Society Horticultural Sci*ence

Miller DE and Burke DW 1975 Effect of soil aeration on Fusarium root rot of beans, *Phytopathology* **65** 519-523

Miller DE and Burke DW 1985 Effect of soil oxygen on Fusarium root rot of beans with respect to seedling age and soil temperature, *Plant Disease* **69** 328-330

Moor U, Karp K and Poldma P 2004 Effect of mulching and fertilization on the quality of strawberries, *Agricultural Food Science* **13**(3) 256-267

Moreno MM 2008 Soil Properties and Production in a Tomato Crop, *Scientia Horticulturae* **116**(3) 256-263

Munshi GD and Sokhi SS 2000 Alternatives to fungicides in the management of diseases of vegetables, *Plant Disease Research* **15**(2) 162-167

Navarro AR, Puerta EOD and Isaza L 1981 Combined effect of population density and chemical control of leaf diseases and yield of bean, *Fitopatologia Colombiana* **10** 7-9

Neilsen GH, Hogue EJ, Forge T and Neilsen D 2003 Mulches and biosolids affect vigor, yield and leaf nutrition of fertigated high density apple, *Horticultural Science* **38**(1) 41-45

Nelsen PE and Wilhelm 1958 Thermal death range of *Nerticillum aboatrum, Phytophthology* **48** 613-616

Nelson EB, Kuter GA and Hoitink HAJ 1983 Effect of fungal antagonists and compost age on suppression of *Rhizoctonia* damping off in container media amended with composted hardwood bark, *Phytopatholoy* **73** 1457-1462

Nesbitt EB, Malajczuk N and G lenn AR 1979 Effect of organic matter on the survival of *Phytophthora cinnamomi* Rands in the soil, *Soil Biology Biochemistry,* **11** 113-136

Newhall AG 1955 Distribution of soil by heat, flooding and fumigation, *Botanical Review* **21** 189- 250

Osnitzkaya NEA 1949 Phytophthora of tomatoes in protected soil, *Orchard and Kitchen Garden* **2** 62-64

Otim NGW and Ingoot D 1987 Effect of cultural practices on the African cassava mosaic disease and its vector, *Bemisia tabaci, Intenational Developmental Research Centre, Canada* **(IDRC-258e)** 105-108

Padmodaya B and Reddy HR 1999 Effect of organic amendments on seedling disease of tomato caused by *Fusarium oxysporum* f.sp. *lycopersici. Journal of Mycology and Plant Pathology* **29**(1) 38-41

Palti J and Katan J 1997 Effect of cultivation practices and cropping system on soil borne diseases, in *Soil borne diseases of tropical plants* (eds) Hillock RJ and Waller JM, UK: CAB International 377-391 pages

Palti J 1981 Cultural practices and infection, crop disease, Berlin: Springer

Phatak SC, Dozier JR, Bateman AG, Brunson KE and Martini NL 2002 Cover crops and conservation tillage in sustainable vegetable production, 401-403. In *Making conservation tillage conventional: building a future on 25 years of research Proceedings of 25th Annual Southern Conservation Tillage Conference for Sustainable Agriculture, Auburn, AL, USA*

Pinkerton J N, Ivors K L, Miller M L, and Moore L W 2000 Effect of soil solarization and cover crops on populations of selected soil borne plant pathogens in western Oregon, *Plant Disease* **84** 952-960

Poll JTK 1996 Effect of soil cover on the yield and quality of vegetables, *Publicatie Proefstation voor de Akkerbouw en de Groenteteelt in de Vollegrond, Lelystad* **81** 103-109

Prasad R, Lallu, and Kumar S 2009 Effect of sowing date and environmental factors on incidence of sclerotinia rot and dry matter of mustard, *Indian Phytopathology* **62** 526-529

Pullman GS and DeVay JE 1977 Control of Verticillium dehliae by plastic tarping, Proceedings of American Phytopathoogical Society **4** 210

Pullman GS, DeVay JE, and Garber RH 1981 Soil solarization and thermal death: a logarithmic relationship between time and temperature for four soil- borne plant pathogens, *Phytopathology* **71** 959-964

Raio A, Zoina A, and Moore LW 1997 The effect of solar heating on natural and inoculated agrobacteria, *Plant Pathology* **46** 320-328

Ramirez VJ and Munnecke DE 1988 Effect of solar heating and soil amendments of cruciferous residues on *Fusarium oxysporum* f. sp. *conglutinans* and other organisms, *Phytopathology* **78** 289 - 295

Rekah Y, Shtinberg D, and Katan J 2000 Disease development following infection of tomato and basil foliage by air borne conidia of the pathogen *Fusarium oxysporum* f. sp. *radicis lycopersici* and *F. oxysporum* f.sp. *basilci, Phytopathology* **90** 1322-1324

Ristaino JB 1991 Influence of rainfall drip irrigation and inoculums density on the development of Phytophthora root rot and trunk rot epidemic and yield in bell pepper, *Phytopathology* **81** 922-929

Robak J 1994 Crop rotation effect on clubroot disease decrease, *Acta Horticulturae* **371** 223-226

Rotem J and Pail J 1969 Irrigation plant decreases, *Annual ReviewPhytophthology* **7** 267-288

Saikia AK and Muniyappa V 1989 Epidemiology and control of tomato leaf curl virus in southern India, *Tropical Agriculture* **66** 350-354

Sainju UM 2005 Cover crops for sustaining vegetable production, improving soil and water qualities and controlling weeds and pests, *Vegetables: Growing Environment Mineral Nutrition* 281-296

Sakai Y, Itho T and Tanaka A 1992 Cultural control of stem blight of asparagus (*Asparagus officinalis* L.), *Bulletin of the Hiroshima Prefectural Agriculture Research Center* **55** 109-119

Saratarato A, Zimmerman MJ, and Yamada T 1988 Web blight in *Cultura de Feijoero Factores que afetam a Productividade* 503-520

Schippers B, Bakker AW and Bakker PAHH 1987 Interaction of deleterious and beneficial rhizosphere microorganisms and the effect of cropping practices, *Annual Review Phytopathology* **25** 339-358

Sharma A and Sharma SK 2005 Effect of oil cakes on the soil borne disease management and soil microflora in apple, *Acta Horticulturae* **696** 341-347

Sharma RC and Sharma JN 2005 Integrated plant disease management *Challenging Problems in Horticultural and Forest Pathology, Solan, India,* xiii + 362 pages

Sharma SL, Chowfla SC and Sohi HS 1976 Control of buckeye rot of tomato by cultural practices, *Indian Journal of Mycology and Plant Pathology* **6** 51-54

Sharma SL, Kainth DS and Khan MA 1983 Cultural control of *Sclerotinia* stalk rot of cauliflower, *Indian Phytopathology* **36**(4) 601-603

Sherbakoff CD 1917 Buckeye rot of tomato fruit, *Phytopathology* **7** 119-129

Shyam KR and Gupta SK 1998 Buckeye rot of tomato, *Research Information Bulletin Directorate of Research,* Dr. Y.S. Parmar University of Horticulture and Forestry, Solan

Silbernagel MJ and Mills LJ 1990 Genetic and cultural control of Fusarium root rot in bush snap beans, *Plant Disease* **74**(1) 61-66

Silva ACF and Althoff DA 2003 Crop rotation for vegetables in southern coastland of Santa Catarina State, *Agropecuaria-Catarinense* **16**(3) 58-65

Singh RS 1983 Organic amendments for root disease control through management of soil microbiota and the host, *Indian Journal of Mycology and Plant Pathology* **13** 1-16

Sood AK 2005 Bacterial wilt of solanaceous vegetable crops and its management in *Challenging Problems in Horticultural and Forest Pathology* 167-182 pages

Spencer S and Benson DM 1982 Pine bark, hardwood bark compost, and peat amendment effects on development of *Phytophthora* spp. and lupine root rot, *Phytopathology* **72** 346 - 351

Stanley CD, Schuster DJ and Jones JB 1988 Plant-row-spacing effect on insect activity, bacterial spot severity and yield for staked-tomato production in west Florida, *Proceedings of Soil Crop Science Society Florida* **47** 212-214

Stapleton JJ and DeVay JE 1982 Effect of soil solarization on the population of selected soil borne microorganism and growth of deciduous fruit tree seedlings, *Phytopathology* **72** 323-326

Steadman JR 1979 Control of plant diseases caused by Sclerotinia disease, *Phytopathology* **69** 904

Stover RH 1962 Fusarium wilt (Panama disease of banana) and *Musa* spp., Phytopathological paper no- 4 CMI Surrey

Subbarao KU 2004 Impact of crop residue amendment on soil borne disease management, 13-16. In *California Conference on Biological control IV,* Berkely California USA

Summer DR, Doupink B Jr. and Boodalis MG 1981 Effect of reduced tillage and multiple cropping on plant diseases, *Annual Review Phytopathology* **17** 167

Sun SK and Huang JW 1985 Formulated soil amendment for controlling other soil diseases, *Plant Disease* **77** 420

Sun SK and Huang JW 1985 Mechanism of control of Fusarium wilt disease by amendments of soil with S-H mixture, *Plant Protection Bullettin (Taiwan)* **25** 127-137

Tamura M and Taketani K 1977 Biology and control of club root of Chinese Cabbage in the Jshikawa Prefecture, *Jshikawa Prefecture Agricultural Experimental Station Bullettin* **9** 1-26

Tan CS and Tu JC 1995 Tillage effect of root rot severity, growth and yields of beans, *Candian Journal of Plant Science* **75** 183-186

Theunissen J 1997 Intercropping in field vegetables as an approach to sustainable horticulture, *Outlook on Agriculture* **26** (2) 95-99

Tinline RD and Spurr DT 1991 Agronomic practices and common root rot in spring wheat: effect of tillage on disease and inoculum in soil, *Canadian Journal of Plant Pathology* **13** 258

Tjamos EC and Faridis A 1980 Control of soil borne pathogens by solar heating in plastic houses, 82-84. In *Proceedings of 5th Congress Mediterranean Phytopathological Union, Hellenic Phytopathological Society, Athens*

Tjamos EC and Paplomatas EJ 1988 Long-term effect of soil solarization in controlling Verticillium wilt of globe artichokes in Greece, *Plant Patholology* **37** 507-515

Trutman P and Kayitare E 1991 Disease control and small multiplication plots improve seed quality and small farm bean yields in Central Africa, *Journal Applied Seed Production* **9** 36-40

Tsao PH and Oster JJ 1981 Relation of ammonia and nitrous acid to suppression of *Phytophthora* in soils amended with nitrogenous organic substances, *Phytopatology* **71** 53 - 59

Upmanyu S, Gupta SK, and Shyam KR 2002 Innovative approaches for the management of root rot and web blight (*Rhizoctonia solani)* of French bean, *Journal of Mycology and Plant Pathology* **32** 317-331

Van Assche C and Uytterbroeck P 1981 The influence of domestic waste compost on plant disease, *Acta Horticulturae* **126** 169-178

Vander Plank JE 1963 *Plant diseases: Epidemic and control,* New York: Academic Press

Varshney JG 2009 Why weed control? *Crop Care* **35**(1) 13-15

Vaughn EK, Roberts AN, and Mellenthin WM 1954 The influence of Douglas fir, saw dust and certain fertilizer elements on the incidence of red stele disease of strawberry, *Phytopathology* **44** 601-603

Verdonck O 1988 Composts from organic waste material as substitutes for the usual horticultural substrates, *Biological Wastes* **26** 325-330

Viana FMP, Kobori RF, Bettiol W, and Athayde Sobrinho C 2000 Control of damping-off in bean plant caused by *Sclerotinia sclerotiorum* by the incorporation of organic matter in the substrate, *Summa-Phytopathologica* **26**(1) 94-97

Voland RP and Epstein AH 1994 Development of suppressiveness to disease caused by *Rhizoctonia solani* in soil amended with composted and non-composted manure, *Plant Disease* **78** 461-466

Volcani Z, Zootra D and Cohen R 1970 Bacterial canker disease of tomatoes and capsicums, *Hassadeh* **50**(9) 1028-1032

Vos JGM, Uhan TS, and Sutarya R 1995 Integrated crop management (ICM) programme for hot pepper (*Capsicum* spp.) under tropical lowland conditions, *Crop Protection* **14**(6) 445-452

Wager VA 1935 Brown rot of tomato fruits due to *Phytophthora parasitica* Dast, *South African Journal Science* **32** 235-237

Webb SE and Linda SB 1992 Evaluation of spunbonded polyethylene row covers as a method of excluding insects and viruses affecting fall-grown squash in Florida, *Journal of Economic Entomology* **85** (6) 2344-2352

White JG and Buczacki ST 1979 Observation on the suppression of club root by artificial or natural heating of soil, *Transactions of British Mycological Society* **73** 271-275

Wilson JD 1956 Comparative control of buckeye rot of tomato by various fungicides, *Phytopathology* **46** 511-512

Xiao CL, Subbarao KV, Schulbach KF, and Koike ST 1998 Effects of crop rotation and irrigation on *Verticillium dahliae* microsclerotia in soil and wilt in cauliflower, *Phytopathology* **88**(10) 1046-1055

Yamada M, Asandhi AA, and Purwati E 2003 Employing one-year crop rotations with three vegetable combinations to control clubroot damage in the West Java highlands, *Research Highlights JIRCAS* 16-17

Zentmyer GA and Bald TG 1977 Management of the environment in *Plant Diseases An Advance Practice* Vol. 1, (eds) JG Horsfall and EB Cowling, New York: Academic Press

Zentmyer GA 1963 Biological control of *Phytophthora* root rot of avocado with alfalfa meal, *Phytopathology* **53** 1383-1387.

2

Compost: Managing Soil Borne Plant Pathogens

Manish Mathur, Ritu Mawar and Satish Lodha

Central Arid Zone Research Institute, Jodhpur - 342 003

Soil-borne plant pathogens like *Pythium, Phytophthora, Sclerotium, Aphanomyces, Thielaviopsis, Macrophomina, Fusarium, Ganoderma,* etc. frequently causes heavy losses in economically valuable crops, horticultural plants and trees through-out the world. These pathogens affect both yield and quality. In intensive agriculture, economically valuable crops are frequently or even continuously planted on the same piece of land. The population of these soil-borne plant pathogens increases with the increased years of cultivation of susceptible crops. The inoculum density of the pathogen in soil is directly proportional to the disease intensity on a crop in the field. The inoculum density of soil borne pathogens can be kept in check with various methods but the compost is one which can reduce the inoculum density of these pathogens in an ecofriendly way and this chapter include how composting inactivate or reduce the density of various soil borne pathogens and keep the ravages of diseases in check.

Management strategies against soil borne pathogens

To reduce population of these pathogens below the economic threshold level, many physical, chemical and biological control approaches have been advocated from many parts of the world. A chemical approach of controlling soil borne pathogens such as use of methyl bromide as a soil fumigant was found environmentally harmful. Chemical soil disinfestations eradicate most microbial residents in the soil. While plants benefit from such conditions for a short time, the microbial vacuum created often leads to a rebounding of parasitic agents, resulting diseases that cause more damage than those originally targeted for control. However, efforts have also been made to evolve environmentally sound methods of management of these pathogen, which includes soil steaming (Runia 1983, 2000), soil solarization (Katan et al 1976), use of organic amendments, particularly cruciferous residues and oil-cake (Ramirez-Villapudua and Munnecke 1988, Gamliel and Stapleton 1993, Lodha et al 1997), bio-controls (Whipps 1997) and crop rotation with non host crops.

Use of organic amendments

An approach that is recently gaining interest in the context of integrated management of soil-borne pathogen(s) is the use of organic amendments such as crop residues, composts or manure (Hoitink and Fahy 1986, Gamliel 2000, Lazarovits et al 2000). Composts have been used with various levels of success in suppression of many soil borne pathogens and the diseases caused by them. This subject has been reviewed by H.A.J Hoitink in a series of articles (Hoitink and Fahy 1986, Hoitink and Keener 1993, Hoitink et al 1997, Hoitink and Boehm 1999). Pathogens that have been reported to be suppressed by incorporation of compost into soil or potting mixes include *Verticillium dahliae, Rhizoctonia solani, Pythium aphanidermatum, Pythium ultimum, Pythium arrhenomanes* and *Macrophomina phaseolina.* The incorporation of crop residue changes the soil characteristic in many ways, including enrichment of nutrients and improvement in microbial activity, water holding capacity, soil aeration and permeability.

Composting is one of the best techniques for bio-conversion of organic residues into inputs for use in sustainable agriculture. Most pathogens are inactivated during the heating phase of composting (Ylimaki et al 1983, Bollen et al 1989). However, a few heat tolerant pathogens may survive this phase. Few approaches have been advocated to minimize the chances of survival of pathogenic propagules soon after composting (Lodha et al 2002). The proper sanitation of composts from plant pathogens is a pre-requisite before their application in soil.

Types of composting

Different methods for preparing composts have been suggested by various workers from India and abroad. Major types of compost and materials used for their preparation are as follows:

Compost type	Developed by	Material utilize	Duration for compost preparation
Indore type	Howard and Ward (1931)	Plant residues, animal dung, wood ash, on-farm weeds, stalks, stems, fallen pruning chaff	6 Months (Temperatures reaches 60°C)
Bangalore type	Acharya (1939)	Plant residues, animal dung, wood ash, on-farm weeds, stalks, stems, and fallen pruning chaff. This method is used for the preparation of compost in scanty rainfall area	It requires more time than the Indore method because turning is not required and in biomass, high temperature is not assured.
Synthetic compost	Hutchinson and Richard (c.f. Gaur, 1982)	Organic nitrogen as dung converted by micro-organisms in to inorganic nitrogen compounds like ammonium sulphate and urea	4-6 months

Leaf compost	Eberhardt and Pipes (1972)	Sewage sludge, urea and grass clippings	6-9 months
Vermi-composting	Tapiador, (1981)	Use of earthworms for composting of organic residues like animal wastes (Poultry, equine, piggery excreta), agricultural residues and industrial waste.	These organic materials undergo complex biochemical changes in the intestines of earthworms

Factors influencing composting process

Certain factors can influence composting process, which must be in an optimum range if aerobic, thermophilic composting is to proceed rapidly and effectively. These are summarized as follows:

1. **C:N ratio:** During composting, microorganisms require carbon for growth and energy, and nitrogen for protein synthesis. Rapid composting is achieved when wastes or mixtures of wastes have C:N ratios of between 15 to 35. Higher ratios can slow the rate of composting.
2. **Moisture content:** The optimum moisture content of organic wastes for rapid aerobic, thermophilic composting ranges from 40 to 60% (by weight). Above 60%, there may be insufficient air space to sustain aerobic decomposition.
3. **Temperature:** As composting proceeds, and if other factors are favourable, microbial activity causes temperature to increase from the mesophilic range (20-40ºC) into the thermophilic range (>40ºC). Optimum temperatures for rapid aerobic composting ranges from 55 to 65ºC.
4. **pH:** Optimum pH for rapid composting of various wastes ranges from 5.0 to 9.0.
5. **Aeration/ Oxygen supply:** A continuous supply of oxygen is required to ensure rapid aerobic, thermophilic composting. There should be at least 30% free space.
6. **Particle size/ Texture:** Grinding, shredding and blending organic wastes can enhance the rate of decomposition during composting by providing a more favourable surface to volume ratio.

Different phases during composting

Hoitink and Boehm (1999) in their comprehensive review described different phases during the process of composting. During the first phase, the temperature increases rapidly to 40-60ºC often reaching 70ºC. During this phase, mesophilic, including beneficial micro-organisms like nitrogen-fixing and biological-control agent, are killed due to heat treatment (Hoitink and Fahy 1986, Bollen 1993). However, certain species of *Bacillus* and other facultative thermophilic organisms survive this high temperature phase (Strom 1985). In the curing phase, the temperature declines to 40ºC and mesophilic microorganism re-colonise the compost from outer low-temperature layers to the center of the pile (Hoitink et al 1991).

Pathogen inactivation

Most pathogens are inactivated during the heating phase of composting (Ylimaki et al 1983, Bollen et al 1989), but certain heat-tolerant pathogens may survive. *Macrophomina phaseolina* is a relatively heat - tolerant pathogen. Concerned with the risk of spreading *M. phaseolina* by amending compost prepared by infected crop residues, experiments were initiated at the Central Arid Zone Research Institute (CAZRI), Jodhpur to bring down *M. phaseolina* propagules in the residues of cluster bean during and after preparation of composts. Studies have shown that density of viable sclerotia of *M. phaseolina* was significantly lower in the partially decomposed cluster bean residue enriched with 4% than 2% urea-N. After preparation of compost 7 day period of summer exposure, resulted in 53 - 61% reduction in counts of *M. phaseolina* in different composts, due to prevailing heat (>45°C).

Release of *M. phaseolina* inoculum from infected residues was significantly reduced by organic composting. Many factors may be involved in reducing population of *M. phaseolina* during composting (i) heat generated in the first phase, (ii) toxicity of conversion products formed during or after the sub-lethal heating process and (iii) microbial antagonism in presence of moisture.

Suppression of soil-borne plant pathogens

Effect of composted sewage sludge on several soil-borne pathogens and diseases was tested by Lumsden et al (1983a). In their study, Aphanomyces root rot of peas; Rhizoctonia root rot of bean, cotton and radish; Sclerotinia drop of lettuce, Fusarium wilt of cucumber and Phytophthora crown rot of pepper were significantly decreased by addition of 10% compost to soil. Likewise, bark compost used as a container medium suppressed four pathogens, *Phytophthora cinnamomi, Pythium* spp, *Rhizoctonia solani* and *Thielaviopsis basicola* (Daft et al 1979, Hoitink and Poole 1980, Nelson and Hoitink 1982, Spencer and Benson 1982). Suppression of *Pythium* diseases have been reported for media amended with composts prepared from different organic wastes (Chen and Hadar 1985, Hadar and Mandelbaum 1986, Chen et al 1988a). Suppression of *Pythium* diseases in these media has been attributed mainly to the biological factors (Chen et al 1987, 1988), and competition for available nutrients has been proposed as the principal mechanism (Chen et al 1987, 1988 a, Elad and Chet 1987). Other proposed mechanisms include hyperparasitism (Hadar et al 1983, Sivan et al 1984), iron competition (Becker and Cook 1988), production of soluble (Lockwood 1977) and volatile (Howell et al 1988) inhibitors or hydrolytic enzymes by soil microflora.

Many types of bio-control agents contribute to the control of Pythium and Phytophthora root rots by composts (Boehm et al 1993, 1997, Hardy and Sivasithamparam 1995, Kim et al 1997). At least, 20% of all bacterial strains recovered on 0.1% strength tryptic soya agar from the rhizosphere of cucumber sown in a composted pine bark-amended potting mix induced biological control of Pythium damping-off when applied as seed treatment (Boehm et al 1993).

Two mechanisms of biological control, based on competition, antibiosis, hyperparasitism and systemic acquired resistance in plants, have been described for compost-amended substrates. Propagules of plant pathogens, including *Pythium* and *Phytophthora* spp. are suppressed through a mechanism known as "general suppression" (Boehm et al 1993, Chen and Inbar 1993). Many types of microorganisms present in compost-amended container media function as biological control agents for diseases caused by *Phytophthora* and *Pythium* spp. (Hardy and Sivasithamparam 1991). The high microbial activity and biomass, caused by the "general soil microflora" in compost-amended substrates prevents germination of spores of the pathogen and infection of the host, presumably through microbiostasis (Mandelbaum and Hadar 1990). In a "specific suppression", a narrow group of microorganisms is capable of eradicating sclerotia of *Rhizoctonia solani* (Hoitink et al 1991). Increased population of antagonistic actinomycetes and lytic bacterial density has been attributed as possible reasons for the suppression of *Macrophomina phaseolina* induced diseases in arid region in compost-amended soil (Lodha et al 2002).

Trankner (1992) reported that composts induce systemic resistance to powdery mildew of wheat. Clulow et al (1995) reported that composts coated onto tubers protect potatoes from late blight. Effective batches of composts increase the activity of the pathogenesis-related proteins β-1,3-glucanase and peroxidase in cucumber and *Arabidopsis* but most of this increased activity does not occur until after the plants become infected with the pathogen (Zhang et al 1996, 1998), suggesting that active batches of composts seem to prime plants to better defend themselves against a given pathogen. Han et al (1998) identified several microorganisms in suppressive composts capable of inducing systemic resistance in radish and cucumber. *Trichoderma hamatum* and *Pantoea agglomerans* applied to roots of radish seedlings induced resistance to foliar bacterial spot caused by *Xanthomonas campestris* pv. *armoraceae*. Antagonistic microorganisms such as *Trichoderma harzianum, Aspergillus versicolor* and *Bacillus* spp. may have contributed to reduced disease incidence. However, the actual mechanism of disease suppression was not elucidated in our study. One possible mechanism of biological disease suppression is the stimulation of germination of sclerotia and subsequent lysis of germ-tubes in soil (Papavizas and Lumsden 1980). Germination of *M. phaseolina* has been found to be stimulated by amendments (Smith 1969), and the resulting germ-tube and hyphae are known to be sensitive to bacteria and actinomycetes leading to lysis of fungal cell walls (Kovoor 1954). Since compost-amended soil held more soil moisture in our field experiments, high population of antagonists estimated in final compost samples might have accelerated antagonism against *M. phaseolina* in the presence of moisture.

Composts have also been used in combination with other effective management strategies to improve control of soil-borne plant pathogens. Gamliel and Stapleton (1993) studied effects of commercial chicken compost, ammonium phosphate fertilizer and soil solarization alone or in combination and found improvement in control of several soil-borne pathogens and yield of lettuce.

Suppressive effect of composts on dry root rot was studied at CAZRI, Jodhpur in a series of experiments. After summer exposure, 4 week old composts of pearl millet, clusterbean, weeds and cauliflower residues (each @ 4 t/ha^{-1}) were incorporated to a depth of 0-30 cm in separate plots and cluster bean crop was grown. Dry root rot severity was significantly reduced and seed yield of cluster bean was enhanced by amendment with different composts (Table 1).

Table 1. Effect of different compost amendments on dry root rot mortality induced by *M. phaseolina* and seed yield of clusterbean

Compost	Mortality (%)	Seed yield (Kg ha $^{-1}$)
Pearl millet	3.2	403
Clusterbean	4.7	436
Weeds	5.5	449
Cauliflower leaves	4.1	461
Control	10.7	313

Soil was amended with 4 t compost per hectare

Amendment of soil with compost, in general, increased the population of antagonistic actinomycetes, lytic bacteria and *Nitrosomonas* and decreased the population of *M. phaseolina* in soil (Table 2). Among composts, increase in antagonistic actinomycetes was significantly different from that achieved in the presence of weed compost.

Table 2. Densities of *M. phaseolina*, associated antagonists and *Nitrosomonas* in compost amended soil

Compost	*M. phaseolina* cfu g^{-1} soil	Antagonistic actinomycetes log cfu g^{-1}	Lytic bacteria log cfu g^{-1}	Nitrosomonas log cfu g^{-1}
Pearl millet	411	11.4	12.5	10.4
Clusterbean	330	11.4	12.7	10.1
Weeds	431	11.5	12.6	9.8
Cauliflower leaves	317	12.0	12.7	10.7
Control	597	11.2	12.3	9.8

cfu – colony forming units

In this study, disease suppression due to addition of composts into the soil could possibly be ascribed to (1) Composts support high levels of total microbial population including antagonists and (2) Composts improve the moisture holding capacity of the soil, which in turn could reduce the pathogenic propagules and disease. These results showed greater potentials of cauliflower compost amendment to reduce *M. phaseolina* density compared to other composts, in which increased density of antagonists brought proportionate reduction in dry root rot incidence. Beneficial effects of incorporating cruciferous residues on growth and yield of watermelon and wheat due to disease suppression and increase in microbes beneficial to plant growth has been well documented (Keinath 1996, Kirkegaard et al 1994). Increased seed yield in all the compost-amended treatments may be a cumulative effect of reduced disease incidence, more retention of soil moisture, availability of nutrients and qualitative and quantitative improvements in microbiological properties. Increased microbial population particularly those of antagonistic actinomycetes and lytic bacteria against *M. phaseolina* in compost amended soil prevents germination of pathogenic propagules and infection of the host, presumably through microbiostasis (Hoitink and Boehm 1999).

In a subsequent study, effect of three composts (pearl millet, weeds, neem) was ascertained against dry root rot pathogen, nitrogen fixation and seed yield of cowpea and clusterbean (Lodha and Burman 2000). Studies have shown that pearl millet and weed composts were significantly superior in reducing plant mortality due to dry root rot and in enhancing seed yield of clusterbean and cowpea (Table 3). Invariably, seed yield also was maximum in pearl millet compost amended plots. Activity of nitrogenase was maximum when both the crops were grown in pearl millet compost amended plots (Table 4).

Table 3. Effect of different composts on dry root rot intensity and seed yield of clusterbean and cowpea

Compost	Mortality (%)	Seed yield (Kg ha^{-1})	Mortality (%)	Seed yield (Kg ha^{-1})
Weed	5.1	583	12.0	423
Neem	6.0	556	14.8	370
Pearl millet	5.2	658	10.2	480
Control	9.8	462	20.0	339

Soil was amended with 4 t compost per hectare

Table 4. Effect of different composts on number of nodules and nitrogenase activity in clusterbean and cowpea

Compost	Clusterbean		Cowpea	
	No. of nodules	Nitrogenases (nmoles 2114/hp)	No. of nodules	Nitrogenases (nmoles 2114/hp)
Weed	4.5	5920	16.0	5978
Neem	2.4	5353	17.8	4684
Pearl millet	2.0	6493	11.1	6751
Control	3.2	5261	16.6	6484

Higher reduction in disease incidence in soil amended with pearl millet compost may be attributed to presence of some fungistatic or fungitoxic compounds. Residues of pearl millet were also found highly effective in bringing down propagules of *M. phaseolina.* Use of weed compost as soil amendment has a potential advantage in nutrient deficient sandy soil, as it was prepared purely from on- and off-season weeds. Many weeds are also host of *M. phaseolina.* Since sclerotia are formed during pathogenesis and released in the soil after disintegration of host tissues, infected weeds may also contribute in increasing inoculum density of *M. phaseolina* in soil. However, heat generated during process of composting inactivate *M. phaseolina* propagules during heat phase to a great extent, but precaution should be taken to use only those weeds, which are tolerant to *M. phaseolina* for preparing composts.

Amendment of soil with different composts or FYM resulted in significant reduction in plant mortality (Table 5) due to charcoal rot compared to non-amended control (Bareja et al 2010). The magnitude of reduction in disease incidence ranged between 26.1-65.7% in compost-amended plots. In general, *Acacia nilotica* compost has shown consistency in reducing plant mortality.

Table 5. Effect of compost-amendments on severity of dry root rot and seed yield of cowpea

Amendment[a]	Dry root rot mortality (%)		Seed yield (Kg ha^{-1})	
	2001	2003	2001	2003
Calotropis procera	7.5 (15.85)[b]	5.0 (12.72)	683	1028
Prosopis juliflora	7.9 (16.28)	3.8 (11.04)	806	1101
Weeds	8.4 (16.82)	6.1 (14.03)	734	1012
Azadirachta indica	4.6(12.36)	8.2 (16.29)	701	968
Acacia nilotica	5.9 (14.02)	5.2 (12.83)	775	960
Farm yard manure	8.5 (16.98)	6.5 (14.78)	729	961
Non-amended soil	11.9 (20.14)	11.1 (18.97)	588	897
LSD (*P*=0.05)	2.27	3.36	37	70

[a]Composts or FYM amended @ 4 t ha^{-1}

[b] Angular transformed values

* Values are mean of three replicates

Microbial activity

Disease suppressive properties of composts rely on a number of factors including microbial activity, microbial population dynamics, nutrient concentrations and other associated chemical and physical factors (Fig 1). The mechanism suggested is that the compost enhances the activity of microorganisms introduced with the compost and stimulates those residents in the soil (Lumsden et al 1983b). High levels of microbial activity in composts were postulated as the primary factor in disease control (Phae et al 1990, Nelson 1991, Nelson and Craft 1992, Craft and Nelson 1996, Nakasaki et al 1998). Addition of mature composts prepared from tree bark (Chen et al 1987), municipal sewage sludge (Lumsden et al 1983a), separate cattle manure and grape pomace (Chen et al 1988) to peat mixes increases microbial activity and induces microbiostasis and suppression to Pythium root rot and damping-off (Mandelbaum et al 1988). Potting mixes prepared with Canadian sphagnum peats varied in suppressiveness to Pythium root rot of Poinsettia caused by *Pythium ultimum.* Root rot and population development of *P. ultimum* were highest in the most decomposed, intermediate in a slightly decomposed peat and suppressed in the least decomposed light peat (Boehm and Hoitink 1992). Microbial activity was highest in the suppressive and lowest in the conducive potting mixes, during production of crop, a decrease in microbial activity was accompanied by an increase in the population density of *P. ultimum* and root rot severity.

Chen et al (1988) established negative correlation of disease severity caused by *Pythium ultimum* on cucumber with microbial activity, based on the rate of hydrolysis of fluorescein diacetate and with microbial biomass. Based on extractible phospholipids phosphate content in the container media, several other workers have also used hydrolysis of FDA as an indicator of microbial activity and the aspect has been reviewed by Hoitink and Boehm (1999). Dehydrogenase activity depends on the metabolic state of soil microorganisms although its use as an indicator of soil microbial activity has been criticized because the enzyme is affected by numerous factors like soil type, pH, etc. (Nannpieri et al 1990, Beyer et al 1992). According to Garcia et al (1994), dehydrogenases activity can be used as an indicator of microbial activity in semi arid soils. Two years after the incorporation of organic wastes, the dehydrogenases activity of the amended soil was significantly greater than that of the amended soil indicating the greater metabolic activity of the microorganisms (Ros et al 2003).

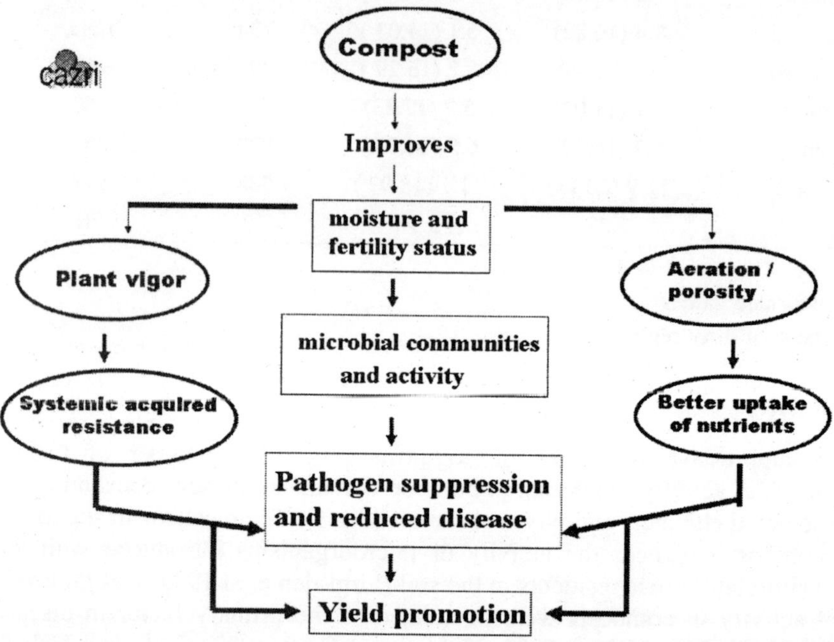

Fig. 1. Mechanism of action for disease suppression and yield promotion by compost

FDA activity seems to provide an image of the quantity and quality of biologically available organic substances in soil required for sustained biological control. FDA activity also affects soil microbial biomass and this has been related to suppression of root rots caused by *Pythium* and *Phytophthora* spp. However, microbial biomass and FDA activity together better predict suppressiveness than either procedure alone (Chen et al 1988, 1988a).

In a study utilizing five compost (*Calotropis procera, Prosopis juliflora*, neem, *Acacia nilotica* and weeds), dehydrogenase activity increased soon after addition of composts in amended compared to non-amended plots, with maximum activity being in *Calotropis* compost-amended soil (Bareja et al 2010). Microbial activity increased in all composts at first sampling date, but greater activity was estimated in *Calotropis* and weed compost-amended soil. The activity declined at second sampling date in *Calotropis, P. juliflora* and weed compost-amended soil but significant increase was estimated in other treatments including non-amended control. At subsequent intervals, a sharp decline in microbial activity was observed in both amended and non-amended soil except *Calotropis* and *P. juliflora* compost-amended soil, where an increase was recorded. On the basis of pooled average, maximum microbial activity, during different stages of crop, was estimated in neem and *Calotropis* followed by *P. juliflora* compost-amended soil, while minimum was in non-amended soil.

Significant negative correlations were established between microbial activity and per cent plant mortality due to dry root rot at 20 days after planting (DAP). At all other intervals, negative correlations between microbial activity and disease level were not significant. Increase in microbial activity in compost compared to non-amended soil is a reflection of increased microbial population and biomass. Negative correlations established between microbial activity and charcoal rot incidence only at 20 DAP indicates that *M. phaseolina* infection in seedlings occurred either at this stage or just before, but was expressed late in the season. In earlier studies, maximum inoculum build-up of *M. phaseolina* and plant mortality due to charcoal rot were recorded, when cowpea seedlings got infection at 15 DAP (Singh and Lodha 1986). Several studies have established that microbial activity and microbial biomass are the main predictors of suppressiveness, where a negative relationship exists (Boehm and Hoitink 1992, You and Sivasithamparam 1995).

Composts for Plant growth

Amendment of soil with compost has shown positive and significant effects on growth parameters of a crop due to increase organic matter, better soil moisture and fertility conditions and increased microbial activity (Fig 1). Cowpea plants grown in amended soil had greater number of nodules compared to non-amended with maximum counts in FYM amended soil, but in *P. juliflora* and *A. nilotica* compost-amended plots these were not significantly different than control. Root length did not differ significantly in amended or non-amended soil, but reduced significantly in weed compost-amended plots (Baraja et al 2010). Except *Calotropis* and weed compost-amended plots, shoot length was greater in other amended plots but significant differences with non-amended soil were recorded only in *P. juliflora, A. nilotica* and FYM amended plots. In fresh and dry shoot weights, consistent trend was not recorded in amended or non-amended plots.

Better nodulation observed in the amended treatments could be ascribed to the availability of soil moisture and nutrients and these factors also enhanced population of beneficial microbes which might have contributed for a better growth of cowpea in *P. juliflora* compost-amended soil. Variations observed in growth parameters of cowpea in different composts can be assigned to differences in biological and chemical properties of the composts (Widmer et al 1998, Abbasi et al 2002).

Improvement in Soil moisture

Soil amendment with compost also improves soil moisture conditions in those soils which are other wise poor in water holding capacity. In study carried out in sandy soils, compost amended plots held more soil moisture (30.4–35.3 mm) compared to corresponding non-amended (29.4 mm) plots (Table 6), (Bareja et al 2010). There was a great variation in different composts at 20 days after planting (DAP), where available soil moisture ranged from 11.9 - 19.6 mm in amended plots compared to 20.2 mm in the non-amended control indicating that cowpea plants effectively utilized soil moisture for the initial growth in amended plots. At 40 DAP, variations in soil moisture conservation were not discernible in different treatments but again FYM amended plots held maximum soil moisture. At fourth sampling date, which coincided with the maturity of cowpea, all the amended plots held more soil moisture than the non amended control.

Table 6. Soil moisture content (mm)* in compost amended and non-amended plots in cowpea field during 2001 crop season

Amendment	Days			
	0	20	40	60
Calotropis procera	33.7±0.9	18.2±0.3	17.9±0.2	8.3±0.6
Prosopis juliflora	35.3±0.4	17.1±0.1	20.6±0.3	9.3±0.4
Weeds	32.2±0.4	17.9±0.3	18.9±0.3	10.3±0.2
Azadirachta indica	29.7±0.2	11.9±0.2	17.9±0.6	8.6±0.2
Acacia nilotica	30.7±0.2	19.2±0.4	18.8±0.4	7.8±0.6
Farm yard manure	30.4±0.2	19.6±0.4	22.5±0.3	6.7±0.3
Non-amended soil	29.4±0.3	20.2±0.4	20.1±0.5	5.2±0.2

*Determined gravimetrically; crop was harvested after 65 days of planting

Availability of more soil moisture in compost-amended compared to non-amended plots can be better utilized for the crop growth. Improved soil moisture conditions, in our study, increased native bacterial population and decreased total fungi. This might have also reduced the population of *M. phaseolina.* In the presence of adequate soil moisture, antagonistic role of some soil bacteria, in reducing the sclerotial population of *M. phaseolina*, is well documented (Dhingra and Sinclair 1975, Lodha 1996).

Improvement in seed yield

All the composts amendment significantly improved seed yield of cowpea compared to non-amended control in a study carried out at CAZRI. Maximum seed yield was recorded in *P. juliflora* compost-amended plots and the yield promotion was significantly greater than in any other compost. Increased seed yield in all the compost-amended treatments may be a cumulative effect of increased soil moisture retention, availability of nutrients, qualitative and quantitative improvements in microbiological properties, increased microbial activity and reduced disease incidence. Beneficial effect of compost-amendment in reducing disease incidence and increasing seed yield has been well documented for sunflower, tomato and cluster bean (Allievi et al 1993, Lodha et al 2002). Improvement in seed yield of cowpea in the year 2003 compared to 2001 is mainly attributed to quantum and distribution of rainfall. In the year 2001, four well distributed rain events (254 mm) occurred during vegetative and flowering stage. Thereafter, no significant rain events occurred leading to severe moisture stress at pod formation stage and onwards till maturity. However, in the year 2003, there were six good evenly distributed rain events (282 mm) till pod formation stage, thus crop experienced moisture stress only after seed formation stage.

Improvement in Micronutrients

Compost also improves availability of micronutrient in soil and as well as in uptake by plants grown in amended soils. Maximum contents of Cu, Fe, and Mn were estimated in plants grown in weed compost-amended soil, while Zn content was higher in those grown in *P. juliflora* amended plots (Table 7) (Bareja et al 2010). Interestingly, Zn content was lower in plants grown in weed compost-amended soil. There was a great variation in Mn content in plants of amended or non-amended soil. Overall, maximum uptake of these nutrients was estimated in plants grown in weed followed by *A. nilotica* compost-amended soil.

Table 7. Micronutrients (ppm)* in compost-amended and non-amended soil and cowpea plants

Amendment	Cu	Fe	Zn	Mn
Calotropis procera	2.14 [d*]	5.22 [f]	0.62 [c]	10.08 [f]
	99**	238	36	47
Prosopis juliflora	2.30 [c]	7.62 [d]	0.52 [d]	13.10 [c]
	101	218	44	56
Weeds	2.52 [a]	8.32 [b]	0.80 [ab]	14.42 [a]
	103	287	35	64
Azadirachta indica	2.32 [c]	7.58 [d]	0.64 [c]	12.13 [e]
	102	205	38	53
Acacia nilotica	2.40 [b]	8.72 [a]	0.80 [ab]	13.16 [b]
	98	256	37	62
Farm yard manure	2.44 [b]	7.84 [c]	0.86 [a]	12.72 [d]
	99	196	36	37
Non-amended soil	2.40 [b]	5.84 [e]	0.76 [b]	9.76 [g]
	92	208	40	56

* Values are mean of three replicates. For each parameter, values in the same column followed by an identical letter are not significantly different according to Fisher's LSD test ($P \leq 0.05$).

** Standard Error Cu =3.6, Fe= 32.6, Zn=3.1 and Mn=9.1.

After two successive crops of cowpea, maximum contents of Cu and Mn were determined in weed compost-amended soil while those of Fe and Zn were in *A. nilotica* and FYM amended soil, respectively. In general, Zn availability remained lower (0.52 - 0.86 ppm) but Mn availability remained higher (9.76-14.42 ppm) in amended and non-amended soils after the harvest of cowpea crop. Enhanced availability of some micronutrients is of great significance not only for plant growth, but also in imparting resistance against soil-borne diseases (Engelhard 1989).

Survival of bio-control agents

Exploitation and recycling of naturally available agricultural waste media have been used for growth and carrier medium for multiplication of various bio-control agents. Research efforts made at CAZRI have shown that utilizing pods of *P. juliflora*, an abundantly grown shrub, with potato improved multiplication of *T. harzianum* in liquid broth (Lodha et al 1999). Thus, utilizing *P. juliflora* plant parts for the preparation of compost can be a better choice for fast multiplication and as a substrate for easy delivery in soil after further experimentation.

Future Thurst

Health and environmental concerns associated with conventional agriculture have raised interests in alternative farming practices that may reduce the adverse effects of conventional production practices and at the same time sustain productivity. Alternative agriculture encompasses various farming systems known as organic, regenerative, ecological, biological and biodynamic. Even though they involve diverse practices, these farming systems pursue a common objective of elimination or reduction of reliance on synthetic fertilizers and pesticides. Instead, they emphasize greater use of biological resources such as animal manure, crop residues, composts and other organic amendments to supply plant nutrients. Among these, use of composts in improving soil fertility, suppressing plant diseases and enhancing productivity has been found not only cost effective but also a practical approach to minimize dependence on chemical sources of fertilizers and pesticides. Where available and when economically feasible, application of this natural, organic resource can provide nutrient supplements as well as inducing disease control. The luxury of dumping our municipal wastes in oceans or in landfills soon will no longer be available should be utilized in compost making and further apply in fields where soil borne diseases are problems.

References

Abbasi PA, Al Dahmani J, Sahin F, Hoitink HAJ, and Miller SA 2002 Effect of compost amendments on disease severity and yield of tomato in organic and conventional production systems, *Plant Disease* 86 156-161

Acharya CN 1939 The hot fermentation process for composting town refuses and other waste materials, *Indian Journal of Agricultural Sciences* 9 741-817

Allievi L, Marchesini A, Salardi C, Piano V and Ferrari A 1993 Plant quality and soil residual fertility six years after a compost treatment, *Bioresource Technology* **43** 85-89

Bareja M, Kumar P and Lodha S 2010 Effect of composts on microbial dynamics and activity, dry root rot severity and seed yield of cowpea in the Indian arid region, *Phytopathologia Mediterranea* **49** 381-392

Becker JO and Cook RJ 1988 Role of siderophores in suppression of *Pythium* species and production of increased growth response of wheat by fluorescent pseudomonads, *Phytopathology* **78** 778-782

Beyer L, Wackendorf C, Balezen FM and Balzer Graf VR 1992 The effect of soil texture and soil management on microbial biomass and soil enzyme activities in arable soils of Northwest Germany, *Agrobiology Research* **45** 276-283

Boehm MJ and Hoitink HAJ 1992 Sustenance of microbial activity in potting mixes and its impact on severity of Pythium root rot of poinsettia, *Phytopathology* **82** 259-264

Boehm MJ, Madden LV and Hoitink HAJ 1993 Effect of organic matter decomposition level on bacterial species diversity and composition in relationship to Pythium damping-off severity, *Applied Environmental Microbiology* **59** 4171- 4179

Boehm MJ, Wu T, Stone AG, Kraakman B and Iannotti DA 1997 Cross-polarized magic-angle spinning ^{13}C nuclear magnetic resonance spectroscopic characterization of soil organic matter relative to culturable bacterial species composition and sustained biological control of Pythium root rot, *Applied Environmental Microbiology* **63** 162-168

Bollen GJ 1993 Factors involved in inactivation of plant pathogens during composting of crop residues, in *Science and Engineering of Composting: Design, Environmental, Microbiological and Utilization Aspects* (eds) HAJ Hoitink and H M Keeners, Worthington, OH: Renaissance Publications, 301-318 pages

Bollen GJ, Volker D and Wijnen AP 1989 Inactivation of soil-borne plant pathogens during small scale composting of crop residues, *Netherland Journal of Plant Pathology* **95 (1)** 19-30

Chen W, Hoitink HAJ and Madden LV 1988a Microbial activity and biomass in container media for predicting suppressiveness to damping-off caused by *Pythium ultimum, Phytopathology* **78** 1447-1450

Chen W, Hoitink HAJ and Schmitthenner AF 1987 Factors affecting suppression of *Pythium* damping-off in container media amended with composts, *Phytopathology* **77** 755-760

Chen W, Hoitink HAJ Schmitthenner AF, and Tuovinen OH 1988 The role of microbial activity in suppression of damping-off caused by *Pythium ultimum, Phytopathology* **78** 314-322

Chen Y and Hadar Y 1985 Composting and use of agricultural wastes in container media, in *Composts: Production, Quality and Use* (eds) De Bertoldi et al, London and New York: Elsvier, 853 pages

Chen Y and Inbar Y 1993 Chemical and spectroscopical analysis of organic matter transformations during composting in relation to compost maturity, in *Science and Engineering of Composting: Design, Environmental, Microbiological and Utilization Aspects* (eds) HAJ Hoitink and HM Keeners, Worthington, OH: Renaissance Publications, 551-600 pages

Clulow SA, Stewart HE, Dashwood EP and Wastie RL 1995 Tuber surface micro-organisms influence the susceptibility of potato tubers to late blight, *Annals of Applied Biology* **126** 33-43

Craft CM and Nelson EB 1996 Microbial properties of composts that suppress damping-off and root rot of creeping bent grass caused by *Pythium graminicola, Applied Environmental Biology* **62** 1550-1557

Daft GC, Poole HA and Hoitink HAJ 1979 Composted hardwood bark: A substitute for steam sterilization and fungicide drenches for control of poinsettia crown and root rot, *Horticulture Science* **14** 185-187

Dhingra OD and Sinclair JB 1975 Survival of *Macrophomina phaseolina* sclerotia in soil: effects of soil moisture, carbon: nitrogen ratios, carbon sources and nitrogen concentrations, *Phytopathology* **65** 236-240

Eberhardt DL and Pipes WO 1972 Illinois composting application, in *Large-Scale composting* (ed) MJ Satriana, Park Bridge, New Jersy: Noyes Data Corporation

Elad Y and Chet I 1987 Possible role of competition for nutrients in bio-control of *Pythium* damping-off by bacteria, *Phytopathology* **77** 190-195

Engelhard AW 1989 Historical highlights and prospects for the future, in *Management of Diseases with Macro and Microelements* (ed) AW Engelhard, The American Phytopathological Society, St. Paul, MN, USA, 9-17 pages

Gamliel A 2000 Soil amendments: A non-chemical approach to the management of soil borne pest *Acta Horticulturae* **532** 39-47

Gamliel A and Stapleton JJ 1993 Effect of chicken compost or ammonium phosphate and solarization on pathogen control, rhizosphere micro-organisms and lettuce growth, *Plant Disease* **77** 886-891

Garcia C, Hernandez T and Costa F 1994 Microbial activity in soils under Mediterranean environmental conditions, *Soil Biology and Biochemistry* **26** 1185-1191

Gaur AC 1982 A manual of rural composting, FAO/UNDP Regional Project, RAS75/004, Field document no 15, FAO, Rome, 102 pages

Hadar Y and Mandelbaum R 1986 Suppression of *Pythium aphanidermatum* damping-off in container media containing composted liquorice roots, *Crop Protection* **5** 88-92

Hadar Y, Harman GE, Taylor AG and Norton JM 1983 Effects of pre germination of pea and cucumber seeds and of seed treatment with *Enterobacter cloacae* on rots caused by *Pythium* spp., *Phytopathology* **73** 1322-1325

Han DY, Coplin DL and Hoitink HAJ 1998 Partial characterization of systemic acquired resistance induced in radish by *Pantoea agglomerans* strain E278Ar, *Phytopathology* **88** S36 (Abstract)

Hardy GESJ and Sivasithamparam K 1991 Sporangial response of *Phytophthora drechsleri* in composted eucalyptus barks mix, *Soil Biology and Biochemistry* **23** 756-765

Hardy GESJ and Sivasithamparam K 1995 Antagonism of fungi and actinomycetes isolated from composted Eucalyptus bark to *Phytophthora drechsleri* in a steamed and non-steamed composted eucalyptus bark-amended container medium, *Soil Biology and Biochemistry* **27** 243-246

Hoitink HAJ and Boehm MJ 1999 Bio-control with in the context of soil microbial communities: A substrate dependent phenomena, *Annual Review of Phytopathology* **37** 427-426

Hoitink HAJ and Fahy PC 1986 Basis for the control of soil-borne plant pathogens with composts, *Annual Review of Phytopathology* **46** 309-320

Hoitink HAJ and Keener HM 1993 Science and Engineering of Composting, in *Design, Environmental, Microbiological and Utilization aspects* (eds) HAJ Hoitink and H M Keener) Worthington, OH: Renaissance Publications, 728 pages

Hoitink HAJ and Poole HA 1980 Factors affecting quality of composts for utilization in container media, *Horticultural Science* **15** 171-173

Hoitink HAJ, Inbar Y, and Boehm MJ 1991 Status of compost amended potting mixes naturally suppressive to soil-borne diseases of floricultural crops, *Plant Disease* **75** 869-873

Hoitink HAJ, Stone AG, and Han DY 1997 Suppression of plant diseases by composts, *Horticultural Science* **32** 184-187

Howard A and Ward YD 1931 *The waste products of agriculture and their utilisation as humus*, London: Oxford University Press, 167 pages

Howell CR, Beier RC and Stipanovie RD 1988 Production of ammonia by *Enterobacter cloacae* and its possible role in the biological control of Pythium pre emergence damping-off by the bacterium, *Phytopathology* **78** 1075-1078

Katan J, Greenberger AH and Grinstein A 1976 Solar heating by polyethylene mulching for the control of diseases caused by soil-borne pathogens, *Phytopathology* **66** 683-688

Keinath AP 1996 Soil amendment with cabbage residue and crop rotation to reduce gummy stem blight and increase growth and yield of watermelon, *Plant Disease* **80** 564-570

Kim KD, Nemec S and Musson G 1997 Effects of composts and soil amendments on soil microflora and *Phytophthora* root and crown rot of bell pepper, *Crop Protection* **16** 165-172

Kirkegaard JA, Gardner PA, Angus JF and Koetz E 1994 Effect of brassica crops on the growth and yield of wheat, *Australian Journal of Agricultural Research* **45** 529-545

Kovoor ATA 1954 Some factors affecting the growth of *Rhizoctonia bataticola* in soil, *Journal of Madras University* **24** 47-52

Lazarovits G, Tenuta M and Conn KL 2000 Utilization of high nitrogen and swine manure amendments for control of soil-borne diseases: Efficacy and mode of action, *Acta Horticulturae* **532** 59-64

Lockwood JI 1977 Fungistatis in soils, *Biological Reviews* **52** 1- 43

Lodha S 1996 Influence of moisture conservation techniques on *Macrophomina phaseolina* population, dry root rot and yield of cluster bean, *Indian Phytopathology* **49** 342-349

Lodha S and Burman U 2000 Efficacy of composts on nitrogen fixation, dry root rot *(Macrophomina phaseolina)* intensity and yield of legumes, *Indian Journal of Agricultural Sciences* **70** (12) 846-849

Lodha S, Sharma SK and Aggarwal RK 1997 Solarization and natural heating of irrigated soil amended with cruciferous residue for improved control of *Macrophomina phaseolina, Plant Patholology* **46** 186-190

Lodha S, Bohra MD and Harsh LN 1999 Evaluation of *Prosopis* pods as a source of carbohydrate for enhancing the growth of soil borne fungi, *Indian Phytopathology* **52** 24-27

Lodha S, Sharma SK and Aggarwal RK 2002 Inactivation of *Macrophomina phaseolina* propagules during composting and effect of composts on dry root rot severity and on seed yield of cluster bean, *European Journal of Plant Pathology* **108** 253-261

Lumsden RD, Lewis JA and Milner PD 1983a Effect of composted sewage sludge on several soilborne pathogens and diseases, *Phytopathology* **73** 1543-1548

Lumsden RD, Lewis JA and Papavizas GC 1983b Effect of organic amendments on soil borne plant pathogens and their antagonists, in *Environmentally sound Agriculture* (ed) W Lockeretz, New York: Praeger Press, 51-70 pages

Mandelbaum R and Hadar Y 1990 Effects of available carbon source on microbial activity and suppression of *Pythium aphanidermatum* in composts and peat container media, *Phytopathology* **80** 794-804

Mandelbaum R, Hadar Y and Chen Y 1988 Composting of agricultural wastes for their use as container media: Effect of heat treatments on suppression of *Pythium aphanidermatum* and microbial activities in substrates containing compost, *Biological Wastes* **26** 261-274

Nakasaki K, Hiraoka S and Nagata H 1998 A new operation for producing disease suppressive compost from grass clippings, *Applied Environmental Microbiology* **64** 4015 - 4020

Nannpieri P, Greco S and Ceccanti B 1990 Ecological significance of the biological activity in soil, in *Soil Biochemistry* (eds) JM Bollag and G Stotzeky Vol. **6**, New York: Marcel Dekker, 293-355 pages

Nelson EB 1991 Introduction and establishment of strains of *Enterobacter cloacae* in golf course turf for the biological control of dollar spot, *Plant Disease* **75** 510- 514

Nelson EB and Craft CM 1992 Suppression of dollar spot on creeping bent grass and annual blue grass turf with compost amended top dressings, *Plant Disease* **76** 954-958

Nelson EB and Hoitink HAJ 1982 Factors affecting suppression of *Rhizoctonia solani* in container media, *Phytopathology* **72** 275-279

Papavizas GC and Lumsden RD 1980 Biological control of soilborne fungal propagules, *Annual Review of Phytopathology* **18** 389-413

Phae CG, Saski M, Shoda M and Kubota H 1990 Characteristics of *Bacillus subtilis* isolated from composts suppressing phytopathogenic microorganisms, *Soil Science and Plant Nutrition* **36** 575-586

Ramirez-Villapudua J and Munnecke DE 1988 Effects of solar heating and soil amendments of cruciferous residues on *Fusarium oxysporum* f.sp. *conglutinans* and other micro-organisms, *Phytopathology* **78** 289-295

Ros M, Hernandez MT, and Garcia C 2003 Soil microbial activity after restoration of a semi arid soil by organic amendments, *Soil Biology and Biochemistry* **35** 463-469

Runia WT 1983 A recent development in steam sterilization, *Acta Horticulturae* **152** 195-200

Runia WT 2000 Steaming methods for soils and substrates, *Acta Horticulturae* **532** 115-123

Singh S and Lodha S 1986 Varietal resistance of cowpea to *Macrophomina phaseolina* (Tassi.) Goid. causing dry root rot and its control, *Indian Journal of Agricultural Sciences* **56** 552-555

Sivan A, Elad Y and Chet Y 1984 Biological control effects of a new isolate of *Trichoderma harzianum* on *Pythium aphanidermatum, Phytopathology* **74** 498-501

Smith WH 1969 Germination of *Macrophomina phaseolina* sclerotia as affected by *Pinus lambertiana* root exudates, *Canadian Journal of Microbiology* **15** 1387-1391

Spencer S and Benson DM 1982 Pinebark, hardwood bark compost, and peat amendment effects on development of *Phytophthora* spp. and lupine root rot, *Phytopathology* **72** 346-351

Strom PF 1985 Effect of temperature on bacterial species diversity in thermophilic solid-waste composting, *Applied Environmental Microbiology* **50** 899-905

Tapiador DD 1981 Vermi-culture and its potential in Thailand and other Asian countries, First National Earthworm Growers convention, Metro Manila, Phillipines

Trankner A 1992 Use of agricultural and municipal organic wastes to develop suppressiveness to plant pathogens, in *Biological control of Plant Diseases* (eds) EC Tjamos, GC Papavizas, and Cook RJ, New York: Plenum Press, 35-42 pages

Whipps JM 1997 Developments in the biological control of plant pathogens, *Advances in Botanical Research* **26** 1-134

Widmer TL, Graham JH and Mitchell JD 1998 Composted municipal waste reduces infection of citrus seedling by *Phytophthora nicotianae, Plant Disease* **82** 683-688

Ylimaki A, Toiviainen A, Kallio H and Tikanmaki E 1983 Survival of some plant pathogens during industrial-scale composting of wastes from a food processing plant, *Annales Agriculturae Fenniae* **22** 77-85

You MP and Sivasithamparam K 1995 Changes in microbial populations of an avocado plantation mulch suppressive of *Phytophthora cinnamomi, Applied Soil Ecology* **2** 33-43

Zhang W, Dick WA and Hoitink HAJ 1996 Compost-induced systemic acquired resistance in cucumber to Pythium root rot and anthracnose, *Phytopathology* **86** 1066-1070

Zhang W, Han DY, Dick WA, Davis KR and Hoitink HAJ 1998 Composts and compost water extract-induced systemic acquired resistance in cucumber and Arabidopsis, *Phytopathology* **88** 450-455.

3

Biological Control of Plant Diseases

S. K. Gupta, A.K. Gupta, Neeraj Sharma and Durga Parasad

Department of Plant Pathology, Dr Y S Parmar University of Horticulture and Forestry, Nauni, Solan - 173 230 (HP)

Biological control, the use of specific organisms that interfere with plant pathogens, is a nature-friendly, ecological approach to overcome the problems caused by commercial chemical methods of plant protection (Chet, 2012). Microorganisms like fungi and bacteria naturally present in their natural habitat, have been parasitizing both cultivated and non-cultivated plants over a million of years. Also they have been interacting to their counterparts and have been governing the infection in plants to a minimum level. The phenomenon of controlling one organism by other is called "Biological control" and this phenomenon attracted the attention of workers quite late during 1920's and 1930's. These attempts were so unsuccessful that research on this phenomenon was virtually ceased for about three decades. The turning point in this area occur in 1963 when contemporary leaders of plant pathology assembled at Berkely to laid down a formidable foundation to the science of biological control of plant pathogens. This international symposium was held on "Ecology of soil borne plant pathogens - prelude to biological control". Since then a variety of biological control are available for use, but further development and effective adoption will require a greater understanding of complex interactions among plants, people and environment.

Definition

The term biological control was coined by late Harry Smith of University of California for insect pest and defined biological control as the "Suppression of insect population by the action of their nature and natural enemies". Biological control was used for the first time in relation to plant pathogens in 1914. Since then a number of definitions were proposed time to time by different workers due to the disagreement among scientists for the biological control. There are some definitions explained below:

DeBach (1964) defined biological control in relation to insect pest and weed as "the action of predators, parasites, or pathogens in maintaining another organism's population density at a lower average than would occur in their absence".

Garret (1956) excluding the role of man in reducing disease and considering antagonism as the central dogma, defined biological control as "any condition under which or practice whereby survival or activity of pathogen is reduced through the agency of any other living organism (except man himself) with the result that there is a reduction in the incidence of the disease caused by that pathogen"

Baker and Cook (1974) stated that the "Biological control is the reduction of inoculums density producing activities of a pathogen or parasite in its active or dormant state, by one or more organisms, accomplished naturally or through the manipulation of the environment, host, or antagonist, or by mass introduction of one or more antagonists".

Deacons (1976) stated that "is a practice in which, or process whereby, the undesirable effects of an organism are reduced through the agency of another organism that is not the host plant, the pest or pathogen, or man".

History of biological control of plant pathogens

W. Roberts (1874) demonstrated the antagonistic action of microorganisms in liquid culture between *Penicillium glaucum* and bacteria and introduced the term "antagonism" used in disease management. Hartley (1921) made the first attempt to control the root diseases with microorganisms introduced in to the soil. Sanford (1926) observed that potato scab (*Streptomyces scabies*) was suppressed by green manuring through the activity of antagonists. Similarly, Millard and Taylor (1927) controlled the scab of potato by adding *Streptomyces praecox* and green grass clipping to sterilize soil. G.B. Sanford and W.C. Broadfoot demonstrated in 1931 a "suppressive effect on take all pathogen (*Gaeumannomyces graminis* var. *tritici*) in pots. Weindling (1932) reported the antagonistic nature of *Trichoderma lignosum* on several plant pathogens. Grossbard (1948-1951), Wright (1952-1957) and other demonstrated that antibiotics were produced in soil by a *Penicillium, Aspergillius, Trichoderma* and *Streptomyces* spp. Rishbeth (1963) inoculated the freshly cut pine stumps with weak pathogen *Peniophthora gigantean* which prevented pathogen invasion by hyphal interference. Since then fungal bio-control agents mainly *Trichoderma* and *Gliocladium* spp. have received much more attention (Chaube and Singh, 1991). Lindberg (1959) found that cultures of *Helminthosporium victoriae* developed a transmissible disease that caused reduced virulance of pathogen on oats. A. Biraghi found in 1950 that cankers caused by *Cryphonectria parasitica* in chestnut trees in Italy were unexpectedly healing.

Different types of interactions involved in biological control

Different types of interactions exist between the pathogen and plant. These interactions significantly affect the plant health in various ways. Odum (1953) proposed the different types of interactions viz., mutualism, protocooperation, commensalisms, neutralism, competition, amensalism, parasitism and predation. These types of interactions can be further divided into two groups.

1. **Positive interactions:** These types of interactions involve those relationships in which both the organisms are benefited. e.g. mutualism, commensalisms, protocooperation etc.

2. **Negative interactions:** These types of interactions involve those relationships in which one organism is benefited at the cost of other. It may be possible that either one or both the organisms are harmed. e.g. antagonism, competition, parasitism and predation etc.
3. **Neutral interactions:** In this type of relationship neither one organism is benefited nor harmed e. g. neutralism.

1. Positive interactions

(i) Mutualism: Mutualism is an association between two or more species where both species derive benefit. Sometimes, it is an obligatory lifelong interaction involving close physical and biochemical contact, such as those between plants and mycorrhizal fungi. However, they are generally facultative and opportunistic. For example, bacteria in the genus *Rhizobium* can reproduce either in the soil or, to a much greater degree, through their mutualistic association with legume plants. These types of mutualism can contribute to biological control, by fortifying the plant with improved nutrition and/or by stimulating host defenses.

(ii) Protocooperation: Protocooperation is a form of mutualism, but the organisms involved do not depend exclusively on each other for survival. Many of the microbes isolated and classified as BCAs can be considered facultative mutualists involved in protocooperation, because survival rarely depends on any specific host and disease suppression will vary depending on the prevailing environmental conditions.

(iii) Commensalism: Commensalism is a symbiotic interaction between two living organisms, where one organism benefits and the other is neither harmed nor benefited. Most plant-associated microbes are assumed to be commensals with regards to the host plant, because their presence, individually or in total, rarely results in overtly positive or negative consequences to the plant.

2. Negative interactions

(i) Antagonism: It is defined as the phenomenon in which one organism releases those substances which are toxic to the growth of other organism.

(ii) Competition: Competition within and between species results in decreased growth, activity and /or fecundity of the interacting organisms. Biocontrol can occur when non-pathogens compete with pathogens for nutrients in and around the host plant. Direct interactions that benefit one population at the expense of another also affect our understanding of biological control.

(iii) Parasitism: Parasitism is a symbiosis in which two phylogenetically unrelated organisms coexist over a prolonged period of time. In this type of association, one organism, usually the physically smaller of the two (called the parasite) benefits and the other (called the host) is harmed to some measurable extent.

3. Neutral interactions

(i) Neutralism: Neutralism describes the biological interactions when the population density of one species has absolutely no effect whatsoever on the other.

Related to biological control, an inability to associate the population dynamics of pathogen with that of another organism would indicate neutralism.

Types of biological control

There are two types of Biological control:

1. Natural biological control
2. Induced biological control

1. Natural biological control

Natural biological control can be defined as the control in which the antagonists are already present in the soil that suppresses almost all root pathogens and can reduce the inoculum. Soil can be classified as conducive and suppressive soils with regard to development of the disease (Table 1). There are many soil borne diseases like root rot (*Phytophthora cinnamoni*), vascular wilt (*Fusarium oxysporum*), damping off (*Pythium* spp.) and take all disease in wheat (*Gaeumannomyces graminis*) which appear in severe form whereas the same diseases develop much less and cause milder effect on their host in other soils, such soils are called suppressive soils. The inhospitality of certain soils to some pathogens is such that either the pathogens cannot establish themselves or they are established, but fail to cause disease or they diminish in severity with continued culture of crop. Thus suppressive soil is an umbrella term encompassing fungistasis, competitive, saprophytic ability and other disease and pathogen interactions, where the defined relationship of reduced disease in the presence of the pathogen and susceptible host exist. The term "Suppressive soils" is opposite to conducive soils.

Table 1. Examples of pathogen inoculums suppressed by the suppressive soils

Pathogen	Disease(s) Caused	Pathogen	Disease (s) caused
Armillaria mellea	Root rot of conifers	*Poria weirii*	Root rot of conifers
Didymella lycopersici	Stem rot of Tomato	*Vertcillium alboatrum*	Wilt of potatoes
F.oxysporum f. sp. *batatos*	Wilt of sweet Potatoes	*Cephalosporium gramineurum*	Stripe of wheat
F. oxysporum f. sp. *dianthi*	Wilt of carnation	*Pythium aphanidermatum*	Root rot of radish
F. oxysporum f. sp. *lycoporsici*	Wilt of Tomato	*Pythium sp*	Root rots
F. oxysporum f. sp. *pisi*	Wilt of Pea	*Sclerotium rolfsii*	Root rot of Tomato
Fusarium udum	Wilt of Pigeon Pea	*Streptomyces scabies*	Scab of Potatoes
Gaeumanomyces graminis var. *tritici*	Take all of wheat	*Fusarium avenaceum*	Root rot of many crops
Phytophnthora cinnamomi	Root rot of crops	*F. oxysporum* f. sp. *cubense*	Wilt of banana
Pseudocercosporella-huspotrichoides	Root rot of cereals	*Pythium ultimum*	Root rots

F. oxysporum f. sp. *lini*	Wilt of flax	*Rhizoctonia solani*	Root rot of many crops
F. oxysporum f.sp *melonis*	Wilt of melon	*Sclerotium cepivorum*	Wilt of onion
Fusarium solani	Root rot of bean	*Heterodera avenae*	Cereal cyst nematode

2. Introduced Biological control

In this, the antagonists are introduced to the soil. This was first attempted by Hartley in 1921 against damping off of coniferous seedlings, common scab of potato (Millard and Taylor, 1927) and in take-all diseases of wheat (Sanford and Broadfoot, 1931, Baker and Snyder, 1965). Introduced biological control can be achieved by the following ways:

(i) Trap Plants: Trap plants have also been suggested for use as biological barriers against spread of *Radiopholus similes*, the burrowing nematode in citrus (Ford, 1968).

If a few rows of rye, corn or other tall plants are planted around a field of beans, peppers, or squash, many of the incoming aphids carrying viruses that attack these crops will first stop and feed on the peripheral taller rows of rye or corn. Most of the aphid borne viruse(s) are non-persistent in the aphids; therefore trap crops reduce the amount of inoculum that reaches a crop. Some plants that are not actually susceptible to certain sedentary plant parasitic nematodes produce exudates that stimulate eggs of this nematode to hatch. The juvenile enter these plants but are unable to develop in to adults and eventually they die. Such plants are also called "Trap crops".

(ii) Antagonistic/Inhibitory plants: African and French marigold (*Tagetes erecta* and *T. patula)* was found to be resistant against root knot nematode (Steiner, 1941).

Fungistasis

Soil fungistasis, a phenomenon of inhibition of the spore germination in soils, is an ecologically important mechanism in maintaining the biological balance in soil. The term fungistasis, originally proposed by Dobbs and Hinson (1953), describes the phenomenon whereby viable propagules, not under the influence of endogenous or constitutive dormancy do not germinate in the soil in conditions of temperature and moisture favourable for germination. There are three stages in fungistatis namely induction, maintenance and release of fungistasis (Watson and Ford, 1972). These stages are controlled by a complex balance of stimulators and inhibitors present in the soil microenvironments. The stimulators are of biotic origin and may act as nutrients while inhibitors are of both biotic and abiotic origin. The organic substances present in the root exudates have important role to play. The sclerotia of *Sclerotium cepivorum* were free from fungistasis by volatile compounds associated with the root of *Allium* species. These substances have been identified as allyl sulfides from allyl cystein sulfoxides originating from *Allium* roots. Similarly, the volatile compounds like alcohols and aldehydes with low molecular weights released from plant residues were also found to be stimulatory to soil fungi. Freeing the fungal propagules from soil fungistasis and allowing a spore germination and lysis are important aspects for those using soil fungistasis as a tool in controlling soil - borne pathogens.

Soil fungistasis and biological control

Soil fungistasis play an important role in controlling plant diseases caused by soil borne pathogens. Fungistasis in natural soil is usually nullified by enriching the soil with energy sources or organic matter. The manipulation of the soil environment by organic amendment to enhance the liberation of volatile fungistatic substances is useful for the biological control. The enhanced production of ammonia was reported from soil amended with chitin, organic matters of crucifers and soyabean, linseed and cotton seed meals, and it was found that ammonia was suppressive to some of the soil - borne fungal pathogens like *Fusarium oxysporum* and *F. solani*, *F. solani* f. sp. *cucurbitae*, *Macrophomina phaseolina* and *Phytophthora cinnamomi* (Chaube and Singh, 1991).

Table 2. Types of different antagonisms involved in the biological control of plant pathogens

Type	Mechanism	Examples
Direct antagonism	Hyperparasitism/ Predation	Lytic/ non lytic mycoviruses *Ampelomyces quisqualis, Lysobacter enzymogenes, Pasteuria penetrans, Trichoderma virens*
Mixed path antagonism	Antibiotics	2,4- diaacteylphoroglucinol, Phenazines, Cyclic lipopeptides
	Lytic enzymes	Chitinases, Glucanases, Proteases
	Unregulated waste products	Ammonia, Carbon dioxide, Hydrogen cyanide
	Physical/ chemical interference	Blockage of soil pores Germination signals consumption Molecular cross talk confused
Indirect antagonism	Competition	Exudated/ leachates consumption Siderophore scavenfing Physical niche occupation
	Induction of host resistance	Contact with fungal cell walls Detection of pathogen- associated, molecular patterns Phytoharmone- mediated induction

(Pal and Gardener, 2006)

Machanism behind biological control

Different types of mechanisms are involved in the biological control (Table 2). Direct antagonism results from the physical contact and/ or a high degree of selectivity for the pathogen by the mechanism(s) expressed by the BCA (s). Hyperparasitism by obligate parasites of the plant pathogens is considered as the most direct type of antagonism because the activites of no other organism would be required to exert the suppressive effect. Indirect antagonism results from the activities that do not involve the sensing or targeting a pathogen by the BCA (s). Stimulation of the host defence pathways by non pathogenic BCA (s) is the most indirect form of antagonism. The most effective BCAs studied to date appear to antagonize pathogens using multiple mechanisms. For instance, pseudomonads known to produce the antibiotic 2,4-diacetylphloroglucinol (DAPG) may also induce host defenses (Iavicoli et al 2003). Additionally, DAPG-producers can aggressively colonize roots, a trait that might further contribute to their ability to suppress pathogen activity in the rhizosphere of wheat through competition for organic nutrients (Raaijmakers and Weller, 2001).

1. Hyperparasitism: Hyperparasitism can be defined as the complex process that involves the tropic growth of the biocontrol agent toward the target organism, coiling, final attack and dissolution of the target organism's cell wall/ cell membrane by the activity of enzymes (Chaube and Singh 1991, Duffy et al 1996, Tewari 1996, Biswas et al 2007). Hyperparasitism is the main mechanism involved in the antagonism of *Trichoderma* as bio- control agent. *Trichoderma harzianum* exhibits the excellent mycoparasitic activity against *Rhizoctonia solani* hyphae whereas *T. virens* relies more on the antibiosis against hyphae of the pathogen (Roberts et al 1997). Nature of antagonism is not only dependent on the antagonist but also on the target pathogen. Parasitism is more common against *R. solani*, while antibiosis is predominant mechanism against *Fusarium* spp. *T. harzanium* also parasitizes the nematodes (Sharma 1996). *T. harzianum* coiled around the second stage juveniles of root knot nematode *Meleoidogyne javanica* and penetrated them by forming haustoria like structure. Egg masses were also penetrated by hyphae of *T. harzianum*. Some species of *Aspergillus* also parasitize directly on the pathogen hyphae (Selvakumar et al 2000). In general there are four classes of hyperparasitism: obligate pathogen, hypoviruses, facultative parasites, and predators. *Pasteuria penetrans* is an obligate bacterial pathogen of root-knot nematodes that has been used as a BCA. There are several fungal parasites of plant pathogens, including those that attack sclerotia (e.g. *Coniothyrium minitans*) while others attack living hyphae (e.g. *Pythium oligandrum*). A single fungal pathogen can be attacked by multiple hyper-parasites e.g. *Acremonium alternatum*, *Acrodontium crateriforme*, *Ampelomyces quisqualis*, *Cladosporium oxysporum* and *Gliocladium virens* are just a few of the fungi that have the capacity to parasitize powdery mildew pathogens (Kiss 2003). Other hyperparasites attack plant-pathogenic nematodes during different stages of their life cycles (e.g. *Paecilomyces lilacinus* and *Dactylella oviparasitica*). In contrast to hyperparasitism, microbial predation is more general and pathogen non-specific and generally provides less predictable levels of disease control. Some BCAs exhibit predatory behavior under nutrient-limited conditions. However, such activity generally is not expressed under typical growing conditions. For example, some species of *Trichoderma* produce a range of enzymes that are directed against cell walls of fungi. However, when fresh bark is used in composts, *Trichoderma* spp. does not directly attack the plant pathogen, *R. solani*. But in decomposing bark, the concentration of readily available cellulose decreases and this activates the chitinase genes of *Trichoderma* spp., which in turn produce chitinase to parasitize *R. solani* (Benhamou and Chet 1997, Biswas et al 2007).

2. Hypovirulence: Inoculation of diseased plant with hypovirulent strains (pathogens with reduced pathogenicity) carrying double stranded RNA or DNA plasmids can result in decreased virulence in the population of the pathogens overall, through transfer of these extrachromosomal elements to the pathogenic strain *in vivo*. Hypoviruses are the examples of hypovirulence. A clssical example is the virus that infects *Cryptonectria parasitica*, a fungus causing chestnut blight has reduced the disease producing capacity of the pathogen. This phenomenon has controlled the chestnut blight in many places (Newhouse et al 1990, Milgroom and Cortesi 2004, Tjamos et al 2010). The introduction of hypovirulent strains in Italy and France has enabled chestnut production to occur in orchard previously infested with pathogen (Griffin, 1986). Hypovirulent *Ophiostoma ulmi* can be used to control the Dutch elm disease (*Ophiostoma ulmi*) in elm (Rogers et al 1986). However the interaction of virus, fungus, host and environment determines the successs or failure of the hypovirulence.

3. Antibiosis: The antagonistic organisms release antibiotics and other metabolites that are harmful to the pathogen and inhibit their growth. Antibiotics are microbial toxins that can, at low concentrations, poison or kill other microorganism (Miroua, 1982). Most microorganisms produce and secrete one or more compounds with antibiotic activity. Many antibiotics have been isolated and characterized from bacterial and fungal bioagents (Table 3). These include gliotoxin and glyoviridin from *T. virens*; viridin alkalye pyrones, isonitriles, peptaibols, susquiterpenes and some steroids from *Trichoderma* spp., compounds of hydroxamate and catecholate group eg: trans and cis-4-2- methoxy butanolide from *Aspergillus* spp. (Chaube el al. 2003). The production of antibiotics usually requires relatively large amounts of carbon while other environmental factors (eg. matric potential, soil reaction) in substrate microsites influence metabolism. Other microorganism may utilize antibiotics as carbon substrates. Efficient strategies must be employed for successful application of antibiotics-producing bioagents for the control of soil borne pathogens. However in the context of the natural environment most described mechanisms of pathogen suppression will be modulated by the relative occurrence of other organism in addition to other pathogen. While many investigations have attempted to establish the importance of specific mechanism of bio-control to particular path systems, all of the mechanisms described below are ideal to the operation to some extent in all natural and managed ecosystems. The most effective BCAs studied to date appear to antagonize pathogen using multiple mechanisms. *Pseudomonades* known to produce the antibiotic 2, 4-diacetylphologlucinol (DAPG) may also induce host defenses. Additionally DAPG- producers can aggressively colonize roots a trait that might further contribute to their ability to suppress pathogen activity in the rhizosphere of the wheat through competition for organic nutrients (Raaimakers and Weller 2001).

Fluorescent pseudomonas produce secondary metabolites with antibiotic activities some of these (Phenazine-1-carboxylic acid (PCA), 2,4-diacetylphloroglucinol (DAPG), oomycin- A, pyocyanine, pyoluteorin and pyrrolnitrin) have been implicated in suppression of soil-borne diseases (Thomshow and Weller 1996). Gene loci involved have been cloned and used to enhance the biological properties of *Pseudomonas* spp. by either over expressing the genes and hence increasing antibiotic production. HCN, a volatile inhibitor, inhibits activities of harmful organisms and reduces plant growth. Transfer of HCN biosynthetic gene clusters to non - HCN producing strains of *Pseudomonas* increased their biocontrol activity against black root rot, as well as against the foliage pathogens *Septoria tritici* and *Puccinia graminis* in wheat (Flaishman et al 1996).

Table 3. Antibiotics produced by different bio-control agents to suppress the growth of plant pathogens

Antibiotic	Nature of antibiotic	Source	Target pathogen	Disease	Reference
2,4-Dicatyl, 2-4 phlorogl-ucinol	Phenol	*Pseudomonas fuorescens* F113	*Pythium* spp.	Damping off	Shanahan et al (1992) Fanton et al (1992)
Agrocin 84	Nucleotide	*Agrobacterium radiobacter*	*Agrobacterium tumefaciens*	Crown gall	Kerr (1980)
Bacillo-mycin D	Amino acid derived	*Bacillus subtilis* AU 195	*Aspergillus flavus*	Aflatoxin con-tamination	Moyne et al (2001)
Bacillonycins fengycin	Amino acid derived	*Bacillus amylo-liquefaciens* FZB42	*Fusarium*	Wilt	Koumoutsi et al (2004)
Zantho-baccin A	Amino acid derived	*Lysobacter* sp. Strain SB-K88	*Aphanomyces cochlioides*	Damping off	Islam et al (2005)
Gliotoxin	Piperazine	*Trichoderma virens*	*Rhizoctonia solani*	Root rots	Wilhite et al (2001)
Herbicolin	-	*Pantoea agglo-merans* C9-1	*Erwinia amylovora*	Fire blight	Sandra et al (2001)
Iturin A	Lipope-ptide	*B. subtilis* QS1713	*Botrytis cinerea* and *Rhizoctonia solani*	Damping off	Paulitz and Belanger (2001), Kloepper et al (2004)
Mycosubtilin	Amino acid derived	*B. subtilis* BBG100	*Pythium apha-nidermatum*	Damping off	Leclere et al (2005)
Phenazines	Phenozine	*P. fluorescens* 2-79 and 30-84	*Gaeumanno-myces graminis* var. *tritei*	Take –all	Thomashow et al (1990)
Pyoluteorin, Pyrrolnitrin	Phenyl pyrol	*P.fluorescens* PF-5	*Pythium ultimum* and *R. solani*	Damping off	Howell and Stipanovic (1980) Mondal and Verma (1998)
Pyrrolnitrin pseudane	Phenyl pyrol	*Burhkolderia cepacia*	*R. solani* and *Pyricularia oryzae*	Damping off and rice blast	Homma et al (1989)
Zwitler micin	Phenyl pyrol	*Bacillus cereus* UW85	*Phytophthora medieagins* and *P. aphanider-matum*	Damping off	Smith et al (1993)

(Pal and Gardener, 2006)

Pseudomonas fluorescens and *P. putida* are reported to suppress several major plant pathogens. The former is well known for its usefulness in the biocontrol of take - all disease of wheat caused by *G. graminis* var. *tritici.* The suppression depends on the ability of the bacterium to colonize the roots and production of an antibiotic Phenazine-1-carboxylic acid, a siderophore called Pyoverdin, and an antifungal factor, iron - regulated, non - siderophore antibiotics may be frequently produced by the fluorescent *pseudomonas* (Thomashow and Weller 1990). To be effective, antibiotics must be produced in sufficient quantities near the pathogen to result in a biocontrol effect. *In situ* production of antibiotics by several different biocontrol agents has been measured (Thomashow et al 2002); however, the effective quantities are difficult to estimate because of the small quantities produced relative to the other, less toxic, organic compounds present in the phytosphere. Several methods have been developed to ascertain when and where biocontrol agents may produce antibiotics (Notz et al 2001), detecting expression in the infection court is difficult because of the heterogenous distribution of plant-associated microbes and the potential sites of infection. In a few cases, the relative importance of antibiotic production by biocontrol bacteria has been demonstrated, where one or more genes responsible for biosynthesis of the antibiotics have been manipulated (Thomashow and Weller 1988). Several biocontrol strains are known to produce multiple antibiotics which can suppress one or more pathogens. For example, *Bacillus cereus* strain UW85 is known to produce both zwittermycin (Silo-Suh et al 1994) and kanosamine (Milner et al 1996). The ability to produce multiple antibiotics probably helps to suppress diverse microbial competitors, some of which are likely to be plant pathogens. The ability to produce multiple classes of antibiotics, that differentially inhibit different pathogens, is likely to enhance biological control. More recently, *Pseudomonas putida* WCS358r strains genetically engineered to produce phenazine and DAPG displayed improved capacities to suppress plant diseases in field-grown wheat (Glandorf et al 2001, Bakker et al 2002).

4. Bacteriocins: Bacteriocins can be defined as the proteinaceous antagonistic substances produced by several species of bacteria and are used to suppress the growth of other strain of closely related bacteria. Bacteriocins are adsorbed to the receptors on the surface of bacterial cells, eliminate the electric charges of the inner membrane and stop the bio-synthesis of adenosine triphosphate. Consequently, the biosynthesis of DNA, RNA, proteins and phospholipids is disrupted and finnaly the growth of the bacterial cell is inhibited and may lead to the death of the bacterial cells. Antibiotic Agrocin-84 produced by *Agrobacterium radiobacter* strain K84 is a bacteriocin and used against *Agrobacterium tumefaciens* (Ryder and Jones 1990).

5. Siderophores: Siderophores are "low molecular weight, virtually ferric specific legends, the biosynthesis of which is carefully regulated by iron and the function of which is to supply iron to the cell". The structural diversity among the different siderophores is quite considerable and depends upon the producing micro-organisms. The ability to sequester iron provides a competitive advantage to micro-organisms and siderophores play an important role in inhibition of one organism by other organism. Pathogens are sensitive to suppression by siderophores for several reasons:

- They produce no siderophore of their own

- They are unable to use the siderophores produced by the antagonists or other microorganisms in their immediate environment
- They produce too little siderophore or a siderophore with a lower affinity for iron than those of antagonists.
- They produce a siderophore that can be used by the antagonist, but they are unable to use the siderophore of the antagonists.

Different fungi produce different siderophores involved in the iron uptake and these are commonly short peptides containing non-protein aminoacids. *Trichoderma virens* produces three types of hydroxamate siderophores; a monohydtoxamate (cis- and trans- fusarinines), a dipeptide of trans fusarinine (dimerum acid), a trimer disdepsipeptide (copragen). *Aspergillus niger* strain AN -27 produced hydroxamate and catecholate groups of siderophores and helps in improving the antagonistic activities, rhizosphere competence and plant growth (Selvakumar et al 2000).

Iron is present as ferric ion (Fe^{3+}) in the microbial environment, which is virtually insoluble in the presence of oxygen and therefore, not available for the microbial growth. Siderophore chelate Fe (II) and microbial receptor proteins specifically recognize and take up the siderophore-Fe-Complex (Mukhopadhyay and Mukherjee, 1998). This results in making iron unavailable to rhizosphere micro-organisms, including pathogens, which produces less siderophores with lower binding coefficient. *Pseudomonas fluorescens* strain WCS 374 induced systemic resistance in host by production of both siderophores (Pseudobactin) and salicylic acid (Bakker et al 2002).

6. Volatile substances: The production of volatile substances by biocontrol agents directly affect the plant pathogens (Table 4). Hydrogen cyanide (HCN) effectively blocks the cytochrome oxidase pathway and is highly toxic to all aerobic microorganisms at picomolar concentrations. The production of HCN by certain fluorescent pseudomonads is believed to be involved in the suppression of root pathogens. *P. fluorescens* CHA0 produces antibiotics, siderophores and HCN, but suppression of black rot of tobacco caused by *Thielaviopsis basicola* appeared to be primarily due to HCN production (Voisard et al 1989). Howell et al (1988) reported that volatile compounds such as ammonia produced by *Enterobacter cloacae* were involved in the suppression of *Pythium ultimum*-induced damping-off of cotton. While it is clear that biocontrol microbes can release many different compounds into their surrounding environment, the types and amounts produced in natural systems in the presence and absence of plant disease have not been well documented and this remains a frontier for discovery. Now the emphasis is on the enhancement of antifungal metabolites productivity of biocontrol through genetic manipulations like directed mutagenesis, protoplast fusion, recombination, transformation or isolation of useful genes from biocontrol fungi without functional sexual stages so that they can tolerate some stress conditions and manage a wide range of pathogens. Production of volatile substances can also be achieved by chemical and physical mutation, sexual hybrids and homokaryons (Mohamed et al 2004 and Palumbo et al 2005).

Table 4. Different volatile metabolites produced by biocontrol agents used in the suppression of plant pathogens

Antagonist	Volatile substances	Pathogen	Reference
Boletus varigatus	Isobutanol, isoamyl alcohol and isobutaric acid	*Phytophthora cinnamonni, Fomes annosus*	Krupa and Nyland, 1972
Enterobacter cloacae	Ammonia	*Pythium ultimum, Rhizoctonia solani, Verticillium dahliae*	Howell et al 1987
Pseudomonas fluorescens strain CHAO	Hydrocyanic acid	*Chalara elegans* *Thielaiopsis basicola*	Voisard et al 1989 Stutz et al 1986

7. Lytic enzymes: Some pathogens depend upon the production of cell wall degrading enzymes to infect the living plants. Proteolytic enzymes produced by biocontrol agents inactivate these enzymes resulting in reduced ability of pathogen to infect the plant. Many microorganisms produce and release lytic enzymes that can hydrolyze a wide variety of polymeric compounds including chitin, proteins, cellulose, hemicellulose and DNA. Expression and secretion of these enzymes by different microbes can sometimes results in the suppression of plant pathogen activiteies directly. Many microorganisms produce and release lytic enzymes that can hydrolyze a wide variety of polymeric compounds, including chitin, proteins, cellulose, hemicellulose and DNA (Table 5). Expression and secretion of these enzymes by different microbes can sometimes result in the suppression of plant pathogen activities directly, e.g. control of *Sclerotium rolfsii* by *Serratia marcescens* appeared to be mediated by chitinase expression (Ordentlich et al 1988) and α β-1,3-glucanase contributes significantly to biocontrol activities of *Lysobacter enzymogenes* strain C3 (Palumbo et al 2005). While they may stress and/or lyse cell walls of living organisms, these enzymes generally act to decompose plant residues and nonliving organic matter. Currently, it is unclear how much of the lytic enzyme activity that can be detected in the natural environment represents specific responses to microbe-microbe interactions. It seems more likely that such activities are largely indicative of the need to degrade complex polymers in order to obtain carbon nutrition. Nonetheless, microbes that show a preference for colonizing and lysing plant pathogens might be classified as biocontrol agents. *Lysobacter* and *Myxobacteria* are known to produce copious amounts of lytic enzymes, and some isolates have been shown to be effective in suppressing fungal plant pathogens (Kobayashi and El-Barrad 1996, Bull et al 2002). So, the lines between competition, hyperparasitism, and antibiosis are generally blurred. Furthermore, some products of lytic enzyme activity may contribute to indirect disease suppression.

Oligosaccharides derived from fungal cell walls are known to be potent inducers of plant host defenses. Interestingly, *Lysobacter enzymogenes* strain C3 has been shown to induce plant host resistance to disease (Kilic-Ekici and Yuen 2003), though the precise activities leading to this induction are not entirely clear. The quantitative contribution of any and all of the above compounds to disease suppression is likely to be dependent on the composition and carbon to nitrogen ratio of the soil organic matter that serves as a food source for microbial populations in the soil and rhizosphere. However, such activities can be manipulated so as to result in greater disease suppression. For example, in postharvest disease control, addition of chitosan can stimulate microbial degradation of pathogens similar to that of an applied hyperparasite (Benhamou, 2004). Chitosan is a non-toxic and biodegradable polymer of beta-1,4-glucosamine produced from chitin by alkaline deacylation. Amendment of plant growth substratum with chitosan suppressed the root rot caused by *Fusarium oxysporum* f. sp. *radicis-lycopersici* in tomato (Lafontaine and Benhamou 1996). Although the exact mechanism of action of chitosan is not fully understood, it has been observed that treatment with chitosan increased resistance to pathogens.

Some bacteria can parasitize fungi and kill them by secreting lytic enzymes like chitinase, β-1,3-glucanases, proteases and lipases. Chitinase produced by *Serratia marcescens* have been associated with biocontrol of fungal disease of pea and bean. The chi A gene was cloned and expressed constitutively in *P. putida*. The chi A + recombinant provided increased protection of radish against *F. oxysporum* f. sp. *redolens* (Dubey 2001).

Table 5. Various lytic enzymes used in the growth inhibition of plant pathogens

Antagonist	Enzyme	Pathogen	Reference
Gliocladium virens	Water extracts with unidentified enzymes	*R. solani*	Lewis and Papvizas (1987)
Talaromyces flavus isolate TF1	Glucose and glucose oxidase	*Rhizoctonia* spp.	Boosalies (1956)
T. flavus	Glucose oxidase	*R. solani*	Boosalies (1956)
Penicillium vermiculatum	Hydrogen peroxidase	*R. solani*	Boosalies (1956)
Trichoderma spp.	Water extracts with unidentified enzymes	*R. solani*	Lewis and Papvizas (1987)
Serratia marcescens	Chitinase	*Sclerotium rolfsii*	Ordentlich et al (1988)

8. Competition: Competition is a classical mechanism of biological control. It involves the competition between antagonist and plant pathogens for space and nutrients (Chaube and Singh 1991). Clark (1965) defined competition in the narrow sense of active demand in excess of the immediate supply of material or condition on the part of two or more organism. Competition among microorganisms in soil and the rhizosphere is primarily for nutrients. *Peniophora gigantea* had been used to control root rot of pine caused by *Heterobasidium annosum*. Conidia of the antagonists *P. gigantea*, are applied usually in chain saw oil to freshly cut stumps, although the primary mechanism is competition for space by physical exclusion of the pathogen, the antagonists also produce an antibiotic like substance against *H. annosum*.

The omnipresence of *Trichoderma* in soil throughout the world proves that it must be an excellent competitor. Elad (2000) reported that biocontrol of *Botrytis cineria* by *T. harzianum* strain T-39 is good example of competition. *B. cinerea* conidia requieextermal nutrients for germination and infection. When conidia of T-39 were applied to leaves, germination of conidia of the pathogen was slower, this is due to the competition (Elad, 1994). The competitive ability of *Trichoderma* was affected by the soil properties. Backman et al (1975) showed that when fungicides those were toxic to the *Trichoderma* were applied to peanut late leaf spot control, there was increase in the disease caused by *Sclerotium rolfsii*, a pathogen susceptible to biocontrol by *Trichoderma*.

In general, soilborne pathogens, such as species of *Fusarium* and *Pythium,* which infect through mycelial contact, are more susceptible to competition from other soil- and plant-associated microbes than those pathogens that germinate directly on plant surfaces and infect through appressoria and infection pegs. Genetic work of Anderson et al (1988) revealed that production of a particular plant glycoprotein called agglutinin was correlated with potential of *P. putida* to colonize the root system. *P. putida* mutants deficient in this ability exhibited reduced capacity to colonize the rhizosphere and a corresponding reduction in Fusarium wilt suppression in cucumber (Tari and Anderson 1988). The most abundant nonpathogenic plant-associated microbes are generally thought to protect the plant by rapid colonization and thereby exhausting the limited available substrates so that none are available for pathogens to grow. For example, effective catabolism of nutrients in the spermosphere has been identified as a mechanism contributing to the suppression of *Pythium ultimum* by *Enterobacter cloacae* (van Dijk and Nelson 2000; Kageyama and Nelson 2003). At the same time, these microbes produce metabolites that suppress pathogens. These microbes colonize the sites where water and carbon containing nutrients are most readily available, such as exit points of secondary roots, damaged epidermal cells and nectaries and utilize the root mucilage. Biocontrol based on competition for rare but essential micronutrients, such as iron, has also been examined. Iron is extremely limited in the rhizosphere, depending on soil pH. In highly oxidized and aerated soil, iron is present in ferric form (Lindsay 1979), which is insoluble in water (pH 7.4) and the concentration may be as low as 10-18 M. This concentration is too low to support the growth of microorganisms, which generally need concentrations approaching 10^{-6}M. To survive in such an environment, organisms were found to secrete iron-binding ligands called siderophores having high affinity to sequester iron from the micro-environment. Almost all microorganisms produce siderophores, of either the catechol type or hydroxamate type (Neilands 1981). Kloepper et al (1980) were the first to demonstrate the importance of siderophore production as a mechanism of biological control of *Erwinia carotovora* by several plant-growth promoting *Pseudomonas fluorescens* strains A1, BK1, TL3B1 and B10. A direct correlation was established *in vitro* between siderophore synthesis in fluorescent pseudomonads and their capacity to inhibit germination of chlamydospores of *F. oxysporum* (Elad and Baker 1985 a and b; Sneh et al 1984). As with the antibiotics, mutants incapable of producing some sidero-phores, such as pyover-dine, were reduced in their capacity to suppress different plant pathogens (Keel et al 1989, Loper and Buyer 1991). The increased efficiency in iron uptake of the commensal microorganisms is thought to be a contributing factor to their ability to aggressively colonize plant roots and an aid to the displacement of the deleterious organisms from potential sites of infection.

9. Induction of host resistance: The induced resistance depends on recognition between host and pathogen. Induction of the resistance response is brought about by the interaction of products of both host and parasite. Localized responses can include biochemical responses, such as induction of phytoalexin biosynthesis in cells and physical responses, such as formation of callus or suberin or lignin as structural barrier around a lesion. The rhizobacteria can reduce plant disease by antagonizing soil-borne pathogenic fungi through production of antibiotic and / or lytic enzymes (Chitinase) and through competition for ecological niche and nutrients as well as by inducing systemic resistance (Table 6).

Systemic acquired resistance (SAR), is mediated by salicylic acid (SA), a compound which is frequently produced following pathogen infection and typically leads to the expression of pathogenesis-related (PR) proteins. These PR proteins include a variety of enzymes some of which may act directly to lyse invading cells, reinforce cell wall boundaries to resist infections, or induce localized cell death. A second phenotype, first referred to as induced systemic resistance (ISR), is mediated by jasmonic acid (JA) and/or ethylene, which are produced following applications of some nonpathogenic rhizobacteria. Interestingly, the SA- and JA- dependent defense pathways can be mutually antagonistic, and some bacterial pathogens take advantage of this to overcome the SAR. e.g. pathogenic strains of *Pseudomonas syringae* produce coronatine, which is similar to JA, to overcome the SA-mediated pathway (He et al 2004). Because the various host-resistance pathways can be activated to varying degrees by different microbes and insect feeding, it is possible that multiple stimuli are constantly being received and processed by the plant. Thus, the magnitude and duration of host defense induction will likely vary over time. Only if induction can be controlled, i.e. by overwhelming or synergistically interacting with endogenous signals, will host resistance be increased. A number of strains of root-colonizing microbes have been identified as potential elicitors of plant host defenses. Some biocontrol strains of *Pseudomonas* sp. and *Trichoderma* sp. are known to strongly induce plant host defenses (Pieterse et al 2002, Harman et al 2004; Haas and Defago 2005). Plant-growth-promoting rhizobacteria (PGPR) inoculations were also found effective in controlling multiple diseases caused by different pathogens, including anthracnose (*Colletotrichum lagenarium*), angular leaf spot (*Pseudomonas syringae* pv. *lachrymans* and bacterial wilt (*Erwinia tracheiphila*) in cucumber. A number of chemical elicitors of SAR and ISR may be produced by the PGPR strains upon inoculation, including salicylic acid, siderophore, lipopolysaccharides, and 2,3-butanediol, and other volatile substances (Van Loon et al 1998; Ongena et al 2004; Ryu et al 2004). Again, there may be multiple functions to such molecules blurring the lines between direct and indirect antagonisms. More generally, a substantial number of microbial products have been identified as elicitors of host defenses, indicating that host defenses are likely stimulated continually over the course of a plant's lifecycle. Excluding the components directly related to pathogenesis, these inducers include lipopolysaccharides and flagellin from Gram-negative bacteria; cold shock proteins of diverse bacteria; transglutaminase, elicitins, and ß-glucans in Oomycetes; invertase in yeast; chitin and ergosterol in all fungi; and xylanase in *Trichoderma* (Numberger et al 2004). These data suggest that plants would detect the composition of their plant-associated microbial communities and respond to changes in the abundance, types and localization of many different signals. The importance of such interactions is indicated by the fact that further induction of host resistance pathways, by chemical and microbiological inducers, is not always effective in improving plant health or productivity under the field conditions (Vallad and Goodman 2004).

ISR is similar to pathogen induced systemic acquired resistance (SAR) as both provide enhanced resistance against challenge pathogen. ISR has been demonstrated against fungi, bacteria, and viruses in several crops. SAR (Systemic Acquired Resistance) and ISR differ in their signaling pathways. SAR is dependent on the synthesis of salicylic acid by the plant that acts as an inducer signal and is associated with the accumulation of novel pathogenesis related proteins, some of which have been shown to possess antifungal activity. Transgenic plants with Nah G gene from *P. putida* which codes for salicylic hydroxylase, causes conversion of salicylic acid to catechol, as a result no SAR develops (Ryals et al 1996). Bacteria that produce salicylic acid as siderophore, under iron limiting condition, have been reported to induce SAR. *P. aeuroginosa* 7 NSK2 has been found to induce SAR in bean and tobacco plants against *Botrytis cinerea* and TMV, respectively (Dubey 2001).

Rhizobacteria mediated ISR signaling pathways does not involve salicylic acid or PRs. Rather it requires jasmonate and ethylene response, and like the SAR, depends on the regulatory protein NPRI. Thus, NPRI differentially regulates ISR and SAR- related gene expression depending on the pathway that is activated. The jasmonate and ethylene dependency of ISR is based on enhanced sensitivity of these hormones, rather than those on the increase in their production. Although the extent of induced resistance attained is similar in both SAR and ISR, the latter usually is less effective.

Table 6. Host resistance induced by biocontrol agents

Antagonist	Plant species	Determinant	Type	Reference
Bacillus mycoides strain Bac J	Sugar beet	Peroxidase, chitinase and B-1,3- glucanase	ISR	Bargabus et al (2002)
Baccilus pumilus 203-6	Sugar beet	Peroxidase, chitinase and B-1, 3-glucanase	ISR	Bargabus et al (2004)
Bacillus subtilis GB03 and IN937a	Arabidopsis	2,3-butanediol	ISR	Ryu et al (2004)
Pseudomonas fluorescens strain CHAO	Tobacco	Siderophore	SAR	Maurhofer et al (1994)
	Arabidopsis	Antibiotica (DAPG)	ISR	Lavicoli et al (2003
Pseudomonas fluorescens strain WCS 374	Radish	Lipopolysacchride	ISR	Leeman et al (1995 a, b)
		Siderophore	ISR	
		Iron regulated factor	ISR	
Pseudomonas fluorescens strain WCS 417	Carnation	Lipopolysacchride	ISR	Van Peer & Schipper (1992)
	Radish	Lipopolysacchride	ISR	Leeman et al (1995 a, b)
	Arabidopsis	Lipopolysacchride	ISR	VanWees et al (1997)
	Tomato	Lipopolysacchride	ISR	Duijff et al (1997)
Pseudomonas putida strain WCS 358	Arabidopsis	Lipopolysacchride	ISR	Meziane et al (2005)
		Siderophore	ISR	Meziane et al (2005)
Pseudomonas putida strain BTP 1	Bean	Z, 3 hexenal	ISR	Ongena et al (2004)
Serratia marcescens 90- 166	Cucumber	Siderophore	ISR	Descafzo and Mauza (1990), Press et al (2001)

(Pal and Gardener, 2006)

Various biocontrol agents used to control various diseases also induces the resistance in host plants against the pathogens (Table 7). e.g. *Trichoderma harzianum* T203 used as a biocontrol agent to control the anthracnose of cucumber (*Colletotrichum orbiculare*). *T. harzianum* T-203 controlled the disease by increasing the resistance in cucumber against the pathogen.

Table 7. Biocontrol agent mediated induced resistance in plants

Diseases	Pathogen	Biocontrol
Anthracnose of cucumber	*Colletotrichum orbiculare*	*Trichderma harzianum* T 203 *Pseudomonas putida* 89B-27
Cucumber mosaic	Cucumber mosaic virus	*P. putida* 89B-27
Bacterial wilt of cucumber	*Erwinia trachaephila*	*P. putida* 89B-27
Vascular wilt of cucumber	*Fusarium oxysporum* f. sp. *cucumerinum*	*P.putida* 89B-27
Wilt of muskmelon	*F. o.* f. sp. *melonis*	*Aspergillus niger* AN27
Vascular wilt of tomato	*F. o.* f. sp. *lycopersici*	*Pseudomonas fluorescens* WCS417
Mottling of tomato	Tomato mottle virus	*Bacillus amyloliquefaciens* 937
Grey mold of bean	*Botrytis cineria*	*T. harzianum* T 39 *P. fluorescens*

(Biswas et al 2007)

Various organisms used as biocontrol agents

(i) Bacterial anrtagonists: Among the major bacterial genera (*Bacillus, Pseudomonas, Pantoea, Arthrobacter* and *Curtobacterium*) are identified as antagonistic to plant pathogen (Table 8). *Pseudomonas* and *Bacillus* are thought to be most appealing candidates for biological control. They are excellent colonizer and widely prevalent in plant rhizosphere. These mediate biocontrol by production of antibiotics, iron- scavenging siderophores, lytic enzymes and microbial cyanide (Mondal and Verma 1998), or by initiating a cascade of events leading to the induction of systemic resistance in host (Mondal and Verma 2002, Daayf et al 2003 and Pieterse et al 1996). Bacteria are extremely important in biological control especially as antagonists of plant pathogen, such as *Fusarium* and certain other which produce germ tubes and attack roots rapidly by multiple infectious. Ten different strains of *Pseudomonas* tested in laboratory against *F. solani* the causal agent of Fusarium wilt of melon. Two, were most antagonistic and their culture filtrates reduced infection under field conditions. Antagonistic effects of saprophytic strain of *Pseudomonas aeruginosa* against *F. solani* and use of *P. putida* and *P. fluorescens* in the control of *Fusarium* wilts have been reported by (Scher and Baker 1982).

B. subtilis was found antagononistic to *F. oxysporum* and *F. solani.* The cell free culture filtrates of AF1 strain of *B. subtilis* has been found to inhibit the growth of wilt fungi viz., *F. udum, F. oxysporum* f. sp. *lycopersici, F. oxysporum* f. sp. *vasinfectum* as reported by Podile and Dube in 1985.

Table 8. Bacterial diseases suppressed by various bacterial antagonist

Cop	Bacterial diseases	Pathogen	Bacterial biocontrol agent	Reference
Potato	Common scab	*Streptomyces scabies*	*Bacillus cereus, B. pumilus, Pseudomonas chlororaphis*	Sturz et al 2004
	Soft rot	*Erwinia carotovora*	*Bacillus thuringeinsis*	Dong et al 2004
	Bacterial wilt/ brown rot	*Ralstonia solanacearum*	*Bacillus cereus, B. subtilis* *Bacillus polymyxa* FU-6 *Pseudomonas fluorescens*	Leclere et al 2005
	Ring rot	*Clavibacter michiganensis* subsp. *sepedonicus*	*Pseudomonas aurefaciens* *P. fluorescens biovar* III	Cruz et al 1992
	Black leg	*Erwinia carotovora* subsp . *carotovora and E. C.* subsp. *atroseptica*	*Curtobacterium* spp. *Pantoea* spp.	Struz and Matheson 1996
Tomato	Bacterial wilt	*Ralstonia solanacearum*	*Pseudomonas fluorescens*	Pardeep et al 2003
	Bacterial speck	*Pseudomonas syringae* pv. tomato	*Pseudomonas* spp. Non pathogenic *P. syringae* strain Cit7	Colin and Chafik,1986 Wilson et al 1998
	Bacterial spot	*Xanthomonas axonopodis* pv. *vesicotoria*	Non pathogenic *P. syringae* strain Cit7	Wilson et al 1998
Cabbage	Black rot	*Xanthomonas campestris* pv. *campestris*	*Bacillus cereus* *B. lentimorbus* *B. pumilus*	Massomo et al 2004
Rajmah bean	Fuscous blight	*Xanthomonas axonopodis* pv. *phaseoli* var. *fuscans*	*Pseudomonas fluorescens* Strain RPB14	Mondal, 2004
Beans	Common bacterial blight	*Xanthomonas axonopodis* pv. *phaseoli*	*Bacillus* spp. *Pseudomonas fluorescens*	Mabagala, 1999
	Halo blight	*Pseudomonas syringae* pv. *phaseolicola*	*Pseudomonas fluorescens* *P. putida* *Streptomyces aureofaciens*	Mahmoud 2002
Soyabean	Leaf spot	*Pseudomonas syringae* pv. *glycinea*	*Pseudomonas syringae* strain 22d/93	Volksch and May 2001
Pea	Bacterial blight	*Pseudomonas syringae* pv. *pisi*	*Pseudomonas* spp. *Pentoea agglomerans* *Arthrobactor* sp. *Curtobacterium* sp.	Elvira 2000
Cucumber	Angular leaf spot	*Pseudomonas syringae pv. lachrymans*	*Bacillus pumilus*INR7 *B. subtilis* strain GB03 *Curtobacterium flaccumfaciens* strain ME1	Raupach and Kloepper 2000

(ii) Actinomycetes as antagonists: Actinomycetes are successful antagonists largely due to their wide spread ability to produce antibiotics. Actinomycetes are successful in biocontrol of *Fusarium* spp., because of their ability to possess ectoenzyme systems to break down the chitinous walls present in fungi. Workers have obtained some biocontrol of *F. solani, F. solani* f. sp. *phaseoli, F. oxyspirum* f. sp. *cubense* and *F. oxysporum* f. sp. *conglutinans, F. oxysporum* f. sp. *cucumerinum* (Sieiketov and Kuldybeav, 1974).

(iii) Fungal antagonists: Fungi have received maximum attention as antagonists possibly because of the case in handling and identification compared to other microbes. However, fungi probably rank "3rd" as potential antagonists in biocontrol, though some situation clearly suggests fungi as most important antagonists (Grossbard 1951).

Table 9. Antagonistic activity exhibited by potential biocontrol agents

Antagonists	Pathogens	Mechanism of action
Aspergillus niger, A. nidulans	*Macrophomina phaseolina, Sclerotium rolfsi, R. solani*	Antibiosis, mycoparasitism, Induced systemic resistance
Acremonium obclavatum	*Puccinia arachidis*	Hyperparasitism
Epicoccum purpurascens	*M. phaseolina*	Antibiosis
Gliocladium roseum	*Sclerotinia sclerotiorum*	Mycoparasitism
Gigaspora margarita	*Heterodera cajani, F. udum*	Antibiosis
Paecilomyces lilacinus	*Rotylenchus reniformis, M. incognita, R. solani*	Mycoparasitism and antibiosis
Penicillium citrinum	*Colletotrichum gloeosporoides*	Parasitism and Antibiosis
P. Pinophilum	*R. solani, Ralstonia solanacearum*	
Trichoderma aureoviride	*R. solani, Meloidogyne javanica,*	Parasitism on mycelium & sclerotia
T. hamatum	*M. phaseolina, Sclerotium rolfsii*	Direct parasitization of 2nd juvenile larva and eggs of nematode

a. *Trichoderma*: The first of brilliant series of papers on parasitism of *Trichoderma viride* appeared in 1932 (Weindling, 1932, Garrett 1956).

b. *Penicillium*: *Penicillium* spp. are very effective and common colonisers. They produce an abundance of conidia and in part, for this reason they are frequently isolated from soil debris. Indication of antagonism to *F. oxysporum* f. sp. *melonis* by *P. commune* and *Penicillium* sp. along with *Harmodendron*, *Pachybasium* and *Fusarium* have been reported by Welvaert, 1961. *Penicillium cyclopium*, the most frequent isolate from the soils has been found to be an active antagonist against *F. avenaceum, F. culmorum* (Kamyshiko et al 1976), and *F. oxysporum* f. sp. *milonis*, *F. oxysporum* f. sp. *lycopersici* and *F. oxysporum* f. sp. *niveum* (Bora et al 1982).

c. *Aspergilllus*: As early as in 1940 *in vitro* lytic activity of *Aspergillus niger* against soil borne plant pathogens like *Macrophomina phaseoli* and *Rhizoctonia solani* was noticed. Further, filtrates of *A. niger* when added along with pathogen filtrates could delay the wilting of plants (Vasudeva and Sikka, 1941). Inhibitory substances are produced by the culture filtrates of *A. niger, A. flavus* and *A. candidus* (Shukla and Dwivedi, 1979). *A. niger* exhibited greatest antagonism both in culture and on host during investigation on damping off disease of egg plant caused by *F. solani, Alternaria alternrta* and *R. solani* (Bora, 1977).

Table 10. Mode of action exhibited by potential biocontrol agents

Antagonist	Pathogen	Reference
Parasitism/Predation		
Bacteria and Actinomycetes		
Arthobacter spp.	*Pythium debayanum*	Mitchell and Hurwitz, 1965
Fungi		
Ampelomyces quisqualis	*S. fuligena*	Phillip, 1985
Gliocadium catenulatum	*Fusarium* spp., *S. sclerotiorum*	Huang, 1978
Pythium nunn	Numerous	Deacon, 1976
P. oligandrum	Numerous	Laing and Deacon, 1990
Trichoderma hamatum	*Pythium* spp., *Rhizoctonia solani*	Chet *et. al.,* 1981
Trichoderma spp.	*R. solani* and *S. rolfsii*	Elad *et. al.,* 1980
Antibiosis		
Bacteria and Actinomycetes		
Agrobacterium radiobacter	*A. tumefaciens*	Kerr and Tate, 1984
Bacillus subtilis	Numerous	Swinburne *et. al.,* 1975
Enterobacter aerogenes	*Phytophthora cactorum*	Gupta and Utkhede, 1987
Streptomyces griseoviridis	*Alternaria brassicicola, R. solani, F.o.* f. sp. *dianthi*	Schippers 1987
Fungi		
Gliocladium virens	*Pythium ultimum, R. solani*	Howell and Stipanovic, 1983
Talaromyces flavus	*Verticillium dahliae*	Fravel *et. al.* 1987
Penicillium oxalicum	*F. o.* f. sp. *conglutinans*	Reyes, 1984
Trichoderma spp.	Numerous	Papavizas, 1985, Clayton et al. 1987
Chaetomium globosum	*Alternaria alternata, F. moniliformae*	Singh and Mishra, 2001
Paecilomyces lilacinus	*Meloidogyne incognita, Rotylenchulus reniformis, R. solani*	Khan and Hussain, 1988
Competition		
Bacteria and Actinomycetes		
Enterobacter spp.	*Pythium* spp	Nelson *et. al.,* 1986
Pseudomonas putida	*F. oxysporum*	Scher and Baker, 1982
Pseudomonas spp.	*F. oxysporum, F. o.* f. sp. *cucumerinum, F. solani, P. aphanidermatum*	Elad and Baker, 1985a and b; Sneh *et. al.,* 1984; Elad and Chet, 1987, Aokiet et al 1991
Fungi		
Pythium nunn	*P. ultimum*	Paulitz and Baker, 1988

P. oligandrum	*P. ultimum*	Martin and Hancock, 1986
T. harzianum	*F.o.* f.sp. *melonis, F.o.* f.sp. *vasinfectum*	Sivan and Chet, 1989
Cross protection		
Bacteria and Actinomyces		
Bacillus subtilis	*F. roseum* f. sp. *dianthi*	Aldrich and Baker, 1970
Pseudomonas syringae	*Colletotrichum lagenarium*	Caruso and Kuc., 1979
Pseudomonas spp.	*F. o.* f. sp . *dianthi*	Van Peer and Schipper, 1991
Fungi		
Colletotrichum lagenarium	*Colletotrichum lagenarium*	Dean and Kuc, 1985
Didymella bryoniae	*D. bryoniae*	Descalzo and Mauza 1990
Phytophthora parasitica	*P. cinnamomi, P. citricola*	Dolan *et. al.*, 1986

(iv) Mycorrhiza - Tool for Biocontrol: Arbuscular mycorrhizal fungi (AMF), a rhizospheric fungi from the phylum Glomeromycota are obligate symbionts of plants forming an association with a majority of them acting as extended root hairs and increasing the area of interaction with soil resulting in benefits for the host ranging from not only enhanced nutrient uptake but drought tolerance and soil improvement leading to higher plant productivity besides imparting resistance to pathogens (Davis and Merge 1980, Charlse and Mosse 1981, Graham and Menge 1982, Krishna and Bagyaraj 1983, Adholeya 2012). The ectomycorrhizal fungus can be seen to penetrate intercelluarly in a network pattern. The vesicular-arbuscular mycorrhizae, on the other hand, lack a fungal mantle. The fungus type penetrates root epidermal cells and proliferates between and within cortical cells. Inside the cells, hyphae branch profusely to produce tree-like structures called "Arbuscules" (Theodorou 1971).

Table 11. Some of commercial bioformulation and their use in disease management

Antagonist	Name of commercial product	Formulation	Application method	Target Disease	Crop
BACTERIA					
Agrobacterium radiobacter strain K-84	Galtrol-A	Pure culture	Applied to seeds, seed-ling, cutting root stem & as soil drench	Crown gall disease	Fruit, nut, ornamental nursery stock etc.
	Nogall	Culture	Root		Tree
	Diegall	Suspension	Cutting dip		
Bacillus subtilis	Kodiak	Dry powder usually applied with chemicals fungicides	Seed treatment	*R.. solani, Fusarium* sp., *Alternaria* sp., *Aspergillus* sp. attacking root	Legumes

	Rhizoplus	Water dispersible granules	Seed treatment, soil drench	*R. solani, Fusarium* sp., *Alternaria* sp. *Streptomyces scabies*	Potato, vegetable & ornamental
	System-3	Dust	Seed treatment	Seedling pathogen	Beans
Burkholoderia cepacia or P. cepacia	Intercept		Seed treatment & spray	*Rhizocotonia* sp., *Fusarium* sp, *Pythium* sp.	Vegetables
Pseudomonas fluorescens	Blight ban A506	Wettable powder	Spray	Frost damage *E. amylovara*	Apple, Peech, Potato, Tomato
Streptomyces griseoviridis	Mycostop	Powder	Drench spray through irrigation water	*Fusarium* sp. *phomopsis* sp, *botrytis* sp., *Phytophthora* sp .	Field vegetable & ornamental crops
FUNGI					
Aspergillus niger AN 27	Kalisena	Powder	Seed treatment & soil application	Seed borne pathogen	Vegetables (Brinjal, Cab-bage, Chiili, Cucumber)
Ampelomyces quisqualis	AQ -10	Water dispersible granules	Spray	Powdery mildew	Cucurbits & Tomato
Coniothyrium minitans	KONI	Granules	Soil application	*S. Sclerotiorum*	Green House crop
Gliocladium virens GL-27	Soil gard	Granule	Soil applic-ation before sowing	Damping off & root rot pathogens	Ornamental & fruit plants
Fusarium oxysporum	Biofox	Dust & alginate granules	Seed treat-ment & soil incorporation	*F. oxysporium, F. moniliforme*	Carnation & tomato
T. harzianum (ATCC 20476)	Binab T	Wettable powder & pellets	Painting on tree wounds, soil incorp-oration	Wilt, take all, roor rot	Vegetables & fruit crops
	Bio-Trek T-22G	Granules	Soil incorpo-ration, spray	*Pythium, Rhizoctonia, Fusarium* sp.	Tomato, Cabbage, Cucumber
	Trichodex	Wettable powder	Spray	*Botrytis* sp, *Colletotrichum, R. stolonifer*	Soyabean, Tomato
Trichoderma sp.	Trichoderma 2000	Granules	Soil incorp-oration	*Rhizoctonia* sp, *S. rolfsii, Pythium* sp.	Nursery & field crop
Talaromyces flavus	Protus	Dust	Seed & soil Application	*Verticillum* sp	Cucurbitaceous vegetable
Mycorrhiza (VAM)	Vam Inoc.	Granules	Soil application	*Pythium & Verticillum* sp	Cucumber

Advantages of biological control over chemical control

Inspite of large number of effective chemicals are present but there is a practical shift toward the biological control because of following reasons:

1. Biological control reduces the exposure to lethal environmental and public safety hazards.
2. Avoid the development of resistance in pathogen toward the chemical.
3. High level of disease control at low cost.
4. Reduces the use of chemical fungicides and nematicides.
5. Safe for the user and farming community.
6. Biological control plays a key role in integrated disease management

Future Thurst

It is now clear that the stage is set for biological control agents to play a greater role in agriculture and horticulture. There is compelling need to identify the potential bioagents and to formulate them for their high level of consistence performance. Importantly, low cost production, corporate commitment, good management and market research will provide rapid success to small and medium size enterprises and also benefit the end user. Bacterial and fungal antagonist can be enhanced to produce antibiotics, microbial cycnide and lytic enzymes by adopting molecular tools. Intrinsic effects of the molecular mechanisms developed recently can be of great use in the successful implementation of biological programme as a whole.

References

Adholeys A 2012 Mycorrhiza: a friendly association with plants towards sustainable food production, *Journal of Mycology and Plant Pathology* **42** 183-188

Aldrich J and Baker R 1970 Biological control of *Fusarium roseum* f. sp. *dianthi* by *Bacillus subtilis, Plant Disease Reporter* **54** 446-448

Anderson AJ, Tari PH, and Tepper CS 1988 Genetic studies on the role of an agglutinin in root colonization by *Pseudomonas putida, Applied Environment Microbiology* **54** 375-380

Aoki M, Uehara K, and Koseki K 1991 An antimicrobial substance produced by *Pseudomonas cepacia* B5 against the bacterial wilt disease pathogen, *Pseuomonas solanacearum, Agricultural and Biological Chemistry* **55** 715-722

Backman PA, Rodriguez-Kabana R, and Williams JC 1975 The effect of peanut leaf spot fungicides on the non-target pathogen *Sclerotium rolfsii, Phytopathology* **65** 773-776

Baker KF and Cook RJ 1974 *Biological Control of Plant Pathogens*, San Francisco: W H Freeman and Company, 433 pages

Baker KF and Snyder WC 1965 *Ecology of Soil-borne Plant Pathogens, Prelude to Biological Control*, Berkeley, Los Angeles: University of California Press, 571pages

Bargabus RL, Zidack NK, Sherwood JW, and Jacobsen BJ 2002 Characterization of systemic resistance in sugar beet elicited by a non-pathogenic, phyllosphere colonizing *Bacillus mycoides*, biological control agent, *Physiology Molecular Plant Pathology* **61** 289-298

Bargabus RL, Zidack NK, Sherwood JW, and Jacobsen BJ 2004 Screening for the identification of potential biological control agents that induce systemic acquired resistance in sugar beet, *Biological Control*, **30** 342-350

Benhamou N and Chet I 1997 Cellular and molecular mechanisms involved in the intersection between *Trichoderma harzianum* and *Pythium ultimum, Applied and Environmental Microbiology* **63** 2095-2099

Benhamou N 2004 Potential of the mycoparasite, *Verticillium lecanii*, to protect citrus fruit against *Penicillium digitatum*, the causal agent of green mold: A comparison with the effect of chitosan, *Phytopathology* **94** 693-705

Biraghi A 1950 Possible active resistance to Endithia parasitica in Castanea sativa in 11th Report of Congress on International Union for Research Organization 643-645 pages

Biswas SK, Chinmay B, and Srivastava SSL 2007 Biological control of plant diseases: Present Status and Future Scope, in *Ecofriendly Management of Plant Diseases* (eds) Ahmad Shahid and Narain Udit, Delhi: Daya Publishing House, 247-264 pages

Boosalis MG 1956 Effect of soil temperature and green manure amendment of unsterilized soil on parasitism of *Rhizoctonia solani* by *Penicillium vermiculatum* and *Trichoderma sp., Phytopathology* **46** 473-478

Bora T, Vildiz M, Akinci C, and Nemli T 1982 Investigations on fungiostasis with respect to wilt disease in important cultiveted soil of the western Aegean Region, *Journal of Turkish Phytopathology* **11** 1-13

Bora T 1977 *In vitro* and *in vivo* investigations on the effect of some antagonistic fungi against the damping off disease of egg plant, *Journal of Turkish Phytopathology* **6** 13-17

Bull CT, Shetty KG, and Subbarao KV 2002 Interactions between myxobacteria, plant pathogenic fungi, and biocontrol agents, *Plant Disease* **86** 889-896

Caruso FL and Kuc J 1979 Induced resistance of cucumber to anthracnose and angular leaf spot of *Pseudomonas lachrimans* and *Colletotrichum lagenarium, Physiological Plant Pathology* **14** 191-201

Chaube HS and Singh US 1991 *Plant Disease Management: Principles and Practices* Bora Raton: CRC Press, pages 319

Chaube HS, Mishra DS, Varshney S, and Singh US 2003 Biological control of plant pathogens by fungal antagonists: Historical Background Present Status and Future Prospects. *Annual Review of Plant Pathology* 2: 1- 42 (eds) TS Thind, HN Gour, and SD Purohit *Indian Society Mycology and Plant Pathology* and Scientific Publishers (India), Jodhpur

Chet I 2012 *Trichoderma* sp.-environmentally friendly biocontrol agent and inducer of plant resistance to diseases, Paper presented at the *3rd Global Conference*, 10-13 Jan, 2012, organized by Indian Society of Mycology and Plant Pathology, Udaipur

Chet I, Harman GE, and Baker R 1981 *Trichoderma hamatum*: Its hyphal interaction with *Rhizoctonia solani* and *Pythium* spp., *Microbial Ecology* **7** 29-38

Clark FE 1965 The concept of competition in microbiology, in *Ecology of Soil-Borne Plant Pathogens* (eds) KF Baker and WC Snyder, Berkeley: University of California Press, 339–347

Clarke CA and Mosse B 1981 Plant growth responses to vesicular-arbuscular mycorrhiza XII, Field inoculation on responses of barley at two soil P levels, *New Phytologist* 87 695-703

Claydon N, Allan M, Hanson JR, and Avent AG 1987 Antifungal alkyl pyrones of *Trichoderma harzianum, Transactions of British Mycological Society* **88** 503-513

Colin JE and Chafik Z 1986 Comparism of biological chemical treatments for control of bacterial speck of tomato under field conditions in Morocco, *Plant Disease* **70** 1048-1050

Cruz AR De La, Poplawsky MV, and Wiese MV 1992 Biological suppression of potato ring rot by fluorescent pseudomonads, *Applied and Environmental Microbiology* **58** 1986-1991

Daayf F, Adam L, and Fernado WGD 2003 Comparative screening of bacteria for biological control of potato late blight (strain US-8), using *in vitro*, detached leaves and whole plant testing systems, *Canadian Journal of Plant Pathology* **25** 276-284

Davis RM and Menge JA 1980 Influence of *Glomus fasciculalus* and soil phosphorus on *Phytophthora* root rot of citrus, *Phytopathology* **70** 447-452

De Bach P 1964 The scope of biological control, in *Biological Control of Insect Pests and Weeds* (ed) P De Bach, London: Chapman and Hall Ltd pages 3-20

Deacons JW 1976 Studies on *Pythium oligandrum* an aggressive parasite of other fungi, *Transactions of British Mycological Society* **66** 383-391

Dean RA and Kuc J 1985 Induced systemic protection in plants, *Trends in Biocontrol* **3** 125-129

Descalzo RC and Mauza B 1990 Comparative efficacy of induced resistance for selected diseases of green house cucumber, *Canadian Journal of Plant Pathology* **12** 16-24

Dobbs CG and Hinson WH 1953 A widespread fungistasis in soil, *Nature London* **172** 197-199

Dolan T, Cohen Y, and Coffey MD 1986 Protection of *Perse* sp. against *Phytophthora* sp.by prior inoculation with a citrus isolate of *Phytophthora parasitica*, *Phytopathology* **76** 194-198

Dong YH, Zhang XF, Xu JL, Zhang LH, Dong YH, and Zhang LH 2004 Insecticidal bacillus thuringiensis silences *Erwinia carotovora* virulence by a new form of microbial antagonism, signal interference, *Applied and Environmental Microbiology* **70** 954-960

Dubey HC 2001 Rhizobactereia in biological control and plant growth promotion, *Journal of Mycology and Plant Pathology* **31** 9

Duffy BK, Simon A, and Weller DM 1996 Combination of *Trichoderma koningii* and fluorescent *Pseudomonads* for control of take-all of wheat, *Phytopathology* **86** 188-194

Duijff BJ, Gianinazzi PV, and Lemanceau P 1997 Involvement of the outer membrane lipopolysaccharides in the endophytic colonization of tomato roots by biocontrol *Pseudomonas fluorescens* WCS417, *New Phytologist* **135** 325-334

Elad Y and Baker R 1985a Influence of trace amounts of cations and siderophore-producing pseudomonads on chlamydospore germination of *Fusarium oxysporum*, *Ecology & Epidemiology* **75** 1047-1052

Elad Y and Baker R 1985b The role of competition for iron and carbon in suppression of Chlamydospore germination of *Fusarium* sp by *Pseudomonas* sp., *Phytopathology* **75** 1053-1059

Elad Y and Chet I 1987 Possible role of competition for nutrients in biocontrol of *Pythium* damping off by bacteria, *Phytopathology* **77** 190-195

Elad Y, Chet I, and Katan J 1980 *Trichoderma harzianum* biocontrol agent effective against *Sclerotium rolfsii* and *Rhizoctonia solani*, *Phytopathology* **70** 119-121

Elad Y 1994 Biological Control of Grape Grey Mold by *Trichoderma harzianum*, *Crop Protection* **13** 35-38

Elad Y 2000 Biological control of foliar pathogens by means of *Trichoderma harzianum* and potential modes of action, *Crop Protection* **19** 709-714

Elvira RM 2000 Sustainable control of pea bacterial blight, In *Approaches for durable genetic resistance and biocontrol by endophytic bacteria* Wageningen, Netherlands: Landbouw-universiteit Wageningen

Fenton AM, Stephens PM, Crowley J, O'Callaghan M, and O'Gara F 1992 Exploitation of gene(s) involved in 2,4-diacetylphloroglucinol biosynthesis to confer a new biocontrol capability to a *Pseudomonas strain*, *Applied and Environmental Microbiology* **58** 3873-3878

Flaishman MAZ, Eyal A, Zilbertein C, and Haas D 1996 Suppression of *Septoria tritici*, blotch and leaf rust of wheat by recombinant cyanide - ptroducing strains of *Pseudomonas putida*, *Molecular Plant Microbe Interaction* **9** 642

Ford HW 1968 Burrowing nematode resistant rootstock as biological barriors in citrus groves, *Citrus Industry* **49** 18-19

Fravel DR, Kim KK, and Papavizas GC 1987 Viability of microsclerotia of *Verticillium dehliae* reduced by a metabolite produced by *Talaromyces flavus*, *Phytopathology* **77** 616-619

Garrett SD 1956 *Biology of Root-Infecting Fungi*, Cambridge, England: Cambridge Univ. Press 293 pages

Glandorf DC, Verheggen P, Jansen T, Jorritsma JW, Smit E, Leefang P, Wernars K, Thomashow LS, Laureijs E, Thomas Oates JE, Bakker PA and Van Loon LC 2001 Effect of genetically modified *Pseudomonas putida* WCS358r on the fungal rhizosphere microflora of field grown wheat, *Applied Environmental Microbiology* **67** 3371-3378

Graham JH and Menge JA 1982 Influence of VAM and soil phosphorus on take all disease of wheat, *Phytopathology* **72** 95

Griffin GJ 1986 Chestnut blight and its control, *Horticultural Reviews* **8** 291

Grossbard E 1951 Antibiotics and microbial antagonism in plant pathology, *Endeavor* **10** 145

Gupta VK and Utkhede RS 1987 Nutritional requitrement of production of antifungal substances by *Enterobacter aerogenes* and *B. subtilis*, antagonists of *Phytophthora cactorum*, *Journal of Phytopathology* **120** 143-158

Haas D and Defago G 2005 Biological control of soil-borne pathogens by fluorescent pseudomonads, *Nature Reviews Microbiology*, **3** 307-319

Harman GE, Howell CR, Viterbo A, Chet I, and Lorito M 2004 *Trichoderma* species opportunistic, avirulent plant symbionts, *Nature Reviews Microbiology* **2** 43-56

Hartley C 1921 Damping-off in forest nurseries, *US Department of Agriculture Bulletin* **934** 1-99

He P, Chintamanani S, Chen Z, Zhu L, Kunkel BN, Alfano JR, Tang X and Zhou JM 2004 Activation of a COI1-dependent pathway in *Arabidopsis* by *Pseudomonas syringae* type III effectors and coronatine, *Plant Journal* **37** 589-602

Homma Y, Sato Z, Hirayama F, Konno K, Shirahama H and Suzui T 1989 Production of antibiotics by *Pseudomonas cepacia* as an agent for biological control of soil-borne plant pathogens, *Soil Biology and Biochemistry* **21** 723-728

Howell CR and Stipanovic RD 1980 Suppression of *Pythium ultimum* induced damping off of cotton seedlings by *Pseudomonas fluorescens* and its antibiotic, Pyoluterin, *Phytopathology* **70** 712-715

Howell CR and Stipanovic RD 1983 Gliovirin, a new antibiotic from *Gliocladium virens* and its role in the biological control of *Pythium ultimum*, *Canadian Journal of Microbiology* **29** 321-324

Howell CR 1987 Relevance of mycoparasitism in the biological control of *Rhizoctonia solani* by *Gliocladium virens*, *Phytopathology* **77** 992-994

Howell CR, Beier RC, and Stipanovic RD 1988 Production of ammonia by *Enterobacter cloacae* and its possible role in the biological control of *Pythium* pre-emergence damping-off by the bacterium, *Phytopathology* **78** 1075-1078

Huang HC 1978 *Gliocladium catenulatum*: hyperparasite of *Sclerotinia sclerotiorum* and *Fusarium* sp., *Canandian Journal of Botany* **56** 2243-2246

Iavicoli A, Boutet E, Buchala A and Métraux JP 2003 Induced systemic resistance in *Arabidopsis thaliana* in response to root inoculation with *Pseudomonas fluorescens* CHA0, *Molecular Plant Microbe Interactions* **16** 851-858

Islam TM, Hashidoko Y, Deora A, Ito T and Tahara S 2005 Suppression of damping-off disease in host plants by the rhizoplane bacterium *Lysobacter* sp. strain SB-K88 is linked to plant colonization and antibiosis against soilborne peronosporomycetes, *Applied Environmental Microbiology* **71** 3786-3796

Kageyama K and Nelson EB 2003 Differential inactiviation of seed exudates stimulation of *Pythium ultimum* sporangium germination by *Enterobacter cloacae* influences biological control efficacy on different plant species, *Applied Environmental Microbiology* **69** 1114-1120

Kamyshiko OP et al 1976 Mikroflora v eksperimental nomsevoohorote I ee antagonistic heskaya activnost k Pochverrnym Fitopatogenam, *Mikology and Fitoputology* **10** 32-36

Keel C, Voisard C, Berling C H, Kahir G and Defago G. 1989. Iron sufficiency is a prerequisite for suppression of tobacco black root rot by *Pseudomonas fluorescnes* strain CHA0 under gnotobiotic contiditions, *Phytopathology* **79** 584-589

Kerr A and Tate ME 1984 Agrocin and biological control of crown gall, *Microbiology Science* **1** 1-4

Kerr A 1980 Biological control of crown gall through production of agrocin 84, *Plant Disease* **64** 25-30

Khan TA and Hussian ST 1988 Studies on the efficacy of *Paecilomyces lilacinus* as biological control agent against a diseases complex caused by onteraton of *Rotylenchulus reniformis, Meloidogyne incognita* and *Rhizoctonia solani* on cowpea, *Nematol Mediterranea* **16** 229-231

Kilic Ekici O and Yuen GY 2003 Induced resistance as a mechanism of biological control by *Lysobacter enzymogenes* strain C3, *Phytopathology* **93** 1103-1110

Kiss L 2003 A review of fungal antagonists of powdery mildews and their potential as biocontrol agents, *Pest Management Science* **59** 475-483

Kloepper JW, Leong J, Teintze M, and Schroth MN 1980 *Pseudomonas* siderophores: A mechanism explaining disease suppression in soils, *Current Microbiology* **4** 317-320

Kloepper JW, Ryu CM, and Zhang S 2004 Induce systemic resistance and promotion of plant growth by *Bacillus* spp, *Phytopathology* **94** 1259-1266

Kobayashi DY and El-Barrad N 1996 Selection of bacterial antagonists using enrichment cultures for the control of summer patch disease of Kentucky bluegrass, *Current Microbiology* **32** 106-110

Koumoutsi A, Chen XH, Henne A, Liesegang H, Gabriele H, Franke P, Vater J, and Borris R 2004 Structural and functional characterization of gene clusters directing non ribosomal synthesis of bioactive lipopeptides in *Bacillus amyloliquefaciens* strain FZB42, *Journal of Bacteriology* **186** 1084-1096

Krishna KR and Bagyaraj DJ 1983 Interaction between *Glomus fasciculatum* and *Sclerotium rolfsii* in Peanut, *Canadian Journal of Botany* **61** 2349

Krupa S and Nylund JE 1972 Studies on ectomycorrhizae of pine. III. Growth inhibition of two root pathogenic fungi by volatile organic constituents of ectomycorrhizal root systems of *Pinus sylvestris* L, *European Journal of Forest Pathology* **2** 88-94

Lafontaine PJ and Benhamon N 1996 Chitosan treatment: an emerging strategy for enhancing resistance of greenhouse tomato to infection by *Fusarium oxysporum* f.sp. *radicilycopersici, Biocontrol Science and Technology* **6** 111-124

Laing SAK and Deacon JW 1990 Aggressiveness and fungal host range of mycoparasitic *Pythium* sp., *Soil Biology & Biochemistry* **22** 905-911

Leclere V, Bechet M, Adam A, Guez JS, Wathelet B, Ongena M, Thonart P, Gancel F, Chollet I M, and Jacques P 2005 Mycosubtilin overproduction by *Bacillus subtilis*

BBG100 enhances the organism's antagonistic and biocontrol activities, *Applied and Environmental Microbiology* **71** 4577-4584

Leeman M, Van Pelt JA, Den Ouden FM, Heinbroek M, and Bakker PAHM 1995a Induction of systemic resistance by *Pseudomonas fluorescens* in radish cultivars differing in susceptibility to *Fusarium* wilt, using novel bioassay, *European Journal of Plant Pathology* **101** 655-664

Leeman M, Van Pelt JA, Hendrickx MJ, Scheffer RJ, and Bakker PAHM 1995b Biocontrol of *Fusarium* wilt of radish in commercial greenhouse trials by seed treatment with *Pseudomonas fluorescens* WCS374, *Phytopathology* **85** 301-1305

Lewis JA and Papvizas GC 1987 Reduction of inoculum of *Rhizoctonia solani* in soil by germlings of *Trichoderma hamatum, Soil Biology & Biochemistry* **19** 195-201

Lindberg GD 1959 A transmissible disease of *Helminthosporium victoriae, Phytopathology* **49** 29-32

Lindsay WL 1979 *Chemical Equilibria in Soils*, New York: John Wiley & Sons, Inc

Loper JE and Buyer JS 1991 Siderophores in microbial interactions of plant surfaces, *Molecular Plant Microbe Interaction* **4** 5-13

Mabagala RB 1999 Epiphytic bacteria from various bean genotypes and their potential for biocontrol of *Xanthomonas axonopodes* pv. *phaseoli, Tanzania Journal of Agricultural Science* **2** 19-26

Mahmoud SM 2002 Evaluation of some biocontrol agents and copper compounds in controlling halo blight diseases of beans caused by *Pseudomonas syringae* pv. *Phaseolicola, Annals Agricultural Science Cairo* **47** 407-415

Martin FN and HancocK JG 1986 Association of chemical and biological factors in soil suppressive to *Pythium ultimum, Phytopathology* **76** 1221-1231

Massomo SMS, Mortensen CN, Mabagala RB, Newman MA, and Hockenhull J 2004 Biological control of black rot in cabbage in Tanzania, with Bacillus strains, *Journal of Phytopathology* **152** 98-105

Maurhofer M, Hase C, Meuwly P, Metraux JP, and Defago G 1994 Induction of systemic resistance to tobacco necrosis virus by the root-colonizing *Pseudomonas fluorescens* strain CHA0: influence of the gacA gene and of pyoverdine production, *Phytopathology* **84** 139-146

Meziane H, Van der Sluis I, Van Loon LC, Hofte M, and Bakker PAH M 2005 Determinants of *Pseudomonas putida* WCS358 involved in inducing systemic resistance in plants, *Molecular Plant Pathology* **6** 177-185

Milgroom MG and Cortesi P 2004 Biological control of chestnut blight with hypovirulence: a critical analysis, *Annual Review of Phytopathology* **42** 311-338

Millard WA and Taylor CB 1927 Antagonism of microorganisms as the controlling factor in the inhibition of scab by green-manuring, *Annals of Applied Biology* **14** 202-216

Milner JL, Silo-Suh L, Lee JC, He H, Clardy JC, Handelsman JC 1996 Production of kanosamine by *Bacillus cereus* UW85, *Applied Environment Microbiology* **62** 3061-3065

Mirova E 1982 *In vitro* study of antagonistic activity of *Trichoderma* spp. toward some soil pathogen, *Gardinarska I Lozarska Nauka* **19** 3-81

Mitchell R and Hurwitz E 1965 Suppression of *Pythium debaryanum* by lytic rhizosphere bacteria, *Phytopatholoy* **55** 156-158

Mohamed HAA, Wafaa HM, and Abo-Aba SM 2004 Influence of salt stress on *Pseudomonas fluorescens* plasmids, some phenotypic traits and antibiosis against Diplodia theobromae, *Journal of Genetic Engineering and Biotechnology* **2** 265-281

Mondal KK and Verma JP 1998 Role of secondary metabolites of Pseudomonas fluorescens in the biocontrol of plant pathogens, In Biological suppression of plant diseases,

Photoparasitic Nematodes and weeds- Present Scenario and Future Thrust, Project Directorate of Biological Control, Bangalore, India, 70-90 pages

Mondal KK and Verma JP 2002 Biological control of cotton diseases in *Biological Control of Crop Diseases,* New York: Marcel Dekker Inc. 480 pages

Mondal KK 2004 Integrated management strategy for fuscus blight and floury leaf spot of rajmash, *Indian Phytopathology* **57** 135-139

Moyne AL, Shelby R, Cleveland TE and Tuzun S 2001 Bacillomycin D: an iturin with antifungal activity against *Aspergillus flavus, Journal of Applied Microbiology* **90** 622-629

Mukhopadhyay AN and Mukharjee PK 1998 Biological control of plant disease: Status in India in biological suppression of plant diseases in *Phytopathogens, Nematodes and Weeds* (eds) SP Singh and SS Hussain, Project Directorate of Biological Control, Banglore, 7-20 pages

Neilands JB 1981 Microbial iron compounds, *Annual Review of Biochemistry* **50** 715-731

Nelson EB, Chao WL, Norton JM, Nash GT, and Harman GE 1986 Attachment of *Enterobacter cloacae* to hyphae of *Pythium ultimum*: Possible role in biological control of *Pythium* pre-mergwence damping- off, *Phytopathology* **76** 327-335

Newhouse JR, Mc Donald WL, and Hoch HC 1990 Virus-like particles in hyphae and conidia of European hypovirulent (ds RNA-containing) strain *of Cryphonectria parasitica, Canadian Journal of Botany* **68** 90-101

Notz R, Maurhofer M, Schnider-Keel U, Duffy B, Haas D, and Defago G 2001 Biotic factors affecting expression of the 2,4-diacetylphloroglucinol biosynthesis gene phlA in *Pseudomonas fluorescens* biocontrol strain CHA0 in the rhizosphere, *Phytopathology* **91** 873-881

Numberger T, Brunner F, Kemmerling B, and Piater L 2004 Innate immunity in plants and animals: striking similarities and obvious differences, *Immunological Review* **198** 249-266

Odum EP 1953 *Fundamentals of Ecology*, W. B. Saunders, Philadelphia, London

Ongena M, Duby F, Rossignol F, Fouconnier ML, Dommes J, and Thonart P 2004 Stimulation of the lipoxygenase pathway is associated with systemic resistance induced in bean by a nonpathogenic *Pseudomonas* strain, *Molecular Plant Microbe Interaction* **17** 1009-1018

Ordentlich A, Elad Y, and Chet I 1988 The role of chitinase of *Serratia marcescens* in the biocontrol of *Sclerotium rolfsii, Phytopathology* **78** 84-88

Pal KK and Gardener BM 2006 Biological Control of Plant Pathogens, *The Plant Health Instructor* DOI: 10.1094/PHI-A-2006-1117-02

Palumbo JD, Yuen GY, Jochum CC, Tatum K, and Kobayashi DY 2005 Mutagenesis of beta-1, 3-glucanase genes in *Lysobacter enzymogenes* strain C3 results in reduced biological control activity toward B*ipolaris* leaf spot of tall fescue and *Pythium* damping-off of sugar beet, *Phytopathology* **95** 701-707

Papavizas GC 1985 *Trichoderma* and *Gliocladium*: Biology, ecology and potential for biocontrol, *Annual Review of Phytopathology* **23** 23-54

Pardeep Kumar, Sood AK and Kumar P 2003 Integration of host resistance, biocontrol agents and soil amendments for the control of bacterial wilt of tomato, *Plant Disease Research* **18** 12-15

Paulitz TC and Baker R 1988 Interation between *Pythium nunn* and *Pythium ultimum* on bean leaves, *Canadian Journal of Microbiology* **34** 947-951

Paulitz TC and Belanger RR 2001 Biological control in greenhouse systems, *Annual Review of Phytopathology* **39** 103-133

Philip DW 1985 Extracellular enzymes and nutritional physiology of *Ampelomyces quisqualis* Ces., hyperparasite of powdery mildew, *in vitro, Phytopathology Zeit* **107** 193-203

Pieterse CMJ, Van Wees SCM, Ton J, Van Pelt JA and Van Loon LC 2002 Signalling in rhizobacteria induced systemic resistance in *Arabidopsis thaliana, Plant Biology* **4** 535-544

Pieterse CMJ, Van Wees SCM, Hoffland E, Pelt JA and Loon LC 1996 Systemic resistance in *Arabidopsis* induced by biocontrol bacteria is independent of salicylic acid accumulation and pathogenesis- related gene expression, *Plant Cell* **8** 1225-1237

Podile AR and Dube HC 1985 Effect of *Bacillus subtilis* on the growth of vascular wilt fungi, *Current Science* **54** 1282-1283

Press CM, Loper JE and Kloepper JW 2001 Role of iron in rhizobacteria mediated induced systemic resistance of cucumber, *Phytopathology* **91** 593-598

Raaijmaker JM and Weller 2001 Combining strains of *Trichoderma harzianum* and solid matrix priming to improve biological seed treatments, *Plant Disease* **73** 631

Raupach GS and Kloepper 2000 Biocontrol of cucumber diseases in the field by plant growth promoting rhizobacteria with and without methyl bromide fumigation, *Plant Disease* **84** 1073-1075

Reyes A 1984 Suppresssive activity of different fungi against cabbage yellows in the soil and *Furarium oxysporum* f. sp. *conglutinans in vitro, Phytopathology* **80** 461-465

Rishbeth J 1963 Stump protection against *Fomes annosus* III inoculation with *Peniophora gigantean, Annals of Applied Biology* **52** 63-77

Roberts DP, Dery PD, Mao W, and Hebrar PK 1997 Use of colonization deficient strain of Escherichia coli in strain combinations for enhanced biological control of cucumber seedling diseases, *Journal of Phytopathology* **145** 461-63

Roberts W 1874 Studies on biogenesis, Philosophical Transactions on the Royal Society of London **164** 466

Rogers HJ, Buck KW and Brasier CM 1986 Transmmission of double-stranded RNA and a disease factor in *Ophiostoma ulmi, Plant Pathology* **35** 277-287

Ryals JA, Neuenschwander UH, Willits MG, Molina A, Steiner HY, and Hunt MD 1996 Systemic Acquired Resistance, *Plant Cell* **8** 1809-18819

Ryder MK and Jones DA 1990 Biological control of crown gall, in Biological Control of Soil-Borne Plant Pathogens (ed) D Hornby, Wallingford, UK: CAB International, 45-63 pages

Ryu CM, Farag MA, Hu CH, Reddy MS, Kloepper JW, and Pare PW 2004 Bacterial volatiles induce systemic resistance in *Arabidopsis, Plant Physiology* **134** 1017-1026

Sandra AI, Wright CH, Zumoff LS, and Steven VB 2001 *Pantoea agglomerans* strain EH318 produces two antibiotics that inhibit *Erwinia amylovora in vitro, Applied and Environmental Microbiology* **67** 282-292

Sanford GB and Broadfoot WC 1931 Studies of the effects of other soil inhabiting micro-organisms on the virulence of Ophiobolus graminis, *Scientia Agricola* **11** 512-528

Sanford GB 1926 Some factors affecting the pathogenicity of actinomyces scabies, *Phytopathology* **16** 525-547

Scher PM and Baker R 1982 Effect of *Pseudomonas putida* and synthetic iron chelator on the induction of soil suppressive to *Fusarium* wilt pathogen, *Phytopathology* **72** 1567-1573

Schippers BA 1987 Interation of deleterious, a beneficial rhizosphere microorganism and the effect of cropping practices, *Annual Review of Phyotopathology* **25** 339

Selvakumar R, Srivastava KD, Rashmi A, Sing DV and Dureja P 2000 Studies on development of *Trichoderma viride* mutants and their effect of *Ustilago segetum tritici, Indian Phytopathology* **53**(2) 185-189

Shanahan P, O'Sullivan DJ, Simpson P, Glennon JD, and O'Gara F 1992 Isolation of 2,4-Diacetylphloroglucinol from a fluorescent pseudomonad and investigation of physiological parameters influencing its production, *Applied and Environmental Microbiology* **58** 353-358

Sharma GI 1996 Studies on integrated management of banded leaf and sheath blight of maize (*Zea mays* L.) caused by *Rhizoctonia solani* (Kuhn). M Sc. (Ag.) Thesis, G B Pant University of Agriculture and Technology, Pant Nagar, India

Shukla AN and Diwivedi RS 1979 Survival of *Rhizoctonia solani* (Kuhn) under the influence of staling growth product of some aspergilli and its growth response to some phenolic substances, *Proceedings of Indian National Science Academy* **45** 269-272

Sieiketov G and Kuldybeav MM 1974 Aktinomitesely antagonisty vozduietelei gribnyk bol ezneiogurtsov, *Trudy Instituta Microbiologii I virusologii Alma Ata* **19** 142-148

Silo-Suh LA, Lethbridge BJ, Raffel SJ, He H, Clardy J, and Handelsman J 1994 Biological activities of two fungistatic antibiotics produced by *Bacillus cereus* UW85, *Applied and Environmental Microbiology* **60** 2023-2030

Singh D, Rana SK, and Singh D 2000 Biocontrol of bacterial wilt /brown rot of potato, *Journal of Mycology and Plant Pathology* **30** 420-421

Singh US and Mishra DS 2001 Induced resistance: Potential as diseases management strategy. In: Biopesticde conference on emerging trends in biological control: BET 2001, held at Chandigarh Feb. 7-9, 2001, 12 page (Abstract)

Sivan A and Chet I 1989 The possible role of competition between *Trichoderma harzianum* and *Fusarium oxysporum* in rhizosphere colonization, *Phytopathology* **79** 198-203

Smith KP, Havey MJ, and Handelsman J 1993 Suppression of cottony leak of cucumber with *Bacillus cereus* strain UW85, *Plant Disease* **77** 139-142

Sneh B, Dupler M, Elad Y, and Baker R 1984 Chlamydospore germination of *Fusarium oxysporum* f. sp. *cucumerinum* as affected by fluorescent and lytic bacteria from *Fusarium* suppressive soils, *Phytopathology* **74** 1115-1124

Steiner G 1941 Nematode parasitic on and associated with roots of marigolds (*Tagetes* hybrids), *Proceedings of the Biological Society of Washington* **54** 31-34

Sturz AV and Matheson BG 1996 Population of endophytic bacteria which influence host resistance to *Erwinia* induced bacterial soft rot on potato tubers, *Plant and soil* **184** 265-271

Sturz AV, Ryan DAJ, Coffin AD, Matheson BG, Arsenault WJ, Kimpinski J, and Christie BR 2004 Stimulating disease suppression in soils: sulphate fertilizer can increase biodiversity and antibiosis ability of root zone bacteria against *Streptomyces scabies*, *Soil Biology and Biochemistry* **36** 343-352

Stutz EW, Defago G, and Kern H 1986 Naturally occurring fluorescent Pseudomonads involved in suppression of black root rot of tobacco, *Phytopathology* **76** 181-185

Swinburne TR, Barr J, and Brown AE 1975 Productiopn of antibiotics by *Bacillus subtilis* and their effect on fungal colonists of apple leaf scars, *Transactions of British Mycological Society* **65** 211-217

Tari PH and Anderson AJ 1988 *Fusarium* wilt suppression and agglutinability of *Pseudomonas putida*, *Applied and Environmental Microbiology* **54** 2037-2041

Tjamos EC, Tjamos SE, and Antoniou PP 2010 Biological management of plant diseases: highlights on research and application, *Journal of Plant Pathology* **92** S4.17 - S4.21

Tewari AK 1996 Biological control of chickpea wilt complex using different formulations of *Gliocladium virens* through seed treatment, Ph.D Thesis, GB Pant University of Agriculture and Technology, Pant Nagar, India 167 pages

Theodorou C 1971 Introduction of mycorrhizal fungi in to soil by spore inoculation of seed, *Australian Forestry* **35** 23-26

Thomashow LS and Weller DM 1988 Role of a phenazine antibiotic from *Pseudomonas fluorescens* in biological control of *Gaeumannomyces graminis* var. *tritici*, *Journal of Bacteriology* **170** 3499-3508

Thomashow LS and Weller DM 1990 Role of antibiotics and siderophores in biological control with mycoherbicides, *Reviews of Weed Science* **2** 1

Thomashow LS and Weller DM 1996 Current concepts in the use of introduced bacteria for biolopgical disease control: mechanism and antifungal metabolites, in *Plant-Microbe Interactions*, (eds) G Stacey and N Keen, New York: Chapman and Hall

Thomashow LS, Bonsall RF, and Weller DM 2002 Antibiotic production by soil and rhizosphere microbes *in situ*, in *Manual of Environmental Microbiology (2nd edition)*, Washington DC: ASM Press, 638-647 pages

Thomashow LS, Weller DM, Bonsall RF, and Pierson LS 1990 Production of the antibiotic phenazine-1-carboxylic acid by fluorescent *Pseudomonas* in the rhizosphere of wheat, *Applied and Environmental Microbiology* **56** 908-912

Vallad GE and Goodman RM 2004 Systemic acquired resistance and induced systemic resistance in conventional agriculture: *Review and Interpretation, Crop Science* **44** 1920-1934

Van Dijk K and Nelson EB 2000 Fatty acid competition as a mechanism by which *Enterobacter cloacae* suppresses *Pythium ultimum* sporangium germination and damping off, *Applied and Environmental Microbiology* **66** 5340-5347

Van Loon LC, Bakker PAHM, and Pieterse CMJ 1998 Systemic resistance induced by rhizosphere bacteria, *Annual Review of Phytopathology* **36** 453-483

Van Peer R and Schipper B 1991 Biocontrol of *Fusarium* wilt by *Pseudomonas* sp. strain WCS417 induced resistance and phytoalexin accumulation, in *Biotic interation and soil borne diseases*, Amsterdam 274-280 pages

Van Peer R and Schippers B 1992 Lipopolysaccharides of plant-growth promoting *Pseudomonas* sp. strain WCS417r induce resistance in carnation to *Fusarium* wilt, *Netherland Journal of Plant Pathology* **98** 129-139

Van Wees SCM, Pieterse CMJ, Trijssenaar A, Van't Westende Y and Hartog F 1997 Differential induction of systemic resistance in *Arabidopsis* by biocontrol bacteria, *Molecular Plant-Microbe Interaction* **10** 716-724

Vasudeva RS and Sikka MR 1941 Effect of certain fungi on the growth of root rot fungi, *Indian Journal of Agricultural Sciences* **3** 422-431

Volksch B and May R 2001 Biological control of *Pseudomonas syringae* pv. *glycinea* by epiphytic bacteria under field conditions, *Microbial Ecology* **41** 132-139

Watson AG and Ford E 1972 Soil fungistasis-a reappraisal, *Annual Review of Phytopathology* **10** 327-348

Weindling R 1932 *Trichoderma lignorum* as a parasite of other soil fungi, *Phytopathology* **22** 837-845

Welvaert W 1961 De fusariose van meloenen enzijn problemen, *Verh. Rijkssta Pl. Ziekt Gent.* **8** 53

Wilhite SE, Lunsden RD, and Strancy D C 2001 Peptide synthetase gene in *Trichoderma virens, Applied and Environmental Microbiology* **67** 5055-5062

Wilson M, Moss WPJP, Wang SY, Dianese AC, Zhang D, Campbell HL, and Defago G 1998 Molecular approaches in the development of biocontrol agents of foliar and floral bacterial pathogens, *Bulletin OILB/SROP* **21** 247-255.

4

PGPR: Their Role in Plant Growth and Disease Management

Rishi Mahajan and C.K. Shirkot

Department of Basic Sciences (Microbiology), Dr. Y.S. Parmar University of Horticulture and Forestry, Nauni Solan, Himachal Pradesh

Plant growth in agricultural soils is influenced by a myriad of abiotic and biotic factors (Nelson, 2004). The global necessity to increase agricultural/horticultural production from a steadily decreasing and degrading land resource base has placed considerable strain on agro ecosystems (Tilak et al 2005). Technologies and strategies not in favor of the sustainability of soil health are today being used at an unprecedented rate to maximize the crop productivity (Fox et al 2007; Kumar et al 2010). Use of chemical fertilizers is credited with nearly fifty per cent increase in agricultural production but they are closely associated with environmental pollution and health hazards.

A few decades ago, growers routinely used physical and chemical approaches to manage the soil environment to improve crop yields and the application of microbial products for this purpose was less common. However, organic growers have actually been promoting the presence of beneficial bacteria without actually knowing it. Addition of compost and the absence of pesticides and more complex organic rotations are likely to promote existing population of beneficial bacteria. During the last couple of decades, the recent biotechnological advancements specifically in agriculture have unlocked new avenues for the augmentation of the agricultural productivity in a sustainable mode and have made possible exploitation of soil microorganisms for improving the crop health (Tank and Saraf 2010).

The region around the root, the rhizosphere is relatively rich in nutrients, due to loss of as much as 40 per cent of plant photosynthates from the roots (Lynch and Whipps 1991). Due to availability of organic nutrients in root exudates, actively growing microbial population forms an important component of the rhizosphere. The microorganisms that colonize the rhizosphere profoundly affect root and plant biology in relation to nutrition, development and health (Sen 2000, Ryan et al 2009). Consequently, the rhizosphere supports large and active microbial populations capable of exerting beneficial, neutral, or detrimental effects on plant growth.

Kloepper and Schroth (1978) introduced the term 'rhizobacteria' to the soil bacterial community that competitively colonized plant roots and stimulated growth and thereby reducing the incidence of plant diseases. Kloepper and Schroth (1981) termed these beneficial rhizobacteria as plant growth-promoting rhizobacteria (PGPR). These encompass all bacteria that inhabit the plant roots and exert a positive effect by various mechanisms, ranging from a direct influence (e.g. increased solubilization and uptake of nutrients or production of plant growth regulators), to an indirect effect (e.g. suppression of pathogen by producing siderophore or antibiosis). Some of these PGPR can also enter root interior and establish as endophytic populations (Kloepper et al 1989, Glick et al 1994, Bashan and de-Bashan 2010).

Cook (2002) considered PGPR as the significant component in the management of agricultural practices with innate genetic potential. The concept of PGPR has now been confined to the bacterial strains that can fulfil at least two of the three criteria such as aggressive colonization, plant growth stimulation and biocontrol (Weller et al 2002, Vessey 2003). PGPR have received prominent attention primarily because of their dual role in plant growth promotion and disease management (Tomcyzak and Lenteren 1999 and Parnwar et al 2004) and thus are being considered as an alternative or a supplemental way of reducing the use of chemicals in agriculture/ horticulture (Gerhardson 2002, Postma et al 2003 and Hameeda et al 2006). The use of PGPR as biofertilizers is one of the most promising biotechnologies to improve primary production with low inputs in fertilizers, through any of the many mechanisms possible: increased solubility of immobile nutrients such as P and Fe (Hayat et al 2010); fixation of atmospheric nitrogen, with previous reports having strongly suggested the positive impacts of microbes on nitrogen uptake involving non legume biological nitrogen fixation (Woo et al 2006, Ardakani et al 2010); phytohormone production (auxins) and eliminating the deleterious effects of plant pathogens on plant growth (Glick et al 1994, Khakhipour et al 2008, Jalili et al 2009). An overview of plant growth promotion mechanisms is outlined in Figure 1.

Use of PGPR has also emerged as a promising component of integrated plant nutrient management because of its manifold role in soil plant system. PGPR inoculants can fulfil diverse beneficial interactions in plants leading to promising solutions for sustainable and environment-friendly agriculture. Way back in 1999, Mishra et al worked on the role of biofertilizer on availability and uptake of nutrient supply mechanisms in plant system. Also inoculation with some PGPR have resulted in P solubilization or enhanced plant uptake of fixed soil P and applied phosphate resulting in higher crop yield (Vessey 2003, Glick et al 2007).

Our understanding on PGPR is now advancing at cellular, genomic and proteomic level. Large numbers of PGPR strains of different bacterial classes and genera with multifunctional traits have, therefore, been described for their potent application in boosting plant activities in modern agriculture. However, it is equally important to study in detail the potentiality of this group of rhizospheric microbiota along with their mechanism of action involved in sustainable crop production. We also need to improve our knowledge for the selection of potent microbial strains colonizing rhizosphere of growing plants for specific restoration programmes. It is important to consider the host plant specificity or adaptation to a particular soil, climatic conditions or pathogens in selecting the isolation conditions and screening assays (Chanway et al 1989, Bowen and Rovira 1999). Other approaches involve selection based on traits known to be associated with PGPR such as root colonization, antibiotic production and siderophore production (Cattelan et al 1999, Carson et al 2000, Giacomodonato et al 2001, Silva et al 2003, El-Tarabily and Sivasithamparam 2006).

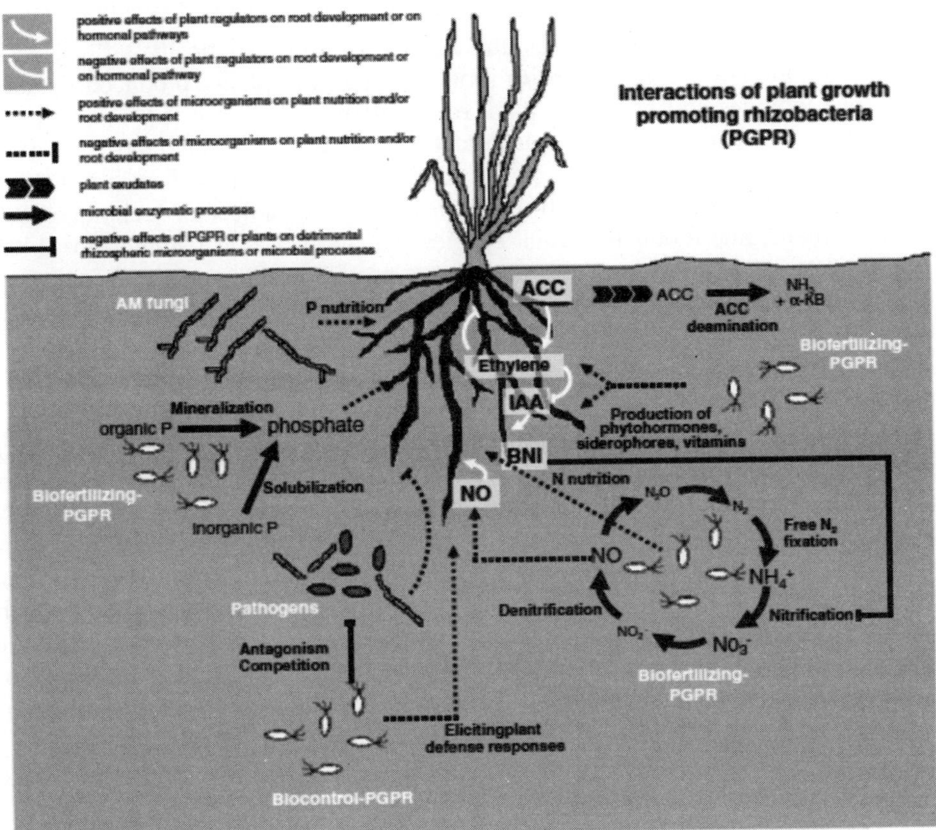

Fig. 1. Plant growth promotion mechanisms associated with PGPR: Biofertilizing-PGPR stimulate plant nutrition by directly increasing the supply of nutrients to plants (e.g. through N fixation, P solubilization and/or mineralization, vitamin and siderophore production) or by increasing the plants access to nutrients due to enhancement of root volume. Promotion of root growth is linked to the ability of PGPR to produce phytohormones (e.g. IAA, ethylene) or by direct influences on plant hormone levels (e.g. deamination of ACC precursor to plant ethylene). Biocontrol-PGPR improve plant heath by inhibiting the growth of plant pathogens or by eliciting plant defense responses. (Richardson et al 2009)

1. PGPR AND HOST INTERACTIONS

1.1 PGPR in Rhizosphere

PGPR can be defined as the indispensable part of rhizosphere biota that when grown in association with the host plants can stimulate the growth of the host. PGPR seemed as successful rhizobacteria in getting established in soil ecosystem due to their high adaptability in a wide variety of environments, faster growth rate and biochemical versatility to metabolize a wide range of natural and xenobiotic compounds (Bhattacharyya and Jha 2012). The degree of intimacy between the PGPR and the host plant can vary depending on where and how the PGPR colonizes the host plant i.e. either rhizospheric colonization or endophytic colonization.

Rhizosphere can be defined as any volume of soil specifically influenced by plant roots and/or in association with roots and hair and plant– produced material. This space includes soil bound by plant roots, often extending a few mm from the root surface and can include the plant root epidermal layer (Andrade et al 1997, Bringhurst et al 2001). This region of soil is much richer in bacteria than the surrounding bulk soil (Hiltner 1904). Studies based on molecular techniques have estimated more than 4,000 microbial species per gram of soil (Montesinos 2003). Plant exudates such as amino acids and sugars provide a rich source of energy and nutrients for the bacteria in rhizosphere, resulting in more microbial populations in the region than outside the region (Haas and Defago 2005).

Root colonization (Fig 2) is the process whereby bacteria survive inoculation onto seed or into soil, multiply in the spermosphere (region surrounding the seed) in response to seed exudates rich in carbohydrates and amino acids (Kloepper et al 1985, 1993), attach to root surface (Weller 1983, Suslow 1982), and colonize the developing root system in soils containing indigenous microorganisms (Weller 1984, Suslow and Schroth 1982, Kloepper et al 1980c). Successful root colonization and persistence of PGPRs in plant rhizosphere are required in order to exert their beneficial effect on the plant (El-liot and Lynch 1984). The intimacy between the plants and the environment in rhizosphere is thus essential for better acquisition of water and nutrients by plants as well beneficial interactions of plants with soil-borne microorganisms (Ryan et al 2009).

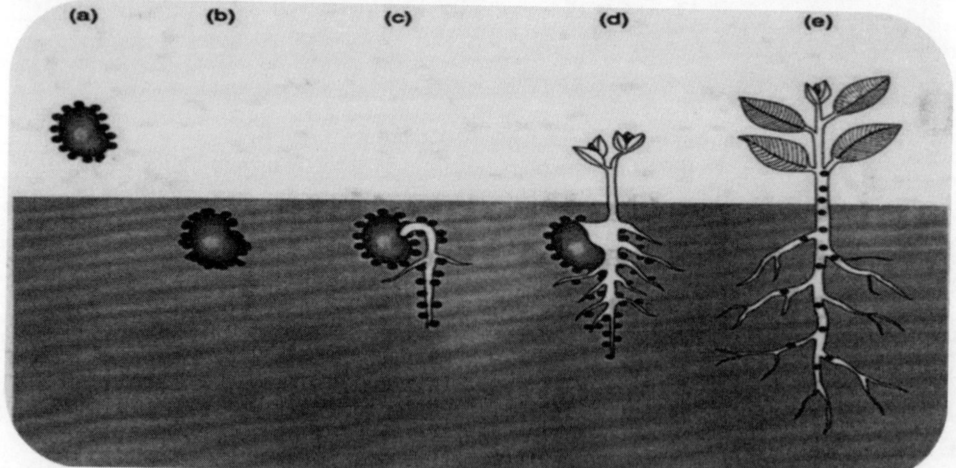

Fig. 2. Steps involved in root colonization of PGPR applied as seed coating (a, b) when planted in soil, bacteria multiply in the spemosphere (c) in response to seed exudates, attach to the root surface (d) and colonize the developing root system (e) some root-colonizing bacteria can colonize the internal regions of the root. (Kloepper et al 1985)

For inoculants that are added in granular form instead of seed coatings, spermosphere colonization would not apply. Therefore, in the root zone, rhizobacteria are efficient microbial competitors that could displace native root colonizing microorganisms (Cook and Baker 1983, Kloepper et al 1999, Vargas-Ayala et al 1999).The nutrient status of rhizosphere determines the nature of root exudates and it can have a direct effect on the composition of the rhizosphere microbial community (Curl and Truelovac 1986, Klein et al 1990, Lynch 1990). Therefore, available nutrients can also probably affect the ability of introduced PGPR to colonize roots and perform their beneficial activity.

Extracellular PGPR (ePGPR) living in the rhizosphere increases plant growth through variety of mechanisms, and include genera such as *Bacillus*, *Pseudomonas*, *Chromobacterium*, *Agrobacterium*, *Azotobacter* and *Azospirillum* (Prithviraj et al 2003 and Gray and Smith 2005). Most rhizosphere microorganisms occur within 50mm of root surface and populations within 10mm of root surface may reach 1.2×10^8 cells g^{-1} soil. Despite large numbers of bacteria in rhizosphere, only 7-15 per cent of the total root surface is generally occupied by microbial cells (Pinton et al 2001 and Gray and Smith 2005).

Selected PGPR must be an aggressive root colonizer and have rapid growth rate. Maize growth promotion and root colonization by strain of *Rhizobium leguminosarum* biovar *phaseoli* and two isolates of fluorescent pseudomonads were found to be the most potential root colonizers (Chabot et al 1998 and Bhatia et al 2005). The most predominant rhizosphere colonizing bacteria belong to the genus *Pseudomonas* and *Bacillus* because of their association with soil organic matter, nutritional diversity and rapid growth rate (Vijaypal et al 1998). Recently it has been reported that soil microorganisms, including free-living as well as associative and symbiotic rhizobacteria belonging to the genera like *Acinetobacter, Alcaligenes, Arthrobacter, Azospirillum, Azotobacter, Bacillus, Burkholderia, Enterobacter, Erwinia, Flavobacterium, Proteus, Pseudomonas, Rhizobium, Serratia, Xanthomonas* in particular, are the integral parts of rhizosphere biota exhibiting successful rhizosphere colonization (Kaymak 2011). A variety of bacterial traits and specific genes are known to contribute to the process of root colonization, but only a few have been identified (Benizri et al 2001, Lugtenberg et al 2001). These include motility, chemotaxis to seed and root exudates, production of pili or fimbriae, production of specific cell surface components, ability to use specific components of root exudates, protein secretion and quorum sensing (Lugtenberg et al 2002 and Oldroyd and Downie 2004).

1.2 PGPR in Endosphere

Certain plant growth promoting rhizobacteria, not only colonize rhizosphere and rhizoplane, but are also reported to be as endophytes. They are localized in the intercellular spaces of the root epidermal cells and vascular tissue (M'Piga et al 1997). Endophytic bacteria live in the plant tissues without causing harm to the host or gaining any benefit other than a non-competitive environment inside the host (Holliday 1989 and Schulz and Boyle 2006) and of the nearly 300000 plant species that exist on the earth, each individual plant is host to one or more endophytes (Strobel et al 2004). An increasing number of observations suggest that root hairs can also play a decisive role in plant root colonization by beneficial bacteria (Rincón et al 2005 and Mattos et al 2008) other than rhizobia with endophytic lifestyle.

Reinhold-Hurek and Hurek (1998) and James (2000) reported that many microorganisms enter into plant roots by three putative pathways i.e. root tips, point of emergence of lateral root and the axils of emerging or developing lateral roots. The endophytes become localized at the point of entry or are able to spread throughout the plant. Root cracks are amply recognized as the main 'hot spots' for bacterial colonization (Sørensen and Sessitsch 2006 and Hardoim et al 2008). These cracks or wounds may be located at the emergence points of lateral roots or may be caused by microbial or nematode activities. Root differentiation and elongation zones and the intercellular spaces in both epidermis and cortex are also proposed to be major sites for bacterial colonization (Reinhold- Hurek and Hurek 2011). Paungfoo-Lonhienne et al 2010 and Prieto et al 2011 have provided evidence of a new site through which some endophytic bacteria enter into the plant roots - the root hairs. Lugtenberg and Kamilova (2009) reported that some plant growth promoting rhizobacteria may stimulate cellular division in roots and increase root hair numbers under soil drought conditions. This then leads to improved water and nutrient uptake, particularly from deeper soil layers (Compant et al 2010).

Bacterial endophytes have been isolated from surface–sterilized plant tissue or extracted from the internal plant tissue (James and Olivares 1998). Endophytic bacteria have been found in numerous plant species, with most members of common soil bacterial genera such as *Pseudomonas*, *Bacillus*, and *Azospirillum* (Chanway 1998). A number of facultative endophytes have been reported from rice (Biswas et al 2000), maize (Rosenblueth and Martinecz 2004), wheat (Cooms and Franco 2003), sorghum (Baldani et al 1997), cotton (Reva et al 2002) and *Arabidopsis* (Sessitsch et al 2004).

Bacterial cells of *Acetobacter diazotrophicus* to localize mostly in the intercellular spaces of roots and stems Sevilla et al (2001). Endophytic and epiphytic bacteria belonging to the genera *Pseudomonas, Ralstonia, Enterobacter, Pantoea* and *Acinetobacter* were isolated from two soybean cultivars. Significant differences were observed in bacterial population densities in relation to season of isolation, soybean growth phase and the tissues from which the isolates were obtained (Sobral et al 2004). The ability to colonize roots has been considered the major factor that determine inoculum efficacy both for crop yield enhancement and disease control. The endophytic bacteria have been considered as potential biocontrol agents as they have the advantage of being relatively protected from the competitive and high-stress environment of the soil (Chen et al 1995 and Blanco and Prieto 2012).

Many studies performed with endophytes have focused on evaluating the colonization pattern of vegetative tissues, as well as the effects of endophytes on plant growth. These studies have been performed by inoculating the host plants with endophytic bacteria, and comparing the inhibition of disease symptoms. Moreover, bacterial endophytes colonize an ecological niche similar to that of phytopathogens, which makes them suitable as biocontrol agents (Chen et al 1995, Benhamou et al 1997, Berg et al 2005). Endophytic bacteria from internal tissues of peanut healthy roots stimulated plant defense reactions and suppressed fungal pathogen infection of peanut tissue Ziedan (2006). The results demonstrated the endophytic bacteria to be useful as biocontrol agent to suppress soil borne diseases of peanut. Diseases of fungal, bacterial, viral origin and in some instances even damage caused by insects and nematodes can be reduced following prior inoculation with endophytes (Kerry 2000, Sturz et al 2000, Ping and Boland 2004, Berg and Hallmann 2006).

It is believed that certain endophyte bacteria trigger a phenomenon known as induced systemic resistance (ISR), which is phenotypically similar to systemic-acquired resistance (SAR). SAR develops when plants successfully activate their defence mechanism in response to primary infection by a pathogen, notably when the latter induces a hypersensitive reaction through which it becomes limited in a local necrotic lesion of brown desiccated tissue (Van Loon et al 1998). Bacterial endophytes and their role in ISR have been reviewed by Kloepper and Ryu (2006). Role of bacterial endophytes in plant–bacterial endophyte interactions that have been studied and their applications have been highlighted in Fig 3.

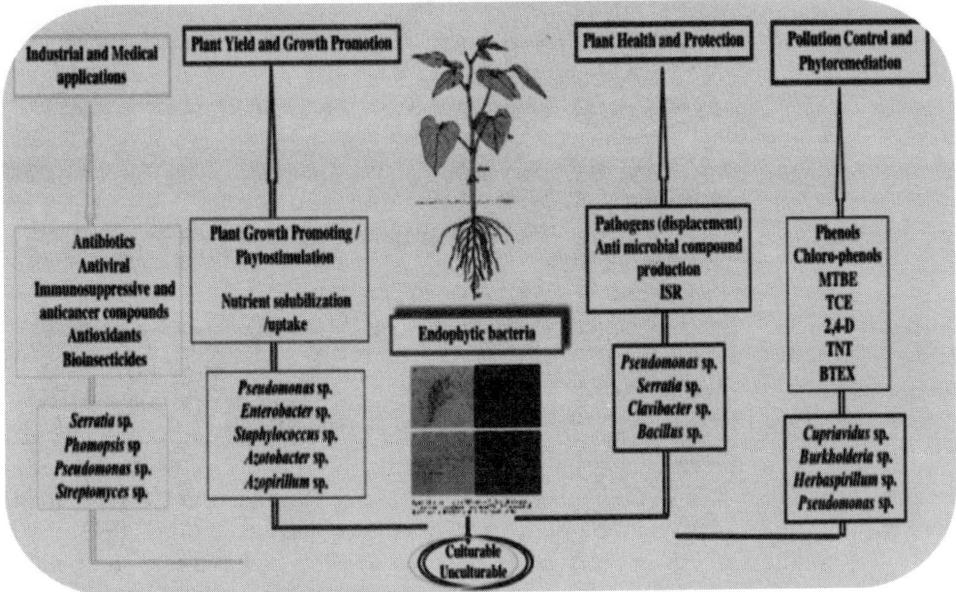

Fig. 3. Schematic diagram of the different plant–bacterial endophyte interactions that have been studied and their applications (Ryan *et al.* 2008)

In some cases, bacterial endophytes have been shown to accelerate seedling emergence, promote plant establishment under adverse conditions and enhance plant growth (Chanway 1997, Bent and Chanway 1998). Endophytes also promote plant growth by a number of mechanisms. These include phosphate solubilisation activity, indole acetic acid production and the production of a siderophore (Costa and Loper 1994, Lee et al 2004, Wakelin et al 2004). Endophytic organisms can also supply essential vitamins to plants (Pirttila et al 2004). Moreover, a number of other beneficial effects on plant growth have been attributed to endophytes and include osmotic adjustment, stomatal regulation, modification of root morphology, enhanced uptake of minerals and alteration of nitrogen accumulation and metabolism (Compant et al 2005). Bacterial endophytes have been shown to prevent disease development through endophyte-mediated *de novo* synthesis of novel compounds and antifungal metabolites. Investigation of the biodiversity of endophytic strains for novel metabolites may identity new drugs for effective treatment of diseases in humans, plants and animals. A summary of natural products that have been derived or produced from various endophytic bacteria is presented in table 1.

Table 1. Natural products that have been derived or produced from various endophytes

Organism	Plant association	Active agent	Activity	Reference
Taxomyces andreane	*Taxus brevifolia*	Taxol	Anticancer	Strobel et al (1993)
Psuedomonas viridiflava	Grass	Ecomycins B and C	Antimicrobial	Miler et al (1998)
Streptomyces griseus	*Kennedia candel*	p-Aminoacetop-henonic acids	Antimicrobial	Guan et al (2005)
Streptomyces NRRL 30562	*Kennedia nigriscans*	Munumbicins	Antibiotic	Castillo et al (2002)
Streptomyces NRRL 30566	*Grevilla pteridifolia*	Kakadumycins	Antibiotic	Castillo et al (2003)
Serratia marcescens	*Rhyncholacis penicillata*	Oocydin A	Antifungal	Strobel et al (2004)
Paenibacillus polymyxa	Wheat, Lodge pine, Canola	Fusaricidin A-D	Antifungal	Beck et al (2003) Li et al (2007) Jensen (2002)
Cytonaema sp.	*Quercus* sp. 103	Cytonic acid A and D	Antiviral	Guo et al (2000)
Streptomyces sp.	*Monstera* sp.	Coronamycin	Antimalarial and antifungal	Ezra et al (2004)

In contrast to the extensive information on the molecular mechanisms of other bacteria-plant interactions (Lugtenberg et al 2002 and Oldroyd and Downie 2004), there is only limited data on the endophyte-host molecular interactions. Endophytes may differentially express genes that are required to enter and colonize the plant to grow and survive within plant tissues and to stimulate plant growth, compete and suppress pathogens, or produce different substances. Endophytic bacteria provide useful and rich models to study the genetic expression of bacteria in their natural niches or habitats (inside plants), which are more structured and variable than culture media under controlled laboratory conditions. Nevertheless, very little work has been done on this. Genomic projects are being performed on some endophytic bacteria, such as *Azoarcus* spp. (Battistoni et al 2005), *Herbaspirillum* sp., *Gluconacetobacter diazotrophicus*, and *Klebsiella* spp., which will certainly be of great help to further understand their molecular interaction with plants. *In vivo* expression technology used to study gene expression in different niches (Rediers et al 2005) including the rhizosphere (Ramos-González et al 2005), may as well be used to study gene expression during endophytic life. Thus, exploitation of endophyte–plant interactions can result in the promotion of plant health and can play a significant role in low-input sustainable agriculture/ horticultural applications.

2. MECHANISM OF ACTION

Kloepper and Schroth (1981) while investigating mechanisms of plant growth promotion by PGPR strains reported that increased plant growth is achieved basically by changing the whole microbial community structure in rhizosphere. PGPR strains can promote plant growth and development either directly and indirectly Castro et al (2008). Direct stimulation includes biological nitrogen fixation, producing phytohormones like auxins, cytokinins and gibberellins, solubilizing minerals like phosphorus and iron, production of siderophores and enzymes and induction of systemic resistance, while indirect stimulation is basically related to biocontrol, including antibiotic production, chelation of available Fe in the rhizosphere, synthesis of extracellular enzymes to hydrolyze the fungal cell wall and competition for niches within the rhizosphere (Van Loon 2007).Thus, based on their mechanism of action, PGPRs can be categorized into three general forms such as biofertilizer, phytostimulator and biopesticide (Table 2).

Table 2. Forms of PGPR and their mechanism of action stimulating plant growth

PGPR forms	Definition	Mechanism of action	References
Biofertilizer	A substance that contains live microorganisms which, when applied on the seed, plant surface or soil, colonizes the rhizosphere and promote plant growth through increased supply of primary nutrients for the host plant	Biological nitrogen fixation	Vessy (2003)
		Utilization of insoluble phosphorus	Somers et al (2004)
Phytostimulator	Microorganism, with the ability to produce phytohormones such as indole acid, gibberellic acid, cytokinins and ethylene	Production of phytohormones	Lugtenberg et al (2002), Somers et al (2004)
Biopesticide	Microorganisms that promote plant growth by controlling phytopathogenic agents	Production of antibi-otics, siderophores, HCN	Somers et al (2004) Chandler et al (2008)
		Production of hydrolytic enzymes	
		Acquired and Induced systemic resistance	

Recent investigations on PGPR revealed that it can promote plant growth mainly by following means; (1) producing ACC deaminase to reduce the level of ethylene in the roots of developing plants (Dey et al 2004) (2) producing plant growth regulators like indole acetic acid (IAA) (Mishra et al 2010), gibberellic acid (Narula et al 2006), cytokinins (Castro et al 2008) and ethylene (3) asymbiotic nitrogen fixation (Ardakani et al 2010) (4) exhibition of antagonistic activity against phytopathogenic microorganisms by producing siderophores, β-1,3-glucanase, chitinases, antibiotics, fluorescent pigment and cyanide (Pathma et al 2011) and (5) solubilization of mineral phosphates and other nutrients (Hayat et al 2010). PGPR may use more than one of these mechanisms to enhance plant growth as experimental evidence suggests that the plant growth stimulation is the net result of multiple mechanisms that may be activated simultaneously.

2.1. PGPR as Biofertilizer

Biofertilizers are the substances, prepared from living microorganisms which, when applied to the seeds or plant surfaces adjacent to soil can colonize rhizosphere or the interior parts of the plants and thereby promotes root growth and thus in turn enhancing nutrient status of host plants. The term, biofertilizer should not be used interchangeably with green manure, manure, intercrop or organic-supplemented chemical fertilizer. Thus, the means by which the PGPRs enhance nutrient status of host plants and thereby act as biofertilizers can be categorized into five distinct areas such as biological N_2 fixation, increasing the availability of nutrients in rhizosphere, increase in root surface area, enhancing beneficial symbioses of the host and finally the combinations of all the above modes of action.

2.1.1 PGPR mediated phosphate acquisition in host plants

Phosphorus (P) is one of the essential macronutrients required for plant growth and development and is intimately involved in a wide range of physiological and biochemical processes. This includes; cellular energy systems, structure and function of nucleic acids, membrane integrity and sugar metabolism (Richardson 2009). However, large areas of soil used for agriculture across the world are deficient in plant-available forms of phosphate which can limit agricultural production. Runge-Metzger (1995) and Cordell et al (2009a) reviewed that of 17.5 million tonnes of P is processed annually from world reserves of rock phosphates, 85 per cent is used in the production of fertilizer and suggested that the world may experience a 'peak' in P supply as early as 2033, despite on-going demand for P to support increased food production. For many countries, particularly developing nations where the cost of P-fertilizers is already preclusive, this puts further strain on the long-term sustainability and economic viability of agriculture and highlights the need to consider possible alternatives and future options (Cordell et al 2009).

In addition to the limited global reserves of P and looming supply issues, it has also long been recognized that the efficiency of P-use from fertilizers is low. Despite most soils containing a large amount of total soil P, only a small proportion is immediately available for plant uptake. The concentration of soluble P in soil is usually very low, normally at levels of 1 ppm or less than 1 ppm due to chemical fixation. This low availability of phosphorous to plants is because of the vast majority of soil P is found in insoluble forms, while the plants can only absorb it in two soluble forms, the monobasic ($H_2PO_4^-$) and the diabasic (HPO_4^{2-}) ions (Glass 1989). A general sketch of phosphorous solubilization in soil is shown in Figure 4. Phosphate solubilizing microorganisms include different types of microorganisms that convert the insoluble form of phosphorus to soluble form through acidification, secretion of organic acids or protons (Richardson et al 2009) and chelation and exchange reactions (Hameeda et al 2008). Bacterial genera like *Azospirillum, Azotobacter, Bacillus, Beijerinckia, Burkholderia, Enterobacter, Erwinia, Flavobacterium, Microbacterium, Pseudomonas, Rhizobium and Serratia* are reported as the most significant phosphate solubilizing bacteria (Sturz and Nowak 2000, Sudhakar et al 2000, Mehnaz and Lazarovits 2006).

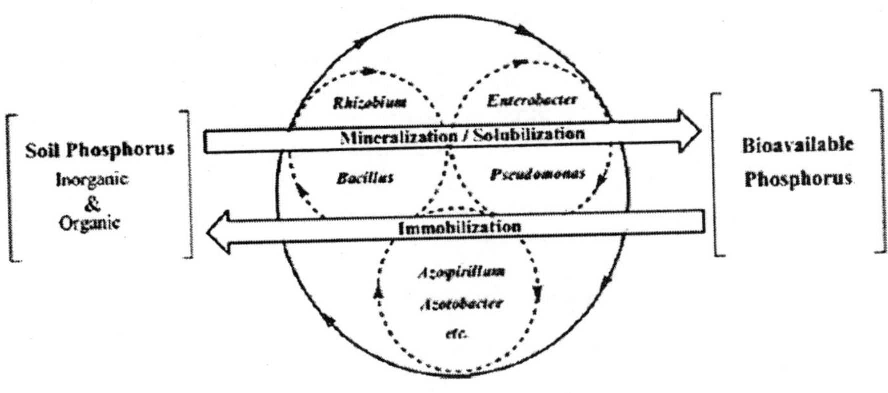

Fig. 4. Schematic representation of solubilisation of soil phosphorus by rhizobacteria (Khan et al 2009)

Biofertilizers have been used as source to improve plant nutrients in sustainable agriculture. Han et al (2006) conducted experiments to evaluate the potential of phosphate solubilizing bacteria (PSB) *Bacillus megaterium* var. *phosphaticum* and potassium solubilizing bacteria (KSB) *Bacillus mucilaginosus* inoculated in nutrient limited soil planted with pepper and cucumber. Their results revealed that rock P and K applied either singly or in combination did not significantly enhance soil availability of P and K, indicated their unsuitability for direct application. Phosphate solubilizing bacteria (PSB) was a more potent P-solubilizer than KSB, and co-inoculation of PSB and KSB resulted in consistently higher P and K availability than in the control without bacterial inoculum and without rock material fertilizer.

Mehta et al (2010) isolated *Bacillus circulans* MTCC 8983 from apple rhizosphere. The strain solubilized tri-calcium phosphate and produced substantial amount of soluble phosphorus (957.3 mg/l) in Pikovskaya's (PVK) broth and exhibited the simultaneous production of indoleacetic acid (IAA) (15.13 µg/ml), siderophore (57.80%) and growth inhibition against *Dematophora necatrix* (46.57%). Regression analysis revealed that phosphate solubilization of *B. circulans* was inversely correlated with pH ($r = -0.98$) and positively correlated with growth ($r = 0.98$), siderophore production ($r = 0.99$), IAA ($r = 0.78$) and antifungal antibiotic activity against *Dematophora necatrix* ($r = 0.87$).

2.1.1.1. Mechanism of phosphate solubilization

There are two components of P in soil, organic and inorganic phosphates. Inorganic P occurs in soil, mostly in insoluble mineral complexes, some of these appearing after the application of chemical fertilizers. Organic matter, on the other hand, is an important reservoir of immobilized P that accounts for 20-80 per cent of soil P.

2.1.1.1.1. Organic Phosphate Solubilization

Organic phosphate solubilization is also called mineralization of organic phosphorus, and it occurs in soil at the expense of plant and animal remains, which contain a large amount of organic phosphorus compounds. The degradability of organic phosphorous compounds depend mainly on the physicochemical and biochemical properties of their molecules, *e.g.* nucleic acids, phospholipids, and sugar phosphates are easily broken down, but phytic acid, polyphosphates, and phosphonates are decomposed more slowly. Phosphorus can be released from organic compounds in soil by three groups of enzymes: (a) Nonspecific phosphatases, which perform dephosphorylation of phospho-ester or phosphoanhydride bonds in organic matter (b) Phytases, which specifically cause P release from phytic acid and (c) Phosphonatases and carbon-phosphorus (C-P) lyases, enzymes that perform C-P cleavage in organophosphonates.

2.1.1.1.2. Inorganic Phosphate Mineralization

Several reports have suggested the ability of different bacterial species to solubilize insoluble inorganic phosphate compounds, such as tri-calcium phosphate, dicalcium phosphate, hydroxyapatite, and rock phosphate (Goldstein 1986). In two thirds of all arable soils, the pH is above 7.0, so that most mineral P is in the form of poorly soluble calcium phosphates (CaPs). Microorganisms must assimilate P via membrane transport, so dissolution of CaPs to Pi (H_2PO_4) is considered essential to the global P cycle. Evaluation of samples from soils throughout the world has shown that, in general, the direct oxidation pathway provides the biochemical basis for highly efficacious phosphate solubilization in Gram-negative bacteria via diffusion of the strong organic acids produced in the periplasm into the adjacent environment. Therefore, the quinoprotein glucose dehydrogenase (PQQGDH) may play a key role in the nutritional ecophysiology of soil bacteria. A number of theories have been proposed to explain the mechanism of phosphate solubilization. Important among them are (a) Acid production theory according to this theory, the process of phosphate solubilization by PSM is due to the production of organic acids which is accompanied by the acidification of the medium (Puente et al 2004) and (b) Proton and enzyme theory According to which, solubilization without acid production is due to the release of protons accompanying respiration or ammonium assimilation (Dighton and Boddy 1989 and Parks et al 1990). In addition, protons arise from produced organic acids which also possess an organic acid anion which is usually capable of forming a complex with metal cation (Hughes and Poole 1991 and Burgstaller and Schinner 1993). Besides these two mechanisms the production of chelating substances (Luo et al 1993)

2.1.1.1.3. Genetic basis of inorganic phosphate solubilization

The production of organic acids is considered to be the principal mechanism for mineral phosphate solubilization, this assumption has been corroborated by the cloning of gene involved in gluconic acid production *viz.* PQQ (4,5-dihydro-4,5-dioxo-1H-pyrrolo-[2,3-] quinoline-2,7,9 tricarboxylic acid) (Fig 5) which is an aromatic, tricyclic ortho-quinone that serves as the redox cofactor for several bacterial dehydrogenases.

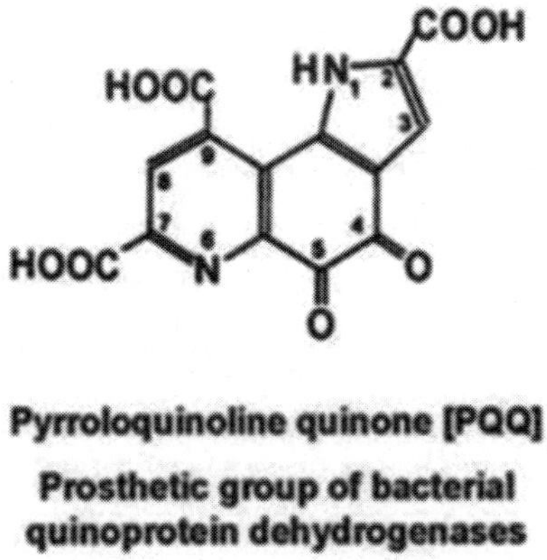

Pyrroloquinoline quinone [PQQ]
Prosthetic group of bacterial quinoprotein dehydrogenases

Fig. 5. Chemical structure of PQQ
(4, 5-dihydro-4, 5-dioxo-1H-pyrrolo- [2, 3-f] quinoline-2, 7, 9-tricarboxylic acid)
(Source: http://www.chris-hony.co.uk/reseach%20pics/pqq.jpg)

PQQ is a prosthetic group required by several bacterial dehydrogenases, including methanol dehydrogenase (MDH) of Gram negative methylotrophs and some glucose dehydrogenases (GDH). PQQ is derived from two amino acids, tyrosine and glutamic acid (Van Kleef and Duine 1988 and Houck et al 1991). PQQ-dependent glucose dehydrogenase (GDH) can oxidize glucose to gluconate. GDH is located in the cytoplasmic membrane and require PQQ for the holoenzyme. Gluconate is further oxidized to 2-ketogluconic acid by gluconate dehydrogenase (GADH).

Sequence analysis of this gene (Liu et al 1992) suggested its probable involvement in the synthesis of the enzyme pyrroloquinoline quinone (PQQ) synthase, which directs the synthesis of PQQ, a co-factor necessary for the formation of the holoenzyme glucose dehydrogenase (GDH)-PQQ. Rodriguez et al (2004) carried out a genetic construction using the broad host range vector pKT230 and plasmid pMCG898, which encodes *Erwinia herbicola* pyrroloquinoline quinone (PQQ) synthase. Their results showed that clones of *Burkholderia cepacia* IS-16 and *Pseudomonas* sp.containing recombinant plasmids produced higher clearing halos in plates with insoluble phosphate as the unique (P) source, in comparison with those of strains without plasmids.

2.1.2 PGPR mediated iron acquisition in host plants

Lankford coined the term siderophore in 1973 as a term to describe low molecular weight molecules that bind ferric iron with an extremely high affinity (Lankford 1973). Term siderophore is derived from a Greek term meaning iron carrier (Ishimaru 1993). The molecular weight of siderphores range from approximately 600 to 1500 daltons and because passive diffusion does not occur for molecules greater than 600 daltons (Fig 6), siderophores must be actively transported (Ishimaru 1993). Once actively transported into the periplasm, the iron siderophore complex is bound to a periplasmic binding protein (Braun and Braun 2002).

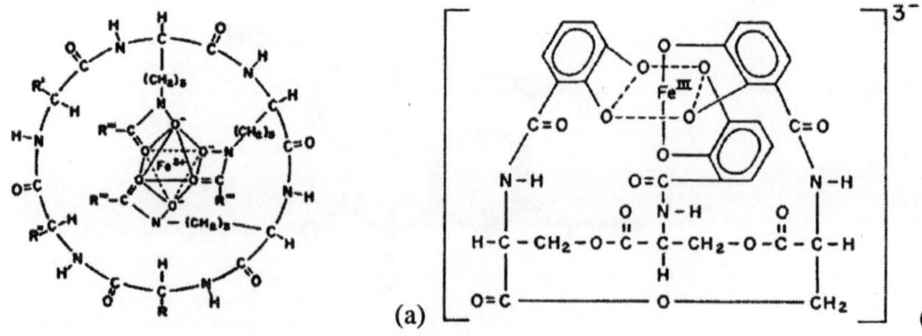

(a) (b)

Fig. 6. General structure of (a) ferrichromes, prototypical hydroxamate
(b) Ferric enterobactin, prototypical catechol-type siderophore (Neilands 1995)

Neilands and Leong (1986) concluded that there are at least three ways by which siderophores could affect plant life. Firstly, the chelaters act in the soil to solubilize and transport Fe (III), a very important mineral in plant nutrition. A second possible effect of siderophores may be via the facilitation of plant disease. Neilands (1982) detected in survey that Gram-negative bacterial phytopathogenes, including species of *Agrobacterium, Erwinia and Pseudomonas* show the presence of siderophores or siderophores-like compounds. Siderophore are by now firmly established as one very important determinant of virulence for infections of animals and plants (Neilands and Leong 1986). The third mechanism is a type of biocontrol in which certain microbial species, such as fluorescent Pseudomonads discourage the growth or metabolic activities of competing microorganisms. Antibiosis, hormone effects, and lytic activity may work in combination with siderophore to afford a microenvironment at the root surface that favors or impedes plant growth.

Plant growth promoting rhizobacteria promote growth of several annual crops by increased uptake of iron through siderophores. Masalha et al (2000) reported that plants grown in non-sterile soil systems show no iron-deficiency symptoms and have fairly high iron levels in roots, in contrast to plants grown in sterile systems. This supported the possible role of microbial community in iron uptake (Walter et al 1994, Yehuda et al 1996). Crowley et al (1992) and Walter et al (1994) also demonstrated the role of microbial siderophore to combat the chlorosis in host plant system (Sharma and Johri 2003, Sharma et al 2003).

PGPR may influence the growth of plant roots through the production of siderophores (Richardson et al 2009) Roots of strategy II type plants (e.g. the Graminaceae, Marschner 1995, Robin et al 2008) secreted phytosiderophores (Fe-chelators) which bind Fe^{3+} and maintain its concentration in soil solution (Lemanceau et al 2009). At the root surface chelated Fe^{3+} can then be directly taken up as a phytosiderophore-Fe complex (strategy II; Lemanceau et al 2009), whereas in non-graminaceous species (strategy I type plants; Marschner 1995, Robin et al 2008), Fe^{3+} must first be reduced to Fe^{2+} prior to being absorbed by the plant (Lemanceau et al 2009). Rhizobacteria (and fungi) also produce siderophores and it has been shown that plants can absorb bacterial- Fe^{3+} complexes, which includes strategy I species possibly by endocytosis (Bar-Ness et al 1991, Vansuyt et al 2007). The capture of these bacterial complexes by plants may play a significant role in nutrition and growth, especially in alkaline and calcareous soils where Fe availability is low (Bar-Ness et al 1991, Masalha et al 2000).

2.1.3. PGPR mediated nitrogen acquisition in host plants

Nitrogen occurs in soil in both organic and inorganic forms and in addition to marked seasonal changes is characterised by a heterogeneous distribution within the soil. Nitrogen inputs through fixation reactions and transformations of N between different pools have important implications for plant growth and for the loss of N from soil systems (Jackson et al 2008).

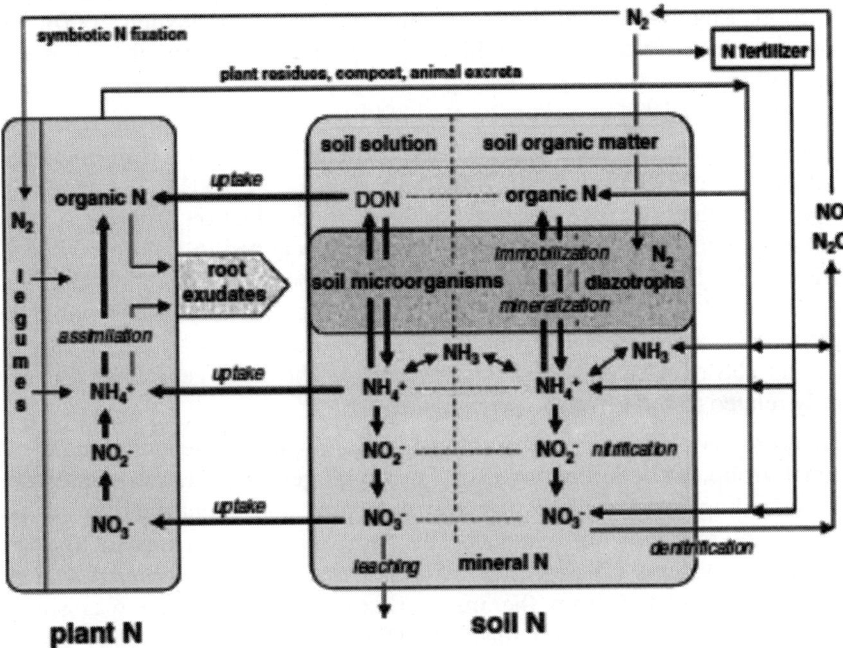

Fig. 7. The plant-soil N cycle and pathways for N transformation mediated by physiological processes (DON = dissolved organic nitrogen) (McNeill and Unkovich 2007, Richardson et al 2009)

Microbial-mediated mineralization of organic forms of N to ammonium (NH_4^+) and its subsequent nitrification to nitrate (NO_3^-) is of major significance to N availability (Fig 7) and has influence on root behaviour and rhizosphere dynamics. Microbial communities are a main component of ecosystems that play critical roles in the biochemical transformations of elements including nitrogen fixation (Madigan et al 2000). Therefore, nitrogen that is available to plants grown for many years without N fertilizers is considered to be due to biological fixation.

Diazotrophs (i.e. N_2-fixing bacteria) are classified as being either symbiotic (rhizobia and Frankia species) or as free-living (associative) and/or root endophytic microorganisms (Lethbridge and Davidson 1983, Rao et al 1998, Cocking 2003). Several groups of symbiotic N_2 fixing bacteria have been identified in soils and flooded systems, such as *Azotobacter*, *Azomonas*, *Beijerinckia*, *Derxia* (aerobic), *Azospirillum*, *Aquaspirillum*, *Thiobacillus*, *Pseudomonas*, *Xanthobacter*, *Rhizobium*, *Methylosinus*, *Mycobacterium* (Microaerobic), *Klebsiella*, *Erwinia*, *Enterobacter*, *Citrobacter*, *Escherichia*, *Bacillus* (facultative anaerobic) and *Desulfovibrio*, *Desulfotomaculum*, *Clostridium* anaerobic, with some being endophytes. Endophytic diazotrophs may have advantage over root-surface associated organisms, as they can colonize the interior of plant roots and establish themselves within niches that are more conducive to effective N_2 fixation and subsequent transfer of the fixed N to host plants (Stewart 1980, Zani et al 2000, Zehr et al 2003, Mirza et al 2006, Bashan and de-Bashan 2010 and Reinhold-Hurek and Hurek 2011).

The process of N_2-fixation, catalyzed by nitrogenase enzymes is essential for maintaining fertility in many ecosystems (Vitousek and Howarth 1991, Bagwell and Lovell 2000, Engelhard et al 2000, Steenhoudt and Vanderleyden 2000). Rhizobia develop symbiotic relationships with host legumes and through atmospheric N_2 fixation within nodules can provide up to 90 per cent of the N requirements of the plant (Höflich et al 1994 and Franche et al 2009).

The *nifH* gene has been widely used to detect nitrogen-fixing bacteria (NFB) (Widmer et al 1999, Zani et al 2000 and Poly et al 2001). It encodes for the dinitrogenase reductase, a key enzyme in the nitrogen fixation process. A substantial molecular diversity of N_2 fixing bacteria has been detected in field grown rice based on retrieval of *nif*H or *nif* D gene fragments from root DNA (Ueda et al 1995).Palus et al (1996) used PCR amplification for the identification of a diazotrophic bacterial endophyte, *Klebsiella*, isolated from the stems of *Zea mays* L. by amplifying portions of *nif*H and 16S rRNA genes from this organism. The *nif* H gene from this organism is closely related to *nif*H from *K. pneumoniae*.

Diazotrophs represent a physiologically and phylogenetically highly diverse functional group, and consequently the functional gene *nifH* (nitrogenase reductase) is the prevailing marker gene for the detection and identification of potential diazotrophs in environmental samples. This approach has been applied to a variety of environmental systems (Diallo et al 2004). In *Bacillus megaterium* C4, a nitrogen fixer originally isolated from the maize rhizosphere *nif*H gene was successfully amplified by polymerase chain reaction (Ding et al 2005). In many diazotroph's nitrogenase activities correspond well to the levels of *nif* H transcription under various conditions (Saito and Minamisawa 2006). Chowdhury et al (2009) studied diversity of *nif*H genes derived from rhizosphere soil and roots of an endemic drought tolerant grass, *Lasiurus sindicus.*

2.2. PGPR as Biostimulator

The most common phytohormones produced by PGPR are auxins, cytokinins, giberellins and to a lesser extent ethylene, with auxins being the most well characterized (Khalid et al 2004, Patten and Glick 1996 and Arshad and Frankenberger 1998). Plant growth stimulation by rhizobacteria is attributed to their ability to produce Indole 3-acetic acid (IAA) (Tsavkelova et al 2007) and thus, IAA production is considered as a key trait for the selection of PGPR (Ali et al 2009, Patten and Glick 1996). Members of the genera *Azospirillum, Pseudomonas, Rhizobium, Acetobacter, Azatobacter, and Bacillus* produce IAA has been effectively used to enhance plant growth and crop productivity (Patten and Glick 1996, Idris et al 2007).

Indole-3-acetic acid (IAA) controls a wide variety of processes in plant development and plant growth and plays a key role in shaping plant root architecture such as regulation of lateral root initiation, root vascular tissue differentiation, polar root hair positioning, root meristem maintenance and root gravitrophism (Aloni et al

2006, Fukaki et al 2007). Production of IAA is widespread among rhizobacteria (Tan and Zou 2001, Khalid et al 2004, Patten and Glick 1996, Spaepen et al 2007, 2008, Mujahid et al 2011). Distribution of IAA biosynthetic pathways among annotated bacterial genomes suggests that 15 per cent (from 369 analysed) contain genes necessary for synthesis of IAA (Spaepen et al 2007).

Idris et al (2007) reported that several routes of tryptophan-dependent IAA biosynthesis (Fig. 8) in Gram negative are known however, no gene products involved in IAA biosynthesis in Gram-positive bacteria were identified, therefore they applied a genomic approach and examined the whole-genome sequence of *B. amyloliquefaciens* FZB42 which revealed three candidate genes with apparent homology to genes previously reported to be involved in IAA metabolism: *ysnE*, encoding a protein similar to the IAA acetyltransferase from *Azospirillum brasilense* (Zimmer et al 1991); *dhaS*, similar to indole-3-acetaldehyde dehydrogenase from *Ustilago maydalis* (Basse et al 1996); and *yhcX*, encoding a putative nitrilase similar to nitrilase2 from *Arabidopsis thaliana* (Hillebrand et al 1998). All three genes had counterparts in the genome of *B. subtilis* 168 (Kunst et al 1997).

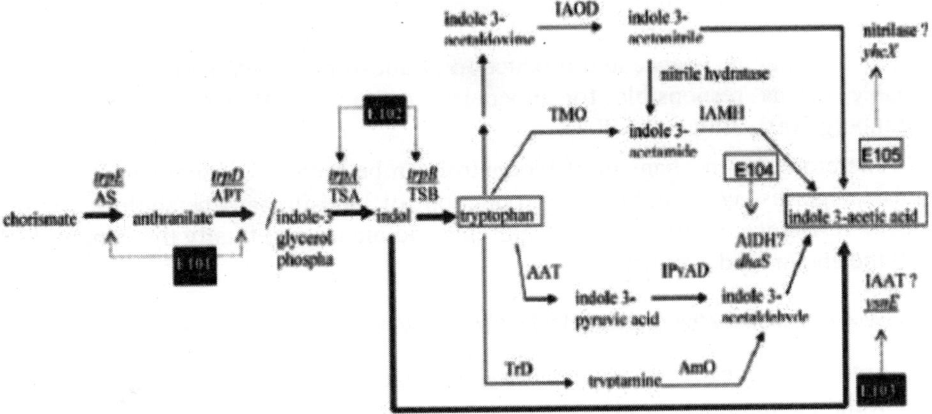

Fig.7. Tryptophan-dependent pathways of bacterial indole-3-acetic acid (IAA) synthesis (Idris et al 2007)

In *A. brasilense*, inactivation of a key enzyme in the IPyA pathway (the ipdC gene, encoding an indole-3-pyruvate decarboxylase) resulted in up to 90% reduction of IAA production (Dobbelaere et al 1999), but mutants were not completely abolished in IAA biosynthesis, suggesting some redundancy in pathways. Irrespective of this, various ipdC mutants displayed altered phenotypes compared to the wildtype strains in their ability to alter wheat root morphology (Dobbelaere et al 1999, Malhotra and Srivastava 2008).

2.3 PGPR as Biocontrol Agent

Soils in which a pathogen causes disease symptoms are called conducive soils. Spontaneous control of plant diseases by bacteria in some fields was discovered at several places around the world. Some soils, called suppressive soils contain bacteria that protect plants against fungal diseases despite the presence of disease-causing pathogens in soil (Schroth and Hancock 1982, Haas and Defago 2005).

Pathogenic micro-organisms affecting plant health are a major and chronic threat to food production and ecosystem stability worldwide. The term "biological control" and its abbreviated synonym "biocontrol" have been used in different fields of biology, but in plant pathology, this term is applied for the use of microbial antagonists (the biological control agent or BCA) to suppress diseases. Most narrowly, biological control refers to the suppression of a single pathogen (or pest) by a single antagonist in a single cropping system. The use of microbes to control diseases, which is a form of biological control, is an environment-friendly approach.

2.3.1. Mechanisms of Biocontrol

Microbial control of plant diseases is a complex phenomenon involving not only the biocontrol microbe, the pathogen, and the plant, but also the indigenous microflora, macrobiota such as nematodes and protozoa, and the plant growth substrate such as soil, stonewool, or vermiculite. Many reviews on biocontrol have been written (Thomashow and Weller 1996, Chin et al 2003, Compant et al 2005a, Haas and Defago 2005).

Competition for nutrients and space, niche exclusion, limiting available Fe supply through production of siderophores, producing lytic enzymes and antibiosis, induced systemic resistance and production of anti-fungal metabolites (AFMs) is the probable means responsible for biocontrol activity of PGPRs (Bloemberg and Lugtenberg 2001, Jing et al 2007).

Therefore the mechanism of biocontrol can be broadly distinguished into three main categories *viz.* antibiosis (production of anti-fungal metabolites AMFs), Induced resistance, competition for nutrients and niches and finally the combinations of all the above modes of action.

2.3.1.1. Antibiosis: Production of Anti-Fungal Metabolites by PGPR

Antibiotics encompass a chemically heterogeneous group of organic, low-molecular weight compounds produced by microorganisms. At low concentrations, antibiotics are deleterious to the growth or metabolic activities of other microorganisms. In the past decades, numerous antibiotics have been isolated from various biocontrol strains representing different bacterial genera (Table 3).

Table 3. Antibiotics produced by bacterial biocontrol agents

Antibiotic[1]	Species/strain[2]	Target Pathogen	Origin/Host
DAPG	*Pseudomonas* spp.		
	Q2-87	Ggt	Wheat, WA, USA
	CHAO	Ggt, Th, Pu	Tobacco, Switzerland
	F113	Pu	Sugar beet, Ireland
	PFM2	St	Wheat, OK, USA
	Pf-5	Pu, Rs	Cotton, Texas,USA
	Q8r1-96	Ggt	Wheat, WA,USA
Phenazines	*Pseudomonas* spp.		
	2-79RN$_{10}$	Ggt	Wheat, WA,USA
	30-84	Ggt	Wheat Kansas, USA

	PGS 12	Fo	Corn, Belgium
	PNA1	Fo, Ps	Chickpea, India
Oomycin A	*Pseudomonas fluorescens* Hv37a	Pu	Barkley, USA
Pyoluteorin	*P. fluorescens*		
	Pf-5	Pu, Rs	Cotton, Texas, USA
	CHAO	Tb, Pu	Tobacco, Switzerland
Pyrrolnitrin	*P. fluorescens* BL915	Rs	Cotton, USA
	B. cepacia B37w	Fs	Potato
	Enterobacter agglomerans	Bc, Rs	Grape, Uzbekistan
	Serratia spp.	Vd, Rs, Ss	Oilseed, rape
DDR	*P. borealis* MA342	Pt,Tc	Wheat, barley
Viscosinamide	*P. fluorescens* DR54	Rs, Pu	Sugar beet, Denmark
Pantocin A and B	*P. agglomerans* EH318	Eh	Apple
Zwittermycin A	*B. cerrus* W85	Pm	Alfalfa
Kanosamine	*B. cerrus* W85	Pm	Alfalfa

[1]DAPG-2,4-diacetylphloroglucinol; DDR-2,3-de-epoxy-2,3-didehydro-rhizoxin; N-BBs-N-Butylbenzenesulphonamide.

[2]Ggt-*Gaemannomyces graminis* vaar *tritici*; Tb- *Thielaviopsis basicolar*; Pu, *Pythium ultimum*; St-*Septoria tritici*; Rs-*Rhizoctonia solani*; Fo-*F. oxysporum*; Ps-*Pythium splendens*; Fs-*Fusarium sambucinum*; Pt-*Pyrenophora terres*; Tc-*Tilletia caries*; Bc-*Botrytis cinerra*; Eh-*Erwinia herbicola*; Pm-*Phytopathora medicaginis*; Vd-V*verticillium dahlia*; Ss-*Sclerotinia sclerotiorum*

Raaijmakers JM, Vlami M and de Souza JT 2002

Most of the PGPRs are recorded to produce AFMs, of which phenazines, pyrrolnitrin, 2, 4-diacetylphloroglucinol (DAPG), pyoluteorin, viscosinamide and tensin are the frequently detected classes (Bloemberg and Lugtenberg 2001 and Bhattacharyya and Jha, 2012). Kanosamine and zwittermycin A have also been reported to be produced by *Bacillus cereus* (Milner et al 1996, Emmert et al 2004). Antibiotics more recently discovered in biocontrol strains are d-gluconic acid (Kaur et al 2006) and 2-hexyl-5-propyl resorcinol (Cazorla et al 2006). Volatiles other than hydrogen cyanide, such as 2,3-butanediol, or blends of volatiles are also produced by *Bacillus* spp. (Ryu et al 2003).

P. fluorescens strain, WCS374 has been recorded to suppress *Fusarium* wilt in radish leading to an average increase of 40 per cent in yield. The individual genes such as *phz*O and *phz*H responsible for the presence of functional group on phenazine compound have been detected (Chin-A-Woeng et al 2001). More recently, informations have been generated on the biosynthesis of pyoluteorin in *Pseudomonas fluorescens* Pf-5 and 2, 4-diacetylphloroglucinol in *P. fluorescens* Q2-87 (Kidarsa et al 2011).

Antibiotic-producing bacteria are too variable in their performance to be successfully used as a common practice in agriculture and horticulture. This inconsistency has been ascribed to a number of factors, including variable expression of genes involved in disease suppression and poor colonization of the host plant by the applied biocontrol agent. Consequently, research has focused on traits involved in the ecological competence of introduced strains and factors that affect antibiotic production.

For a PGPR to be suitable for biocontrol, firstly it must not only synthesize and release the anti-fungal metabolites, but also compete successfully with other organisms for nutrients from the root and for niches on the root to deliver the metabolite along the whole root system. Secondly, the PGPR should produce the anti-fungal metabolite in the right microniche on the root surface. Finally, bacterial mutants (defective in structural genes involved in synthesis of anti-fungal metabolite) should be biocontrol negative (Chin et al 2000, Pliego et al 2008).

2.3.1.2 Induced resistance

Induced resistance is a physiological "state of enhanced defensive capacity" elicited by specific environ stimuli, whereby the plant's innate defenses are potentiated against subsequent biotic challenges (Van Loon et al 1998). The two most clearly defined forms of induced resistance are SARand ISR (Fig. 8), which can be differentiated on the basis of the nature of the elicitor and the regulatory pathways involved, as demonstrated in model plant systems.

Interaction of some bacteria with the plant roots can result in plants resistant to some pathogenic bacteria, fungi, and viruses. This phenomenon is called induced systemic resistance (ISR) (Van Loon et al 1998, Lugtenberg and Kamilova 2009). ISR acts through a different signaling pathway to that regulating systemic acquired resistance (SAR). Bacterial determinants that are claimed to produce ISRs include siderophores, the *O*-antigen of lipopolysacharide and salicylic acid. Presence of salicylic acid even in nanogram amounts has been indicated to cause an ISR when present (DeMeyer et al 1999, Bloemberg and Lugtenberg 2001, Kloepper et al 2004, Van Loon 2007, Lugtenberg and Kamilova 2009).

The mechanism of ISR elicited by PGPR sometimes, overlaps partly with that of pathogen-induced systemic acquired resistance (SAR). Both ISR and SAR represent a state of enhanced basal persistence of the plant that depends on the signaling compounds jasmonic acid and salicylic acid (Van Loon 2007). Berg (2009) highlighted that pathogens are differently sensitive to the resistance activated by these signaling pathways. These interactions are highly specific on each component: the host plant, the pathogen, as well as the PGPR strain. They recognize each other by chemical signaling: root exudates as well as microbial metabolites can have this function. The mechanisms of ISR include (1) developmental—escape: linked to growth promotion, (2) physiological—tolerance: reduced symptom expression, (3) environmental: associated with microbial antagonisms in the rhizosphere, and (4) biochemical— resistance: induction of cell wall reinforcement, induction of phytoalexins, induction of pathogenesis-related proteins, and "priming" of defense responses (resistance).

Recently, interactions between *Bacillus* spp. and plants with special reference to induced systemic disease resistance have been reported by Choudhary and Johri (2009). Several species of *Bacillus* like *B. amyloliquefaciens, B. subtilis, B. pasteurii, B. cereus, B. pumilus, B. mycoides and B. sphaericus* (Ryu et al 2004) are presently recorded to elicit significant reduction in disease incidence on diversity of hosts.

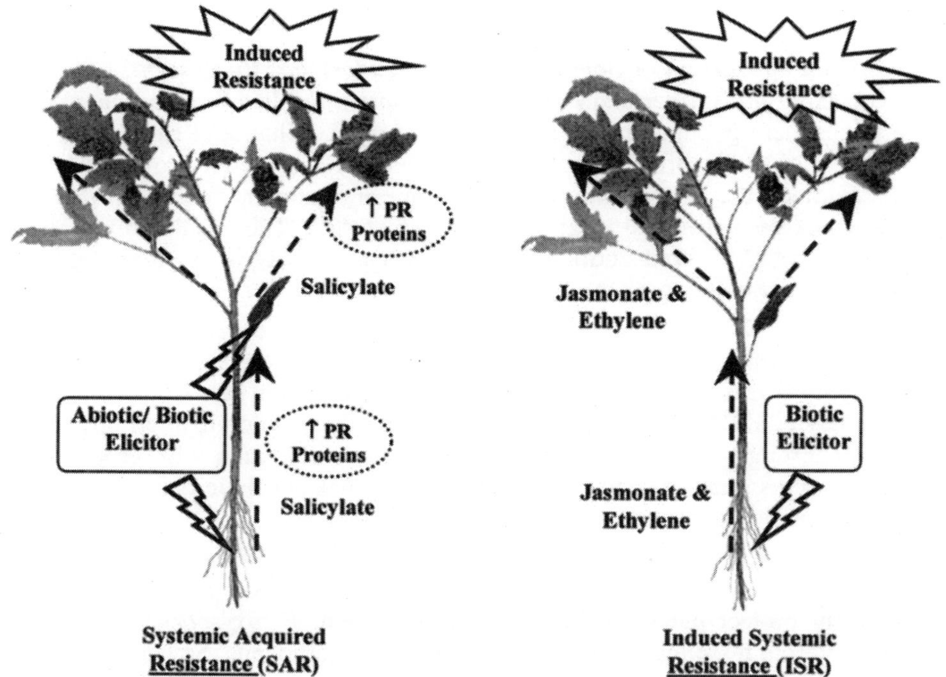

Fig. 8. A pictorial comparison of the two best characterized forms of induced resistance in plants, both which lead to similar phenotypic responses (Vallad and Goodman 2004)

2.3.1.3 Competition for nutrients and niches

Rhizobacteria can suppress the growth of various phytopathogens in variety of ways like competing for nutrients and space, limiting available Fe supply through producing siderophores, producing lytic enzymes and antibiosis (Jing et al 2007). Chitinases are particularly useful in agriculture as biocontrol agents against fungal phytopathogens because of their ability to hydrolyse the chitinous fungal cell wall (Chaiharn and Lumyong 2008, Maisuria et al 2008, Suresh et al 2010, Wahyudi et al 2011 and Kumar et al 2012).

Competition of biocontrol bacteria with the pathogen for nutrients and niches in the rhizosphere (CNN) has been suggested for decades as a possible mechanism of biocontrol, but experimental proof was lacking. Kamilova et al (2008) argued that if this mechanism exists such biocontrol strains can be selected. To this end, they applied a mixture of rhizosphere strains onto surface-sterilized seeds, which were subsequently allowed to germinate in a gnotobiotic system (Simons et al 1996). After one week, the root tip, which contained the best competitive root colonizers, was removed from the seedling and the bacterial content was briefly allowed to multiply and subsequently applied onto fresh seeds for a new enrichment cycle. After three of such cycles, the isolated bacteria were as good as, or even better biocontrol agents. Mutant studies confirmed the proposed mechanism (Kamilova et al 2005, Validov 2007).

3. PGPR INOCULATION AND DISEASE MANAGEMENT

Root dip application of *Bacillus subtilis, Pseudomonas fluorescens, Aspergillus awamori, Aspergillus niger* and *Penicillium digitatum* resulted in significant decline in the rhizosphere population of *Fusarium oxysporum.* (Khan and Khan 2002). The root systems of plants offer different micro-habitats for bacterial growth. Cho et al (2003) studied the endophytic colonization of balloon flower (*Platycodon grandiflorum*) by antifungal strain *Bacillus* sp. C422. The results revealed that *Bacillus* sp. C422 showed antibiotic activities against several phytopathogens by producing the antibiotic iturin A.

Han et al (2006) identified an antagonistic strain of *Bacillus* sp. against *Streptomyces scabiei* and observed that the culture broth had a suppressive effect on common scab disease in a pot assay, decreasing the infection rate from 75 to 35 per cent. Nikolay et al (2006) reviewed microbially mediated solubilization of insoluble phosphates through the release of organic acids in combination with production of other metabolites such as siderophores, phytohormones and lytic enzymes that take part in biological control against soil borne phytopathogens.

Kavroulakis et al (2007) isolated a strain Fs-K of *Fusarium solani* from endophytic root tissues of tomato plant, which was able to colonize root tissues and subsequently protect plants against the root pathogen. *F. oxysporum* elicited and induced systemic resistance against the tomato foliar pathogens. Seven plant growth promoting rhizobacteria, isolated from rhizophere were able to inhibit a wide range of fungal species and were found to be effective against *Fusarium* sp., *Didymella bryoniae* and *Myrothecium* species Lokesh et al (2007). The inoculation with phosphate solubilizing bacteria was useful in controlling *Phytophthora* crown blight of pepper under field conditions and increased the yield (Akgul and Mirik, 2008).

Senthilkumar et al (2008) isolated 137 bacterial isolates from surface sterilized root, stem and nodule tissues of soyabean and screened them for antifungal activity. It was observed that *Paenibacillus* sp. (HKA-15) and *Bacillus* sp. (HKA-121) were potential candidates for biocontrol of charcoal rot fungus as well as for soyabean plant growth promotion. Recep et al (2009) studied the effect of plant growth promoting rhizobacteria as biocontrol agents against *Fusarium* dry root rot of potatoes under storage conditions. *B. cepacia* strain OSU-7 resulted in an inhibition zone ranging from 35.33 to 47.37 mm.

The capability of a novel bio-organic fertilizer (BIO2) that integrated the biocontrol agent *Bacillus subtilis* N11 and mature composts to control *Fusarium* wilt of banana in pot experiments was investigated. The results showed that the application of the BIO2 significantly decreased the incidence rate of *Fusarium* wilt compared to the control. The colonization of the natural biocontrol agent on banana roots using a GFP marker was also studied (Zhang et al 2011). Liang et al (2011) isolated and evaluated *Bacillus megaterium* L8, a plant growth-promoting rhizobacterium (PGPR) for its ability to induce resistance in cucumber against seedling damping-off caused by *Pythium aphanidermatum. Bacillus megaterium* L8 treatment of cucumber roots effectively suppressed seedling damping-off than the control as judged by seedling survival rate (83.45 versus 31.68% at 28 days). Ramyasmruthi et al (2012) isolated 18 bacterial isolates from the rhizosphere of brinjal, capsicum, chilli and screened them for the production of various PGPR activities. Dual plate assay against *Alternaria alternata* OTA36; *Alternaria brassicola* OCA1; *Alternaria brassiceae* OCA3; *Collectotrichum gleosporidose* OGC1 revealed anti-fungal activity by isolate R. The isolate R was identified as *Pseudomonas fluorescens* by biochemical test. Chilli seeds inoculated with *Pseudomonas* showed 100 per cent germination index and almost 50% reduction in disease incidence by *C. gloeosporiodes* OGC1.

Shehata et al (2012) conducted a pot and field experiment to evaluate the effect of *Pseudomonas fluorescens* and *Bacillus subtilis*. The pot experiment evaluated the probable suppressive effect of rhizobacteria as bioagents against *Macrophomina phaseolina*, *Rhizoctonia solani* and *Sclerotium rolfsii* under artifically infested soil. Results showed that co-inoculation of soybean with rhizobacteria led to a significant decrease in pre and post emergence damping-off caused by all pathogens besides enhancing the nodulation.

FUTURE PROSPECTS

PGPR are an environmentally friendly alternative to chemical fertilizers and pesticides, the use of which is regulated and sometimes forbidden, the market for bioinoculants is therefore expanding. PGPR inoculants can fulfil diverse beneficial interactions in plants leading to promising solutions for sustainable and environment-friendly agriculture/horticulture. Application of PGPR in agricultural and horticultural crops has established considerable promises in both the laboratory and greenhouse experiment and to some extent in field as well. However, inconsistent response of PGPR in different environments and on various hosts remains as a significant impediment to their widespread development and application. Paramount to further success of PGPR there is need for better understanding of the ecology of microorganisms, either indigenous or introduced, within the rhizosphere both as individual organisms and through their interaction and directly with host plants. Recent progress in molecular biology and biotechnology can further help in the understanding of rhizobacterial interactions with crop plants that has encouraged a suitable area of research in PGPR mechanisms relating to rhizosphere colonization. PGPR can be classified as biofertilizers, phytostimulators and biopesticides with certain bacteria having overlapping applications. Therefore, the need of the hour is identify naturally occurring rhizobacterial strains with multifunctional PGP traits, which have a coordinated expression of genes for enhancing the plant growth, yield, nutrient uptake of various crop plants and alleviating the attack of both pests and diseases in field crops in different agro ecosystems. The success and commercialization of PGPR strains is another aspect that needs to be taken into account, which depend on the linkages between the scientific organizations and industries. Commercial success of PGPR strains requires economical and viable market demand, consistent and broad spectrum action, safety and stability, longer shelf life, low capital costs and easy availability of career materials. Quality control in this step is crucial to retain the confidence of farmers on the efficacy of the antagonistic strain.

References

Akgul D S and Mirik M 2008 Biocontrol of *Phytophthora capsici* on pepper plants by *Bacillus megaterium* strain, *Journal of Plant Pathology* **90**(1) 29-34

Ali B, Sabri AN, Ljung K, and Hasnain S 2009 Auxin production by plant associated bacteria: impact on endogenous IAA content and growth of *Triticum aestivum* L., *Letters Applied Microbiology* **48** 542–547

Aloni R, Aloni E, Langhans M, and Ullrich CI 2006 Role of cytokinin and auxin in shaping root architecture: regulating vascular differentiation, lateral root initiation, root apical dominance and root gravitropism, *Annals of Botany* **97** 883-893

Andrade G, Esteban E, Velascol L, Maria JL, and Bedmar EJ 1997 Isolation and identification of N_2- fixing microorganisms from the rhizosphere of *Capparis spinosa* (L.), *Plant and Soil* **197** 19-23

Ardakani SS, Heydari A, Tayebi L, and Mohammedi M 2010 Promotion of cotton seedlings growth characteristics by development and use of new bioformulations, *International Journal Botany* **6(2)** 95–100

Arshad M and Frankenberger WT 1998 Plant growth regulating substances in the rhizosphere, Microbial production and functions, *Advances in Agronomy* **62** 46-151

Bagwell CE and Lovell CR 2000 Persistence of selected *Spartina alterniflora* rhizoplane diazotrophs exposed to natural and manipulated environmental variability, *Applied Environmental Microbiology* **66** 4625-4633

Baldani JI, Caruso L, Baldani VLD, Gori SR and Dobereiner J 1997 Recent advances BNF with non-legume plants, *Soil Biology and Biochemistry* **29** 911-922

Bar-Ness E, Chen Y, Hadar Y, Marschner H, and Römheld V 1991 Siderophores of *Pseudomonas putida* as an iron source for dicot and monocot plants, *Plant and Soil* **130** 231-241

Bashan Y and de-Bashan LE 2010 How the plant growth-promoting bacterium *Azospirillum* promotes plant growth–a critical assessment, *Advances in Agronomy* **108** 77-136

Basse CW, Lottspeich F, Steglich W and Kahmann R 1996 Two potential indole-3-acetaldehyde dehydrogenases in the phytopathogenic fungus *Ustilago maydis*, *Europian Journal Biochemistry,* **2242** 648-656

Battistoni F, Bartels D, Kaiser O, Marie Reamon-Buettner S, Hurek T, and Reinhold-Hurek B 2005 Physical map of the *Azoarcus* sp. strain BH72 genome based on a bacterial artificial chromosome library as a platform for genome sequencing and functional analysis, *FEMS (Fedration Europian Microbiology Society) Microbiology Letters* **249** 233-240

Beatty PH and Jensen SE 2002 *Paenibacillus polymyxa* produces fusaricidin-type antifungal antibiotics active against *Leptosphaeria maculans*, the causative agent of blackleg disease of canola, *Canadian Journal of Microbiology* **48** 159-169

Beck HC, Hansen AM, and Lauritsen FR 2003 Novel pyrazine metabolites found in polymyxin biosynthesis by *Paenibacillus polymyxa*, *FEMS Microbiol Letters* **220** 67–73

Benhamou N, Kloepper JW, and Tuzun S 1997 Induction of resistance against *Fusarium* wilt of tomato by combination of chitosan with an endophytic bacterial strain: ultrastracture and cytochemistry of the host response, *Planta* **204** 153-168

Benizri E, Baudoin E, and Guckert A 2001 Root colonization by inoculated plant growth promoting rhizobacteria, *Biocontrol Science Technology* **11** 557-574

Bent E and Chanway CP 1998 The growth-promoting effects of a bacterial endophyte on lodgepole pine are partially inhibited by the presence of other rhizobacteria, *Canadian Journal Microbiology* **44** 980-988

Berg G and Hallmann J 2006 Control of plant pathogenic fungi with bacterial endophytes, in *Microbial Root Endophytes* (eds) BJE Schulz, CJC Boyle and TN Sieber, Berlin: Springer-Verlag, 53-69pages

Berg G, Eberl L, and Hartmann A 2005 The rhizosphere as a reservoir for opportunistic human pathogenic bacteria, *Environmantal Microbiology* **7** 1673-1685

Berg G 2009 Plant-microbe interactions promoting plant growth and health: perspectives for controlled use of microorganisms in agriculture, *Applied Microbiology Biotechnology* **84** 11-18

Bhatia S, Dubey RC, and Maheshwari DK 2005 Enhancement of plant growth and suppression of collar rot of sunflower caused by *Sclerotium rolfsii* through fluorescent *Pseudomonas, Indian Phytopathology* **58** (1) 17-24

Bhattacharyya PN and Jha DK 2012 Plant growth-promoting rhizobacteria (PGPR): emergence in agriculture, *World Journal of Microbiology Biotechnology* **28** 1327–1350

Biswas JC, Ladha JK, and Dazzo FB 2000 Rhizobia inoculation improves nutrient uptake and growth of lowland rice, *Soil Biology and Biochemistry* **64** 1644-1650

Blanco JM and Prieto P 2012 Bacterial endophytes and root hairs, *Plant and Soil* DOI 10.1007/s11104-012-1212-9

Bloemberg GV and Lugtenberg BJJ 2001 Molecular basis of plant growth promotion and biocontrol by rhizobacteria, *Current Opinion Plant Biology* **4** 343-350

Bowen GD and Rovira AD 1999 The rhizosphere and its management to improve plant growth, *Advances in Agronomy* **66** 1-102

Braun V and Braun M 2002 Active transport of iron and siderophore antibiotics, *Current opinion in Microbiology* **5** 194-201

Bringhurst RM, Cardon ZG and Gage DJ 2001 Galactosides in the rhizosphere: utilization by *Sinorhizobium meliloti* and development of a biosensor, *Proceedings of the National Academy of Sciences USA,* **98** 4540- 4545

Burgstaller W and Schinner F 1993 Leaching of metals with fungi, *Journal of Biotechnology* **27** 91-116

Carson KC, Meyer JM, and Dilworth MJ 2000 Hydroxamate siderophores of root nodule bacteria, *Soil Biology Biochemistry* **32** 11–21

Castillo U, Harper JK and Strobel GA 2003 Kakadumycins, novel antibiotics from *Strepto-myces* sp. NRRL 30566, an endophyte of *Grevillea pteridifolia, FEMS Microbiol Letters* **224** 183-190

Castillo UF, Strobel GA, and Ford EJ 2002 Munumbicins, wide-spectrum antibiotics prod-uced by Streptomyces NRRL 30562, endophytic on Kennedia nigriscans, *Microbiology* **148** 2675-2685

Castro RO, Cantero EV, and Bucio JL 2008 Plant growth promotion by *Bacillus megaterium* involves cytokinin signaling, *Plant Signal Behaviour* **3(4)** 263–265

Cattelan AJ, Hartel PG, and Fuhrmann JJ 1999 Screening for plant growth-promoting rhizobacteria to promote early soybean growth, *Soil Science Society American Journal* **63** 1670-1680

Cazorla FM, Duckett SB, Bergstrom ET, Noreen S, and Odijk R 2006 Biocontrol of avocado *Dematophora* root rot by the antagonistic *Pseudomonas fluorescens* PCL1606 correlates with the production of 2-hexyl 5-propyl resorcinol, *Molecular Plant Microbe Interaction* **19** 418-28

Chabot R, Beauchamp CJ, Kloepper JW, and Antoun H 1998 Effect of phosphorus on root colonization and growth promotion of maize by bioluminescent mutants of phosphate sloublizing *Rhizobium leguminosarum biovar phaseoli, Soil Biology and Biochemistry* **30** 1615-1618

Chaiharn M and Lumyong S 2008 Phosphate solubilization potential and stress tolerance of rhizobacteria from rice soil in Northern Thialand, *World Journal of Microbiology and Biotechnology* **25** 305-314

Chandler D, Davidson G, Grant WP, Greaves J, and Tatchell GM 2008 Microbial biopesticides for integrated crop management: an assessment of environmental and regulatory ustainability, *Trends Food Science Technology* **19** 275–283

Chanway CP, Nelson LM and Holl FB 1989 Cultivar-specific growth promotion of spring wheat (*Triticum aestivum* L.) by co-existent *Bacillus* species, *Canadian Journal of Microbiology* **34** 925-929

Chanway CP 1997 Inoculation of tree roots with plant growth promoting soil bacteria: an emerging technology for reforestation, *Forest Science* **43** 99-112

Chanway CP 1998 Bacterial endophytes: ecological and practical implications, *Sydowia* **50** 149-170

Chen C, Bauske EM, Musson G, Rodriguez KR, and Kloepper JW 1995 Biological Control of *Fusarium* wilt on cotton by use of endophytic bacteria, *Biological Control* **5** 83-91

Chin-A-Woeng TF, Thomas-Oates JE, Lugtenberg BJ, and Bloemberg GV 2001 Introduction of the phzH gene of *Pseudomonas chlororaphis* PCL1391 extends the range of biocontrol ability of phenazine-1-carboxylic acid producing *Pseudomonas* spp. strains, *Molecular Plant Microbe Interaction* **14**(8) 1006-1015

Chin-A-Woeng TFC, Bloemberg GV, and Lugtenberg BJJ 2003 Mechanisms of biological control of phytopathogenic fungi by *Pseudomonas* spp., in *Plant-Microbe Interactions*, (eds) G Stacey and NT Keen, St. Paul, MN: American Phytopathology Society **6:**173-224

Chin-A-Woeng TFC, Bloemberg GV, Mulders IHM, Dekkers LC, and Lugtenberg BJJ 2000 Root colonization is essential for biocontrol of tomato foot and root rot by the phenazine-1-carboxamide-producing bacterium *Pseudomonas chlororaphis* PCL1391, *Molecular Plant Microbe Interaction* **13** 1340-1345

Cho So J, Lim WT, Hong SY, Park SR, and Yun HD 2003 Endophytic colonization of balloon flower by antifungal strain *Bacillus* sp. C422, *Bioscience Biotechnology Biochemistry* **67**(10) 2132-2138

Choudhary DK and Johri BN 2009 Interactions of *Bacillus* spp. and plants-with special reference to induced systemic resistance (ISR), *Microbiology Research* **164** 493-513

Chowdhury SP, Schmid M, Hartmann A, and Tripathi AK 2009 Diversity of 16S rRNA and *nif* H genes derived from rhizosphere soil and roots of an endemic drought tolerant grass, *Lasiurus sindicus, European Journal of Soil Biology* **45** 114-122

Cocking EC 2003 Endophytic colonization of plant roots by nitrogen-fixing bacteria, *Plant and Soil* **252** 169-175

Compant S, Duffy B, Nowak J, and Barka EA 2005a Use of plant growth-promoting bacteria for biocontrol of plant diseases: principles, mechanisms of action, and futureprospects, *Applied and Environmental Microbiology* **71** 4951-4959

Compant S, van der Heijden MGA and Sessitsch A 2010 Climate change effects on beneficial plant-microorganism interactions, *FEMS Microbial Ecology* **73** 197-214

Cook RJ and Baker KF 1983 The nature and practice of biological control of plant pathogens, St. Paul, Minnesota: American Phytopathogical Society

Cook RJ 2002 Advances in plant health management in the twentieth century, *Annual Review Phytopathology* **38** 95-116

Cooms JT and Franco CMM 2003 Visualization of an Endophytic *Streptomyces* spp. in wheat seed, *Applied and Environmental Microbiology* **69** 4260-4262

Cordell D, Drangert JO, and White S 2009a. The story of phosphorus: Global food security and food for thought. *Global Environment Change* **19** 292–305

Cordell D, Schmid-Neset T, White S, and Drangert JO 2009b Preferred future phosphorus scenarios: A framework for meeting long-term phosphorus needs for global food demand, 23–43 *In* (eds) D Mavinic, K Ashley and F Koch *International conference on nutrient recovery from wastewater streams*, IWA Publishing, London, UK

Costa JM and Loper JE 1994 Characterization of siderophore production by the biological-control agent *Enterobacter cloacae*, *Molecular Plant Microbe Interaction* **7** 440-448

Crowley DE, Römheld V, Marschner H, and Szaniszlo PJ 1992 Root-microbial effects on plant iron uptake from siderophore and phytosiderophores, *Plant and Soil* **142** 1–7

Curl EA and Truelovac B 1986 *The Rhizosphere,* Berlin: Springer-Verlag

DeMeyer G, Capieau K, Audenaert K, Buchala A, Metraux JP, and Höfte MN 1999 Amounts of salicylic acid produced by the rhizobacterium *Pseudomonas aeruginosa* 7NSK2 activate the systemic acquired resistance pathway in bean, *Molecular Plant Microbe Interaction* **12** 450-458

Dey R, Pal KK, Bhatt DM, and Chauhan SM 2004 Growth promotion and yield enhancement of peanut (*Arachis hypogaea* L.) by application of plant growth-promoting rhizobacteria, *Microbiology Research* **159** 371-394

Diallo MD, Willems A, Vloemans N, Cousin S, Vandekerckhove TT, and de Lajudie P 2004 Polymerase chain reaction denaturing gradient gel electrophoresis analysis of the N_2-fixing bacterial diversity in soil under *Acacia tortilis* ssp *raddiana* and *Balanites aegyptiaca* in the dryland part of Senegal, *Environmental Microbiology* **6** 400-415

Dighton J and Boddy L 1989 Role of fungi in nitrogen, phosphorus and sulfur cycling in temperate forest ecosystems, in *Nitrogen, phosphorus and sulfur ultilization by fungi* (eds) L Boddy, R Marchant and D Read, Cambridge: Cambridge University Press

Ding Y, Wang J, Liu Y and Chem S 2005 Isolation and identification of nitrogen- fixing *Bacilli* from plant rhizospheres in Beijing region, *Journal of Applied Microbiology* **99** 1271-1281

Dobbelaere S, Croonenborghs A, Thys A, Vande Broek A and Vanderleyden J 1999 Phytostimulatory effect of *Azospirillum brasilense* wild type and mutant strains altered in IAA production on wheat, *Plant and Soil* **212** 155-164

El-liot LF and Lynch JM 1984 Pseudomonads as a factor in the growth of winter wheat (*Triticum aestivum* L.), *Soil Biology Biochemistry* **16** 69-71

El-Tarabily KA and Sivasithamparam K 2006 Non-streptomycete actinomycetes as biocontrol agent of soil-borne fungal plant pathogens and as plant growth promoters, *Soil Biology Biochemistry* **38** 1505–1520

Emmert EA, Klimowicz AK, Thomas MG and Handelsman J 2004 Genetics of zwittermicin A production by *Bacillus cereus, Applied and Environmental Microbiology* **70** 104-113

Engelhard M, Hurek T, and Reinhold-Hurek B 2000 Preferential occurrence of diazotrophic endophytes, *Azoarcus* spp., in wild rice species and land races of *Oryza sativa* in comparison with modern races, *Environmental Microbiology* **2** 131-141

Ezra D, Castillo UF, and Strobel GA 2004 Coronamycins, peptide antibiotics produced by a verticillate *Streptomyces* sp. (MSU-2110) endophytic on *Monstera* sp., *Microbiology* **150** 785-793

Fox JE, Gulledge J, Engelhaupt E, Burow ME and McLachlan JA 2007 Pesticides reduce symbiotic efficiency of nitrogen-fixing rhizobia and host plants, *Proceedings of National Academy of Sciences USA* **104** 10282-10287

Franche C, Lindström K and Elmerich C 2009 Nitrogen-fixing bacteria associated with leguminous and non-leguminous plants, *Plant and Soil* doi:10.1007/S11104-008-9833-8

Fukaki H, Okushima Y, and Tasaka M 2007 Auxin-mediated lateral root formation in higher plants, *International Review of Cytolology* **256** 111-137

Gerhardson B 2002 Biological substitutes for pesticides, *Trends in Biotechnology* **20** 338-343

Giacomodonato MN, Pettinari MJ, Souto GI, Mendez BS, and Lopez NI 2001 A PCR-based method for the screening of bacterial strains with antifungal activity in suppressive soybean rhizosphere, *World Journal of Microbiology and Biotechnology* **17** 51-55

Glass ADM 1989 *Plant nutrition: an introduction to current concepts*, Boston: Jones and Bartlett Publishers

Glick BR, Jacobson CB, Schwarze MMK, and Pasternak JJ 1994 1-Aminocyclopropane1-carboxylate deaminase mutants of plant growth promoting rhizobacteria *Pseudomonas putida* GR 12-2 do not stimulate canola root elongation, *Canadian Journal of Microbiology* **40** 911-915

Glick BR, Todorovic B, Czamy J, Chang Z, Duan J and McConkey B 2007 Promotion of plant growth by bacterial ACC deaminase, *Critical Review of Plant Science* **26** 227-242

Goldstein AH 1986 Bacterial solubilization of mineral phosphates: historical perspective and future prospects, *American Journal of Alternative Agriculture* **1** 51-57

Gray EJ and Smith DL 2005 Intracellular and extracellular plant growth promoting rhizobacteria: commonalities and distinctions in the plant bacterium signaling processes, *Soil Biology and Biochemistry* **37** 395- 415

Guan SH, Sattler I, Lin WH, Guo DA and Grabley S 2005 p-Aminoacetophenonic acids produced by a mangrove endophyte: Streptomyces griseus subspecies, *Journal National Proceedings* **68** 1198-1200

Guo B, Dai JR, Ng S, Huang Y, Leong C, Ong W, and Carte BK 2000 Cytonic acids A and B: novel tridepside inhibitors of hCMV protease from the endophytic fungus *Cytonaema species, Journal National Proceedings* **63** 602-604

Haas D and Defago G 2005 Biological control of soil-borne pathogens by fluorescent pseudomonads, *National Review Microbiology* **3** 307-319

Hameeda B, Harini G, Rupela OP, Wani SP, and Reddy G 2008 Growth promotion of maize by phosphate-solubilizing bacteria isolated from composts and macrofauna, *Microbiology Research* **163** 234-242

Hameeda B, Reddy YHK, Rupela OP, Kumar GN and Reddy G 2006 Effect of carbon substrates on rock phosphate solubilization by bacteria from composts and macrofauna, *Current Microbiology* **53** 298-302

Han HS, Supanjani, and Lee KD 2006 Effect of co-inoculation with phosphate and potassium solubilizing bacteria on mineral uptake and growth of pepper and cucumber, *Plant Soil and Environment* **52**(3) 130-136

Hardoim PR, Van Overbeek LS, and Van Elsas JD 2008 Properties of bacterial endophytes and their proposed role in plant growth, *Trends in Microbiology* **16** 463-471

Hayat R, Ali S, Amara U, Khalid R, and Ahmed I 2010 Soil beneficial bacteria and their role in plant growth promotion: a review, *Annual Microbiology* **60(4)** 579-598

Hillebrand H, Bartling D, and Weiler EW 1998 Structural analysis of the *nit2/nit1/nit3* gene cluster encoding nitrilases, enzymes catalyzing the terminal activation step in indole-acetic acid biosynthesis in *Arabidopsis thaliana, Plant Moecular Biology* **36** 89-99

Hiltner L 1904 Uber neuere Erfahrungen und Probleme auf dem Gebiet der Bodenbakteriologie und unter besonderer Berucksichtigung der Grundungung und Brache, *Arbeiten der Deutschen Landwirtschaftlichen Gesellschaft* **98** 59-78

Höflich G, Wiehe W, and Kühn G 1994 Plant growth stimulation by inoculation with symbiotic and associative Rhizosphere microorganisms, *Experientia* **50** 897-905

Holliday P 1989 *A Dictionary of Plant Pathology*, Cambridge: Cambridge University Press

Houck DR, Hanners JL, and Unkefer CJ 1991 Biosynthesis of pyrroloquinoline quinine, Biosynthetic assembly from glutamate and tyrosine, *Journal of American Chemistry Soceity* **113** 3162-3166

Hughes MN and Poole RK 1991 Metal speciation and microbial growth-the hard and soft facts, *Journal of General Microbiology* **137** 725-734

Idris EE, Iglesias EJ, Talon M and Borriss R 2007 Tryptophan dependent production of indole-3-acetic acid (IAA) affects level of plant growth promotion by *Bacillus amyloliquefaciens* FZB42, *Molecular Plant Microbe Interaction* **20** 619–626

Ishimaru C A 1993 Biochemical and genetic analysis of siderophores produced by plant associated *Pseudomonas* and *Erwinia* species, in *Iron chelation in plants and soil microorganism* (eds) LB Barton and BC Hemming, Academic Press Inc

Jackson LE, Burger M, and Cavagnaro TR 2008 Roots, nitrogen transformations, and ecosystem services, *Annual Review of Plant Biology* **59** 341-363

Jalili F, Khavazi K, Pazira E, Negati A, Asadi Rahmani H, Rasuli Sadaghiani H, and Miransari M 2009 Isolation and characterization of ACC demainase producing fluorescent pseudomonads to elevate salinity stress on Canola (*Brassica napus* L.) growth, *Journal of Plant Physiology* **166** 667-674

James EK and Olivares FL 1998 Infection and colonization of sugarcane and other graminaceous plants by endophytic diazotrophs, *Critical Review of Plant Science* **17** 77-119

James EK 2000 Nitrogen fixation in endophytic and associative symbiosis, *Field Crops Res*earch **65** 197-209

Jing YD, He ZL, and Yang XE 2007 Role of soil rhizobacteria in phytoremediation of heavy metal contaminated soils, *Journal Zhejiang University Science* **8**(3) 192–207

Kamilova F, Lamers G, and Lugtenberg B 2008 Biocontrol strain *Pseudomonas fluorescens* WCS365 inhibits germination of *Fusarium oxysporum* spores in tomato root exudate as well as subsequent formation of new spores, *Environmental Microbiology* **10** 2455-2461

Kamilova F, Validov S, Azarova T, Mulders I, and Lugtenberg B 2005 Enrichment for enhanced competitive plant root tip colonizers selects for a new class of biocontrol bacteria, *Environmental Microbiology* **7** 1809–17

Kaur R, Macleod J, Foley W and Nayudu M 2006 Gluconic acid, an antifungal agent produced by Pseudomonas species in biological control of take-all, *Phytochemistry* **67** 595–604

Kavroulakis NN, Besi MI, Katsou P, Damaskinou A, Ehaliotis C, Zervakis GI, and Papadopoulou KK 2007 Antagonistic bacteria of composted agro-industrial residues exhibit antibiosis against soil-borne fungal plant pathogens and protection of tomato plants from *Fusarium oxysporum* f.sp. *radicis-lycopersici, Plant and Soil* **333** 233-247

Kaymak HC 2011 Potential of PGPR in Agricultural Innovations, in *Plant growth and health promoting bacteria* (ed) DK Maheshwari, *Microbiol monographs* Berlin: Springer, **18** 45-79

Kerry BR 2000 Rhizosphere interactions and the exploitation of microbial agents for the biological control of plant-parasitic nematodes, *Annual Reviewof Phytopathology* **38** 423-441

Khakhipour N, Khavazi K, Majallali H, Pazira E and Asadirahmani H 2008 Production of auxin hormone by florescent pseudomonads, *American European Journal of Agriculture Environmental Science* **4** 687-692

Khalid A, Arshad M, and Zahir ZA 2004 Screening plant growth-promoting rhizobacteria for improwing growth and yield of wheat, *Journal of Applied Microbiology* **96** 473–480

Khan AA, Jilani G, Akhtar MS, Naqvi SMS, and Rasheed M 2009 Phosphorus solubilizing bacteria: occurrence, mechanisms and their role in crop production, *Journal of Agricultural Biologial Sciences* **1** 48-58

Khan MR and Khan SM 2002 Effects of root-dip treatment with certain phosphate solubilizing microorganism on the fusarial wilt of tomato, *Bioresource technology* **85** 213-215

Kidarsa TA, Goebel NC, Zabriskie TM, and Loper JE 2011 Phloroglucinol mediates cross-talk between the pyoluteorin and 2, 4 diacetylphloroglucinol biosynthetic pathways in *Pseudomonas fluorescens* Pf-5, *Molecular Microbiology* **81**(2) 395-414

Klein DA, Salzwedel JL, and Dazzo FB 1990 Microbial colonization of plant roots, in *Biotechnology of Plant –Microbe Interactions,* (eds) JP Nakas and C Hagedon, New York: Mc. Graw-Hill, 1989-1225 pages

Kloepper JW and Ryu CM 2006 Bacterial endophytes as elicitors of induced systemic resistance, in *Microbial Root Endophytes* (eds) BJE Schulz, CJC Boyle and TN Sieber, Berlin: Springer-Verlag, 33-52 pages

Kloepper JW and Schroth MN 1981 Relationship of in vitro antibiosis of plant growth promoting rhizobacteria to plant growth and the displacement of root microflora, *Phytopathology* **71** 1020-1024

Kloepper JW and Schroth MN 1978 Plant growth-promoting rhizobacteria on radishes, 879-882 *In Proceings of the 4th International Conference on Plant Pathogenic Bacteria* Vol. 2, Station de Pathologie Vegetale et Phytobacteriologie, INRA, Angers, France

Kloepper JW and Schroth MN 1981 Relationship of *in vitro* antibiosis of plant growth promoting rhizobacteria to plant growth and the displacement of root microflora, *Phytopathology* **71** 1020-1024

Kloepper JW, Lifshitz R, and Zablotowicz R M 1989 Free-living bacterial inocula for enhancing crop productivity, *Trends in Biotechnology* **7** 39-44

Kloepper JW, Rodriguez Ubana R, Zehnder GW, Murphy JF, Sikora E, and Fernadez C 1999 Plant root bacterial interactions in biological control of soil borne diseases and potential extension to systemic and foliar diseases, *Australian Plant Pathology* **28**(1) 21-26

Kloepper JW, Ryu CM, and Zhang S 2004 Induced systemic resistance and promotion of plant growth by *Bacillus* spp., *Phytopathology* **94** 1259-1266

Kloepper JW, Schroth MN, and Miller TD 1980 Effect of rhizosphere colonization by plant growth promoting rhizobacteria on potato development and yield, *Phytopathology* **70** 1080-1082

Kloepper JW 1993 Plant growth promoting rhizobacteria as biocontrol agent, in *Soil Microbial Ecology: Applications of agricultural and envrionmental management* (ed) FB Metting, New York, USA: Jr. Marcel Dekker Inc 255-274 pages

Kloepper LW, Scher FM, Laliberte M, and Zaleska I 1985 Measuring the spermospher colonizing capacity (spermosphere competence) of bacterial inoculants, *Canadian Journal of Microbiology* **31** 926-929

Kucey RMN 1988 Effect of *Penicillium bilaji* on the solubility and uptake of P and micronutrients from soil by wheat, *Canadian Journal of Soil Science* **68** 261-270

Kumar A, Kumar A, Devi S, Patil S, Payal C, and Negi S 2012 Isolation, screening and characterization of bacteria from Rhizospheric soils for different plant growth promotion (PGP) activities: an *in vitro* study, *Recent Research in Science and Technology* **4**(1) 1-5

Kumar JIN, Bora A, and Amb MK 2010 Chronic toxicity of the triazole fungicide tebuconazole on a heterocystous nitrogen-fixing rice paddy field cyanobacterium, *Westiellopsis prolifica Janet Journal of Microbiology and Biotechnology* **20** 1134-1139

Kunst F, Ogasawara N, Moszer I, Albertini AM, Alloni G, Azevedo V, Bertero MG, Bessières P, Bolotin A, Borchert S, Borriss R, Boursier L, Brans A, Braun M, Brignell SC, Bron S, Brouillet S, Bruschi CV, Caldwell B, CapuanoV, and Carter NM 1997 The complete genome sequence of the gram-positive bacterium *Bacillus subtilis*, *Nature* **390** 249-256

Lankford CE 1973 Bacterial assimilation of iron, *Critical Review of Microbiology* **2** 273-331

Lee S, Flores-Encarnacion M, Contreras-Zentella M, Garcia- Flores L, Escamilla JE, and Kennedy C 2004 Indole-3-acetic acid biosynthesis is deficient in *Gluconacetobacter diazotrophicus* strains with mutations in cytochrome C biogenesis genes, *Journal of Bacteriology* **186** 5384-5391

Lemanceau P, Kraemer SM, and Briat JF 2009 Iron dynamics in the rhizosphere as a case study for analyzing interactions between soils, plants and microbes, *Plant and Soil* **321** 513-535

Lethbridge G and Davidson MS 1983 Root-associated nitrogen fixing bacteria and their role in the nitrogen nutrition of wheat estimated by ^{15}N isotope dilution, *Soil Biology and Biochemistry* **15** 365-374

Li J, Beatty PK, Shah S, and Jensen SE 2007 Use of PCR-targeted mutagenesis to disrupt production of Fusaricidin-type antifungal antibiotics in *Paenibacillus polymyxa*, *Applied and Environmental Microbiology* **73** 3480-3489

Liang J, Tao R, Hao Z, Wang L and Zhang X 2011 Induction of resistance in cucumber against seedling damping-off by plant growth-promoting rhizobacteria (PGPR) *Bacillus megaterium* strain L8, *African Journal of Biotechnology* **10**(36) 6920-6927

Liu TS, Lee LY, Tai CY, Hung CH, Chang YS, Wolfram JH, Rogers R, and Goldstein AH 1992 Cloning of an *Erwinia carotovora* gene necessary for gluconic acid production and enhanced mineral phosphate solubilization in *Escherichia coli* HB101: Nucleotide sequence and probable involvement in biosynthesis of the coenzyme pyrroloquinoline quinine, *Journal of Bacteriology* **174** 5814-5819

Lokesh S, Bharath BG, Raghavenra VB, and Govidappa M 2007 Importance of plant growth promoting rhizobacteria in enhancing the seed germination and growth of watermelon attacked by fungal pathogens, *Acta Agronomica Hungariaca* **55**(2) 243-249

Lugtenberg B and Kamilova F 2009 Plant-Growth-Promoting Rhizobacteria, *Annual Review of Microbiology* **63** 541-556

Lugtenberg BJJ, Chin-A-Woeng TFC, and Bloemberg GV 2002 Microbe-plant interactions: Principles and mechanisms, *Antonie vanLeeuwenhoek* **81** 373-383

Lugtenberg BJJ, Dekkers L, and Bloemberg GV 2001 Molecular determinants of rhizosphere colonization by *Pseudomonas, Annual Review of Phytopathology* **38** 461-490

Luo AC, Sun X, and Zhang US 1993 Species of insoluble phosphate solubilizing bacteria in rhizosphere and their effect on yield and nutrient uptake of wheat crop, *Plant and Soil* **57** 223-230

Lynch JM 1990 Some consequences of microbial rhizosphere competence for plant and soil, in *The Rhizosphere*, (ed) JMLynch, New York: Welly Interscience, 1-10pages

Lynch LM and Whipps JM 1991 Substrate flow in rhizosphere, in *The rhizosphere and plant growth* (eds) DL Keister and B Cregan, Beltsville Symposium, *Agricultural Research* 14, Kluwer, Dordre Cht, The Netherlands

M'Piga P, Belanger RR, Paulitz TC, and Benhamou N 1997 Increased resistance to *Fusarium oxysporum* f. sp. *radicis-lycopersici* in tomato plants treated with the endophytic bacterium *Pseudomonas fluorescens* strain 63-28, *Physiology Molecular Plant Pathology* **50** 301-320

Madigan MT, Martinko JM, and Parker J 2000 Brock Biology of Microorganisms, New Jersy, Prentice Hall: Upper Saddle River

Maisuria VB, Gohel V, Mehta AN, Patel RR, and Chhatpar HS 2008 Biological control of *Fusarium* wilt of pigeonpea by *Pantoea dispersa*, a field assessment, *Annals of Microbiology* **58** 177-185

Malhotra M and Srivastava S 2008 An ipdC gene knock-out of *Azospirillum brasilense* strain SM and its implications on indole-3-acetic acid biosynthesis and plant growth promotion, *Antonie vanLeeuwenhoek* **93** 425-433

Marschner H 1995 *Mineral nutrition of higher plants* 2nd edition, London: Academic Press

Masalha J, Kosegarten H, Elmaci Ö, and Mengel K 2000 The central role of microbial activity for iron acquisition in maize and sunflower, *Biology and Fertility of Soils* **30** 433-439

Mattos KA, Pádua VLM, Romeiro A, Hallack LF, Neve BC, Ulisses TMU, Barros CF, Todeschini AR, Previato JO, and Mendonça-Previato L 2008 Endophytic colonization of rice (*Oryza sativa* L.) by the diazotrophic bacterium *Burkholderia kururiensis* and its ability to enhance plant growth, *An Academy Brazilian Ciências* **80** 477493

McNeill AM and Unkovich M 2007 The nitrogen cycle in terrestrial ecosystems, in *Nutrient cycling in terrestrial ecosystems* (eds) P Marschner and Z Rengel, Berlin: *Springer-Verlag*, 37-64 pages

Mehnaz S and Lazarovits G 2006 Inoculation effects of *Pseudomonas putida*, *Gluconacetobacter azotocaptans*, and *Azospirillum lipoferum* on corn plant growth under greenhouse conditions, *Microbiology Ecology* **51(3)** 326-335

Mehta P, Chauhan A, Mahajan R, Mahajan P K and Shirkot C K 2010 Strain of *Bacillus circulans* isolated from apple rhizosphere showing plant growth promoting potential, *Current Science* **98** 538-542

Miller CM, Miller RV, Garton-Kenny D, Redgrave B, Sears J, Condron MM, Teplow DB and Strobel GA 1998 Ecomycins, unique antimycotics from *Pseudomonas viridiflava*, *Journal Applied Microbiology* **84** 937-944

Milner J, Silo-Suh L, Lee JC, He H, Clardy J, and Handelsman J 1996 Production of kanosamine by *Bacillus cereus* UW85, *Applied and Environmental Microbiology* **62** 3061-3065

Mirza MS, Mehnaz S, Normand P, Prigent-Combaret C, Moënne-Loccoz Y, Bally R, and Malik K A 2006 Molecular characterization and PCR detection of a nitrogen-fixing *Pseudomonas* strain promoting rice growth, *Biology and Fertility of Soils* **43** 163-170

Mishra D, Verma DP, and Sharma A K 1999 Sustainable agriculture: a need of the day, *Farmer and Parliament* **34**(8) 9-10

Mishra M, Kumar U, Mishra PK, and Prakash V 2010 Efficiency of plant growth promoting rhizobacteria for the enhancement of *Cicer arietinum* L. growth and germination under salinity, *Advances in Biological Research* **4(2)** 92-96

Montesinos E 2003 Plant-associated microorganisms: a view from the scope of microbiology, *International Microbiology* **6** 221-223

Mujahid M, Sasikala C, and Ramana CV 2011 Production of indole-3-acetic acid and related indole derivatives from L-tryptophan by *Rubrivivax benzoatilyticus* JA2, *Applied Microbiology and Biotechnology* **89** 1001-1008

Narula N, Deubel A, Gans W, Behl RK, and Merbach W 2006 Paranodules and colonization of wheat roots by phytohormone producing bacteria in soil, *Plant Soil Environment* **52(3)** 119-129

Neilands JB and Leong and Saliy A 1986 Siderophores in relation to plant growth and disease, *Annual Review of Plant Physiology* **37** 187-208

Neilands JB 1982 Microbial envelope proteins related to iron, *Annual Review Microbiology* **36** 185-309

Neilands JB 1995 Siderophores: Structure and Function of Microbial IronTransport Compounds, *The journal of Biological Chemistry* **45** 26723-26726

Nelson LM 2004 Plant growth promoting rhizobacteria (PGPR): Prospects for new inoculants, Online, *Crop Management* doi:10.1094/CM-2004-0301-05-RV

Nikolay V, Maria V, and Lana N 2006 Stimulation of phosphate solubilizing and biocontrol activity of microorganism: potential and future trends, *Journal of Applied Microbiology and Biotechnology* **71**(2) 137-144

Oldroyd GED and Downie JA 2004 Calcium, kinases and nodulation signaling in legumes, *National Review Molecular Cell Biology* **5** 566-576

Palus JA, Borneman J, Luddon PW, and Triplett EW 1996 A diazotrophic bacterial endophyte isolated from stems of *Zea mays* L. and *Zea luxurians* Iltis and Doebley, *Plant and Soil* **186** 135-142

Parks EJ, Olson GJ, Brinckman FE, and Baldi F 1990 Characterization by high performance liquid chromatography (HPLC) of solubilization of phosphorus in iron ore by a fungus, *Journal of Indian Microbiology* **5** 183-190

Parnwar S, Sunaina V, Tomar DK, Sharma A, and Singh L 2004 Evaluation of *Entarobacter* spp. KH-7 for biocontrol of *Fusarium* dry rot of potato, *Plant Archives* **4** (1) 49-51

Pathma J, Kennedy RK, and Sakthivel N 2011 Mechanisms of fluorescent pseudomonads that mediate biological control of phytopathogens and plant growth promotion of crop plants, in *Bacteria in agrobiology: plant growth responses* (ed) DK Maheshwari, Berlin: Springer, 77-105 pages

Patten CL and Glick BR 1996 Bacterial biosynthesis of indole-3-acetic acid, *Canadian Journal of Microbiology* 42 207-220

Paungfoo-Lonhienne C, Rentsch D, Robatzek S, Webb RI, Sagulenko E, Näsholm T, Schmidt S, and Lonhienne TGA 2010 Turning the table: plants consume microbes as a source of nutrients *PLoS One* **7** e11915

Perret X and Broughton WJ 1998 Rapid identification of *Rhizobium* strains by targeted PCR fingerprinting, *Plant and Soil* **204** 21-34

Ping L and Boland W 2004 Signals from the underground: bacterial volatiles promote growth in *Arabidopsis*, *Trends in Plant Science* **9** 263-266

Pinton R, Varanini Z, and Nannipieri P 2001 The rhizosphere as a site of biochemical interactions among soil components, plant and microorganisms, in *The rhizosphere* (eds) R Pinton, Z Varanini, and P Nannipieri, New York, USA: Mercel Dekker Inc 1-18 pages

Pirttila A, Joensuu P, Pospiech H, Jalonen J, and Hohtola A 2004 Bud endophytes of Scots pine produce adenine derivatives and other compounds that affect morphology and mitigate browning of callus cultures, *Plant Physiology* **121** 305-312

Pliego C, DeWeert S, Lamers G, De Vicente A, and Bloemberg G 2008 Two similar enhanced root-colonizing *Pseudomonas* strains differ largely in their colonization strategies of avocado roots and *Rosellinia neatrix* hyphae, *Environmental Microbiology* **10** 3295-3304

Poly F, Ranjard L, Nazaret S, Gourbière F Atlas, and Jocteur Monrozier L 2001 Comparison of *nifH* gene pools between soils and between soil microenvironments of contrasting properties, *Applied and Environmental Microbiology* **67** 2255-2267

Postma J, Montanari M, and van den Boogert PHJF 2003 Microbial enrichment to enhance the disease suppressive activity of compost, *European Journal of Soil Biology* **39** 157-160

Prieto P, Schilirò E, Maldonado-González M, Valderrama R, Barroso-Albarracín JB, and Mercado-Blanco J 2011 Root hairs play a key role in the endophytic colonization of olive roots by *Pseudomonas* spp. with biocontrol activity, *FEMS Microbial Ecology* **62** 435-445

Prithviraj B, Zhou X, Souleimanov A, Kahn WM, and Smith DL 2003 A host- specific bacteria- to- plant signal molecule (Nod factor) enhances germination and early growth of diverse crop plants, *Planta* **21** 437-445

Puente ME, Bashan Y, Li CY, and Lebsky VK 2004 Microbial populations and activities in the rhizoplane of rock-weathering desert plants, root colonization and weathering of igneous rocks, *Plant Biology* **6** 629-642

Raaijmakers JM, Vlami M, and de Souza JT 2002 Antibiotic production by bacterial biocontrol agents, *Antonie van Leeuwenhoek* **81** 537-547

Ramos-Gonzalez MI, Campos MJ, and Ramos JL 2005 Analysis of *Pseudomonas putida* KT2440 gene expression in the maize rhizosphere: *In vivo* expression technology capture and identification of root-activated promoters. *Journal of Bacteriology* **187** 4033-4041

Ramyasmruthi S, Pallavi O, Pallavi S, Tilak K, and Srividya S 2012 Chitinolytic and secondary metabolite producing *Pseudomonas fluorescens* isolated from Solanaceae rhizosphere effective against broad spectrum fungal phytopathogens, *Asian Journal of Plant Science and Research* **2** (1) 16-24

Rao VR, Ramakrishnan B, Adhya TK, Kanungo PK, and Nayak DN 1998 Current status and future prospects of associative nitrogen fixation in rice, *World Journal on Microbiology and Biotechnology* **4** 621-633

Recep K, Fikrettin S, Erkol D, and Cafer E 2009 Biological control of the potato dry rot caused by *Fusarium* species using PGPR strains, *Biological Control* **50** 194-198

Rediers H, Rainey PB, Vanderleyden J, and De Mot R 2005 Unraveling the secret lives of bacteria: Use of *in vivo* expression technology and differential fluorescence induction promoter traps as tools for exploring niche-specific gene expression, *Microbiology and Molecular Biology Review* **69** 217-261

Reinhold-Hurek B and Hurek T 2011 Living inside plants: bacterial endophytes, *Current Opinion in Plant Biology* **14** 435-443

Reinnhold HB and Hurek T 1998 Life in grasses: diazotrophic endophytes, *Trends in Microbiology* **61** 139-144

Reva ON, Smirnov VV, Pattersson B, and Priest FG 2002 *Bacillus endophyticus* sp. nov., isolated from the inner tissues of cotton plants (*Gossypium* sp.), *International Journal of Systemic and Evolutionary Microbiology* **52** 101-107

Richardson AE, Barea JM, McNeill AM, and Combaret CP 2009 Acquisition of phosphorus and nitrogen in the rhizosphere and plant growth promotion by microorganisms, *Plant and Soil* **321** 305-339

Richardson AE 2009 Regulating the phosphorus nutrition of plants: molecular biology meeting agronomic needs *Plant and Soil* **322** 17-24

Rincón A, Ruíz-Díez B, García-Fraile S, Lucas García JA, Fernández-Pascual M, Pueyo JJ, and de Felipe MR 2005 Colonisation of *Pinus halepensis* roots by *Pseudomonas fluorescens* and interaction with the ectomycorrhizal fungus *Suillus granalatus*, *FEMS Microbiol Ecology* **51** 303-311

Robin A, Vansuyt G, Hinsinger P, Meyer JM, Briat JF, and Lemanceau P 2008 Iron dynamics in the rhizosphere: consequences for plant health and nutrition, *Advances in Agronomy* **99** 183-225

Rodriguez H, Gonzalez T, Goire I, and Bashan Y 2004 Gluconic acid production and phosphate solubbilization by the plant growth promoting bacterium *Azospirillum* spp., *Naturwissenschaften* **91** 552-555

Rosenblueth M and Martinez-Romero E 2004 *Rhizobium etli* and populations and their competitiveness for root colonization, *Archives of Microbiology* **181** 337-344

Runge-Metzger A 1995 Closing the cycle: obstacles to efficient P management for improved global security in *Phosphorus in the global environment* (ed) H Tiessen, Chichester, UK: Wiley, 27-42 pages

Ryan PR, Dessaux Y, Thomashow LS, and Weller DM 2009 Rhizosphere engineering and management for sustainable agriculture, *Plant and Soil* **321** 363-383

Ryan RP, Germaine K, Franks A, Ryan DJ, and Dowling DN 2008 Bacterial endophytes: recent developments and applications, *FEMS Microbiol Letters* **278** 1-9

Ryu CM, Farag MA, Hu C-H, Reddy MS, and Wie HX 2003 Bacterial volatiles promote growth of *Arabidopsis, Proceedings of National Academic Science USA* **100** 4927-4932

Ryu CM, Farag MA, Hu CH, Reddy MS, Wei HX, and Kloepper JW 2004 Bacterial volatiles induce systemic resistance in *Arabidopsis, Plant Physiology* **134** 1017-1026

Saito A and Minamisawa K 2006 Evaluation of nitrogen-fixing ability of endophytic clostridia based on acetylene reduction and reverse transcription-PCR targeting the *nif H* transcript and ribosomal RNA, *Microbes and Environment* **21** 23-35

Schroth MN and Hancock JG 1982 Disease-suppressive soil and root-colonizing bacteria, *Science* **216** 1376-1381

Schulz B and Boyle C 2006 What are endophytes? in *Microbial Root Endophytes* (eds) BJE Schulz, CJC Boyle and T N Sieber, Berlin: Springer-Verlag, 1-13pages

Sen B 2000 Biological control: A success story, *Indian Phytopathology* **53** 243-249

Senthilkumar M, Swarnalakshmi K, Govindaswami V, Lee YK, and Annapurna K 2008 Biocontrol potential of soyabean bacterial endophytes against charcoal rot fungus *Rhizoctonia bataicola*, *Current Microbiology* **58**(4) 288-293

Sessitsch A, Reiter B, and Berg G 2004 Endophytic bacterial communities of field-grown potato plants and their plant-growth-promoting and antagonistic abilities, *Canadian Journal of Microbiology* **50 (4)** 239-249

Sevilla M, Burris RH, Gunapala N, and Kennedy C 2001 Comparison of benefit to sugarcane plant growth and $^{15}N_2$ incorporation following inoculation of sterile plants with *Acetobacter diazotrophicus* wild type and *Nif* mutant strains, *Molecular Plant Microbe Interaction* **4** 358-366

Sharma A and Johri BN 2003 Growth promoting influence of siderophore-producing *Pseudomonas* strains GRP3A and PRS9 in maize (*Zea mays* L.) under iron limiting conditions, *Microbiology Research* **158** 243-248

Sharma A and Johri BN 2003 Combat of iron-deprivation through a plant growth promoting fluorescent *Pseudomonas* strain GRP3A in mung bean, *Microbiology Res*earch **158** 77–81

Sharma A, Johri BN, Sharma AK, and Glick BR 2003 Plant growth-promoting bacterium *Pseudomonas* sp. Strain GRP3 influences iron acquisition in mung bean (*Vignaradiata* L. Wilzeck), *Soil Biology Biochemistry* **38** 887–894

Shehata HS, Hamed ER, and Eleiwa ME 2012 Potential for improving healthy and productivity of Soybean by plant growth promoting rhizobacteria, *Australian Journal of Basic and Applied Sciences* **6**(4) 98-107

Silva HSA, Romeiro RS and Mounteer A 2003 Development of a root colonization bioassay for rapid screening of rhizobacteria for potential biocontrol agents, *Journal of Phytopathology* **15**142-46

Simons M, van der Bij AJ, Brand I, de Weger LA, Wijffelman CA, and Lugtenberg BJJ 1996 Gnotobiotic system for studying rhizosphere colonization by plant growth-promoting *Pseudomonas* bacteria, *Molecular Plant Microbe Interaction* **9** 600-607

Sobral JK, Arauja WL, Mendes R, Geraldi IO, Kleiner AAP, and Azevedo JL 2004 Isolation and characterization of soybean- associated bacteria and their potential for plant growth promotion, *Environmental Microbiology* **6**(12) 1244-1251

Somers E, Vanderleyden J, and Srinivasan M 2004 Rhizosphere bacterial signalling: a love parade beneath our feet, *Critical Review in Microbiology* **30** 205–240

Sorensen J and Sessitsch A 2006 Plant-associated bacteria lifestyle and molecular interactions, in *Modern soil microbiology* (eds) JD van Elsas, JK Jansson, and JT Trevors, CRC Press, 211-236pages

Spaepen S, Dobbelaere S, Croonenborghs A, and Vanderleyden J 2008 Effects of Azospirillum brasilense indole-3-acetic acid production on inoculated wheat plants, *Plant and Soil* **312** 15-23

Spaepen S, Vanderleyden J,and Remans R 2007 Indole-3-acetic acid in microbial and microorganism-plant signaling, *FEMS Microbiology Reviews* **31** 425–448

Steenhoudt O and Vanderleyden J 2000 *Azospirillum*, a free-living nitrogen-fixing bacterium closely associated with grasses: genetic, biochemical and ecological aspects, *FEMS Microbiology Review* **24** 487-506

Stewart WD 1980 Systems involving blue-green algae (cyanobacteria), in *Methods for evaluating biological nitrogen fixation* (ed) FJ Bergersen, John Wiley and Sons, Chichester, 583-635pages

Strobel G, Daisy B, Castillo U, and Harper J 2004 Natural products from endophytic microorganisms, *Journal of National Procedings India* **67** 257-268

Strobel GA, Stierle A, Stierle D, and Hess WM 1993 Taxomyces andreanae a proposed new taxon for a bulbilliferous hyphomycete associated with Pacific yew, *Mycotaxonomy* **47** 71-78

Sturz AV and Nowak J 2000 Endophytic communities of rhizobacteria and the strategies required to create yield enhancing associations with crops, *Journal of Applied Soil Ecology* **15** 183-190

Sturz AV, Christie BR, and Nowak J 2000 Bacterial endophytes: potential role in developing sustainable systems of crop production, *Critical Review Plant Sciences* **19** 1-30

Sudhakar P, Chattopadhyay GN, Gangwar SK, and Ghosh JK 2000 Effect of foliar application of *Azotobacter, Azospirillum* and *Beijerinckia* on leaf yield and quality of mulberry (*Morus alba*), *Journal of Agricultural Science* **134** 227–234

Suresh P, Pallavi P, Srinivas V, Praveen Kumar S, Jeevan Chandra, and Ram Reddy S 2010 Plant growth promoting activities of fluorescent Pseudomonads associated with some crop plants, *African Journal of Microbiology Research* **4(14)** 1491-1494

Suslow TV and Schroth MN 1982a Rhizobacteria of sugar beets: Effect of seed application and root colonization on yield, *Phytopathology***72** 199-206

Suslow TV 1982 Role of root colonizing bacteria in plant growth, in *Phytopathogenic Prokaryotes* (eds) MS Mount and GH Lacy, London: Acdemic Press, 187-223pages

Tan RX and Zou WX 2001 Endophytes: A rich source of functional metabolites, *Natural Product Reports* **18** 448-459

Tank N and Saraf M 2010 Salinity resistant plant growth promoting rhizobacteria ameliorates sodium chloride stress on tomato plants, *Journal of Plant Interactions* **5** 51-58

Thomashow LS and Weller DM 1996 Current concepts in the use of introduced bacteria for biological disease control: mechanisms and antifungal metabolites, in *Plant-Microbe Interact*ion (eds) G Stacey and NT Keen, **1**:187–235. New York: Chapman & Hall

Tilak KVBR, Ranganayahi N, Pal KN, De R, Saxena AK, Nautiyal CS, Mital S, Tripathi AK, and Joshi BN 2005 Diversity of plant growth and soil health supporting bacteria, *Current Science* **89**(1) 136-150

Tomcyzak A and Lenteren JCV 1999 The use of plant growth promoting rhizobacteria (PGPR) to decrese susceptibility of cucumber to spider mites, *Bulletin of OILB-SROP* **22**(1) 251-254

Tsavkelova EA, Cherdyntseva TA, Botina SG, and Netrusov AI 2007 Bacteria associated with orchid roots and microbial production of auxin, *Microbiology Research* **162** 69–76

Ueda T, Suga Y, Yahiro N and Matsuguchi T 1995 Remarkable N_2-fixing bacterial diversity detected in rice roots by molecular evolutionary analysis of *nif H* gene sequences, *Journal of Bacteriology* **177** 1414-1417

Validov S 2007 Biocontrol of tomato foot and root rot by *Pseudomonas* bacteria in stonewool, PhD thesis, Leiden University http://hdl.handle.net/1887/12480

Vallad GE and Goodman RM 2004 Systemic Acquired Resistance and Induced Systemic Resistance in Conventional Agriculture, *Crop Science* **44** 1920-1934

Van Kleef MAG and Duine JA 1988 L-tyrosine is the precursor of PQQ biosynthesis in Hyphomicrobium X, *FEMS Microbiology Letters* **237** 91-97

Van Loon LC, Bakker PA, and Pieterse CMJ 1998 Systemic resistance induced by rhizosphere bacteria, *Annual Review of Phytopathology* **36** 453-483

Van Loon LC 2007 Plant responses to plant growth-promoting rhizobacteria, *European Journal Plant Pathology* **119** 243-254

Vansuyt G, Robin A, Briat JF, Curie C, and Lemanceau P 2007 Iron acquisition from Fe-pyoverdine by *Arabidopsis thaliana, Moleuular Plant Microbe Interaction* **20** 441-447

Vargas-Ayala R, Rodriguez-Kabana R, Morgan-Jones G, McInroy JA, and Kloepper JW 1999 Microbial shifts in soils and rhizosphere induced by vevletbean (*Mucuna duringiana*) in cropping systems to control root-knot nematodes, *Biological Control* **34** 468-473

Vessey JK 2003 Plant growth promoting rhizobacteria as biofertilizers, *Plant and Soil* **255** 571-586

Vijaypal I, Jalalipal V, and Jalali I 1998 Rhizosphere bacteria for biocontrol of plant diseases, *Indian Journal of Microbiology* **38** 187-204

Vitousek PM and Howarth RW 1991 Nitrogen limitation on land and on the sea, How can it occur? *Biogeochemistry* **13** 87-115

Wahyudi AT, Astuti RP, Widyawati A, Meryandini A, and Nawangsih A 2011 Characterization of *Bacillus* sp. strains isolated from rhizosphere of soybean plants for their use as potential plant growth for promoting rhizobacteria, *Journal of Microbiology and Antimicrobials* **3**(2) 34-40

Wakelin S, Warren R, Harvey P, and Ryder M 2004 Phosphate solubilization by *Penicillium* spp. closely associated with wheat roots, *Biology and Fertility of Soils* **40** 36-43

Walter A, Römheld V, Marschner H, and Crowley DE 1994 Iron nutrition of cucumber and maize: Effect of *Pseudomonas putida* YC3 and its siderophore, *Soil Biology and Biochemistry* **26** 1023-1031

Weller DM 1984 Distribution of a take-all suppressive strain of *Pseudomonas fluorescens* on seminal roots of winter apple, *Applied and Environmental Microbiology* **48** 897-899

Weller DM, Raaijmakers JM, Gardener BB, and Thomashow LS 2002 Microbial populations responsible for specific soil suppressiveness to plant pathogens, *Annual Review of Phytopathology* **40** 309-348

Weller DM 1983 Colonization of wheat roots by a fluorescent pseudomonad suppressive to take-all, *Phytopathology* **73** 1548-1553

Whitelaw Ma 2000 Growth promotion of plants inoculated with phosphate-solubilizing fungi *Advances in Agronomy* **69** 100-151

Widmer F, Shaffer BT, Porteous LA, and Seidler RJ 1999 Analysis of *nifH* gene pool complexity in soil and litter at a Douglas fir forest site in the Oregon cascade mountain range, *Applied and Environmental Microbiology* **65** 374-380

Woo SL, Scala F, Ruocco M, and Writo M 2006 The molecular biology of the interactions between *Trichoderma* spp. phytopathigenic fungi and plants, *Phytopathology* **96** 181-185

Yazdani M, Bahmanyar MA, Pirdashti H, and Esmaili MA 2009 Effect of phosphate solubilization microorganisms (PSM) and plant growth promoting rhizobacteria (PGPR) on yield and yield components of corn (*Zea mays* L.), *World Academy of Science, Engineering and Technology* **49** 90-92

Yehuda Z, Shenker M, Römheld V, Marschner H, Hadar Y, and Chen Y 1996 The role of ligand exchange in uptake of iron from microbial siderophores by graminaceous plants, *Plant Physiology* **112** 1273–1280

Zani S, Mellon MT, Collier JL, and Zehr JP 2000 Expression of *nif* H genes in natural microbial assemblages in lake George, New York, detected by reverse transcriptase PCR, *Applied and Environmental Microbiology* **66** 3119-3124

Zehr JP, Jenkins BD, Short SM, and Steward GF 2003 Nitrogenase gene diversity and microbial community structure: a cross-system comparison, *Environmental Microbiology* **5** 539-554

Zhang N, Wu K, He X, Li S, Zhang ZH, Shen B, Yang XM, Zhang RF, Huang QW, and Shen QR 2011 A new bioorganic fertilizer can effectively control banana wilt by strong colonization with *Bacillus subtilis* N11, *Plant and Soil* **344** 87-97

Ziedan EHE 2006 Manipulating endophytic bacteria for biological control of soil borne diseases of peanut, *Journal of Applied Sciences Research* **2**(8) 497-502

Zimmer W, Aparicio C, and Elmerich C 1991 Relationship between tryptophane biosynthesis and indole-3-acetic acid production in *Azospirillum*: Identification and sequencing of a *trpGDC* cluster, *Molecular Genetics and Genomics* **229** 41-51.

5

Management of crown gall in stone fruits in the realm of biological control

A.K. Gupta, Sapna Devi and Rishi Mahajan

Department of Plant Pathology, Dr. Y. S. Parmar University of Horticulture and Forestry, Nauni, Solan- 173 230, Himachal Pradesh, India

Stone fruits (Cherry, Peach, Plum, Apricot and Nectarines) are grown in different parts of the Country and fruits are available from May to August. During cultivation these crops are infected with several diseases caused by fungi, bacteria, viruses and nematodes which not only affect the quantity but also impair the quality of the produce thus making the cultivation of these crops less remunerative. Among the various diseases affecting the stone fruits, crown gall (*Agrobacterium tumefaciens*) has emerged as major threat in the successful production and establishment of disease free stone fruit plants in nurseries. In Himachal Pradesh alone, the incidence of this disease ranged from 70 to 100 per cent. The plants with tumour are not saleable and if such plants are planted in orchard they die in a year or two. The disease can also affect the well-established trees in the orchard. Crown gall pathogen is a genetic parasite as it incorporates its tumour inducing principle (TIP) in to the host genome and modifies the host cell, due to this property it hampers the development of resistant cultivars against the pathogen. The disease has also been observed in areas which were previously free from this disease owing to the movement of apparently looking healthy plants carrying incipient infection of crown gall.

Host range and geographical distribution

The host range of *Agrobacterium tumefaciens* is extremely broad in contrast to other plant bacterial pathogens and affects many woody and herbaceous plants belonging to 140 genera of more than 60 families (Agrios 2005). However, it causes manifold losses to grapevine, stone and pome fruit, nut trees and a few ornamentals (e.g. rose) on which most economic damage is done (Kerr 1992).

Smith and Townsend (1907) laid the landmark finding in the field of Bio science- the association of bacterium with trademark plant disease "Crown gall". They quoted it as a plant tumour of bacterial origin. It has been reported as the most serious disease of stone fruits in Australia (Htay and Kerr 1974), Oregon (Moore 1976), South West Ontario, Canada (Dhanvantari 1976), California (Schroth and Moller, 1976), United Kingdom (Crosse et al 1976), South Africa (Matthee et al 1977), Italy (Bazzi and Mazzucchi 1978), Switzerland (Grimm 1987), British Columbia (Utkhede and Smith 1993), Northern areas of Pakistan (Wicks et al 1993), Hungary (Sule 1994), Albania (Isufi and Myrta 1996) and Jordan (Hamed 2004). Crown gall was frequently observed on bitter almond in Tunisia (Rhouma et al 2005).

The disease in India was observed in early 1940's by Singh (Jindal et al 1990) on pome fruits in Kumaon Hills of Uttaranchal, Punjab and Kashmir. Jindal and Sharma (1988) and Gupta et al (2002 and 2005b) observed the occurrence of biovar-1 of *A. tumefaciens* infecting almond, peach and cherry in Himachal Pradesh. Sharma et al. (2004) reported the occurrence of crown gall in the warmers areas of India. Devi et al (2008) first observed the occurrence of biovar-2 on peach from India.

Symptoms

Crown gall is characterized by the formation of tumour or galls of varying shape and size on roots and stems. Hedgcock (1910) observed that the disease first appears as small overgrowth on the stem and roots particularly near the soil line. In early stages of gall development, the galls are more or less spherical, white and soft (Cook 1923). As the tumour enlarges, thin surface of gall becomes much more convoluted (Fig 1). Later on, the tissues become dark brown or black, due to death and decay of the peripheral cells. The tumour lies outside but close to the outer surface of the host being connected only by a narrow neck of tissues (Agrios 2005). Plants with tumours at their crown or on their main roots, grow poorly and result in reduced yield (Ark and Bingham 1958). Under severe conditions infected plants and vines eventually die (Burr et al 1995). Crown gall incidence also resulted in the replant problems in various peach cultivars (Singh and Dhiman 2004). Crown gall infection also caused reduction in chlorophyll content (40.20%), leaf area (28.97%) and plant biomass (11.99%) in peach (Gupta et al 2005b).

Fig. 1: Tumour formation at the crown and root portion on cherry rootstock-Colt

Causal organism

Crown gall is a bacterial disease and caused by *Agrobacterium tumefaciens* (Smith and Townsend) Conn. (Syn. *Rhizobium radiobacter* (Beijerinck and Van Deldan) Young et al). The bacterium is peritrichous (Romanenko and Perepnikhatka 1984), Gram negative, off white in colour, ovate to oblong, smooth, glistening, opaque to translucent, convex, squishy-squashy smooth margin colonies. The individual cell is rod shaped, hyaline, about 1-3 μm long by 0.4 to 0.8 μm in width (Jindal et al 1990, Kamal, 2004)

Classification

Kingdom :	Prokaryotae
Division:	Gracilicutes
Class:	Proteobacteria
Family:	Rhizobiaceae
Genus:	*Agrobacterium*
Species:	*tumefaciens*

Biovar differentiation

Agrobacterium can be differentiated into three biovars. Biovar I and biovar II are generally found in tumours of various plants belonging to Rosaceae family, while biovar III is restricted to grapevine (Panagopoulos et al 1978).There are different biochemical tests to differentiate between biovar of *Agrobacterium* (Table 1).

Table 1. Different biochemical tests for the differentiation of biovars of *Agrobacterium*

Biochemical test (s)	**Biovar**			**Reference**
	I	**II**	**III**	
3-Ketolactose test	+	-	V	Dhanvantari (1978) Jindal and Sharma (1988) Benjama et al (2001)
Growth in 2% NaCl	+	-	+	Kaene et al (1970)
Growth at 35°C	+	V⁻	V	Anderson et al (1977)
Action on litmus milk	Alk	Ac	Alk	Gupta et al (2005b)
Acid production from erythritol	-	+	-	Hayward (1964)
Acid production fron ethanol	-	+	-	Jindal and Sharma (1988)
Ferric ammonium citrate	+	-	-	Hendrickson et al (1934)
Oxidase reaction	+	V⁻	V	Kovacs (1956) Kerr and Panagopoulos (1977) Moore et al (1980)
Citrate utilization	V⁻	+	+	Kerr and Panagopoulos (1977) Moore et al (1980)
L-tyrosine utilization	-	+	-	Kerester et al (1973)

+ Positive reaction, - Negative reaction, V - Variable reaction, Alk – Alkaline reaction, Ac – Acidic reaction

Pathogenicity

Smith (1939) reported the susceptibility of Cupressaceae family to crown gall infection upon artificial inoculation. Hard and woody plants take long time for the appearance of crown gall symptoms. Therefore several herbaceous plants such as tomato, tobacco, carrot disc and other as indicator plants were used to differentiate *A. tumefaciens* from *A. radiobacter* (Manigault and Kurkdjian 1967, Moore 1976). A large number of dicotyledonous plants were found to respond positively to *A. tumefaciens* upon artificial inoculation for gall development (Burr et al 1995). Four week old tomato seedlings (*Lycopersicon esculentum* Mill.) remained the most popular material among the several workers for proving the pathogenicity (Kerr 1969, New and Kerr 1972, Cooksey and Moore 1980, Jindal and Sharma 1988, Kamal 2004, Gupta and Khosla 2007). However, apart from tomato, sunflower (*Helianthus annus* L.), *Kalanchoe diagremantiana* and grape (*Vitis vinifera* L.) have also been extensively used for pathogenicity test upon artificial inoculation (Burr et al 1995).

Mechanism of tumour development

After numerous attempts to demonstrate the mechanism of gall formation in plants, Zaenen et al (1974) reported the presence of large plasmid (Ti plasmid) in all tumourigenic strains of *Agrobacterium* examined. The cells around the wound produce various phenolic compounds and attract *Agrobacterium.* The bacteria attaches itself to cell walls in response to phenolic compounds (acetosyringone) and other signals, they become activated and begin the processing of DNA in their Ti plasmid (tumor-inducing plasmid). Acetosyringone plays a further role in the infection process, because at higher concentrations (about 10^{-5} to 10^{-4} Molar) than those that cause chemotaxis it activates the virulence genes (Vir genes) on the T_i plasmid .These genes coordinate the infection process by two ways;

- By the production of proteins (permeases) that are inserted in the bacterial cell membrane for uptake of compounds (opines) that will be produced by the tumours.
- By the production of an endonuclease - a restriction enzyme - that excises part of the *Ti* plasmid termed the T-DNA (transferred DNA).

The excised T-DNA is released by the bacterium and enters the plant cells, where it integrates with the plant chromosomes. The actual mechanism of transfer is still unclear, but it seems to require a conditioning process, perhaps mediated by the production of cytokinins (plant hormones) by the bacterium. The tzs (transzeatin) gene on the T_i plasmid codes for the hormone. Only a small part of the plasmid (the T-DNA) enters the plant; the rest of the plasmid remains in the bacterium to serve futher roles. After integration with the plant genome, the genes on the T-DNA code for:

- production of cytokinins
- production of indoleacetic acid
- synthesis and release of novel plant metabolites - the opines and agrocinopines.

The plant hormones upset the normal balance of cell growth, leading to the production of galls and thus to a nutrient-rich environment for the bacteria. The opines are unique amino acid derivatives, different from normal plant products, and the agrocinopines similarly are unique phosphorylated sugar derivatives. All these compounds can be used by the bacterium as the sole carbon and energy source. Since they are absent in normal plants and they provide Agrobacterium with a unique food source that other bacteria cannot use. Different strains of A. tumefaciens contain different types of T_i plasmid which code for production of different types of opines. One of the most common types of T_i plasmid found in strain C58 of A. tumefaciens codes for production of nopaline, and for agrocinopine A. The part of the plasmid that remains in the bacterium codes for the uptake and catabolism of these compounds (the Noc gene and the Acc gene). The other common type of *Ti plasmid* codes for synthesis of octopine and agropine. Tumour cells subsequently grow and divide independently of the bacteria, and their organization, rate of growth, and rate of division can no longer be controlled by the host plant. It was subsequently shown that genes controlling the virulence of bacterium (Van Larebeke et al 1975; Watson et al 1975) and the catabolism of octopine (Chilton et al 1976) or nopaline (Watson et al 1975) as well as induction of synthesis of these compounds in tumours (Montoya et al 1977) are all plasmid coded traits. Johnson and Das (1998) reported that the unusual amino acid sugar conjugates (opines) are not normally synthesized in the plants. These observations supported the notion that bacterial plasmid DNA is transferred to and expressed in host cell. Finally, the physical conformation of the presence (Chilton et al 1977) and transcription (Drummond et al 1977) of plasmid DNA sequence was observed in the nucleic acid extracted from tobacco tumour tissues that has been maintained axenically for several years.

Disease management

Crown gall disease is an economically important disease and its management still remains a great challenge despite of the fact that many attempts have been made by several workers worldwide because virulent strains of *A. tumefaciens* contain a large Ti-plasmid which enters the susceptible host through relatively fresh wounds and integrates with the plant genome (Watson et al 1975). The plants once infected can not be cured of the disease as it is Ti-plasmid responsible for the gall development and it can not be simply destroyed there. More over infection can not be detected unless the gall development starts (Banfield 1934).

Several antibacterial compounds (Klemmer et al 1955, Deep 1958a, Deep and Young 1965) and fumigants (Kapshuk 1933, Dickey 1962, Deep et al 1968) were tested and found ineffective. Biological control (Kerr 1972) against crown gall has come up from several places. Strain K-84 of *A. radiobacter* gave promising results in reducing the crown gall development in several countries (Kerr 1972). The different approaches for the management of disease are discussed below:

1. Cultural control

Cultural practices play a significant role in minimizing the incidence of crown gall (Wicks et al 1993). Susceptible nursery stock should not be planted in the infested field (Smith and Townsend 1907). Infested fields should be planted with corn and other grain crops for several years (Agrios 2005). Since the bacterium enter through wounds, therefore wounding of crown and roots should be avoided during transplantation and root chewing insects should be controlled to reduce the incidence of crown gall. There are other practices such as destruction of infected plant material, proper field sanitation, rotation of nursery site and use of well pitted rootstocks. Gupta et al (2005a) observed reduced crown gall incidence when one third dose of P and K along with normal dose of N and farm yard manure were applied. Soil amendments with 1.5 per cent mustard cake and Eucalyptus leaves were also effective in reducing the incidence.

2. Soil solarization

Soil solarization involves the use of transparent polyethylene sheet to capture the energy from sun to heat the soil for controlling various soil pathogens (Katan 1981). The soil solarization technique for targeting soil pathogens was used for the first time by Jones et al (1966). Polyethylene (more commonly known as polythene) is most popular material used in agriculture, was first introduced on commercial scale in 1939 (Byrdson 1970). Among different polyethylene mulches, transparent mulch was able to transmit most of the solar radiations that heat the soil (Macfayden 1967). He also suggested that mulching should be carried out during the period of high temperature. Thinnest polyethylene mulch (25 to 30 μm) have been found to be more effective in heating due to better radiation transmittance than the other one (Horowitz and Regev 1980).

Israel has made a landmark in soil solarization technology. Their experience over a period of time has revealed that mulching the soil with transparent polyethylene sheet in hot season proved effective against soil borne pathogens and weeds (Katan et al 1976, Jacobson et al 1980). Studies on soil solarization have also been carried out in several other countries like Jordon (Al-Raddad 1979), England (White and Buczacki 1979), Greece (Tjamos and Faridis 1980, Tjamos and Paplomatas 1988), California, USA (Pullman and DeVay 1977, Stapleton and DeVay 1982), Australia (Kaewruang et al 1989) and India (Chauhan et al 1988). Soil solarization and thermal death point of different soil pathogens was observed and developed a logarithmic relationship between time and temperature of soil borne plant pathogens and reported that soil solarization is dependent on thermal dose (Pullman et al 1981). It was reported that there was a decrease in population densities of *Agrobacterium* at lower depth due to cumulative effects of sub lethal temperatures on the pathogen (Tjamos and Paplomatas 1988). The decrease in the population densities of *Agrobacterium* at lower depth due to the cumulative effect of sub lethal temperatures was much more in irrigated than in non irrigated solarized plots (Raio et al 1997). Longer duration of soil solarization (90 days) was much more conducive in achieving desired thermal doses and its effectiveness was also correlated to soil type.It was observed that incidence of crown gall on "Mazzard" cherry rootstock planted in solarized beds having silt clay soil was reduced significantly (Pinkerton et al 2000). Longer solarization duration (90 days) was much more conducive in achieving desired thermal doses approximately 195 h above 40ºC and 62.5 h above 45ºC in the upper 8cm soil layers. At 15 and 30cm soil depths, thermal doses were

196.5 and 93 h above 35°C during 2001 and 2002, respectively which resulted in significant reduction of crown gall on cherry rootstock-Colt (Gupta and Khosla 2007).

3. Chemical control

A number of fungicides, antibiotics and other chemicals such as fumigants, soil sterilizing agents, herbicides have been used against the emerging crown gall problem (Ark 1941, 1942) but the promising results were obtained with chloropicrin because it penetrated to the depth of 20 cm when applied to soil and killed *Agrobacterium* effectively (Kapshuk 1933). However, it was experimentally showed that mercuric chloride, copper sulphate and chloride of lime were useless as disinfectant of the soil naturally or artificially infested with *A. tumefaciens*. The effectiveness of aureomycin against crown gall was also proved (Blanchard 1951). There was more than 90 per cent reduction in *A. tumefaciens* in artificially infested soil at all depths of 3, 9 and 15 inches by methyl bromide (Dickey 1962). Several authors used different antibiotics against *A. tumefaciens* and reported the efficacy of terramycin (Deep 1958a), penicillin and streptomycin (Hampton 1948, Deep 1958b). Later on, Ark and Bingham (1958) reported that neither streptomycin nor tetracycline root dip (60 minutes) gave a successful control of crown gall in either pear or cherry rootstocks having the disease in incipient form. Deep and Young (1965) applied a large number of fungicides in the soil prior to sowing of seed in nursery beds and observed a higher incidence of crown gall and this may be due to the reduction in soil fungal micro floral population leading to less competition of *A. tumefaciens* which increased the disease incidence (Deep and Young 1965). Deep et al (1968) tested 7 fumigants for their effectiveness in crown gall on Mazzard cherry. They reported that methyl bromide; chloropicrin-propurgyl bromide appeared to give some control of crown gall. Ogg and Graf (1974) tested 8 herbicides as pre-plant application both in greenhouse and field conditions on apple, pear, cherry and plum. They reported increase in the incidence of crown gall and hairy root on sweet cherry by the use of trifluraince application @ 1.7 kg/h. The fumigation of infested soil with vortex and antagonist *Agrobacterium radiobacter* HLB-2 both reduced the occurrence of infection and tumour development on grape plants (Pu Xin and Goodman 1993).

Utkhede and Smith (1993) evaluated the biological and chemical treatment for the control of crown gall on young apple trees in British Columbia. Soil sterilization with formaldehyde (4%) resulted in minimum disease incidence (9.77%) Bleaching powder (0.3%) also reduced the disease incidence to 23.33%, while application of herbicides as pre -plant treatment of oxyfluorfen (0.5 L/ha) reduced the disease incidence to 30.33% and application of antibacterial compounds like copper sulphate (1%) as root dip resulted in minimum disease incidence to 30.17 per cent. Streptocycline also reduced the disease incidence to 34.90% in cherry (Kamal 2004). The application of streptocycline (0.1%) and copper sulphate (0.3%) were effective to some extent and controlled crown gall disease by 40 to 50 per cent on cheery rootstock–Colt (Gupta et al 2005a). The application of formaldehyde (4%) as pre-plant application reduced the crown gall disease incidence by 81 per cent, whereas oxyfluorfen (0.5 l/ha) also reduced the incidence by 50 per cent on cherry rootstock-Colt (Gupta and Kamal, 2006).

4. Biological control

The credit for pioneering work in the field of biological control of crown gall goes to New and Kerr (1972) who for the first time found that a pathogen to non pathogen ratio of 1:1 completely inhibited the crown gall on peach when *Agrobacterium radiobacter* strain K-84 was inoculated on to the peach seeds sown in soil infested with *A. tumefaciens*. Since 1972, *A. radiobacter* strain K-84 has been effectively and widely used for the control of crown gall across the world as seed treatment, root dip treatment and soil drench in infested soil to control the crown gall on peach and other stone fruits (Kerr 1972). High efficacy of *A. radiobacter* strain K-84 was observed on almond in Greece (Panagopoulos and Psallidas 1973). Stone fruit trees developed galls when the ratio of non pathogenic to pathogenic exceeded the 1:1 ratio of non pathogenic to pathogenic *Agrobacterium* (Kerr 1974). The excellent control of crown gall on apricot, almond, peach and plum trees were obtained when inoculated injured young trees were dipped in strain K-84 suspension containing 10^6 cells/ ml for five seconds (Moller and Schroth 1976, Schroth and Moller 1976). *A. radiobacter* strain K-84 has been widely used as soil drench; seed treatment, tomato assay and root dip on peach, cherry, almond and grape in Canada (Dhanvantari 1976). This strain has been widely used in Hungary on raspberry (Sule and Kollanyi 1977, Sule 1994), peach, cherry and apple (Grimm et al 1982); in Oregon and Washington on boysenberry, cherry and rose (Moore 1976, Johnson and Dileone 1999), pecan (Bouzar et al 1983); in Italy on cherry, Myrobalan and peach rootstocks (Bazzi et al 1978; Zoina et al 1987), in United Kingdom on cherry rootstocks F12/1 and hybrid selection FB-22 and Colt (Garrett 1978, 1979, Garrett et al 1985); in Belgium on apple and pear (Rysheuvels et al 1984); in Australia on almond (Shim et al 1987); in Spain on stone fruits and roses (Lopez et al 1987); in Switzerland (Grimm 1987); in Spain on stone fruits (Lopez et al 1989) and in Egypt on greenhouse grown peach plants (Tawfik 1990). Successful control of crown gall was also obtained in Italy on stone fruit trees (Raio and Zoina 1992); in France on plum (Pierronnet and Eyquard 1993); in Korea on stone fruits (Yang et al 1991); in Poland on Mazzard cherry, Mahaleb cherry and Myrobalan plum (Sobiczewski et al 1991); on peach in China (Wang et al 1995) and in South Australia on almond (McClure et al 1998). Gupta et al (2010) after screening 510 isolates found that among these, twelve isolates of *A. radiobacter* were able to inhibit crown gall on both peach and cherry rootstock-Colt and isolate UHFBA-11 on peach as root dip minimized the incidence (5.37%) of crown gall as compared to 54.17 per cent incidence in untreated peach plants. The same efficacy was observed on cherry rootstock-Colt where it resulted in 23.61 per cent incidence of crown gall as compared to 71.11 per cent incidence in control.

Breakdown of strain K -84 of *Agrobacterium radiobacter*

Several strains of *Agrobacterium tumefaciens* were insensitive to *A. radiobacter* strain K-84 and there was no antagonistic activity of strain K-84 against *A. tumefaciens* under *in vitro* conditions (Kerr and Htay 1974, Kerr and Panagopoulos 1977, Schroth and Moller 1976). *Agrobacterium radiobacter* contains a plasmid, termed pAgK84, which codes of agrocin 84 production. It also contains another plasmid, pNOC, which codes for nopaline uptake and catabolism. The agrocin plasmid is not a conjugative plasmid, but pNOC (also present in the biocontrol strain) is conjugative and it can mobilize the agrocin plasmid during its own transfer (Johnson and Das 1998). The breakdown of control of crown gall by strain K-84 is due to the transfer of pAgk84 plasmid to pathogens (*A. tumefaciens)* and due to this transfer they become resistant to agrocin-84 (Ellis and Kerr 1979, Panagopoulos et al 1979, Penyalver and Lopez 1999). After numerous attempts to develop another strain of K-84, Jones et al (1988) and Jones and Kerr (1989) developed strain K- 1026, mutant of strain K-84 which is transfer deficient (Tra^-) to avoid this breakdown. There was no significant difference in the crown gall incidence between the plants treated with K-1026 or K-84 indicating that strain K-1026 is as effective as strain K-84 in controlling the crown gall. Wang et al (2000) obtained 67-99 per cent control of the crown gall by using genetically modified strain K 1026 of strain K-84 on flowering cherry in China. Benjama et al (2002) tested 206 Moroccan isolates of pathogenic *A. tumefaciens* under *in vitro* conditions for their sensitivity against *A. radiobacter* strain K-84 and K-1026 and obtained that strain K-1026 of *A. radiobacter* was more effective than strain K-84. Fortunately all *A. tumefaciens* isolates from Himachal Pradesh and Uttarakhand were nopaline utilizing and were inhibited by *A. radiobacter* strain K-84 (Gupta et al 2011, unpublished).

Mechanism of biological control of strain K-84 and strain K-1026 of *A. radiobacter*

Lopez et al (1989) suggested that mechanisms such as root colonization, blockage and competition for infection site could be involved in biological control of *A. tumefaciens* in field. Since then strain K-1026 and its parent strain K-84 have been used successfully in controlling crown gall on stone fruits and ornamental plants. The main mechanism involved in the biological control of crown gall is the bio-synthesis and secretion of agrocin-84 by strain K-84 because it is a antibiotic like substance which specifically inhibit the tumourigenic *Agrobacterium* (Fig. 2a) strains (Kerr and Htay 1974, Kerr and Roberts 1976). Agrocin 84 is an adenine nucleotide (Fig. 2b) with a 6 glucofuran and a methylated pentamide attached (fraudulent nucleotide).

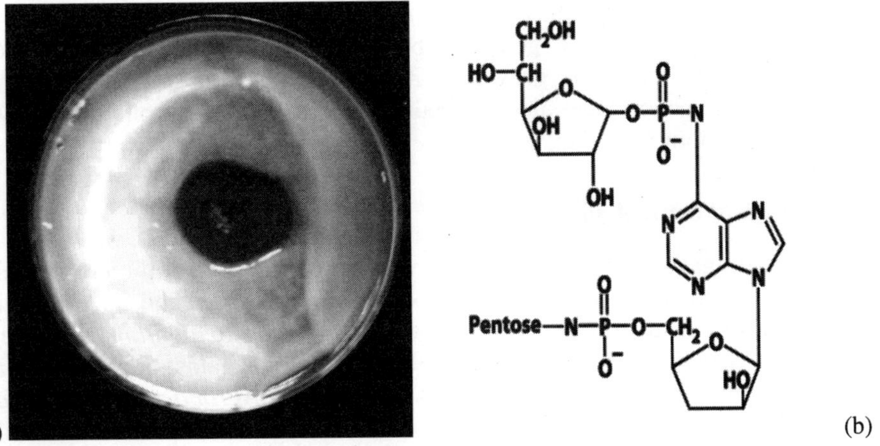

Fig 2. (a) Detection of agrocin production in *A. radiobacter* isolate (b) Structure of agrocin K-84

Handelsam and Stabb (1996) proved that colonization and survival of antagonists in the rhizosphere is essential for the biological control of soil borne pathogens. *A. radiobacter* strain K-84 and strain K-1026 have proved to be good colonizer of the root system in different host (Vicedo et al 1993). Pathogenic strains of A. tumefaciens with a *Ti* plasmid that codes for nopaline production also cause the plant to produce agrocinopines. The plasmid codes for an agrocinopine permease, which is inserted in the bacterial membrane. The inhibitor, agrocin 84, is recognised by this permease, enters the pathogen cells and there it blocks DNA synthesis.

Work done at Dr. Y. S. Parmar University

Native *Agrobacterium radiobacter* isolate tested against crown gall in stone fruit nurseries controlled crown gall by 89.1 per cent as compared to 78 per cent by strain K-84 as seed treatment on peach. The disease incidence in untreated plants was 72.6 per cent. On the basis of 16S rRNA sequence analysis and with the help of universal primers –VirD2 and ipt of *A. tumefaciens*, 141-1A-1, Cherry1A-8-2 and Cherry2E-2-2 were identified as *A. radiobacter*. The results of non-amplification with specific agrocin primers of K-84 also confirmed that all our native isolates are producing either analogue of agrocin or other novel compounds, on account of which these isolates are giving better control of crown gall in stone fruit nurseries than world over used strain K-84. To further identify the relatedness of strains/ isolates of native *A. radiobacter* and *A. tumefaciens* with standard isolates - *A. radiobacter* strain K-84 and *A. tumefaciens* C-58, oligonucleotides FGPS6 (5'GGAGAGTTAGATCTTGGCT-CAG3') and FGPL132 (5'CCGGGTTT-CCCCATTCGG3') were used to amplify a ribosomal region containing 1,479 bp (99.5%) of the 16S rRNA, the intergenic spacer region between 16S and 23S rRNA genes of the seventeen isolates belonging to native *A. radiobacter* and *A. tumefaciens* along with standards. Amplicons were digested with *Taq*I, *Rsa*I and *Ava*I. The studies revealed that native *A. radiobacter* isolates viz., 139-1A-2, 139-1A-5-1, 1401A-7, Cherry 2E-3-2, Cherry 2E-2-2 and strain K-84 are of the same group, whereas *A. tumefaciens* strain C-58, Peach 2E-1, Peach 2E-3 and Peach 2E-10 have similarity with each other in isolate Cherry 1A-7-2 identified as *A. tumefaciens* (on the basis of 16S rRNA sequence analysis) a unique phenomenon where loss of Ti plasmid with acquisition of genes responsible for production of antibacterial compounds/agrocin inhibiting growth of *A. tumefaciens*, was observed. Results of this unique phenomenon were also ascertained by non-amplification with vir-*D2* and *ipt* primers of tumour producing *A. tumefaciens*.

Other biological control agents

Several fungal and bacterial isolates were tested to control the crown gall and some level of control was achieved with species of *Penicillium* and *Bacillus*. Disease incidence was reduced from 61.9 to 4.9 per cent by using *Penicillium* (Cooksey and Moore 1980). *Bacillus subtilis* strains were very effective against nopaline strain C58 and AR125 and also against octopine strain B6 of *Agrobacterium tumefaciens* (Hayman and Farrand 1988). However, *in vitro* experiments, there was no inhibition zone recorded with reference antagonist strain K-84 and strain K-1026 against B6 strain of *A. tumefaciens*. These results indicated that antibacterial compounds of *Bacillus subtilis* strains were different from those K-84 and K-1026. It was found that 77 isolates of *Pseudomonas fluorescens* inhibited the growth of *A. tumefaciens* in culture (Canfield and Moore 1991). Strain 1-1-4 of *Pseudomonas fluorescens* has been reported to provide more than 90 per cent control of crown gall on peaches (Zhang et al 1991). Strain K-84 produces a second antibacterial compound called agrocin 434 (Donner et al 1993), but this antibacterial compound affects only biovar-2 strains of *Agrobacterium tumefaciens* (Fajardo et al 1995). Strain K-84 is also able to produce a third antibacterial substance called ALS-84 which inhibit many tumourigenic *Agrobacterium* strains *in vitro* conditions (Penalver et al 1994). Silva and Romeiro (2001) reported 14 actinomycetes effective against crown gall on rose under greenhouse conditions. *Bacillus subtilis* was again found effective against crown gall on cherry rootstock-Colt and it was even more effective when combined with soil solarization (Gupta and Khosla 2007). A non-pathogenic strain of *Agrobacterium vitis* VAR03-1 was found effective for the biological control of crown gall of grapevine (*Vitis vinifera*), rose (*Rose multiflora*), and tomato (*Lycoper sicon esculentum*) and this is the first report of non-pathogenic strain of *Agrobacterium vitis* VAR03-1 that effectively control the crown gall caused by tumourigenic *A. vitis, A. rhizogenes* and *A. tumefaciens*, respectively (Kawaguchi et al 2008).

Future prospects

Crown gall is very important disease of stone and pome fruits and causes major losses in terms of money and labour all around the world wherever these fruits are cultivated. There is no chemical which provide the complete control of crown gall, therefore major emphasis is given on biological control by antagonistic strain of *A. tumefaciens*. All European countries have *Agrobacterium radiobacter* strain K-84 which is antagonistic to *A. tumefaciens* and provide nearly complete control of crown gall but this strain is not available in India. The success of *A. radiobacter* strain K-84 in Australia, USA, Canada, New Zealand and many European countries has encouraged the Indian workers to isolate the native antagonistic non pathogenic *A. tumefaciens* (=*A. radiobacte*r) which is currently not available in India. Five *A. radiobacter* isolates have now been identified most effective as it reduced the crown gall incidence by 51 to 75 per cent on peach and cherry rootstocks (Gupta et al 2011, unpublished).

Reference

Agrios GN 2005 *Plant Pathology*, New York: Academic Press

Al-Raddad AMM 1979 Soil disinfection by plastic tarping, M.Sc. thesis, University of Jordon

Anderson AR 1977 Taxonomy and host specificity of the genus *Agrobacterium*, Ph.D. thesis, Oregan State University, USA

Ark PA and Bingham DJ 1958 Response of pear and cherry rootstocks to streptomycin and tetracycline when applied to control crown gall, *Plant Disease Reporter* **42** 673-674

Ark PA 1941 Chemical eradication of crown gall on almond trees, *Phytopathology* **31** 556-579

Ark PA 1942 Control of crown gall of peach in the nursery, *Phytopathology* **32** 826

Banfield WM 1934 Life history of crown gall organism in relation to its pathogenesis on the red raspberry, *Journal of Agricultural Research* **48** 761

Bazzi C and Mazzucchi U 1978 Biological control of crown gall in cherry, myrobalan and peach rootstocks in Italy, **1** 252-254. *In Proceeding of 4th International Conference on Plant Pathogenic Bacteria*, Angers, France

Bazzi C, Baldo L, Sacchetti G, and Mazzucchi U 1978 Complete success in the biological control of bacterial tumours of stone fruits in Emilia, *Informatore Agrario* **34** 1571-1572

Benjama A, Redwane A, El Gadda M, El Medraoui L, Ouzrit M, Boussad L, E L-Quard R, and Nesme X 2002 First result on biological control with strains K84 and K1026 of *Agrobacterium radiobacter var. tumefaciens* causing crown gall on fruit trees and rose in Morocco, *Bulletin OEPP* **32** 309-314

Benjama A, Redwane A, Nesme X , El-Gadda, El-medraoui L, Mourabit K, Aaglane A, and Al-Mai S 2001 Simple and rapid biochemical tests to characterize *Agrobacterium* causing crown gall of stone fruits and rose, *Al- Awamia* **103** 83-90

Blanchard FA 1951 Aureomycin chemotherapy of crown gall in tomatoes, *Phytopathology* **41** 954-958

Bouzar H, Moore LW, and Schaad NW 1983 Crown gall of pecan: A survey of *Agrobacterium* strains and potential for biological control in Georgia, *Plant Disease* **67** 310-312

Burr TJ, Reid CL, Yoshimura M, Momol EA, and Bazzi C 1995 Survival and tumorigenicity of *Agrobacterium vitis* in living and decaying grape roots and canes in soil, *Plant Disease* **79** 677-682

Byrdson JA 1970 *Plastic Materials,* London: Iliffe Books

Canfield ML and Moore LW 1991 Isolation amd characterization of opine utilizing strains of *Agrobacterium tumefaciens* and fluorescent strains of *Pseudomonas* spp. from rootstocks of *Malus*, *Phytopathology* **81** 440-443

Chauhan YS, Nene YL, Johonson C, Hawara MP, Sexsena NP, Singh S, Sharma SB, Sahrawat KL, Burford JR, Rupela OP, Kumar JV, Rao DK, Sithanamthan S, and Singh S 1988 Effect of soil solarization on pigeonpea and chickpea, *Research Bulletin* ICRISAT **11** 16

Chilton MD, Drummond MH, Merlo DJ, Sciaky D, Montoya AL, Gordon MP, and Nester EW 1977 Stable incorporation of plasmid DNA into higher plant cells: the molcecular basis of crown gall tumorigenesis, *Cell* **11** 263-271

Chilton MD, Farrand SK, Levin R, and Nester EW 1976 RP4 promotion of transfer of a large *Agrobacterium* plasmid which confers virulence, *Genetics* **83** 609-618

Cook MT 1923 Early stages of crown gall, *Phytopathology* **13** 476-482

Cooksey DA and Moore LW 1980 Biological control of crown gall with fungal and bacterial antagonists *Phytopathology* **70** 506-509

Crosse JE, Garrett CME, Martins JMS, and Billing E 1976 Bacterial diseases, in *Plant Pathology*, (ed) JE Crosse, East Malling Research Station Report, 107-110 pages

Deep IW and Young RA 1965 The role of pre-planting treatments with chemicals in increasing the incidence of crown gall, *Phytopathology* **55** 212-216

Deep IW, Ray A, McNeilan A, and Mac Swan IC 1968 Soil fumigants tested for control of crown gall, *Plant Disease Reporter* **52** 102-105

Deep IW 1958a Crown gall chemotherapy with terramycin, *Plant Disease Reporter* **42** 1210-1213

Deep IW 1958b Reduction in incidence of crown gall of Mazzard cherry following antibiotic treatments, *Plant Disease Reporter* **42** 476-480

Devi S, Gupta AK, Bhardwaj SS, and Khosla K 2008 First report on the occurrence of biovar-II of *Agrobacterium tumefaciens* causing crown gall on peach, *Indian Phytopathology* **61** 386

Dhanvantari BN 1976 Biological control of crown gall of peach in South-Western Ontario, *Plant Disease Reporter* **60** 549-551

Dhanvantari BN 1978 Characterization of *Agrobacterium* isolates from stone fruits in Ontario, *Canadian Journal of Botany* **56** 2309-2311

Dickey RS 1962 Efficacy of five fumigants for the control of *Agrobacterium tumefaciens* at various depths in the soil, *Plant Disease Reporter* **46** 73-76

Donner SC, Jones DA, Mcclure NC, Rosewarme GM, Tate ME, Ker A, Fajardo NN, and Clare BG 1993 Agrocin 434, a new plasmid encoded agrocin from the biocontrol Agrobacterium strains K-84 and K-1026, which inhibit biovar 2 Agrobacteria, *Physiological and Molecular Plant Pathology* **42** 185-194

Drummond MH, Gordon MP, Nester EW, and Chilton MD 1977 Foreign DNA of bacterial plasmid origin is transcribed in Crown gall tumours, *Nature* **263** 731-738

Ellis JG and Kerr A 1979 Transfer of agrocin 84 production from strain 84 to pathogenic recipient: A comment on the previous paper, in *Soil Borne Plant Pathogens*, (eds) B Schippers and W Gams, London: Academic Press, 579-583, pages

Fajardo NN, Tate ME, and Clare BG 1995 Agrocin 434: an additional biological control component of crown gall, in *Improving Plant Productivity with rhizosphere bacteria* (eds) MH Ryder, Stephens and GD Bowen, CSIRO Division of soil, Australia, 128-130 pages

Garrett CMC, Fletcher DA, Trowell SD, Adams AN, Stickels JE, Barbara DJ, and Blake VSA 1985 Diseases of cherries, *Report of East Malling Research Station* 163-164 pages

Garrett CME 1978 Biological control of crown gall, *Agrobacterium tumefaciens, Annals of Applied Biology* **89** 96-97

Garrett CME 1979 Biological control of crown gall in cherry rootstock propagation, *Annals of Applied Biology* **91** 221-226

Grimm R, Sule S, and Lozano JC 1982 Control of crown gall (*Agrobacterium tumefaciens* Smith and Townsend) in nurseries, 531-537 *In Proceedings of V th International Conference on Plant Pathogenic Bacteria*

Grimm R 1987 Control of crown gall in Swiss apple nurseries, *Bulletin OEPP* **17** 269-272

Gupta AK and Kamal B 2006 Pre-planting application of soil sterilents and herbicides for the management of crown gall disease on "Colt" cherry rootstock, *Indian Journal of Agricultural Sciences* **76** 426-429

Gupta AK, Kamal B and Khosla K 2005a Effect of fertilizers, soil amendments and antibacterial compounds on the incidence of crown gall on cherry rootstock-Colt, *Journal of Mycology and Plant Pathology* **33** 286-287

Gupta AK, Kamal B and Kumar S 2005b Biovar characterization of *Agrobacterium tumefaciens* isolates from stone fruits and effect of crown gall disease on growth parameters on peach and cherry rootstock-Colt, *Indian Phytopathology* **58** 486-488

Gupta AK, Kamal B and Verma KD 2002 Isolation and characterization of *Agrobacterium tumefaciens* causing crown gall in stone fruits, *Indian Phytopathology* **55** 400

Gupta AK and Khosla K 2007 Integration of soil solarization and potential native antagonist for the management of crown gall on cherry rootstock colt, *Scientia Horticulturae* **112** 51-57

Gupta AK, Khosla K, Bhardwaj SS, Thakur A, Devi S, Jarial RS, Sharma C, Singh KP, Srivastava DK, and Lal Rup 2010 Biological control of crown gall on peach and cherry rootstock Colt by native *Agrobacterium radiobacter* isolates, *The Open Horticulture Journal* **3** 1-10

Hamed K 2004 Response of some stone fruit rootstocks to crown gall disease in Jordan, *Dirasat-Agricultural Sciences* **31** 289-295

Hampton JE 1948 Cure of crown gall with antibiotics, *Phytopathology* **38** 11-12

Handelsam J and Stabb EV 1996 Biocontrol of soil borne plant pathogens, *Plant Cell* **8** 1855-1869

Hayman GT and Farrand SK 1988 Characterization and mapping of agrocinopine agroiin 84 locus on the nopaline Ti plasmid pTiC58, *Journal of Bacteriology* **170** 1759-1767

Hayward AC 1964 Characteristics of *Pseudomonas solanacearum*, Journal of Applied Bacteriology **27** 265-277

Hedgcock G 1910 Field studies of the crown gall of grape, *USDA Bulletin* **183** 40

Hendrickson AA, Baldwin IL, and Riker AJ 1934 Studies on certain physiological characters of *Phytomonas tumefaciens*, *Phytomonas rhizogenes* and *Bacillus radiobacter* II, *Journal of Bacteriology* **28** 597-618

Horowitz M and Regev Y 1980 Mulching with plastic sheets as a method of weed control, *Hassadeh* **60** 395-399

Htay K and Kerr A 1974 Biological control of crown gall: Seed and root inoculation, *Journal of Applied Bacteriology* **37** 525-530.

Isufi E and Myrta M 1996 The defense of fruit trees in Albania: Problems and Prospects, *Informator Fitopatologico* **46** 33-36

Jacobson R, Greenberger A, Katan J, Levi M, and Alon H 1980 Control of Egyptian broomrape (*Orobanche aegyptica*) and other weeds by means of solar heating of the soil by polyethylene mulching, *Weed Science* **28** 312-316

Jindal KK and Sharma RC 1988 Occurrence of biovar 1 of *Agrobacterium tumefaciens* on almond in India, *Indian Phytopathology* **41** 614-615

Jindal KK, Ved Ram, and Gupta GK 1990 Occurence of crown gall on pecan nut plants, *Plant Disease Research* **5** 216-217

Johnson KB and Dileone JA 1999 Effect of antibiosis on antagonist dose plant disease response relationships for the biological control of crown gall of tomato and cherry, *Phytopathogloy* **89** 974-980

Johnson TM and Das A 1998 Organization and regulation of expression the *Agrobacterium* genes, in *Rhizobiaceae* (eds) HP Spaink, A Condorosi and PJJ Hooykaas, London: Kluwer Academic Publishers, 566 pages

Jones DA and Kerr A 1989 *Agrobacterium radiobacter* strain K1026, a genetically engineered deviative of strain K84, for biological control of crown gall, *Plant Disease* **73** 15-18

Jones DA, Ryder MH, Clare BG, Farrand SK, and Kerr A 1988 Construction of tra˙ deletion mutant of pAgk84 to safeguard the biological control of crown gall, *Molecular General Genetics* **212** 207-214

Jones JP, Overman AG, and Geraldson CM 1966 Effect of fumigants and plastic films on the control of several soil borne pathogens of tomato, *Phytopathology* **56** 929-932

Kaewruang W, Sivastithamparam K, and Hardy GE 1989 Use of soil solarization to control root rots in gerberas (*Gerbera Jamesonii*), *Biology and* Fertility *of Soil* **8** 38-47

Kamal Brijesh 2004 Studies on crown gall of cherry and its management, Ph.D. thesis, Dr. Y.S. Parmar University of Horticulture and Forestry, Nauni, Solan

Kapshuk AA 1933 Bacteriological study of crown gall of fruit trees, in Russian English Summary, *Bull. N. Caucasian Inst. For Plant Protection, Rostoff-on Do* **2** 69-78

Katan J, Greenberger A, Alon H, and Feinstein A 1976 Solar heating by polyethylene mulching for the control of disease caused by soil borne pathogens, *Phytopathology* **66** 683-688

Katan J 1981 Solar heating (solarization) of soil for control of soil borne pests, *Annual Review of Phytopathology* **19** 211-236

Kawaguichi A, Inoue K, and Ichinose Y 2008 Biological control of crown gall of grape vine, rose, and tomato by non- pathogenic *Agrobacterium radiobacter* strain VAR03-1, *Phytopathology* **98**1218-1225

Keane PJ, Kerr A, and New PB 1970 Crown gall of stone fruit II. Identification and nomenclature of *Agrobacterium* isolates, *Australian Journal of Biological Science* **23** 585-595

Kerr A and Htay K 1974 Biological control of crown gall through bacterioin production, *Physiological Plant Pathology* **4** 37-44

Kerr A and Panagopoulos CG 1977 Biotype of *Agrobacterium radiobacter* var. *tumefaciens* and their biological control, *Phytopathology Z* **90** 172-179

Kerr A and Roberts WP 1976 *Agrobacterium*: Correlations between and transfer of pathogenicity, octopine and nopaline metabolism and bacterioin 84 sensitivity, *Physiological Plant Pathology* **9** 205-211

Kerr A 1969 Crown gall of stone fruit I. Isolation of *Agrobacterium tumefaciens* and related species, *Australian Journal of Biological Science* **22** 111-116

Kerr A 1972 Biological control of crown gall seed inoculation, *Journal of Applied Bacteriology* **35** 493-497

Kerr A 1974 Soil microbiological studies of *Agrobactium radiobacter* and biological control of crown gall, *Soil Science* **118** 168-172

Kerr A 1992 *The Prokaryots*, 2nd edition., Vol. III, (ed.)Balows K *et al* Berlin: Springer-Verlag, 2214-2235 pages

Kersters K, De Ley J, Sneath PHA, and Sackin M 1973 Numerical taxonomic analysis of *Agrobacterium, Journal of General Microbiology* **78** 227-239

Klemmer HW, Riker AJ, and Allen ON 1955 Inhibition of crown gall by selected antibiotics, *Phytopathology* **45** 618-625

Kovacs N 1956 Identification *of Pseudomonas pyocyana* by the oxidase reaction, *Nature* **178** 703

Lopez MM, Gorris MT, Salcedo CI, Montojo AM, and Miro M 1989 Evidence of biological control of *Agrobacterium tumefaciens* strains sensitive and resistant to agrocin 84 by different *Agroabcterium radiobacter* strains on stone fruit trees, *Applied and Environmental Microbiology* **5** 741-746

Lopez MM, Gorris MT, Temprano FJ, and Drive RJ 1987 Results of seven years of biological control of *Agrobacterium tumefaciens* in Spain, *Bulletin OEPP* **17** 273-279

Macfayden A 1967 Thermal energy as a factor in the biology of soils, in *Thermology* (ed) AH Rose, New York: Academic Press, 653 pages

Manigault P and Kurkdjian A 1967 Influence de la lumiere sur le development des tumeurs de la plantule de pois, *Pisum sativum* L., induites par *Agrobacterium tumefaciens* (Smith et Town) Conn, C.R. Academic Science, Paris, Ser D **264** 2304-2306

Matthee FN, Thomas AC, and Plessis HJ 1977 Biological control of crown gall, *Deciduous Fruit Grower* **27** 303-307

McClure NC, Ahmadi AR, and Clare BG 1998 Construction of a range of derivatives of the biological control strain *Agrobacterium rhizogenes* K84 a study of factors involved in biological control of crown gall disease, *Applied and Environmental Microbiology* **64** 3977-3982

Moller WJ and Schroth MN 1976 Biological control of crown gall, *California Agriculture* **30** 8-9

Montoya AL, Chilton MD, Gordon MP, Sciaky D, and Nester EW 1977 Octopine and nopline metabolism in *Agrobacterim tumefaciens* and crown gall tumour cells: role of plasmid genes, *Journal of Bacteriology* **129** 101-107

Moore LW, Anderson A and Kado CI 1980 Gram negative bacteria-*Agrobacterium*, in *Laboratory guide for identification of plant pathogenic bacteria*, (ed) NW Schaad, St. Paul, Minnesota: The American Phytopathogical Society, 17-25 pages

Moore LW 1976 Research findings of crown gall and its control, *American Nurseryman* **144** 8-9, 68-69, 122-128

New PB and Kerr A 1972 Biological control of crown gall field: measurements and glasshouse experiments, *Journal of Applied Bacteriology* **35** 279-287

Ogg AG and Graf GT 1974. Effect of herbicides in fruit tree nursery stock, *Circular* 8 pages, College of Agriculture Research Centre, Washington State University No. 576

Panagopoulos CG and Psallidas PG 1973 Characteristics of Greek isolates of *Agrobacterium tumefaciens*, *Journal of Applied Bacteriology* **36** 233-240

Panagopoulos CG, Psallidas PG, and Alivizatos AS 1978 Studies on biotype-3 of Agrobacterium radiobacter var. tumefaciens, In Proceeding of 4th international Conference on Plant Pathogenic Bacteria, Angers **1** 221-228

Panagopoulos CG, Psallidas PG, and Alivizatos AS 1979 Evidence of a breakdown in the effectiveness of biological control of crown gall, In *Soil Borne Plant Pathogens (*eds*)* B Schippers and W Gams, London: Academic Press, 569-578 pages

Penalver R and Lopez MM 1999 Co-colonization of the rhizosphere by pathogenic strains K-84 and K-1026, used for crown gall bio-control, *Applied Environmental Microbiology* **65** 1936-1940

Penalver RB, Vicedo CI, Salcedo, and Lopez MM 1994 *Agrobacterium radiobacter* strain K 84, K 1026 and K 84 produced an antibiotic-like substance, active *in vitro* against *A. tumefaciens* and phytopathogenic *Erwinia* and *Pseudomonas*, *Biocontrol Science and Technology* **4** 259-267

Pierronnet A and Eyquard JP 1993 Prunus rootstocks and crown gall, *Arboriculture- Fruitiere* **466** 37-41

Pinkerton JN, Ivors KL, Miller ML, and Moore LW 2000 Effect of soil solarization and cover crops on populations of selected soil borne plant pathogens in western Oregon, *Plant Disease* **84** 952-960

Pu Xin-An and Goodman RN 1993 Effects of fumigation and biological control on infection of indexed crown gall free grape plants, *American Journal of Enology and Viticulture* **44** 241-248

Pullman GS and DeVay JE 1977 Control of Verticillium dehliae by plastic tarping, Proceedings of American Phytopathological Society **4** 210

Pullman GS, DeVay JE, and Garber RH 1981 Soil solarization and thermal death: a logarithmic relationship between time and temperature for four soil- borne plant pathogens, *Phytopathology* **71** 959-964

Raio A and Zoina A 1992 Analysis of *Agrobacterium* populations in central and southern Italy and search for natural antagonists, *Bulletin DILB-SROP* **15** 136-138

Raio A, Zoina A, and Moore LW 1997 The effect of solar heating on natural and inoculated agrobacteria, *Plant Pathology* **46** 320-328

Rhouma A, Boubaker A, Nesme X, and Dessaux Y 2005 Susceptibility of some stone and pome rootstocks to crown gall, *Phytopathologia Mediterranea* **44** 275-284

Romanenko VM and Perepnikhatka VI 1984 Flagellae of *Agrobacterium tumefaciens* cells, *Mikrobiologilicherkii Zhurnal* **46** 66-68

Rysheuvels P, Laroche M, and Verhoyen M 1984 Study of the applicability of biological control of *Agrobacterium radiobacter* var. *tumefaciens* by *Agrobacterium radiobacter* var. *radiobacter* (Kerr's strain 84) on fruit crops, *Parasitica* **40** 183-196

Schroth MN and Moller WJ 1976 Crown gall Controlled in the field with non-pathogenic bacterium, *Plant Disease Reporter* **60** 275-278

Sharma RC, Kamal B, and Gupta AK 2004 Occurrence of crown gall of stone fruits in warmer areas of India and its management, *Acta Horticulturae* **662** 411-417

Shim JS, Farrand SK, and Kerr A 1987 Biological control of crown gall: construction and testing of new biocontrol agents, *Phytopathology* **77** 463-466

Silva HSA and Romerio Rda S 2001 Isolation and selection of actinomycetes antagonistic to *Agrobacterium tumefacien, Revista Ceres* **48** 285-291

Singh J and Dhiman JS 2004 Peach replant problem: assessment and investigation, *Acta Horticulturae* **662** 419-422

Smith CO 1939 Susceptibility of species of cupressaceae to crown gall as determined by artificial inoculation, *Journal of Agricultural Research* **12** 919-925

Smith EF and Townsend CO 1907 A plant tumour of bacterial origin, *Science* **25** 671-673

Sobiczewski P, Karczewski J, and Berczynski S 1991 Biological control of crown gall *Agrobacterium tumefaciens* in Poland, *Fruit Science Reports* **18** 125-132

Stapleton JJ and DeVay JE 1982 Effect of soil solarization on populations of selected soil borne microorganisms and growth of deciduous fruit tree seedlings, *Phytopathology* **72** 323-326

Sule S and Kollanyi L 1977 Biological control of *Agrobacterium* crown gall on raspberries, *Novenyvedelem* **12** 244

Sule S 1994 Biological control of *Agrobacterium* - caused root crown gall of stone fruits, *Novenyvedelem* **30** 568

Tawfik AE 1990 Comparison of biological ad chemical soil treatments for controlling crown gall in peach, *Agricultural Research Review* **68** 555-561

TJamos EC and Faridis A 1980 Control of soil borne pathogens by solar heating in plastic houses, 82-84, *Proceedings of 5th Congress Mediterrean Phytopathology,* Union, Hellenic *Phytopathological Society* Athens

Tjamos EC and Paplomatas EJ 1988 Long-term effect of soil solarization in controlling Verticillium wilt of globe artichokes in Greece, *Plant Pathology* **37** 507-515

Utkhede RS and Smith EM 1993 Evaluation of biological control and chemical treatments for control of crown gall on young apple trees in Kopotnay Valley of British Columbia, *Journal of Phytopathology* **137** 265-271

Van Larebeke N, Genetello C, Schell J, Schilperoort RA, Hermans AK, Hernalsteens JP, and Van Montagu M 1975 Acquisition of tumour inducing -ability by non-oncogenic agrobacteria as a result of plasmid transfer, *Nature* **255** 742-743

Vicedo B, Penlaver R, Asins MJ, and Lopez MM 1993 Biological control of *Agrobacterium tumefaciens* colonization, and pAgK84 transfer of the Ti plasmid of *Agrobacterium radiobacter* K84 and Tra$^-$ mutant strain K1026, *Applied and Environmental Microbiology* **59** 309-315

Wang HM, Liang YJ, Wang JH, and Li JQ 1995 Study on control of root crown gall disease in stone fruit trees with antigall agents, *Plant Protection* **1** 24-26

Wang HY, Wang HM, Wang JH, and Ng TB 2000 The pathogen of crown gall disease on flowering cherry and its sensitivity to strain K1026 *European Journal of Plant Pathology* **106** 475-479

Watson B, Currier TC, Gordon MP, Chilton MD, and Nester EW 1975 Plasmid required for virulence of *Agrobacterium tumefaciens*, *Journal of Bacteriology* **123** 255-264

White JG and Buczacki ST 1979 Observation on the suppression of club root by artificial or natural heating of soil, *Transaction of British Mycological Society* **73** 271-275

Wicks TJ, Haque MI, and Doolan DW 1993 Crown gall on stone fruits in the Northern areas of Pakistan, *FAO Plant Protection Bulletin* **41** 23-28

Yang DC, Yang DC, and Choi KT 1991 Isolation and identification of *Agrobacterium* spp. and transformation of *Nicotiana glauca* by selected strains, *Journal of Korean Tabacco Science* **13** 43-52

Zaenen I, Van Larebeke N, Teuchy H, Van Montagu M, and Schell J 1974 Super-coiled circular DNA in crown gall inducing *Agrobacterium* strains, *Journal of Molecular Biology* **86** 109-127

Zhang CM, Lu SH, Wei CM, and Zhao JH 1991 Bacterium strain 1-1-4 *Pseudomonas fluorescens* for biocontrol of peach crown gall disease, *Acta Agriculturae Shanghai* **7** 65-69

Zoina A, Aloj B, and Noviello C 1987 Tumorigenic *Agrobacterium* from peach trees in Campania and Calabria regions: biovars and sensitivity to agrocin 84, *Annali della Facolta di Scienze Agrarie della Universita degli Studi di Napoli Portici* **21** 39-50.

6

AM Fungi and Their Role in Plant Disease Management

Narender K. Bharat and Praveen Mehta

Department of Plant Pathology, Dr YS Parmar University of Horticulture and Forestry, Nauni, Solan - 173 230 (HP)

The term mycorrhiza came from Greek words **'myco'** means fungus and **'rhiza'** means root which literally means fungus root and refers to the relationship that many plant species have with fungi. Mycorrhiza have been described as perhaps the ultimate in reciprocal parasitism (symbiosis) where in the fungus supplies the higher plant partner with mineral nutrients and plant growth hormones and protects the roots against pathogens. Higher plant supplies the fungus with energy substrate. These fungi are unique as they are partly inside or outside of the host. The study of complex mycorrhizal symbiosis with plants is one of the important areas of ecological research.

Approximately 92 per cent of plant families have mycorrhiza and in 80 per cent of their species have arbuscular-mycorrhiza which is the ancestral and predominantal form and indeed most prevalent symbiotic association found in plants at all. Up to 80 per cent of phosphorus and 25 per cent of nitrogen is obtained through these associations and their growth and development depend upon stimulation taken by root exudates where up to 20 per cent of carbon can be obtained by fungus. Arbuscular-mycorrhizal (AM) fungi, earlier known as vesicular arbusicular - mycorrhizal (VAM) fungi are the most common form of mycorrhiza occurring in association of most of the plant species (Hennigan 2009).

While looking into the history, around 400 million years old 'Rhynie Chart' contained the earliest fossil assemblage yielding plants preserved in sufficient detail to detect mycorrhizae. They were indeed observed in the stem of *Aglaophyton major*. The major historical landmarks in development and understanding of the mycorrhizal fungi and their association with plants are as follows:

- Frank (1885) first used the term *Mycorhizen* (Mycorhizae) meaning *fungus root* to describe the symbiotic association of fungal and plant root tissue. He also distinguished between ectotrophic and endotropic mycorrhizae.

- Mosse (1962) made critical studies of mycorrhizal phenomenon and understood their most important function, material exchange which is dependent on union of tissues of both the components.
- Nicolson (1967) proposed the term universal symbiosis to describe vesicular-asbuscular mycorrhizal (VAM) fungal plant association. He said that VAM are totally dependent on the host for their nutrients.
- Mosse (1973) described that these fungi are ubiquitous in soil and are found in the roots of many Angiosperms, Gymnosperms, Pteridophytes and Thallophytes.
- Bagyaraj (1984) explained how mycorrhizal associations play important role in plant growth and nutrient uptake.
- Barea (1991) found that roots of most flowering plants live in mutual symbiosis with mycorrhizae which biotrophically colonize the root cortex and extra metrical mycelia help in the plants to obtain plant nutrient from soil.
- Verma and Hock (1995) explained how VAM fungi can be used to promote growth of biotechnologically developed plants in soil.

Types of mycorrhizae

Initially Frank (1985) distinguished two main types of mycorrhizae viz., Ectomycorrhiza and Endomycorrhiza. Presently seven different types of mycorrhizae are known: i) Ectomycorrhiza, ii) Endomycorrhiza, iii) Arbutoid mycorrhiza, iv) Monotropoid mycorrhiza, v) Orchidaceous mycorrhiza, vi) Ericoid mycorrhiza and vii) Ectendomycorrhiza (Gaur, 2006).

1. Ectomycorrhiza: Ectomycorriza is formed due to symbiosis between plant roots and fungi belonging to the phyla Basidiomycota and Ascomycota. There are more then 5000 species of ectomycorrhizal forming fungi (Wilcox, 1991). Basidiomycetes belonging to families Amanitaceae, Boletaceae, Cortinariaceae, Russiclaceae, Tricholomataceae, Rhizophogonaceae and Schlerodermataceae, are the fungal symbionts often encountered in these associations. Tree species belonging to the Families Pinaceae, Salicaceae, Betulaceae, Fagaceae, Tibiaceae, Rosaceae, Leguminoceae and Juglandaceae are the hosts (Gaur, 2006). These are commonly found on the roots of forest trees like pine, oak, birch and eucalyptus. These form a thick mantle (covering) out side the roots. Ectomycorrhiza is visible to the naked eye as fuzzy covering on the roots. Inside the roots they are present in intercellular space and form a characteristic structure called hartig net. The hartig net may sometime become totally absent at a stage in development and this stage is known as perirhizic mycorrhiza. Ectomycorrhiza improves growth of forest seedlings and the fungal partner of ectomycorrhiza can be grown on defined synthetic media.

2. Ectendomycorrhiza: In this form of mycorrhiza, sheath may be much reduced or even absent. The hartig net is well developed and hyphae also penetrate the host cell. Here mycobiont formation is similar to the ectomyorrhiza e.g. *Boletus* associated with *Monotropha indica.*

3. Arbutoid mycorrhiza: In this form there is well developed hartig net. The sheath is usually much reduced even absent. The fungal hyphae penetrate into the host cell. In addition there is intercellular penetration. These form fungal coil in the cells. These fungi belong to class Basidiomycetes and host plants belong to order Ericales.

4. Monotropoid mycorrhiza: These have well developed fungal sheath enclosing the roots. They form hartig net, hyphae enter the cell of the epidermis or cortex. Specialized unbranched haustoria penetrate the epidermal cell and go to the complicated development as the host plant grows. The fungus is often associated with the mycorrhizal roots of adjacent plants. The monotropoid mycorrhiza are found in roots of achlorophyllous plants of Monotropaceae.

5. Orchidaceous mycorrhiza: These are formed by association of Basidiomycetes fungi with roots of orchids. The fungus enters the protocorms (germinating seed) and stimulates the growth of protocorm mainly through carbon and phosphorus nutrition. Orchids are generally propagated through tissue culture and it is observed that there is build up of mycorrhizal infection. The fungus commonly associated with orchid is *Tulasnella* spp. It can be cultured on artificial medium.

6. Ericoid mycorrhiza: In plants of Ericaceae and related families hair like roots are enmeshed in an extensive external weft of hyphae which also penetrate the cortical cells of the root lets. Most of these fungal symbionts are the members of Ascomycetes, however, Basidiomycetes genus *Clavaria* has also been reported to form the symbiotic association but normally do not form the sheath. Ericoid mycorrhizae are associated with plants like blue berries, *Azaleas* sp., *Rhododendron* etc. The fungus involved is *Peziella ericae* which can be cultured in the laboratory.

7. Endomycorrhiza or Arbuscular mycorrhiza: Endomycorrhizae, also known as arbuscular mycorrhizae occur in plants growing in different ecological regions (Hayman, 1982). They are commonly associated with most of agricultural and horticultural crops, shrubs and tropical trees. Arbuscular mycorrhizae have been reported in thousand genera of plants covering 200 families. There are at least 300 thousand receptive plant hosts in the world and there are 120 species of arbuscular mycorrhizal fungi found associated with these plant species (Mukerji, 1996). Plants (80-90%) ranging from bryophytes to flowering plants have dynamic association of AM fungi. However, there are few families of plants viz., Pinaceae, Betulaceae, Orchidaceae, Fumariaceae, Commelinaceae, Urticaceae, Salicaceae, Sagaceae and Caesalpiniaceae which do not form AM association with roots. Few vesicular plants in families such as Cruciferae, Chinopodiaceae, Cyperaceae and Junaceae are not capable of forming AM associations.

The fungi forming arbuscular mycorrhizae are the naturally occurring fungal component of soil biota in most terrestrial ecosystem and also represent an obligate symbiotic group distinct form the rest of soil microbial biomass. AM fungal symbiosis is dynamic process and their interaction affects all physiological aspect of the host and these associations occur abundantly in phosphorus and other nutrient deficient soils.

Arbuscular mycorrhiza are the type of endomycorrhiza in which the fungus penetrates cortical cell wall of the host. The hypha grows both inter as well intracellularly into the cortical cell by penetrating the wall and causing invagination of the plasma membrane. These produce coil, highly branched haustorium like structure called arbuscule, which swells at the end producing roundish bodies called sporangioles. Arbuscules are known to be the site for nutrient exchange (Pannori, 2002). In some cases vesicles which are globose sac like structures are formed as a result of intercalary or terminal swellings of the fungal hyphae (Plate 1). Vesicles contain high concentration of triglycerides at maturity and are thought to be storage and survival organs (Gaur, 2006). Arbuscular mycorrhizae (AM) were previously named as vasicular- arbuscular mycorrhizae (VAM) but research (Hennigan, 2009) uncovered that some of the fungi did not form vesicles, so they are now referred to as arbuscular mycorrhizae (AM).

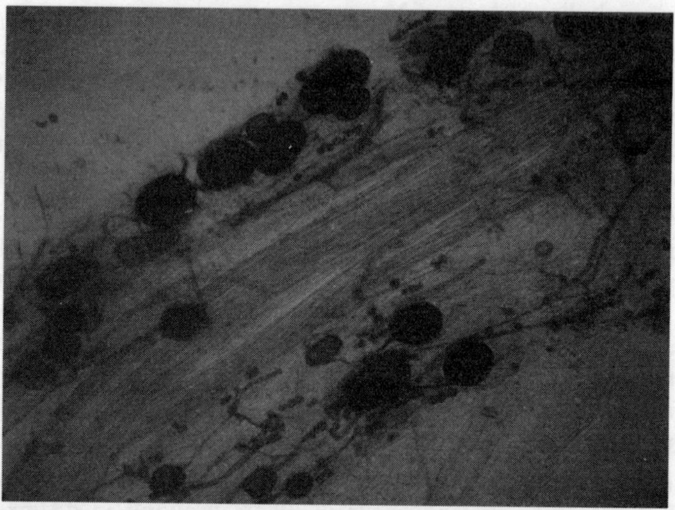

Plate 1. Colonization of apple roots by AM fungi with hyphae, arbuscules and vesicles.

There is no specificity between plants and fungus in AM fungi. However, it is important to distinguish between specificity or ability to colonize effectiveness (plant response to colonization) and infectiveness (inability to colonize) because AM are widely different in these abilities depending on the environment. They do have wide host range, and are capable of long term relationship with many different plants. It is well known that AM fungi are not host specific. However, the degree of AM colonization and its effect can differ with different host-endophytic combinations for example plants like maize and soybean appear to be more mycorrhizal than crops like potato. However, AM fungi do not normally exhibit any species or strain specificity to the particular host plant except perhaps some host preference. The same plant may be infected by more than two species at a time. Extent to which a species or variety of mycorrhizal fungi colonize is determined by background of the host plant or kind of root exudates of host plant or other barriers to AM infection in tree crops (Gaur, 2006).

Classification of AM fungi

The first monograph on AM fungi has been prepared by Thaxter (1922). These fungi were included in the family Endogonaceae of order Mucorales (Zygomycetes). Fungi of this family were located in four genera producing spores in sporocarps:

Endogone Link: Fries, *Glaziella* Berk., *Sclerocystis* Berk. & Br. and *Sphaerocreas* Sacc. & Ellis. Thaxter (1922) and Godfrey (1957) considered chlamydosporic species to be anamorphs of those producing zygospores following the finding of both types of spores in sporocarps of *Glomus fasciculatum* (at that time known as *Endogone fasciculata* and *E. microcarpa*). In 1974, Gerdemann and Trappe revised the Endogonaceae, 44 species in seven genera were characterized. Among them, many taxa were redefined, and two genera (*Acaulospora*, *Gigaspora*) and 12 species were described as new. In 1979, Ames and Schneider erected a new genus in the Endogonaceae, *Entrophospora*. In 1986, Walker and Sanders separated the genus *Scutellospora* with fungi forming spores having at least one inner germination wall from the genus *Gigaspora*, containing species with spores lacking an inner germination wall having no physical contact with their main, structural wall. In 1990, Morton and Benny located soil-borne fungi forming arbuscules in roots of terrestrial plants in a new order, Glomales, consisting of two suborders, Glomineae J.B. Morton & Benny and Gigasporineae J.B. Morton & Benny. The former suborder consisted of the type family Glomaceae, with the genera *Glomus* and *Sclerocystis*, and the Acaulosporaceae fam. nov., containing the genera *Acaulospora* and *Entrophospora*. The latter suborder was proposed to include the Gigasporaceae fam. nov. with the genera *Gigaspora* and *Scutellospora*. In 1990, Almeida and Schenck concluded that a continuum of morphological properties exists between sporocarpic *Glomus* species and the members of the genus *Sclerocystis*. As a result, the five-species genus *Sclerocystis* was reduced to a single-species one. In 2000, Redecker *et al* utilizing both morphological and molecular data, transferred *S. coremioides* to the genus *Glomus* and, thereby, eliminated the genus *Sclerocystis* from the Kingdom Fungi. Based on data from molecular, morphological and biochemical investigations, Morton and Redecker (2001) erected two new families in the order Glomales (now Glomerales), i.e., Archaeosporaceae and Paraglomaceae (now Paraglomeraceae). Each of these families was phylogenetically distant from each other and from other glomalean families, despite similarities in mycorrhizal morphology. The family Archaeosporaceae contained one genus, *Archaeospora*, with three species forming a typical *Acaulospora*-like spores from the neck of a sporiferous saccule. Two of these species, *Archaeospora gerdemannii* and *A. leptoticha*, were considered to dimorphic, forming *Glomus*-like spores as well. The genus *Paraglomus* in the family Paraglomaceae (now Paraglomeraceae) consisted of two species producing spores indistinguishable from those of *Glomus* species.

The classification of arbuscular fungi presented here in this chapter is that of Schüßler *et al.* (2001) with emendations of Oehl and Sieverding (2004), Walker and Schüßler (2004), Sieverding and Oehl (2006), Spain *et al.* (2006) and Walker *et al.* (2007a,b). In this system of classification, fungi forming or considered to form arbuscular mycorrhizae are placed in four orders, i.e., Archaeosporales, Diversisporales, Glomerales, and Paraglomerales, comprising ten families and thirteen genera, belonging to the class Glomeromycetes of the phylum Glomeromycota (Fig. 1).

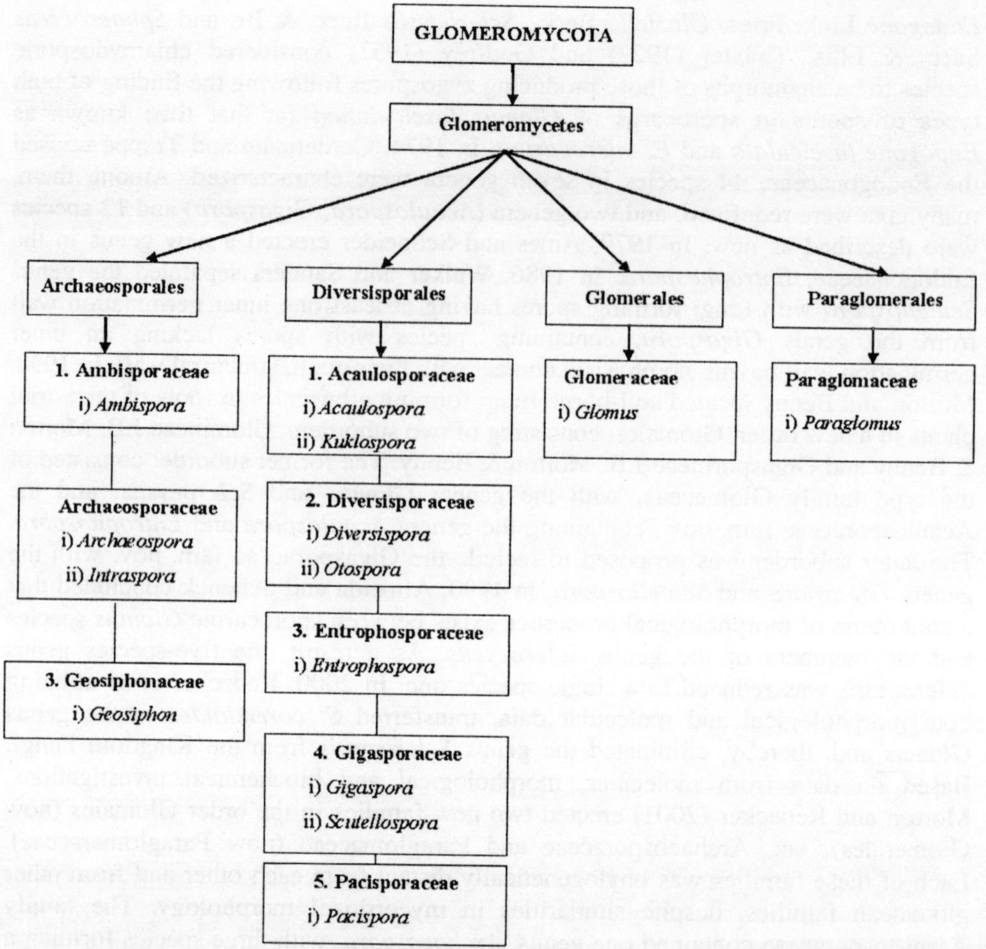

Fig. 1. Classification of Arbuscular-mycorrhizal fungi

Important genera of AM fungi

i. Glomus: Chlamydospores are produced singly or in loose to tight aggregates in soil less commonly in roots, form vesicles. Spores are formed on a straight or recurvate and cylindrical subtending hypha.

ii. Gigaspora: Germ tube produced directly through spore wall, inner flexible wall group absent, axillary cells finely papillate or echinulate. Spores are formed terminally on a bulbous subtending hypha.

iii. Acaulospora: Spore arises laterally form the neck of sporiferous saccule with funnel shaped stack of large inflatenned hyphae. Spores are sessile, i.e., no pedicel (a short branch of the sporiferous saccule neck) is formed.

iv. Scutellospora: Germ tube form germination shield, inner flexible wall group always present, axillary cell knobby broadly papillate or smooth. Spores are formed terminally on a bulbous subtending hypha. (Plate 2).

Glomus sp. *Acaulospora* sp. *Gigaspora* sp. *Scutellospora* sp.

Plate 2. Chlamydospores of important Genera AM fungi

Functions of AM fungi

The main role of arbuscular mycorrhizal (AM) associations in plant developments are:

1. Plant growth and nutrition
2. Plant hormone, water deficits
3. Plant micropropagation system
4. Plant disease management

1. Plant growth and nutrition: AM fungi help in nutrient uptake from soil. The benefits from symbiosis occur because AM fungal hyphae extend beyond the root hair and increase absorptive surface area of colonized roots of host plant. The growth of the host plant is enhanced because of the improved nutrient status in plant tissues.

a. Phosphate nutrition: AM enhances plant growth by improving mineral nutrition. Available evidences suggested that AM have significant effect on phosphorus (P) uptake by host plants. Greater soil exploration by mycorrhizal roots as a mean of increasing phosphorus uptake is well known. In phosphate deficient soil, immobile phosphate ions develop a phosphate depletion zone around the roots. Hyphae reach beyond this zone and directly translocate P from soil to root cortex. Phosphorus is transferred into plant cell in exchange of carbon in the arbuscules with in the colonized cells. Fungal hyphae are known to absorb phosphorus from lower concentration (Hayman, 1983).

b. Nitrogen nutrition: AM improve growth, nodulation and nitrogen fixation by *Rhizobium* and *Frankia*. AM increase nitrogen nutrition by different mechanisms like improvement in symbiotic nitrogen fixation by supplying phosphate and molybdenum to Azotrophs. In legume *Rhizobium* requires phosphate supply and restricted root system of legumes compared to grasses leads to poor competition for soil phosphate. Hence mycorrhizal infection of legumes roots stimulate N fixation. There may be direct uptake of combined nitrogen by AM fungi. These associations also facilitate nitrogen transfer, a process by which a part of nitrogen is fixed by nodulated plant also benefits the non N-fixing plant nearby.

2. Plant hormones, water deficit: Arbuscular mycorrhizae directly affect the level of plant hormones such as cytokinines, and gibberelline-like substance(s). Beneficial effects of AM fungi have been reported under drought and saline conditions. The important feature is that under drought conditions the mycorrhizal roots have ability to tap additional water sources unavailable to non-mycornizal plant roots. Such effects are apparent under phosphate limitations as additional phosphate to non-mycorrhizal plants boosted plant performance under drought (Gaur, 2006).

3. Micropropagation system: Traditional micropropagation techniques allow rapid production of high quality disease free and uniform planting material (Kapoor *et al* 2008). In plant tissue culture if plantlets are artificially inoculated with AM fungi the growth and survival of these inoculated plants is enhanced as compared to un-inoculated ones. The beneficial effects of AMF inoculation to micro-propagated plantlets are: i) development of superior root system, ii) increased photosynthetic efficiency, iii) increased water conducting capacity, iv) enhanced nutrient uptake, v) avert attack by harmful soil born pathogens, and vi) alleviate environmental stresses. *In vitro* mycorrhization refers to the inoculation of AMF to the roots of micropropagated plantlets growing on Agar medium (Pons *et al* 1983). Although technique of mycorrhization is of importance for the development of micropropagated plantlets, contamination of the inoculums, behavior of the host *in-vitro* and obligate nature of endophyte are some major hurdles in the establishment of mycorrhizal host symbiosis *in vitro* (Kapoor *et al* 2008). To circumvent this problem an alternative method, *ex vitro* inoculation or *in vivo* mycorrhization can be used which is relatively early, naturally feasible and easily implementable method and show no technical problem associated with *in vivo* mycorrhization. In *ex vitro* inoculation, the AM fungal spores are inoculated to the micropropagated plantlets after transplanting in to pots. The longevity of effective time of mycorrhization depends on the type of AMF inoculum used. For example when the inoculum is being applied in the form of mycelium, the roots should be encountered with in a short period of time (hours/day) as compared to three week when inoculum is based on AMF spores. It has been shown that the spore dormancy of AMF differ among species and genera. So it is advisable to use germinated AMF spores for effective mycorrhization. There are several examples to show the beneficial effects of AMF on improving growth and development of several *in vitro* raised horticultural crops like apple, peach, grapes, coffee, strawberry, lilium, rose and capsicum (Kapoor *et al* 2008).

4. Plant disease management

Arbuscular mycorrhizal associations were initially thought to be symbiotic associations which help the plant to absorb nutrient more efficiently from soil. It is now well recognized that the incidence of root diseases is reduced upon mycorrhizal inoculation either by some direct action on pathogen or indirectly by improving the resistance of the host plant (Dehne, 1982; Aguilar and Barea, 1996). In case of ectomycorrhizae - pathogen system, several mechanisms have been put forward to explain disease protection of feeder roots of host tree, which could broadly be categorized as: i) utilization of surplus carbohydrates in roots by mycorrhizal endophytes, eventually restricting the quantum of nutrients available to the pathogens, ii) formation of mechanical barrier in the form of fungal mantle against ingress by the pathogens, iii) release of antibiotics inhibitory to the pathogens, iv) formation of protective microbial rhizosphere population, and v) production of metabolites released by mycorrhizal host cortical cells functioning as inhibitor to infection and spread of the pathogens. It has been time and again demonstrated that fungal mantle of ectomycorrhizal roots are formidable physical barriers that restrict penetration by several root pathogens. The hartig net may function as supplementary physical barrier since progression of the invading pathogen get blocked in this region. It has been observed that a mycorrhizal strain of *Laccaria laccata*, protected primary roots of Douglas fir from root rot infection by *Fusarium oxysporum* (Sylvia and Sinclair, 1983).

The arbuscular mycorrhizae - pathogen interactions have, however, different mechanisms by which the mycorrhizae accomplish reduction in plant disease incidence. The important mechanisms as discussed by Aguilar and Barea (1996) are as follows:

1. Improved nutrient status of the host plant: Increased nutrient uptake in host plants by AM symbiosis results improved growth thereby more vigorous plants. Such plant itself thus will be more resistant to pathogen attack. Increased status of phosphorus and other nutrition in plants could account for the higher tolerance to pathogen(s) attack.

2. Damage compensation: It has been suggested that AM fungi increased host tolerance of pathogen attack by compensating for the loss of root biomasses or function caused by pathogens (fungi and nematodes). This represents an indirect contribution through conservation of root system function, both by fungal hyphae growing out into the soil and increasing absorbing surface of the root and by maintenance of the root cell activity through formation of arbuscules.

3. Competition for host photosynthates: It has been proposed that the growth of both AM fungus and root pathogen depends on host photosynthates and so they compete for carbon compounds reaching the roots. When AM fungi have primary access to photosynthates, pathogen gets less carbon which may result growth inhibition in pathogen.

4. Competition for infection/colonization sites: Fungal root pathogen and AM fungus although colonize the same host tissues, usually develop in different cortical cells, indicating some sort of competition for space. Both localized and non localized mechanisms could exist, probably depending upon the pathogen, e.g. *Phytophthora* development is reduced in AM fungal colonized and adjacent uncolonized region of AM root system and that in the former the pathogen does not penetrate arbuscule containing cells. Even in localized mechanisms some non localized mechanisms are also found in host tissue (Dehne, 1982).

5. Anatomical and morphological changes in the root system: It has been demonstrated that AM colonization induces remarkable changes in the root system morphology as well as in meristematic and nuclear activity of root cells. This might affect rhizosphere interactions particularly the development of pathogen infection. However, its significance in plant disease management has not yet been sufficiently considered.

6. Microbial changes in the rhizosphere: AM formation induces changes in host physiology that can be decisive for root exudation pattern, which consequently cause qualitative and quantitative alternation in microbial population in the rhizosphere, there is evidence that microbial shifts occur in the rhizosphere and resulted microbial equilibria could influence the growth of plants. Although this effect is not sufficiently evaluated as mechanism for AM associated biological control but there are the indications that such mechanism could operate. Changes in soil microorganism population induced by AM formation may lead to stimulation of certain component of microbiota, which in turn may be antagonistic to root pathogen.

7. Activation of plant defense mechanisms: Activation of specific plant defense mechanisms as the response to AM colonization is the obvious basis for the protective capacity of host plants. The elicitation by AM symbiosis of specific plant defense reaction could predispose the plant to an early response to attack by a root pathogen (Aguilar and Barea, 1996). It is only a weak or very local, transient activation of plant defense mechanisms that occur during AM formation. During the life cycle, plants involve a number of defense responses elicited by various signals, including those associated with pathogen attack. Among the compounds involved in plant defense studied in relationship to AM formation are phytoalexins, enzymes of the phenyl propanoid pathway, chitinase, β,1-3-glucanase, peroxidases, pathogenesis related (PR) proteins, callose, hydroxy proline-rich glycoproteins (HRGP) and phenolics. Activation of the phenyl propanoid related enzyme especially phenylalanine ammonium lyase (PAL) increased in amount when plant roots were colonized with AM fungi. PR protein and HRGP are synthesized only locally and in very low amounts. Particularly PR-b proteins are usually synthesized in cell containing arbuscules. Accumulation of these compounds may reduce the pathogen infection. In summary only weak responses to AM infection have been observed regarding some activity like production of phytoalexins, PR proteins, HRGP, and phenyl propanoid related proteins, indicating that AM fungi do not elicit typical defense to subsequent pathogen infection but these can sensitize the root or pathogen and enhance the mechanisms of defense to subsequent pathogen infection.

8. Increased lignification of root epidermal cells: The increased lignification of root epidermal cells induced by AM colonization has also been suggested as a mechanism of reduced infection of mycorrhizal plants. This deserves further investigation as process would make penetration of pathogen into root tissue more difficult.

In literature, a wide variety of reports in recent years with experimental evidences have indicated that a host plant previously inoculated with a AM fungal symbiont increased tolerance to a wide range of root diseases in plants. Inoculation of host plants with AM fungi have been shown to exert probiotic influence in the management of diseases like wilts and root rots. It is speculated that infection with mycorrhizal fungi elicits a resistance mechanism by the host which may suppress the subsequent infection by fungal pathogens. Disease caused by soil borne fungi can be influenced by the formation of AM fungi in the root system. In general, mycorrhizal plants suffer less damage by the pathogens, incidence of diseases is decreased and pathogen development is inhibited (Dehne, 1982). Bharat and Bhardwaj (2001) found reduction in the severity of white root rot (*Dematophora necatrix*) of apple after inoculation with native isolate of *Glomus* spp. (Table 1). Similar results were obtained by Raj and Sharma (2009) when they inoculated apple nursery plants with native isolate of AM fungi. They also observed that the effect of mycorrhizal inoculation increased with the inoculation of phosphate solubilizing bacterium (*Azotobacter* spp.) Mehta (2010) and Bharat *et al.* (2011) have reported that inoculation of apple seedling root stock with a potent AM fungal isolate resulted in increased growth of the seedlings when grown in apple replant disease (ARD) soil (Table 2). The artificial inoculation with AM fungi has been found to reduce disease incidence in various crop plants as listed in Table 3.

Table 1. Effect of VAM fungi on severity of white root rot of apple caused by *Dematophora necatrix*

VAM	Disease severity (%)	VAM root colonization (%)	No. of spores/ 100g soil
Glomus mosseae	37.50 (37.75)	26.00 (30.90)	152
G. fasciculatum	38.00 (38.00)	27.00 (31.05)	148
Acualospora laevis	32.50 (34.74)	37.00 (37.46)	191
Gigaspora gilmorei	56.54 (48.74)	21.0 (27.55)	123
Glomus spp. (local isolate)	27.50 (31.61)	40.00 (39.00)	234
Control	100.00 (90.00)	-	-
C. D. 0.05	1.54	1.13	5

(Bharat and Bhardwaj, 2001)

Table 2. Effect of AMF inoculation and other treatments on growth parameters of apple seedlings grown in replant disease soil

Treatment	Increase in seedling height (cm)	Increase in stem diameter (mm)	Intrenodal length (cm)	Leaf area (cm^2)	Shoot fresh weight (g)	Root fresh weight (g)	Shoot dry weight (g)	Root dry weight (g)
1. Apple replant diseased (ARD) soil	12.89^{e}	1.33^{e}	1.39^{b}	15.57^{d}	14.80^{c}	6.25^{e}	8.00^{d}	3.75^{c}
2. Virgin soil (Non apple soil)	14.85^{d}	1.62^{c}	1.41^{b}	15.50^{d}	14.45^{c}	7.10^{de}	8.80^{cd}	4.50^{b}
3. ARD soil + AMF inoculation	14.07^{d}	1.45^{d}	1.43^{b}	15.65^{d}	17.30^{b}	8.10^{d}	9.40^{c}	4.95^{b}
4. Virgin soil+ AMF inoculation	19.98^{b}	1.72^{b}	1.44^{ab}	18.56^{c}	15.75^{bc}	8.00^{d}	9.60^{bc}	5.15^{b}
5. ARD soil+ formaldehyde fumigation	19.60^{b}	1.75^{b}	1.43^{b}	20.03^{b}	20.07^{a}	9.50^{c}	11.70^{ab}	5.45^{ab}
6. Virgin soil+ formaldehyde fumigation	18.08^{c}	1.55^{c}	1.44^{ab}	18.56^{c}	19.20^{ab}	9.35^{cd}	10.65^{b}	5.50^{ab}
7. ARD soil+ formaldehyde fumigation + AMF inoculation	23.90^{a}	1.92^{a}	1.51^{a}	20.32^{b}	21.45^{a}	12.55^{a}	12.0^{a}	5.80^{a}
8. Virgin soil+ formaldehyde fumigation + AMF inoculation	24.40^{a}	1.92^{a}	1.50^{a}	24.42^{a}	20.85^{a}	10.85^{b}	12.10^{a}	6.00^{a}
$CD_{(0.05)}$	1.04	0.08	0.07	0.82	3.10	1.33	1.10	0.66

Figures in each column are mean value of five replicates

Values within a column with different letters are significantly different at $P = 0.05$ (Mehta, 2010)

Table 3. Other successful application of AM fungi in plant disease management

Pathogen	Host Plant
Olpidum brassicae	Tobacco, Lettuce
Phytopthora cinnamoni	Avocado
P. palmivora	Papaya
P. parasitica	Citrus
Fusarium oxysporum f.sp. *cepae*	Onion
F. oxysporum f.sp. *cucumerinum*	Cucumber
F. oxysporum f.sp. *lycopersici*	Tomato
Phoma terrestris	Onion
Phyrenochaeta terrestris	Onion
Thielaviopsis bassicola	Citrus, Onion

(Dehne, 1982)

Role of AM fungi in the management of diseases caused by nematodes

Mycorrhizae effect a broad range of soil borne plant pathogens including plant parasitic nematodes especially endophytic parasitism of plants by nematodes can be influenced by the establishment of VA mycorrhizae. Penetration rate of parasitic nematodes can be decreased, their development inside the root may be retarded, or degree of damage caused by the nematodes may be lowered. There are some plant parasitic nematodes (Table 4) whose infection rate, reproduction and damage can be controlled in the presence of AM fungi.

Table 4. Some successful application of AM fungi in management of plant nematodes

Host	Nematode
Carrot	*Meloidogyne hapla*
Grape	*M. arenaria*
Tomato	*M. incognita*
Cucumber	*M. incognita*
Citrus	*Rodopholus similes*
Bean	*Rotylenchus reniformis*
Cucumber	*Rotylenchus reniformis*
Cotton	*Rotylenchus reniformis*
Tomato	*Rotylenchus reniformis*
Cucumber	*Pratylenchus penetrans*

(Dehne, 1982)

Mass production of AM fungi

The role of AM fungi in improving growth of plant and their role in plant disease management has been studied but the mass production of these symbionts for field application has been limited due to their inability to grow on artificial media. Development of technique for the production of large quantities of inoculum is important for field application. There are some important commonly used methods for mass production of AM Fungi as listed below (Verma and Hock, 1995; Gaur, 2006; Kapoor et al 2008):

1. Soil based inocula

(a) Pot culture: It is most widely used standard and conventional method of maintaining AMF culture around the world. In this method AMF spores are inoculated to the root(s) of trap plant raised on sterilized soil. Though the usual substrate used in pot culture is sterilized sand-soil mixture, some time inorganic inert materials like peat, perlite and vermiculite can also be used as substrate. Trap plants commonly used for pot culture are *Sorghum helipanse*, *Panicum* sp., *Zea mays, Trigonella sp.* etc. The inoculum so produced, consists of mixture of soil, spores, hyphal segments and infected root pieces and generally takes around 3-4 months to multiply.

(b) Inoculum rich soil pellets: A technique of AMF inoculum production in which soil pellets are enriched with a AMF inoculum was introduced by Hall and Kelson (1981). The pellets had an average dry weight of 1.55g and measured 12x12x6 mm. These dry pellets can be glued with seed and can easily be broadcasted like other fertilizers and spread during the seed sowing or transplantation.

2. Soil free inocula

(a) Aeroponic culture: Apart from soil based pot culture, aeroponic culture is most widely used method for AMF inoculum production. Now-a-days for physio-logical genetic studies and for *in-vitro* mycorrhization, focus has shifted towards alternative soil-less culture for mass production for clean and pure AMF propagules. In aeroponic culture, pure and viable spore of particular fungus are selected to inoculate the culture plant which are later transferred into a controlled aeroponic chamber, where nutrient solution is provided in the form of mist.

(b) Root organ culture: In this system we grow a plant *in vitro* on MS nutrient media. Then spores of AM fungi are collected from soil or culture from pot and these spores are disinfected with Tween-20 and rinsed thoroughly with streptomycin-gentamycin antibiotic solution. All the steps must be done on ice to maintain dormancy of spores and spores are placed on root very carefully for germination.

(c) Nutrient film technique (NFT): In this technique plant roots are provided with a shallow layer of rapidly flowing nutrient solution. As a result of this root mats are formed as upper layer and above the liquid. Pre-inoculated seedlings are transferred into NFT unit. Inoculum produced by this method is ideal for the production of easily harvestable solid mats of roots with more concentrated less bulky form of inoculum than that produced by plants grown in soil based other solid media.

Inoculation methods

Various methods have been tried and tested for introducing AM fungi in field crops (Gaur, 2006):

1. Transplanted crop: Seedlings grown in small beds can be inoculated with AM culture and can be transplanted when AM colonization has been received. This can be done in crops like chilli, tomato and horticultural crops like citrus, mango, apple, etc.

2. Directly sown plants

(a) Coating of seed with AM inoculum: Seeds are coated with AM fungi with some adhesive such as methyl cellulose. It is expected that inoculum would stick to the seed and subsequently infect the emerging radicals. It has been proved in large seed plants such as citrus in field nurseries.

(b) Mycorrhizae pellets: Seeds are incorporated into an inoculum to form multi-seeded pellets which are mixed with seed or directly broadcasted in the field.

(c) Fluid drilling: Mixing of seed and inoculum into a uniform suspension containing 40 per cent (w/v) methyl cellulose and then slurry is applied in furrows.

(d) Inoculation in furrows: Inoculum is placed under or beside seed sown in furrows which may be effective method for field crops.

(e) Pre-cropping: Population of AM propagules can be raised *in situ* by growing strongly mycorrhizal host plants and leaving the infected roots and associated spores in the soil to infect the field crops. Judicious crop rotation might be practiced as a mean of manipulative raising of *in-situ* AM propagules.

Constraints

1. Arbuscular mycorrhizae are obligate symbiont, so pure culture preparation is also a difficult task.
2. AM fungi can be affected by several other microorganisms in the rhizosphere.
3. AM fungi vary in effect with host, environment and soil nature.
4. Most studies have been carried out at greenhouse or in pots.

Future Thrusts

The nature has blessed us by providing a lot of beneficial microorganisms which are in their own ways benefitting us directly or indirectly. Now we have to understand their ecology and exploit these microbial forces to the benefit of agriculture and environment as a whole. Arbuscular- mycorrhizal (AM) fungi are one of them. These symbiotic fungi have potential to improve growth and suppress soil borne pathogens in most of crops particularly grown in nutrient deficient soils under organic agriculture. It is important to coordinate comprehensive research on disease control mechanisms with AM fungi. Studies are needed to identify the compatibility of AM with others management practices. Efforts to mass multiply mycorrhizae in large quantity using lighter carrier material should be intensified. Molecular techniques like polymerase chain reaction (PCR) should be standardized to identify mycorrhizal colonization within the plant root using different species specific primers. Efforts to make AM fungi as a potent biological control agent for plant diseases caused by fungi, bacteria and nematodes should be done in fields. Possible role of AM fungi in biological control must be considered and exploited in plant breeding programmes aimed at selecting resistant cultivars. Efforts to use AM fungi in tissue culture so that tissue culture plants inoculated with AM fungi with more resistance to soil borne pathogens, temperature and any others factor are produced.

References

Aguilar CA and Barea JM 1996 Arbuscular mycorrhiza and biological control of soil borne plant pathogen- an overview of mechanism involved, *Mycorrhiza* **6** 457-464

Almeida RT and Schenck NC 1990 A revision of the genus *Sclerocystis* (Glomaceae, Glomales), *Mycologia* **82** 703-714

Ames RN, Schneider RW 1979 *Entrophospora*, a new genus in the Endogonaceae, *Mycotaxon* **8** 347-352

Bagyaraj DJ 1984 Biological interaction with VAM fungi, in *VA Mycorrhizae* (eds) CL Powell and DJ Bagyaraj, USA: CRC Press, 131-153 pages.

Bharat NK, Mehta P and Sharma JN 2011 Effect of indigenous arbuscular- mycorrhizal fungi on growth of apple seedlings grown in replant soil, *Indian Phytopathology* **64** (Supplementary issue) 25

Bharat NK and Bhardwaj LN 2001 Interaction between VA-Mycorrhizal fungi and *Dematophora necatrix* and their effect on apple seedling, *Indian Journal of Plant Pathology* **19** 47-51

Barea JM 1991 Vesicular arbuscular mycorrhizae as modifiers of soil fertility, *Advances in Soil Science* **15** 1-14

Dehne HW 1982 Interaction between VAM fungi and plant pathogen, *Phytopathology* **72** 1115-1119

Frank AB 1885 Veber nene Mykorrhiza-formen. *Berichte der Deutschen Botanischen Gesellschaft* **5** 395

Gaur AC 2006 Biofertilizer in sustainable agriculture, New Delhi: ICAR

Gerdemann JW and Trappe JM 1974 The Endogonaceae in the Pacific Northwest, *Mycologia Memoirs* **5** 1-76

Godfrey RM 1957 Studies of British species of *Endogone*. I. Morphology and taxonomy, *Transactions of the British Mycological Society* **40** 117-135

Hall LA and Kelson A 1981 An improved technique for the production of endomycorrhizal infestated soil pellets. *New Zealand Journal of Agricultural Research* **24** 221-222

Hayman DS 1982 Influence of soils and fertility on activity and survival of vesicular-arbuscular mycorrhizal fungi, *Phytopathology* **72** 1119-1125

Hayman DS 1983 The physiology of vesicular arbuscular endomycorrhizal symbiosis. *Canadian Journal of Botany* **61** 944-963

Hennigan T 2009 Towards an understanding of Arbuscular Mycorrhizal fungi, *Answers Research Journal* **2** 21-28

Kapoor R, Sharma D and Bhatnagar AK 2008 Arbuscular mycorrhiza in mycopropagation system and there potential application, *Scientia Horticulturae* **116** 227-239

Mehta P 2010 Role of arbuscular-mycorrhizal fungi in apple replant problem, M.Sc. thesis, Dr YS Parmar University of Horticulture and Forestry, Nauni, Solan, HP

Morton JB and Benny GL 1990 Revised classification of arbuscular mycorrhizal fungi (Zygomycetes): a new order, Glomales, two new suborders, Glomineae and Gigasporineae, and two new families, Acaulosporaceae and Gigasporaceae, with an emendation of Glomaceae, *Mycotaxon* **37** 471-491

Morton JB and Redecker D 2001 Two families of Glomales, Archaeosporaceae and Paraglomaceae, with two new genera *Archaeospora* and *Paraglomus*, based on concordant molecular and morphological characters, *Mycologia* **93** 181-195

Mosse B 1962 The establishment of vesicular- arbuscular mycorrhiza under aseptic condition, *Journal of General Microbiology* **27** 509-520

Mosse B 1973 Advances in the study of vesicular-arbuscular mycorrhizae. *Annual Review of Phytopathology* **11** 171-196

Mukerji KG 1996 Taxonomy of endomycorrhizal fungi, in *Advances in Botany* (eds) KG Mukerji, B Mathur, BP Chambola and P Chitralekha, New Delhi: APH Publication Corporation, 213-221 pages

Nicolson TH 1967 Vesicular arbuscular mycorrhiza: a universal plant symbioses, *Science Progress* **55** 561-581

Oehl F and Sieverding E 2004 *Pacispora*, a new vesicular arbuscular mycorrhizal fungal genus in the Glomeromycetes, *Journal of Applied Botany* **78** 72-82

Pannori R 2002 Fundamental of Micorrhizal Biology and Biotechnology, Udaipur: Agrotech Publishing Academy

Pons F, Gianinazzi-Pearson V, Gianinazzi S and Navatel JC 1983 Studies of VA Mycorrhizae *invitro*: micorrhizal synthesis of axenically propagated wild cherry (*Prunus avium* L.) plants, *Plant and Soil* **71** 217-221

Raj H and Sharma SD 2009 Investigation of soil solarization and chemical sterilization with beneficial microorganisms for control of white root rot and growth of nursery apple, *Scientia Horticulturae* **119** 126-131

Redecker D, Morton JB and Bruns TD 2000 Molecular phylogeny of the arbuscular mycorrhizal fungi *Glomus sinuosum* and *Sclerocystis coremioides*, *Mycologia* **92** 282-285

Schüßler A, Schwarzott D and Walker C 2001 A new fungal phylum, the Glomeromycota: phylogeny and evolution, *Mycological Research* **105** 1413-1421

Sieverding E and Oehl F 2006 Revision of *Entrophospora* and description of *Kuklospora* and *Intraspora*, two new genera in the arbuscular mycorrhizal Glomeromycetes, *Journal of Applied Botany and Food Quality* **80** 69-81

Spain JL, Sieverding E and Oehl F 2006 *Appendicispora*: a new genus in the arbuscular mycorrhiza-forming Glomeromycetes, with a discussion of the genus *Archaeospora*, *Mycotaxon* **97** 163-182

Sylvia DM and Sinclair WA 1983 Phenolic compounds and resistance to fungal pathogens induced in primary roots of Douglas-Fir seedlings by the ectomycorrhizal fungus *Laccaria laccata, Phytopathology* **73** 390-397

Thaxter R 1922 A revision of the Endogonaceae, *Proceedings of the American Academy of Arts and Sciences* **57** 291-351

Verma A and Hock B 1995 Mycorrhiza: Structure, function, molecular biology and biotechnology, Germany: Spinger Verlag

Walker C and Sanders FE 1986 Taxonomic concepts in the Endogonaceae: III. The separation of *Scutellospora* gen. nov. from *Gigaspora* Gerd. & Trappe, *Mycotaxon* **27**169-182

Walker C and Schüßler A 2004 Nomenclatural clarifications and new taxa in the Glomeromycota, *Mycological Research* **108** 979-982

Walker C, Vestberg M, Demircik F, Stockinger H, Saito M, Sawaki H, Nishmura I and Schüßler A 2007a Molecular phylogeny and new taxa in the Archaeosporales (Glomeromycota): *Ambispora fennica* gen. sp. nov., Ambisporaceae fam. nov., and emendation of *Archaeospora* and Archaeosporaceae, *Mycological Research* **111** 137-13

Walker C, Vestberg M and Schüßler A 2007b Nomenclatural clarifications in Glomeromycota *Mycological Research* **111** 253-255

Wilcox HE 1991 Mycorrhizae, in *Plant Roots: The Hidden Half* (eds) Y Waisel, A Eshel and KV Kafi, New York: Marcel Dekker Publication, 731-764 pages.

7

Citrus Decline in Darjeeling Hills and Potential Microorganisms for its Management

B.N. Chakraborty, U. Chakraborty, K. Rai, S. Allay and U. De

Immuno-Phytopathology Laboratory, Department of Botany, University of North Bengal, Siliguri 734013, West Bengal, India

E-mail:bncnbu@gmail.com

Citrus is the second (after grapes) most widely cultivated fruit crop of the world. It is an ancient commercial crop being cultivated in Darjeeling-Sikkim hills. Indian accession of citrus is numbered as 457 while eight south and south eastern Asian countries have 555 accessions of citrus. The world wide accession of citrus is estimated as 6000. These germplasms could be profitably used as horticultural tools and developing new varieties. *Citrus reticulata* (Blanco) (Rutaceae) is the principal fruit crop of Darjeeling hills known as 'Darjeeling mandarin'. It is considered as the most important cash crop of the marginal families of the hills (Thapa 2007). But recently its decline is posing a threat to the citrus farmers. The symptoms are defoliation of young shoots and twigs and ultimately drying up of the whole plant. This results in decline of general plant health and decreased fruit production attributed to a broad spectrum of pathogens. Severely affected orange orchards of Darjeeling hills are in Mirik valley, Tshering gaon and Simana busty. A multidisciplinary attempt has been made to assess the conditions of the orchards and analyse the relative abundance of the contributing factors for the citrus decline (Mukhopadhyay et al 1985). Lack of proper knowledge on agro-techniques of orchard management of citrus cultivation in terms of nursery management, proper seedling selection, soil health status, plant protection measures, amount of manuring and fertilizer required for the plant, irrigation in winter, mulching practice, pruning practice, intercropping (suitable and unsuitable crop) etc. after the establishment of orchard singly or in combination leads to citrus decline in Darjeeling hills. It was observed that chronic infection of foot and root rot alone or coupled with the infection of *Fusarium, Macrophomina* etc. were associated to large number of diebacked plants in combination with nutrient deficiencies. It is a well-known fact that heavy phosphorus application causes zinc deficiency in plant tissue (Marschner 1993, Cooper and Tinker 1978) and Lambert et al (1979) showed that high phosphorus depressed mycorrhizal inoculation, thus plants were not able to get more zinc.

Rhizosphere soil of plants contains a wide variety of microorganisms- bacteria, fungi and actinomycetes which are both beneficial and harmful. Among the beneficial microorganisms, two groups are very important agriculturally - the mycorrhizal fungi (AMF) and biocontrol fungi. Since microorganisms do not live in isolation in the soil, their effects on plant growth, nutrition and protection against diseases are due to their interactions. Mycorrhizal fungi colonize plant roots and extend the root system into the surrounding soil. The relationship is beneficial because the plant enjoys improved nutrient and water uptake, disease resistance and superior survival and growth. The dual activity of AMF and BCA is generating ample evidence to act as bioprotector as well as biofertilizer which play a significant role to sustain agriculture. Phosphorus (P) is a major essential macronutrient for biological growth and development. Phosphorus in soils is immobilized or becomes less soluble either by absorption, chemical precipitation, or both. A survey of Indian soils revealed that 98% of these need phosphorus fertilization either in the form of chemical or biological fertilizer. Although phosphorus content in an average soil is 0.05%, only 0.1% of the total phosphorus present is available to the plants because of its chemical fixation and low solubility. Application of chemical phosphatic fertilizers is practiced though a majority of the soil phosphorus reaction products are only sparingly soluble (Tilak et al 2005). Under such conditions, AMF solubilize the insoluble inorganic phosphorus of soil and make it available to the plants. Chakerborty and co workers (unpublished data) determined the population of different species of AM fungi isolated from the rhizosphere of mandarin from three different regions namely Mirik valley, Tshering gaon and Simana busty (Table 1).

Table 1. Population of AM fungi in the rhizosphere of Mandarin plants collected from Mirik, Darjeeling

Genus & species	**% of AM spores from different regions**		
	Mirik valley	**Tshering gaon**	**Simana busty**
Glomus mosseae	36	29	31
G. badium	2	-	31
G. fasciculatum	10	11	16
G. intraradices	-	3	-
G. versiforme	7	9	-
Gigaspora margarita	20	21	23
G. rosea	5	-	8
G. gigantean	6	13	15
Acaulospora spinosa	5	5	3
A. bireticulata	6	9	-
Scutellospora sp.	3	-	-

Among the mycorrhizal fungi isolated from the rhizosphere of Mandarin, *Glomus mosseae* could be determined as the predominant one, followed by other genera such as *Gigaspora*, *Acaulospora* and *Scutellospora*. Hyphal and spore characters of each AMF has been illustrated in Table 2 and Fig.1 (A-O).

Table 2. Description of AM Fungi found in rhizosphere of mandarin

AMF spore	Colour	Size (μm)	Shape	No. of wall layers	Other descriptions
Glomus mosseae	Brown to orange-brown	200	Globose to sub-globose	3	Hyphae double layered
G. badium	Brownish orange to reddish brown	200-500 x 290-260	Mainly ovoid to irregular sometimes globose to sub-globose	3	Spores formed blastically at the tip of hyphae developing from a hyphal plexus positioned in the centre of sporocarp
G. fasciculatum	Pale yellow to bright brown	70-120	Globose to sub-globose	3	Spores produced directly with one or more subtending hyphae
G. intraradices	White pale cream to yellow brown	40-140	Globose to sub-globose	3	Outer two layers only present in young spores, both thin and degrade with spore maturation.
G. versiforme	Orange to re-brown in sporocarp, white to white cream-coloured to pale orange yellow in pot cultures	60-160	Globose to subglobose sometimes ovoid	2	Recurved septum positioned at the innermost sublayer of the laminate layer, giving the appearance of being inserted.
Gigaspora gigantea	Greenish yellow	260-318	Globose to subglobose sometimes ovoid	2	Spores formed from bulbous sporogenous cell.
G. margarita	Yellowish white to sunflower yellow	320-370	Globose to subglobose sometimes ovoid	2	Spores produced singly at the tip of bulbous sporogenous cell.
G. rosea	Pale cream with pale pink tint in new healthy spores	160-280	Globose to subglobose	3	Auxiliary cells in aggregates of 4-20 borne on tightly coiled hyaline hyphae.
Acaulospora spinosa	Cream to pale orange brown	140-220	Globose to subglobose	2	Spores formed from sporiferous saccule.
Acaulospora bireticulata	Brownish	280-410	Globose	3	Spores borne laterally from neck of sporiferous saccule
Scutellospora *sp.*	Dark orange brown to red brown	340-640	Globose, subglobose often elliptical	2	Spores develop blastically from a hyphal tip which swells to form the sporogenous cell

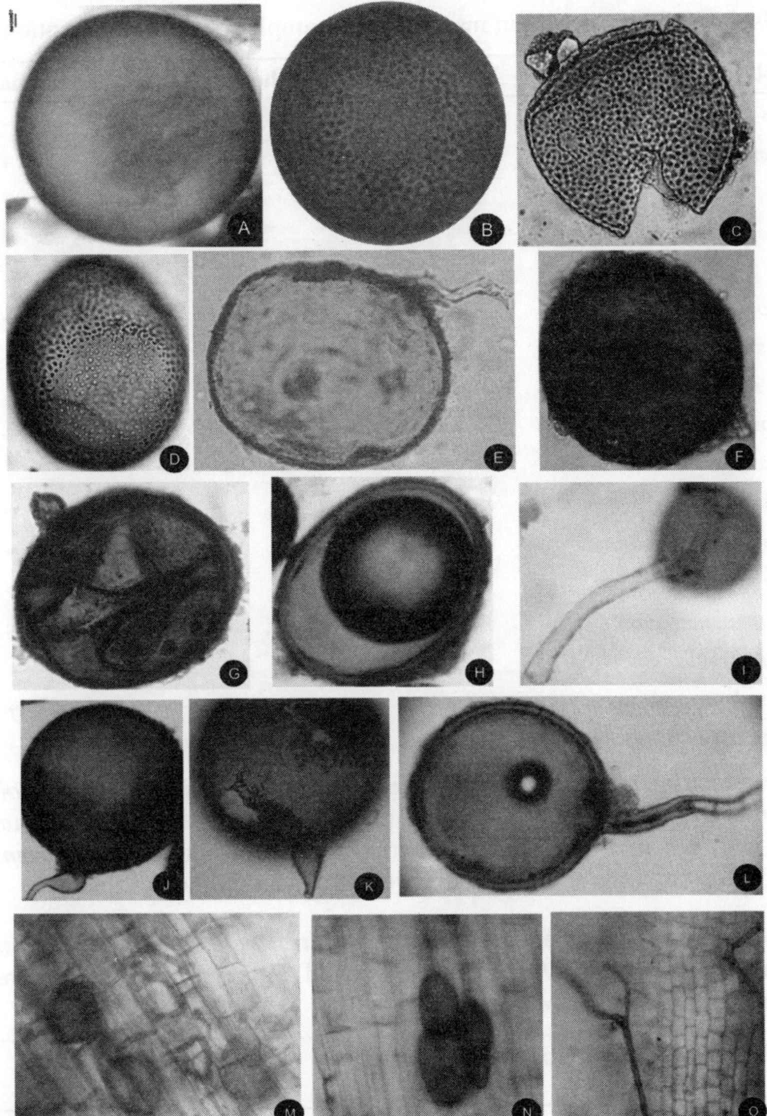

Fig. 1. AMF spores and root colonization in *Citrus reticulata* (A. Glomus, B & C. *Acaulospora bireticulara,* D. *Acaulospora spinosa,* E. *Glomus mosseae,* F. *Glomus constrictum,* G. *Gigaspora gigantea,* H. *Scutellospora rubra;* I. *Glomus badium,* J & K. *Gigaspora margarita,* L. *Glomus fasciculatum,* M, N and O. Colonization of *Citrus reticulata* roots with AM fungi

Root colonization was 70% in Tshering gaon, 72% in Simana busty and 72% in Mirik Valley. Percentage of AM spores determined from different regions showed maximum of different *Glomus* sp. The top six inches of fertile soil contains over two tons of fungi and bacteria per acre. Isolation of soil fungus from rhizosphere of mandarin revealed many plant growth promoting fungi like *Trichoderma* and some pathogenic fungi (Table 3).

Table 3. Isolated fungi present in mandarin soil samples of different locations

Fungal isolates	Mirik Valley	Tshering Gaon	Simana Busty
Aspergillus clavatus	+	+	-
A. fumigates	+	-	+
A. mellus	+	+	+
A. niger	+	+	+
A. oryzae	-	+	-
A. versicolor	-	+	+
Alternaria sp.	-	+	+
Emericella nidulans	+	-	-
Fusarium oxysporum	+	+	+
Fusarium solani	+	+	+
Gonronella butlerii	+	-	+
Macrophomina phaseolina	+	+	-
Penicillium sp.	-	+	+
Sporotrichum pruinosum	+	-	-
Trichoderma harzianum	+	+	+
Trichoderma hamatum	+	+	+
Trichoderma asperellum	+	+	+

Pathogenicity test showed that isolates of *M. phaseolina, Fusarium solani* and *F. oxysporum* greatly affected 2 year old plants. *Macrophomina phaseolina* caused maximum root rot in Simana busty whereas *F. solani* and *F. oxysporum* caused maximum wilt symptoms and loss in dry weight in Mirik valley (Table 4).

Table 4. Screening for resistance of Mandarin (*Citrus reticulata*) against *Macrophomina phaseolina, Fusarium solani* and *F. oxysporum* on different root samples of *Citrus reticulata*

	*Root rot index	**Loss in dry mass of root, %		***Colour intensity		
	M. phaseolina	*F. solani*	*F. oxysporum*	*M. phaseolina*	*F. solani*	*F. oxysporum*
Mirik valley	56.8	82	42.5	++++	++++	++
Tshering gaon	39.7	56.8	39.7	++++	+++	++
Simana busty	87.5	78.6	28.5	+++	++++	+

* On the basis of root area affected; 0-10% (0.10); 11-25% (0.25); 26-50% (0.50); 51-75% (0.75); 76-100% (1.0).

** In relation to control on the basis of 3 separate trials of 50 plants each.

*** + Light brown, ++ Deep brown, +++ Blackish brown, ++++ Black

Biochemical and molecular biology studies carried out to explore the mechanisms involved in biological control revealed that *Trichoderma* is a rather specific mycoparasite. Lectins were found to be involved in the recognition between *Trichoderma* and its host fungi, whereas chitinase is involved in the degradation of the host cell wall (Dev et al 2010). The fungal hyphae in the soil absorb nutrients by active metabolic means and translocate them throughout their mycelium. This process is especially important for the absorption of nutrients such as phosphate, ammonium, potassium, etc. which are relatively immobile and have low movement coefficients in the soil (Sanders and Tinker 1973).

There are several studies which have focused on mycoparasitic nature of *Trichoderma* species and its contribution to plant health (Chet 1987, Ousley et al 1994, Harman 2000, Egberongbe et al 2010). Defense responses of mandarin plants were demonstrated during early stages of root colonization by *T. harzianum* and *G. mosseae* (Chakerborty and co-workers, unpublished data) observed marked reduction in disease development following dual inoculations of *G. mosseae* and *T. harzianum*. These observed root rot reduction following separate and dual application of *G. mosseae* and *T. harzianum* may be correlated with increase accumulation of defense enzymes. Sundaresan et al (1993) reported that in cowpea plants which had mycorrhizal association, accumulation of phytoalexins was much higher. Increased activity of chitinase, β-1,3-glucanase and peroxidase were also determined in tea plants following treatments with Josh- a bioformulations of AMF (Chakraborty et al 2007). Allay and Chakraborty (2010) also demonstrated the existence of isoforms of peroxidase (PR9) constitutively present in healthy roots and leaves of mandarin plants. New isoforms were detected in mandarin roots inoculated with *G. mosseae*. However, new isoforms were also noticed in mandarin leaves following inoculation of roots with *G. mosseae* and *T. hamatum,* singly or jointly. Interestingly dual application of *G. mosseae* and *T. hamatum* induced additional isozyme. The induction of systemic resistance was confirmed in their earlier study as evident in immunological assays. Chakarborty and coworkers (unpublished data) observed a significant inhibition of growth of mycelia of these root pathogens (*F. oxysporum*, *F. solani* and *M. phaseolina*) in dual cultures with *T. harzianum, T. hamatum* and *T. asperellum*. Maximum inhibition occurred in *F. solani* (Fig. 2) followed by *F. oxysporum* and minimum in *M. phaseolina*.

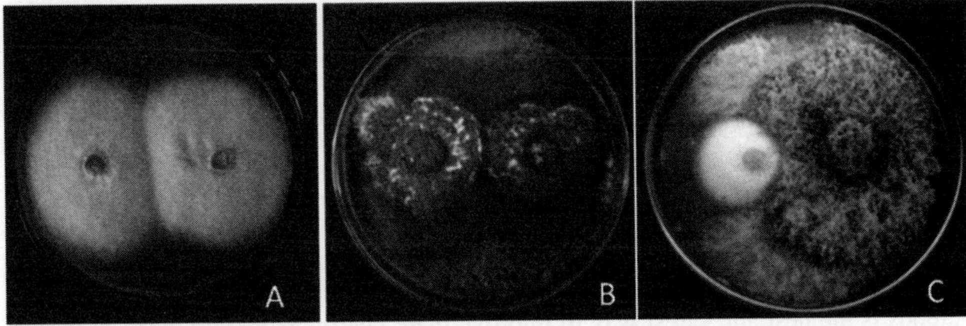

Fig. 2. In vitro interaction of *Trichoderma harzianum* with *Fusarium solani* [A- *F. solani*; B- *T. harzianum*; C- *F. solani* + *T. harzianum*]

Antibiosis to *F. solani* by biocontrol agent (*T. harzianum*) was evaluated *in vivo* condition. The application of *T. harzianum* to the soil as a biocontrol agent, in glasshouse conditions, not only resulted in reduced disease severity but also enhanced plant growth. Application of *G. mosseae* in the rhizosphere of *Citrus* plants led to an increase in the growth of seedlings in terms of increase in height and number of leaves. Joint inoculation with both the microorganisms (*G. mosseae* and *T. harzianum*) gave most significant results (Fig. 3).

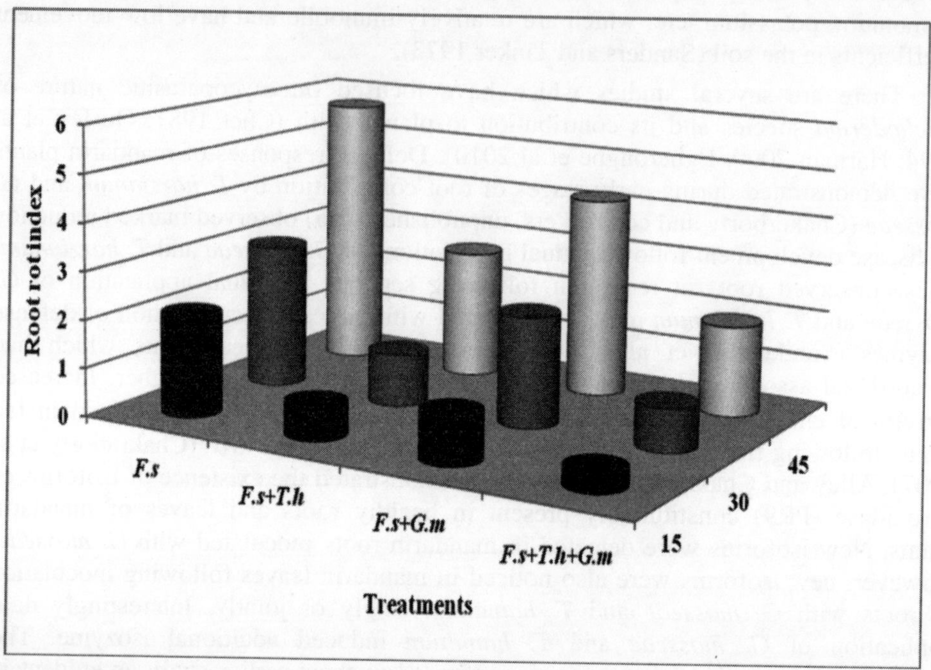

Fig. 3. Effect of application of *T. harzianum* and *G. mosseae* on the development of root rot disease caused by *Fusarium solani*

Controlling plant diseases by biological methods is the best and safest way to improve plant health. Implementation of mycorrhizal colonization and induction establishment should be brought about quickly in order to check wide spread use of fungicides.

Future Thrust

Mandrin (*Citrus reticulate*) is a principal fruit crop of Darjeeling hills. For the conservation of *Citrus reticulata* (Darjeeling mandarin), a multidisciplinary approach including motivation to farmers for cultivation of this crop giving priority to plant protection, crop protection augmentation and market chain improvement is required to be coordinated not only for table and processed fruit but also for the production of secondary products. Controlling plant diseases particularly soil borne by biological methods is the best and safest way to improve plant health. More and more emphasis should be given to isolate native micro-organisms with multiple activities which not only control the disease(s) but will also be safer to the environment.

References

Allay S and Chakraborty BN 2010 Activation of defense response of mandarin plants against *Fusarium* root rot disease using *Glomus mosseae* and *Trichoderma hamatum*, *Journal of Mycology and Plant Pathology* **40 (4)** 499-511

Chakraborty BN, Chakraborty U, Barman BC, Bhutia LP, and Ghosh PK 2007 Induction of systemic resistance in tea plants against root rot pathogen upon field application of VAM, phosphate solubilizing fungus and bacterium, in *Rhizosphere Biotechology: Plant Growth Retrospect and Prospect* (eds) AK Roy, BN Chakraborty, DS Mukadam and Rashmi, Jodhpur: Scientific Publishers, 61-71 pages

Chakraborty U and Chakraborty BN 1989 Interaction of *Rhizobium leguminosarum* and *Fusarium solani* f. sp. *pisi* on pea affecting disease development and phytoalexin production, *Canadian Journal of Botany* **67** 1698-1701

Chet I 1987 *Trichoderma*- application, mode of action and potential as a biocontrol agent of soilborne plant pathogenic fungi, in "Innovative approaches to plant disease control" New York: Edition John Wiley and Sons, 137-160 pages

Cooper KM and Tinker PB 1978 Translocation and Transfer of Nutrients in Vesicular-Arbuscular Mycorrhizas- II: Uptake and Translocation of Phosphorus, Zinc and Sulfur, *New Phytol*ogist **81** 43-52

Daniels BA and Skipper HD 1982 Methods for the recovery and quantitative estimation of propagules from soil, in *Methods and Principles of mycorrhizal Research* (ed) N Schenck, American Phytopathological Society, St. Paul, Minnesota, 29-35pages

Dev Neha and Dawande AY 2010 Biocontrol of soil borne plant pathogen *Rhiozoctonia solani* using *Trichoderma* spp. and *Pseudomonas fluorescens*, *Asian Journal of Biotechnology Resources* **1** 39-44

Egberongbe HO, Akintokun AK, Babalola OO, and Bankole MO 2010 The effect of *Glomus mosseae* and *Trichoderma harzianum* on proximate analysis of soybean (*Glycine max* (L.) Merrill.) seed grown in sterilized and unsterilized soil, *Journal of Agricultural Extension and Rural Development* **2** 54-58

Gerdemann JW and Nicolson TH 1963 Spores of mycorrhizal *Endogone* species extracted from soil by wet sieving and decanting, *Transactions of British Mycological Society* **46** 235-244

Giovannetti M and Mosse B 1980 An evaluation of techniques for measuring vesicular arbuscular mycorrhizal infection in roots, *New Phytologist* **84** 489-500

Harman GE 2000 Myth and dogmas of biocontrol changes in perceptions derived from research on *Trichoderma harzianum* T-22, *Journal of Plant Diseases* **84** 377-393

Lambert DH, Baker DE, and Cole H 1979 The Role of mycorrhizea in the interaction of phosphorus with zinc, copper and other elements, *Soil Science Society of America Journal* **38** 477-479

Marschner H 1993 Zinc uptake from soils, in *Zinc in Soil and Plants* (ed) AD Robson, Dordrecht, Netherlands: Kluwer Academic Publishers, 59-77 pages

Mukhopadhyay S, Mukhopadhyay AK, Mukherjee N, Ghosh MR, Bose TK, Chowdhury AK, Chakrabarti NK, Dutta Ray S, Konar A, and Gurung A 1985 The state of mandarin orange cultivation in Darjeeling district, *Review Tropical Plant Pathology* **2** 95-105

Ousley MA, Lynch M, and Whipps JM 1994 Potential of Trichoderma as consistent plant growth stimulation, *Biology and Fertilizer of Soils* **17** 85-90

Sanders FE and Tinker PB 1973 Phosphate flow into mycorrhizal roots, *Pesticide Science* **4** 385-395

Schenck NC and Perez Y 1990 *Manual for the Identifiation of VA Mycorrhizal Fungi*, 2nd edition, INVAM- University of Florida, Gainesville, USA

Sundaresan P, Raja UN, and Gunasekaran P 1993 Induction and accumulation of phytoalexins in cowpea roots infected with a mycorrhizal fungus *Glomus fasciculatum* and their resistance to *Fusarium* wilt disease, *Journal of Biosciences* **18** 291-301

Thapa KK 2007 Citrus: An indigenous crop of Darjeeling hills, *Asian Agri-History* **11**(2) 133-137

Tilak KVBR, Ranganayaki N, Pal KK, De R, Saxena AK, Shekhar NC, Mittal S, Tripathi AK, and Johri BN 2005 Diversity of plant growth and soil health supporting bacteria, *Current Science* **89** (1) 136-150

Walker C 1992 Systematics and taxonomy of the arbuscular endomycorrhizal fungi (Glomales)-a possible way forward, *Agronomie* **12** 877-897.

8

Plant Disease Resistance: Concepts

R.K. Dogra, Naval Kishore* and K. Kumar

Department of Fruit Science, Dr. Y.S.P.U.H.F., Nauni, Solan (H.P.) 173 230
** CSKHPKV, HAREC, Bajaura, Distt. Kullu (H.P.) 175 125*

Plant diseases are one of the major limiting factors in crop production. Plants are attacked by many disease-causing organisms including bacteria, fungi, viruses and nematodes. These pathogens cause large crop losses and probably since the beginning of agriculture have contributed to human hunger and malnutrition. In 1858, Julius Kuhn, in Germany had defined plant disease as abnormal changes in the physiological processes which disturb the normal activity of plant organs. Agrios (1997) defined disease as an abnormal condition in the plant produced by an organism or an environmental factor. More precisely, disease may be defined as the series of invisible and visible responses of plant cells and tissues to a pathogenic microorganism or an environmental factor that result in adverse changes in form, function or integrity of plants and may lead to partial impairment or death of plant or its parts. The plant affected by a disease is known as host, and the organism that produces disease is known as pathogen. Nature has gifted the crop plants with an inbuilt mechanism for defending themselves from the attack of pathogens. Plants in nature are generally resistant to most pathogens. This inherent inbuilt mechanism restricts the invasion and proliferation of pathogens and is termed as resistance. The ability of a pathogen to cause disease in a host plant is usually the exception not the rule. This is because plants have a natural ability to recognize potential invading pathogens and to mount successful defenses. Successful pathogens cause disease because they are able to evade recognition or suppress host defense mechanism or both. We lose 12% of total crop yields to pathogen infection– equivalent to nine hundred million tons worldwide annually (Krimsky and Wrubel, 1996). Since the beginning of 20th century, classical breeding for disease resistance in plants has been a major method for controlling plant diseases. The control of plant diseases is thus of fundamental importance and is a major objective of plant-breeding and pathology programs and the agricultural chemical industry.

Disease Development

Disease outcome is determined by the three-way interaction of the pathogen, the plant, and the environmental conditions, an interaction known as the disease triangle. The "Disease triangle" is a central concept of Plant Pathology. The availability of susceptible host is a prerequisite for an invasion by a virulent race/pathotype of the given pathogen. Presence of a virulent pathogen is essential for invading a resistant strain of the concerned host. The interaction between a pathogen and its host is controlled by their genetic make-up and is in accordance to gene-for-gene relationship. The functions of host and pathogen are dependent on the prevailing biotic and abiotic factors. It is the tendency of the plant not to be infected by particular pathogen. Plant disease resistance is crucial to the reliable production of food. However, disease control measures are also successful for most crops. Plant disease resistance derives both from pre-formed defences and from infection-induced responses mediated by the plant immune system. Relative to a disease-susceptible plant, disease resistance is often defined as reduction of pathogen growth on or in the plant, while the term disease tolerance describes plants that exhibit less disease damage despite similar levels of pathogen growth. Defence-activating compounds can move cell-to-cell and systemically through the plant vascular system, but plants do not have circulating immune cells so most cell types in plants retain the capacity to express a broad suite of antimicrobial defences. Although obvious qualitative differences in disease resistance can be observed when some plants are compared (allowing classification as "resistant" or "susceptible" after infection by the same pathogen strain at similar pathogen inoculum levels in similar environments), a gradation of quantitative differences in disease resistance is more typically observed between plant lines or genotypes. Plants are almost always resistant to certain pathogens but susceptible to other pathogens; resistance is usually pathogen species-specific or pathogen strain-specific.

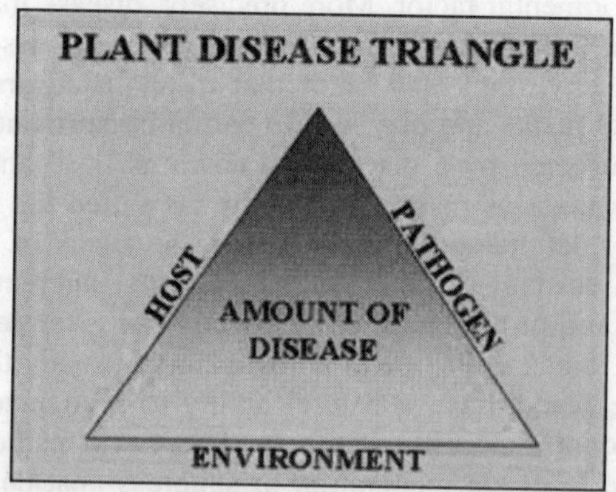

The disease triangle

Types of Resistance

The terms vertical and horizontal resistance were first used by Vanderplank (1963). Vertical resistance is generally determined by major genes. It is also known as race-specific, pathotype specific or specific resistance. Horizontal resistance is known as race-non specific, pathotype non-specific and is partial or general resistance. It is controlled by polygenes and is pathotype non-specific. It does not prevent the development of disease but it slows down the rate of spread of disease in the population. In horizontal resistance, pathogenic ability is entirely independent of each other in genetic terms (Robinson, 1976). Vanderplank published a classic book on Plant Diseases: Epidemics and Control in 1963. He used the term *vertical resistance* to describe the single-gene resistance, and the term *horizontal resistance* to describe the many-gene resistance. Vertical resistance is normally qualitative resistance in the sense that it is either present or absent, and there are no intermediates. However, there are a few exceptions to this rule. Horizontal resistance is the resistance of the biometricians. It is quantitative resistance in the sense that it can occur at every level between a minimum and a maximum.

With vertical resistance, there are single genes for resistance in the host plant, and there are also single genes for parasitic ability in the parasite. This is a very important phenomenon known as the gene-for-gene relationship, and it is the definitive character of vertical resistance. The gene-for-gene relationship was discovered by H.H. Flor (1942), while working on flax rust. Flor showed that, for every resistance gene in the host, there was a corresponding or matching gene in the parasite. It is now realised that the gene-for-gene relationship evolved in plants to operate as a system of locking. Each resistance gene in the host corresponds to a tumbler in a lock. And each parasitism gene in the parasite corresponds to a notch in a key. An individual plant host may have several of these resistance genes, these tumblers, which collectively constitute a biochemical lock. And an individual parasite may have several of these parasitism genes, these notches, which collectively constitute a biochemical key. When a parasite individual is infecting a host individual, its biochemical key either does, or does not, fit the biochemical lock. If the key fits, the infection is described as a matching infection, and it is a successful infection, because the "door" of resistance has been unlocked and "opened". When this happens, the vertical resistance is described as having broken down. If the parasite key does not fit the host lock, the infection is described as a non-matching infection. It fails because the "door" of resistance remains "locked and barred", and the parasite is denied entry. This system of locking is the definitive characteristic of the gene-for-gene relationship.

Perhaps the best way of understanding horizontal resistance is to think of it as the resistance which invariably remains after a vertical resistance has been matched. When a parasite succeeds in unlocking a vertical resistance, it then comes up against a second line of defence which is the horizontal resistance. What is so economically important about horizontal resistance is that it operates equally against all strains of the parasite, regardless of what biological keys they may have. Horizontal resistance begins to function at the moment a matching infection occurs, and at the moment the vertical resistance breaks down. This means that horizontal resistance cannot be matched, in the way that vertical resistance is matched, and it cannot break down, in the way that vertical resistance breaks down. This is the main practical difference between the two kinds of resistance. Vertical resistance operates only against *non-matching* strains of the parasite. Because some matching always occurs, vertical resistance is certain to break down sooner or later. It is temporary resistance. Finally, we come to an odd biological fact. Not all species of plant have vertical resistance (s). Furthermore, species of host plant which do have vertical resistances have them against only some of their species of parasite. This has been the bane of Mendelian plant breeding. Before their resistance breeding can start, Mendelian breeders must find a gene for resistance. If this genetic source of resistance cannot be found, for the simple reason that it does not exist, Mendelian plant breeders cannot breed for resistance. The breeding cannot even begin. Conversely, every plant has horizontal resistance to every one of its parasites. This is one of the main advantages of this kind of resistance. The biometricians can breed for resistance to any species of plant parasite.

Genetics of Host Pathogen Interaction

Interaction between the disease inciting organism and the host is conventionally termed as "host parasite interaction" but more appropriate term is "host pathogen interaction". Among thousands plant pathogenic microorganisms, only some can infect a broad range of plant species. The term host resistance is used when a pathogen species can be pathogenic on the host species but certain strains of that plant species resist certain strains of the pathogen species. There can be overlap in the causes of host resistance and non-host resistance. Pathogen host range can change quite suddenly if, for example, the capacity to synthesize a host-specific toxin or effector is gained by gene shuffling /mutation or by horizontal gene transfer from a related or relatively unrelated organism. The first successful attempt to understand the genetics of host pathogen interaction was made by H. H. Flor. He published his conclusion in 1955 in the form of the "gene for gene" hypothesis which stated that "for each gene conditioning the resistant reaction in the host, there is a specific gene conditioning rust pathogenecity in the pathogen." Flor worked with flax and flax rust caused by *Melampsora lini*, where the virulence of the pathogen is recessive to avirulence and resistance is dominant to susceptibility in host. In this situation, resistant reaction is observed only when the host with dominant gene is attacked by a pathogen with dominant gene. In other cases it is a susceptible reaction. For a plant to exhibit resistance to (or incompatibility with) a pathogen, a R gene must be present in the plant and a corresponding avirulence (avr) gene must be present in the pathogen. An absence of either leads to disease (or compatibility). R genes encode receptors that enable plants to detect the ingress of pathogens whose avirulence genes cause them to produce the corresponding ligands. Thus, R gene products might be expected to have two functions: first, molecular recognition; and second, activation of plant defence upon recognition.

Disease resistance genes enable plants to recognize specific races of pathogens and to mount an effective localized defence response (Table 1). Most encode a leucine-rich repeat motif, which might provide a basis for recognitional specificities and their evolution. The first R gene to be isolated was *Hm1* from maize that confers resistance to *Cochliobolus carbonum.* This gene encodes an enzyme that detoxifies *a C. carbonum* toxin. The tomato *Pto* gene, which confers resistance to races of *Pseudomonas syringae* that carry the *avrPto* gene, was the first race-specific R gene to be isolated. *Pto* gene is polygenic in nature. R-proteins are variably expressed within a species, thus, allowing its distinct members to differ in their resistance to a particular organism. R protein RPMl-interact with protein 4 (RIN4), which confers resistance against strains of *Pseudomonas syringae* that carry the avirulence factors AvrRpml or AvrB. The interaction of RIN4 with AvrRpml and AvrB lead to RIN4 phosphorylation, while the interaction of RIN4 with AvrRpt2 results in the degradation of RIN4, as well as the R proteins RPM1 and RPS2. (Kim *et al* 2009)

Table 1. Cloned plant disease resistance genes

Class	Gene	Plant	Pathogen	Infection Type/ Organ Attacked	Predicted Features of R protein*	Race Specific	Year Isolated	References
1	Hm1	Maize	*Helminthosporium maydis (race 1)*	Fungal necrotroph/ leaf	Detoxifying enzyme Hc toxin reductase	Yes	1992	van't Slot and Knogge (2002)
2	Asc-1	Tomato	*Alternaria alternata* f.sp. *lycopersici* (AAL toxin)	Fungal necrotroph/ leaf	TM helix-LAG1 motif	No	2000	Brandwagt *et al.* (2000)
3A	Pto	Tomato	*Pseudomonas syringae p.v. phaseolicola (avrPphB)*	Extracellular bacteria/ leaf	Different subfamily	Yes	2001	van't Slot and Knogge (2002)
3B	PSB1	Arabidopsis	*Pseudomonas syringae p.v. .phaseolicola (avrPphB)*	Extracellular bacteria/ leaf	Different subfamily	Yes	2001	Swiderski and Innes (2000)
4A	RPS2	Arabidopsis	*Pseudomonas syringae p.v. maculicola (avrRpt2)*	Extracellular bacteria/ leaf	CC-NB-LRR Intracellular protein	Yes	1994	van't Slot and Knogge (2002)
	Mla1/ Mla6	Barley	*Blumeria graminis* f.sp. hordei (resp. race 1, race 6)	Biotrophic intracellular fungus with haustoria/leaf		Yes	2001	Zhou *et al.* (2001)
	R1	Potato	*Phytophthora infestans (race1)*	Biotrophic intracellular Oomycete with haustoria/ leaf and tuber		Yes	2002	Ballvora *et al.* (2002)
	RPP8	Arabidopsis	*Peronospora parasitica*	Biotrophic intracellular Oomycete with haustoria/leaf		Yes	1998	van't Slot and Knogge (2002)
4B	N	Tobacco	Mosaicvirus	Intracellular virus/leaf and phloem	TIR-NB-LRR Intracellular protein	Yes	1994	van't Slot and Knogge (2002)
	RPP4	Arabidopsis	*Peronospora parasitica*	Biotrophic intracellular Oomycete with haustoria/leaf		Yes	2002	Kawchuk *et al.* (2001)

4C	Bs2	Pepper	*Xanthomonas campestris pv. vesicatoria (avrBs2)*	Extracellular bacteria/ leaf	NB-LRR Intra-cellular protein	Yes	1999	van't Slot and Knogge (2002)
	Dm3	Lettuce Bremia	lactuca	Biotrophic intracellular Oomycete with haust-oria/leaf		Yes	2002	Shen *et al.* (2002)
4D	RRS-1	Arabid-opsis	*Ralstonia solanacearum (race1)*	Extracellular bacteria/ leaf	TIR-NB-LRRNLS-WRKY	Yes	2002	Deslandes *et al.* (2002)
4E	Pi-ta	Rice	*Magnaporthe grisea (avrPita)*	Hemibiotrop hic intrace-llular fungus without haustoria/leaf	NB-LRD	Yes	2000	van't Slot and Knogge (2002)
5A	Cf-9	Tomato	*Cladosporium fulvum (Avr9)*	Biotrophic extracellular fungus without haustoria/ leaf	eLRR-TM-sCT Extracellul ar protein with single membrane-spanning region and short cyto-plasmic C terminus	Yes	1994	van't Slot and Knogge (2002)
5B	Ve1 Ve2	Tomato	*Verticillium albo-atrum*	Extracellular vascular wilt fungus without haus-toria/root and stem	CC-eLRR-TM-ECS eLRR-TM-PEST-ECS	Yes	2001	Kawchuk *et al.* (2001)
6	Xa-21	Rice	*Xanthomonas oryzae p.v. oryzae* (all races)	Extracellular bacteria/ leaf	eLRR-TM-kinase	Yes	1995	van't Slot & Knogge (2002)
	FLS2	Arabid-opsis	Multiple bacteria (flagellin)	Extracellular bacteria/ leaf		No	2000	van't Slot & Knogge (2002)
7	RPW 8.1 RPW 8.2	Arabid-opsis	Multiple powdery mildewspecies	Biotrophic intracellular fungus with haustoria/ leaf	Small, pro-bable me-mbrane protein with CC domain	No	2001	Xiao *et al.* (2001)
8	Rpg1	Barley	*Puccinia graminis f.sp.tritici*	Biotrophic intracellular funguswith haustoria/ stem	Receptor kinase-like protein with 2 tan-dem kinase domains	No	2002	Brueggeman *et al.* (2002)

***ECS**= endocytosis signal; **LAG1**= longevity assurance gene; **LR**= leucine-rich domain; **PEST**= Pro-Glu-Ser-Thr; **sCT**= single cytoplasmic tail; **TM**= transmembrane; Source: Khan (2009)

Molecular Basis of Host Pathogen Interaction

The estimated loss caused by pathogens is typically around 10 to 20 per cent (Boyer, 1982). The appropriate response of plant emerges from the perception of an extracellular signal and its transduction between and within plant cells. Specificity of the interactions between plants and pathogens is still an incomprehensible phenomenon with a complicated hierarchy of biological organization. Elucidation of this phenomenon represents an important task of contemporary plant pathology (Scheel, 1998; Nimchuk *et al* 2001). At the molecular level, defence systems often depend on the combination of a specific set of dominant R genes in the plant and a corresponding set of dominant avirulence (Avr) genes in the pathogen (Keen, 1990). This gene-for-gene resistance strategy underlines the molecular basis of defence systems in plants. According to the terminology involved in gene-for-gene resistance, when the plant is resistant, the pathogen is said to be avirulent and the interaction is said to be incompatible. When the plant is susceptible, the pathogen is said to be virulent and the interaction is said to be compatible. In a given plant-pathogen interaction, it is often possible that more than one specific combination of Avr and R genes are operating at the same time, and such different combinations are often co-dominant. These multiple combinations reflect the complexity of defence mechanisms of the plant during pathogen attack. Physiological features such as K^+/H^+ exchange, rapid oxidative burst, hypersensitivity at the site of infection, cross linking of plant cell walls, synthesis of antimicrobial compounds known as phytoalexins, and induction of pathogen-related proteins such as chitinases and glucanases (Lamb *et al* 1989) provide disease resistance.

Defence Mechanism

Understanding the mechanistic basis of why a pathogen causes disease in one host plant and not in other has long intrigued and motivated the researchers. Plants in nature are generally resistant to most pathogens. When a pathogen comes in contact with a host or enters into the host, it sometimes fails to develop successful infection due to predominant defence mechanism in the host. This ability is under genetic control. This inherent capacity or ability of host to defend against pathogen is called resistance and opposite of it is susceptibility. Upon infection by a particular avirulent pathogen, a corresponding R gene recognizes the *avr* product and triggers the defence mechanism.

A. Morphological or structural defence

Several morphological and structural barriers in host plants restrict the entry of pathogen:

(a) Pre existing structural defence

(i) Waxes and Cuticles

(ii) Structure of epidermal cell wall: Lignification or deposition of salicylic acid gives additional strength to cell wall.

(iii) Structure and number of natural openings: small size of stoma surrounded by raised broad lipped structures that prevent pathogens' entry.

(b) Post infectional structural defence

(i) Formation of cork layers: Below area of infection multilayered cork cells produced.

(ii) Formation of tyloses: Protrusions of cell walls. These structures obstruct the movement of fluid and abstruct the pathogen in xylem.

(iii) Formation of absission layer: It is represented by an empty space between the healthy and diseased tissue due to dissolution of middle lamella.

(iv) Swelling of cell walls: swell due to lignification.

B. Biochemical defence mechanisms

a) Pre existing biochemical defence mechanisms

(i) Release of inhibitor compounds by plant in its environment: e.g. onion is resistant to *Colletotrichum circinans* attack due to release of catechol and protocatechuic acid.

(ii) Presence of toxic chemicals in plant: e.g. Tannic acid present in tobacco plants inhibits TMV.

(iii) Absence of essential nutrients for growth of pathogens.

b) Post infectional

(i) Synthesis of phenols: Increase in level of phenol in resistant varieties eg. Phenols increase in resistant cabbage infected by *Botrytis cinerea.*

(ii) Release phytoalexins: Inhibit development of pathogens. e.g. Rishitin in case of late blight of potato caused by *Phytophthora infestans.*

(iii) Defence through induced synthesis of proteins and enzymes: Resistant varieties synthesize proteins and enzymes e.g. ethylene.

(iv) Defence due to hypersensitive reaction.

C. Hypersensitive reaction

As the pathogen comes in contact with the host, the host cells at the site of invading pathogen are killed. The mechanisms of this process are still mysterious, though a caspase-3–like activity has been reported to function in plant hypersensitive cell death. Nuclei of adjoining cells move towards pathogen and disorganize to form brown granules. Finally adjoining cells are killed forming a trap of dead cells so that pathogen fails to invade e.g. hypersensitive reaction observed in wart disease of potato, late blight of potato, rusts, powdery mildews and viral diseases. Interaction between the Avr gene-product and the R-gene product triggers a series of signalling responses. These result in the biosynthesis of salicylic acid which acts as a central signalling intermediate in plant defence. Hatsugai *et al.* (2009) examined the hypersensitive response in *Arabidopsis thaliana* to bacterium *Pseudomonas syringae* infection. The plant cell large central vacuole fuses to the plasma membrane, releasing the vacuole contents to the outside of the cells. This extracellular fluid can inhibit bacterial growth and also promote cell death. This is due to the presence of hydrolytic enzymes that degrade the plant cell wall. The authors then examined the requirements for this membrane fusion process, finding that inhibitors of caspase-3 activity or proteasome inhibitors could abolish this effect. In fact, they determined that the plant's caspase-3 activity belongs to PBA1, the β^1 catalytic subunit of the plant proteasome. This work thus identifies the plant caspase-3 activity and determines its role in activating membrane fusion to discharge vacuolar content.

Systemic acquired resistance

It is a secondary response which is systemic in nature and provides broad-range resistance that leads to Pathogenesis-Related (PR) gene expression.

D. Harmonal pathways in plant defence

Phytohormones such as ethylene and jasmonate have been found to synergistically induce the expression of plant-defence genes in response to different pathogen attacks (Penninckx *et al* 1998). Exogenous application of methyl jasmonate to Arabidopsis plants reduces disease development after infection by several fungi, such as *Alternaria brassicicola*, *Botrytis cinerea*, and *Plectospherella cucumerina* (Thomma *et al* 2000). Lorenzo *et al.* (2003) have demonstrated that ERF1 (Ethylene response factor 1), a downstream component of the ethylene-signaling pathway (Solano *et al* 1998), is an early ethylene- and jasmonate-responsive gene. They suggest that ERF1 may be a common component of both ethylene and jasmonate signaling pathways. In order to investigate the involvement of ERF1 in the regulation of the expression of pathogenesis-related genes, Lorenzo *et al.* (2003) performed a transcriptome analysis of Col-0 and 35S-ERF1 transgenic plants using Arabidopsis Gene Chip from "Affymetrix", which allows the simultaneous monitoring of 8000 genes. They reported about 164 genes, which were constitutively expressed, 77 genes inducible and 70 genes repressed after 6 h of treatment with jasmonate and ethylene (Lorenzo *et al* 2003). Most research has focused on understanding how relevant genes are upregulated during pathogen-defence mechanism. However, plant responses to pathogen attack also involve the down regulation of several genes. In potato, the mRNA and protein levels of Rubisco are drastically reduced by pathogen infection or elicitor treatment. In parsley, the expression of several genes involved in cell proliferation and cell-cycle regulation as well as flavanoid biosynthesis, are repressed to a large extent during defence responses (Somssich and Hahlbrock, 1998). Reports of Lorenzo *et al.* (2003) confirmed the combination of positive and negative regulatory mechanisms of pathogen-acquired resistance in plants. The high number of defence-related genes whose expression was upregulated by ERF1 is fully consistent with their previous results (Berrocal-Lobo *et al* 2002) demonstrating the over expression of ERF1 in transgenic plants that confers resistance to several pathogens. In summary, different types of phyto-hormonal signalling pathways are involved in plant defence responses during pathogen attack. These results indicate that hormonal transcription factors are key elements in the integration of signal transduction for the regulation of defence-related genes.

E. Breeding for Disease Resistance

Plant breeders focus a significant part of their effort on selection and development of disease-resistant plant lines. Plant diseases can also be partially controlled by use of pesticides, and by cultivation practices such as crop rotation, tillage, planting density, use of disease-free seeds and cleaning of equipment, but plant varieties with inherent (genetically determined) disease resistance are generally the first choice for disease control. Breeding for disease resistance has been underway since plants were first domesticated, but it requires continuous efforts. This is because pathogen populations are often under natural selection for increased virulence, new pathogens can be introduced to an area, cultivation methods can favour increased disease incidence over time, changes in cultivation practice can favour new diseases, and plant breeding for other traits can disrupt the disease resistance that was present in older plant varieties. A plant line with acceptable disease resistance against one pathogen may still lack resistance against other pathogens.

There are numerous instances across the world and human hardship caused by sudden outbreak of epidemics. The more devastating epidemics are sudden. To be unprepared is to maximize the epidemic (NAS, 1972). The outbreak of epidemics is occasional; the major epidemics come as a surprise to both growers and breeders. A few epidemics resulted in human misery are potato famine of 1840's in Ireland, coffee rust of 1870's in Ceylon, sugarcane red rot of 1838-39, Bengal rice famine of 1942, downy mildew epidemic of pearl millet of 1970-71 in India; and southern corn blight of 1970 in the United States. The defence mechanisms are principally based on avoidance, resistance or tolerance. Avoidance operates before parasitic contact between host and parasite is established and decreases the frequency of incidence. After parasitic contact has been established the host may resist the parasite by decreasing its growth, or tolerate its presence by suffering relatively little damage. Avoidance is mainly active against animal parasites and includes such diverse mechanisms as volatile repellents, mimicry and morphological features like hairs, thorns and resin ducts. Resistance is usually of a chemical nature. Little is known of tolerance; it is very difficult to measure and is usually confounded with quantitative forms of resistance. Parasites classified as fungi, bacteria, viruses or viroids are considered disease inciting parasites or pathogens. Resistance mechanisms are by far the most important defence mechanisms employed by host plants, including our crops, against pathogens. Avoidance and tolerance play a minor role here. In the never-ending arms race between plant and pathogen, the latter have developed widely different host ranges. Pathogens such as some *Pythium* species, *Rhizoctonia solani*, and *Sclerotinia sclerotiorum* have a wide host range; they are non-specialized, polyphagous pathogens or generalists. The latter one, for instance, has been reported to attack hundreds of plant species belonging to at least 64 families of flowering plants and gymnosperms (Parlevliet, 1989). A large proportion of the pathogens, however, have a narrow host range restricted to a few closely related plant species; they are specialized, monophagous pathogens or specialists. *Puccinia hordei* Otth. and *Phytophthora phaseoli*, pathogenic on barley (*Hordeum vulgare* L.) and lima beans (*Phaseolus lunatus* L.), respectively are typical specialists.

Breeding Methods for Disease Resistance

Plant introduction means taking a genotype or a group of genotypes of plants into a new environment where they were not being grown before. It may involve new varieties of a crop already grown in an area, wild relatives of the crop species or a totally new crop species for the area. Generally materials are introduced from other countries or continents, but it may be from one environment to another within the same country. This is an easy and rapid method of developing disease resistant variety. The resistant variety may be introduced and after testing, if found suitable can be released in the disease prone area.

(i) Selection

In crop improvement, selection is very important because it allows the selected genotypes to reproduce, while the undesirable genotypes are eliminated. Thus the breeder is able to improve the various characteristics by selecting the desirable types. In the asexually propagated crops, clonal selection is the ideal method for incorporating disease resistance.

(ii) Hybridization

Hybridization is used when resistant genes are either available in the germplasm or in wild species of crop plants. After hybridization the breeding material is handled by different methods depending upon the mode of pollination of the crop. For isolating disease resistant plants in self pollinating crops mass selection and pureline, pedigree selection methods are practiced. Mass selection of seed from the resistant plants surviving in natural conditions is a simple but time consuming method. In pureline or pedigree selection, individual highly resistant plants and their progenies are propagated separately and are tested for resistance under artificial conditions. The pedigree method is used when the resistance is governed by polygenes and the resistant variety is an adopted one which also contributes some desirable genes. In cross pollinated crops, there is no control on the pollen source, so recurrent selection or back cross methods are used. A desirable but susceptible variety of a crop is crossed with a resistance source (cultivated or wild relative).Progeny is evaluated for resistance, and the resistance individuals are backcrossed to the desirable variety and is repeated several times until the resistance is stabilized in the genetic background of the desirable variety . The backcross method is used when the resistance is governed by oligogenes and the resistant parent is unadapted type or the resistance gene is to be transferred from wild species.

By making the use of F_1 hybrids of two different but homozygous lines carrying different genes for resistance enables to make use the phenomenon of heterosis or hybrid vigor. Mutations both spontaneous and induced are also good sources of disease resistance. Each of the above steps can be difficult to successfully accomplish, and many highly refined methods are used to increase the effectiveness and reduce the cost of resistance breeding.

(iii) Exploitation of vertical resistance

Vertical resistance refers to resistance of a host to specific race of a pathogen. The breeder should make best use of available genes for specific resistance to prevalent races. Vertical resistance can be exploited by developing cultivars with individual major genes, developing multilines, gene pyramiding and gene deployment.

(iv) Varieties with individual major gene

In this method for the prevalent race of pathogen, an individual major gene is transferred into well adapted variety through backcrossing. Resistant plants are selected from segregating populations and backcrossed to the recurrent parent. As a result, the genotype of the backcross progeny becomes similar to that of the parent to which the backcrosses are made. After 5-6 backcrosses, the progeny would be almost identical with the parents used for backcrossing with added resistance to specific race of pathogen. Although it helps to control the prevalent race of pathogen, but sometimes major genes may become susceptible to minor prevalent races.

(v) Development of multilines

Multiline varieties are mixtures of several purelines of similar height, flowering and maturity dates, seed colour and agronomic characteristics, each of which has a different genes for resistance to the given disease. The seeds of genotypes with individual major genes are mixed together. This approach is used in self-pollinated species. The resistant genes to different races of pathogen are transferred into different genotypes by separate backcross programme. This is done by transferring disease resistance genes from several donor parents into a single recurrent parent. Each donor parent is used in a separate backcross programme and should have a different gene for resistance to the given disease, so that each pureline obtained from backcross programme has a different resistance gene. During the given period of time, the prevalence of pathogen races in the area would determine the lines to be mixed to produce a multiline. If any component line becomes susceptible to the pathogen is immediately replaced by a new resistant line. The breeder should keep several resistant lines in store for future use. There are two main disadvantages of this method. First, several backcross programmes have to be launched to transfer each major resistant gene in a separate background. Second, the recurrent parent limits the agronomic performance of the multiline. Also, it requires extensive testing against specific races.

(vi) Gene pyramiding

Gene pyramiding is defined as method aimed at assembling multiple desirable genes into a single genotype. It differs from multilines, where each major gene is incorporated into a separate genotype and the resistant genotypes are mixed together. In pyramiding approach, all major genes are incorporated in a single cultivar which provides resistance against several new races that may develop in the pathogen. It increases the durability of disease resistance and also broadens the genetic base of released cultivars. This is associated with the fact that presence of target trait genes must be confirmed by phenotyping mostly at the individual level and that individual phenotypic performance is a good indicator of the genotype only if genes have a major effect on phenotypic performance and the error of phenotyping is minimal. Besides this, other factors influencing the success of gene pyramiding are the inheritance model of the genes for the target traits, linkage and/or pleiotropism between the target trait and other traits. Allelic genes can not be combined in the same genotype. The effect conferred by a recessive gene can not be evaluated on heterozygous individuals and progeny testing is required. If the target gene is tightly linked to genes with large negative effects, these undesirable genes may be transferred together with the target gene into the recipient line and result in reduced performance of other trait (linkage drag). Therefore, any improvement in the knowledge of the trait genetics and technique for inferring genotype-phenotype relationship will be useful. With the help of modern plant breeding techniques such as marker assisted selection and genetic engineering, the efficiency of gene pyramiding can be increased.

(vii) Gene deployment

It refers to planned or strategic use of major genes in development of resistant cultivars for various geographical areas. This approach helps in preventing disease and maintaining the genetic diversity .It requires a number of genes with similar effectiveness for the control of prevalent races of the pathogen. Main drawback of gene deployment approach is difficulty in finding different genes with similar effectiveness.

(viii) Tissue culture and molecular techniques

Advances in plant tissue culture include meristem tip propagation, callus and single cell culture, haploid plant production and protoplast isolation, culture transformation, fusion and regeneration into whole plants. These advances have opened up a whole new array of possibilities and methodologies for plant improvement including improvement of plant resistance to infection by pathogens. The potentials of these techniques is further augmented by combination with molecular technologies (genetic engineering). Molecular techniques allow the detection, isolation, modification, transfer and expression of single genes or groups of related genes from one generation to another. Several tissue culture techniques e.g. regeneration of whole plant from callus, single cell, protoplast and microspore or pollen, lead by themselves to plants showing greater variability in many characteristics including resistance to disease. Selection of the best among such plants and subsequent application of classical breeding techniques make possible the production of improved plants with greater efficiency and at a much greater rate. It is generally expected that breeding for disease resistance will quickly profit greatly from the application of techniques in molecular genetics and tissue culture.

(ix) Engineering Plant Resistance to Pathogens

The cloning of R genes has opened doors for producing disease-resistant crop plants. Transgenic plants that resist pathogen attacks can be engineered either by insertional mutagenesis or by map-based cloning methodology. Using the insertional mutagenesis method of transformation, a fragment of DNA is inserted into the coding region or regulatory region of a gene, which results in disruption of gene expression. After this, one proceeds to the cloning of the plant DNA that flanks the integrated insertional mutagen by plasmid rescue. The cloned plant DNA can then be used as a hybridization probe to isolate the gene by screening a lambda or cosmid library constructed from the wild-type plant. The final test is to introduce the cloned gene by transforming sensitive plants and examining them in order to determine whether they have become disease-resistant. The most commonly used DNA molecules for insertional mutagenesis are transposons (Baker *et al* 1986) and T-DNA from the Ti plasmid. In the map-based cloning which is also known as positional cloning, one needs to determine the chromosomal location of the gene of interest. The restriction fragment length polymorphism (RFLP) is the most frequently used method for linking the chromosomal position to a particular trait. In this case, co-inheritance of the disease resistance trait with one specific RFLP DNA probe defines the approximate location of the R gene on the chromosome. The linkage between several specific probe as a starting point to seek, or to land on the R gene. The gene will be identified in the next step after screening for coding sequences (e.g., cDNA) within the specific region using either a yeast artificial chromosome (YAC) vector or a bacterial artificial chromosome (BAC) vector. The final test can be done after transferring the candidate cDNA fragments into susceptible plants by looking for the acquisition of disease resistance in the transgenic plants.

Resistance of plants to pathogen attacks entails a concerted action of many gene products, and thus, alteration in the expression of multiple genes is required. In this respect, transcription factors may be excellent targets of genetic engineering for improving disease resistance in plants. Transcription factors generally regulate a group of genes rather than a single gene. Therefore, manipulating a single transcription factor could have the same effect as manipulating a set of specific genes within the plant. As highlighted above, transgenic plants allow the targeted expression of pathogen-related genes *in vivo* and are therefore an excellent system to assess the function of/and tolerance conferred by the encoded proteins. Another purpose for using transgenic plants is to improve disease resistance in agronomically valuable plants.

Future Thrusts

Plant diseases results in nearly 10-20 per cent of total crop yields losses. Cloning an R gene is an intensive and time-consuming task. Knowing the gene structure will lead to a better understanding of the pathogen-host interactions and the availability of cloned R genes will increase the potential for engineering disease resistance in plants. Unfortunately, up to date, the functions of many proteins and other molecules that interact with R genes during defence mechanisms are largely unknown. The genetic approaches utilized so far to isolate and characterize disease-related genes may need to be complemented by biochemical studies to understand fully their functions (Kotchoni and Shonukan, 2002). The production of mutants using an antisense-RNA approach could turn out to be a powerful technique in understanding the potential role of a gene product during defence mechanisms (Kotchoni and Bartels, 2001).

However, tremendous opportunities for crop improvement are likely to arise, as the complete sequencing of Arabidopsis plant genome (Arabidopsis Genome Initiative, 2000), which is a molecular research model, become a reality. Arabidopsis carries around 150 resistance (R) genes, whereas many crop plants may have two to three times this number (Arabidopsis Genome Initiative, 2000; Xiao and Jang, 2000). Many R genes, which confer resistance to various plant species against a wide range of pathogens, have been isolated. However, the key factors that switch on and switch off these genes during plant defence mechanisms remain poorly understood (Glazebrook, 2001). The population of India has peaked to 1.21 billion (2011 census), and the foodgrain production has not exceeded 230 MT during the last decade. By 2050, India needs to double its food grain production. The need of the hour is to develop resistant varieties to overcome food hunger and provide quality food. The crops needs to be grown in more sustainable agricultural systems. The use of chemicals to control the diseases has to be limited in the context of global warming and climate change. From the breeding point of view, there is a need to continuously develop durable disease resistance varieties. This can be made possible by exploiting quantitative disease resistance and transforming genes or gene constructs encoded for resistance into crops by using modern genetic engineering techniques. A combination of novel approaches including molecular techniques and genetics will provide insights into pathogen-defence mechanism and subsequent disease resistance in years to come.

References

Agrios GN 1997 Plant Pathology, 4th edition, New York: Academic Press

Arabidopsis Genome Initiative 2000 Analysis of the genome sequence of the flowering plant *Arabidopsis thaliana, Nature* **408** 796-813

Baker B, Schell J, Lorz H, and Fedoroff, N 1986 Transposition of the maize controlling element "Activator" in tobacco, *Proceedings of national Academy of Sciences of* USA **83** 4844-4848

Ballvora A, Ercolano MR, Weiss J, Meksem K, Bormann CA, Oberhagemann P, Salamini F, and Gebhardt C 2002 The R1 gene for potato resistance to late blight (*Phytophthora infestans*) belongs to the leucine zipper/NBS/LRR class of plant resistance genes, *Plant Journal* **30** 361-371

Berrocal-Lobo M, Molina A, and Solano R 2002 Constitutive expression of ethylene-response-factor 1 in Arabidopsis confers resistance to several necrotrophic fungi, *Plant Journal* **29** 23-32

Boyer JS 1982 Plant productivity and environment, *Science* **218** 443-448

Brandwagt BF, Mesbah LA, Takken FLW, Laurent PL, Kneppers TJA, Hille J, and Nijkamp, HJ 2000 A longevity assurance gene homolog of tomato mediates resistance to *Alternaria alternata* f. sp. *lycopersici* toxins and fumonisin B1, *Proceedings of National Academy of Sciences of USA* **97** 4961-4966

Brueggeman R, Rostoks N, Kudrna D, Kilian A, Han F, Chen J, Druka A, Steffenson B, and Kleinhofs A 2002 The barley stem rust-resistance gene Rpg1 is a novel disease-resistance gene with homology to receptor kinases, *Proceedings of National Academy of Sciences of USA*, **99** 9328-9333

Deslandes L, Olivier J, Theulieres F, Hirsch J, Feng DX, Bittner-Eddy P, Beynon J, and Marco Y 2002 Resistance to *Ralstonia solanacearum* in *Arabidopsis thaliana* is conferred by the recessive RRS1-R gene, a member of a novel family of resistance genes. *Proceedings of National Academy of Sciences of USA*, **99** 2404-2409

Flor HH 1942 Inheritance of pathogenicity in *Melompsora lini, Phytopathology* **32** 653-669

Flor HH 1955 Host-parasite interaction in flax rust: its genetics and other implications, *Phytopathology* **45** 680-685

Glazebrook J 2001 Genes controlling expression of defence responses in Arabidopsis, *Current Opinion on Plant Biology* **4** 301-308

Hahn MG 1996 Microbial elicitors and their receptors in plants, *Annual Review of Phytopathology* **34** 387-412

Hatsugai N, Iwasaki S, Tamura K, Kondo M, Fuji K, Ogasawara K, Nishimura M, and Hara-Nishimura I 2009 A novel membrane fusion-mediated plant immunity against bacterial pathogens, *Genes and Development* **23** 2496-2506

Kawchuk LM, Hachey J, Lynch DR, Kulcsar F, Rooijen G, Waterer DR, Robertson A, Kokko E, Byers R, Howard RJ, Fischer R, and Prüfer D 2001 Tomato *Ve* disease resistance genes encode cell surface-like receptors. *Proceedings of National Academy of Sciences of USA*, **98** 6511-6515

Keen NT 1990 Gene-for-gene complementarity in plant-pathogen interactions, *Annual Review of Genetics* **24** 447-463

Khan MA 2009 Importance and use of resistance gene analogs, *The Open Plant Science Journal* **3** 54-60

Kim MG, Geng X, Lee SY, and Mackey D 2009 The Pseudomonas syringae type III effector AvrRpm1 induces significant defenses by activating the Arabidopsis nucleotide-binding leucine-rich repeat protein RPS2, *Plant Journal* **57** 645-653

Kotchoni OS and Bartels D 2001 Plant Biotechnology: a new hope for coping with environmental stresses in Africa, *Grenzenlos Forschen*, **42** 300-302

Kotchoni OS and Shonukan OO 2002 Regulatory mutations affecting the synthesis of cellulase in *Bacillus pumilus, World Journal of Microbiology and Biotechnology* **18** 487-491

Krimsky S and Wrubel RP 1996 *Agricultural biotechnology and the environment: science, policy and social issues,* Urbana-Champaign, IL: University of Illinois Press, 294 pages

Kuhn J 1858 *Die Krankheiten Der Kulturgewachse,* Kessinger Publishing Company

Lamb CJ, Lawton MA, Dron M and Dixon RA 1989 Signal and transduction mechanics for activation of plant defences against microbial attack, *Cell* **56** 215-224

Lorenzo O, Piqueras R, Sanchez-Serrano JJ and Solano R 2003 Ethylene response factor 1 integrates signals from ethylene and jasmonate pathways in plant defence, *Plant Cell* **15** 165-178

NAS 1972 *Genetic vulnerability of major crops,* Washington, DC: National Academy of Sciences

Nimchuk Z, Rohmer L, Chang JH and Dangl JL 2001 Knowing the dancer from the dance: R-gene products and their interactions with other proteins from host and pathogen, *Current Opinion on Plant Biology* **4** 288-294

Parlevliet JE 1989 Identification and evaluation of quantitative resistance, In *Plant Disease Epidemiology, Vol. 2: Genetics, Resistance and Management* (eds) KJ Leonard and WE Fry, New York: McGraw-Hill Publishing Company, 215-248 pages

Penninckx IAMA, Thomma BPHJ, Buchala A, Metraux JP, and Broekaert WF 1998 Concomitant activation of jasmonate and ethylene response pathways is required for induction of a plant defence, *Plant Cell* **10** 2103-2113

Robinson RA 1976 *Plant Pathosystems,* Berlin, Heidelberg, New York: Springer-Verlag

Scheel D 1998 Resistance response physiology and signal transduction, *Current Opinion on Plant Biology* **1** 305-310

Shen KA, Chin DB, Arroyo-Garcia R, Ochoa OE, Lavelle DO, Wroblewski T, Meyers BC, and Michelmore RW 2002 Dm3 is one member of a large constitutively expressed family of nucleotide binding site leucine-rich repeat encoding genes. *Molecular Plant Microbe Interaction* **15** 251-261

Solano R, Stepanova A, Chao QM, and Ecker JR 1998 Nuclear events in ethylene signaling: A transcriptional cascade mediated by ethylene-insensitive 3 and ethylene-response-factor 1, *Gene Development* **12** 3703-3714

Somssich I and Hahlbrock K 1998 Pathogen defence in plants: a paradigm of biological complexity, *Trends in Plant Science* **3** 86-90

Swiderski MR and Innes RW 2001 The Arabidopsis PBS1 resistance gene encodes a member of a novel protein kinase subfamily, *Plant Journal* **26** 101-12

Thomma BPHJ, Eggermont K, Broekaert WF, and Cammue BPA 2000 Disease development of several fungi on Arabidopsis can be reduced by treatment with methyl jasmonate, *Plant Physiology and Biochemistry* **38** 421-427

Van't Slot KAE and Knogge W 2002 A dual role for microbial pathogen derived effector proteins in plant disease and resistance, *Critical Review on Plant Sciences* **21** 229-271

Vanderplank JE 1963 *Plant Diseases: Epidemics and Control,* New York and London: Academic Press

Xiao S, Ellwood S, Calis O, Patrick E, Li T, Coleman M, and Turmer JG 2001 Broad spectrum powdery mildew resistance in Arabidopsis thaliana mediated by RPW8, *Science* **291** 118-120

Xiao W and Jang JC 2000 F-box proteins in Arabidopsis, *Trends in Plant Sciences* **5** 454-457

Zhou F, Kurth J, Wei F, Elliotta C, Valèa G, Yahiaouic N, Kellerc B, Somervilled S, Wiseb R, and Schulze-Lefert P 2001Cell-autonomous expression of barley Mla1 confers race specific resistance to the powdery mildew fungus *via* a Rar1-independent signaling pathway, *Plant Cell* **13** 337-350.

9

Acquired Resistance in Plants

Jagdish Kumar

Principal Scientist and Head, IARI, Regional Station, Wellington, Distt. Nilgiris, Tamil Nadu - 643 231

Acquired resistance may be defined as the enhancement of resistance of plants being otherwise susceptible in response to an extrinsic stimulus without alterations of the genome. The inducing agent can be of biotic or abiotic nature. It exists in two different forms. It may be localized at the site of the inducing treatment or it may be systemic. The latter is commonly called SAR (systemic acquired or activated resistance) and is effective in some, all or newly emerging plant parts distant from the site of induction. With abiotic inducers it is more difficult to prove whether they induced localized or systemic resistance. Two mechanisms are possible to explain how the systemic nature of induced resistance is achieved: the message from the site of induction may be translocated by the actual inducer molecule or by some intermediary signal. For plant protection in practice the systemic type of induced resistance will be more important than induced localized resistance.

Most of the conventional chemical and biological plant protection procedures aim for direct control of pests and diseases by elimination of the parasites. These procedures generate problems especially in monoculture crops, because they enhance the risk for the development of pesticide - resistant population. Although disease resistance deserves special interest, this biological procedure also presents problems. Several examples exist showing that the resistance genes of plants have been overcome soon after their introduction and resistant cultivars often do not fulfil the necessary requirements for yield and quality. The natural protection of resistant plants against pathogens or chewing insects is partly based on a variety of constitutive barriers already present in the plant before actual attack. The combined effect of all these barriers is referred to as constitutive resistance. In addition, plants can activate protective mechanisms upon contact with invaders or after treatment with certain chemicals (non antimicrobial): this is termed induced or acquired resistance. This type of resistance may be distinguished both from cross - protection, in that the resistance is not specific for the virus, and from antagonism, in that an active plant process is involved in the mechanism.

Acquired immunity encompasses several independent processes, including viral cross-protection, bacterial and fungal antagonism and a response later known as Induced Resistance (IR). This type of resistance is activated by a prior pathogen infection (biological induced resistance) or by treatment with a chemical (chemical induced resistance: CIR), and may be distinguished both from cross-protection, in that the resistance is not specific for the virus, and from antagonism, in that an active plant process is involved in the mechanism. IR is thought to play an important role in the preservation of plants in nature. Although several types of biological IR have been identified, the one that has been studied extensively is systemic acquired resistance or SAR (Ross 1961; Ryals et al 1994). SAR is induced by pathogens causing necrotic reactions that can range from single cell hypersensitive response (HR) to necrotic disease lesions. The response is characterised by the induction of long-lasting, systemic protection that is often effective against viral, bacterial, and fungal pathogens.

Viral cross protection in plants is known as an acquired immunity phenomenon, where a mild virus isolate/strain can protect plants against economic damage caused by a severe challenge strain/isolate of the same virus. Mild strain cross protection (MSCP) has been used extensively to control losses caused by a few major virus diseases in some parts of the world. So far, none of the many proposed mechanisms can fully explain the intact process of MSCP. In fact, it may be that different mechanisms are involved in MSCP against different viruses, even when different research approaches are used for the same virus, different mechanisms could be proposed. The molecular detail of MSCP still remains unclear, although several lines of evidence imply that the resistance is protein and/or RNA mediated. Some data to date have shown that a minimum time (a few days to less than a month) is required for the mild virus strain to establish MSCP. To investigate interference among virus strains and the plant host at an early stage of MSCP at a subcellular level, we developed a rapid micro-extraction method for the preparation of total nucleic acid (TNA), combined with other molecular methods, to monitor the interaction of virus strains at short time intervals in young plants. This method was initially developed to further study the mechanism of MSCP against Citrus tristeza virus, but has potentially widespread application to other viruses after having been efficiently used to extract over 50,000 TNA samples of citrus viruses, viroids, and bacteria (Zhou and Zhou, 2012).

Systemic acquired resistance is a broad-spectrum, systemic, plant defense response that is triggered by exposing a plant to certain natural or synthetic chemical compounds or to a weak (hypovirulent) strain of a specific pathogen. These inducers or elicitors do not have a direct effect on pathogens. Rather, they are more like an intruder setting off the house alarm. Depending on the plant species and its genetic makeup, the response to a perceived intruder might include one or more of the following: signal the distant plant cells (for example, salicylic acid), chemical warfare (for example, enzymes), build stronger barriers (for example, lignin), or localized suicide (hypersensitive reaction).

In view of the applicability of the phenomenon in future plant protection measures it was critical to identify chemicals that elicit the same type of resistance response. With the description of such compounds, an example being the benzo(1,2,3)thiadiazole-7-carbothioic acid S-methyl ester (BTH) the potential of IR as a new plant protection strategy has been tremendously boosted (Gorlach et al 1996).

Plants are constantly attacked by pathogens, and as a result plants have evolved a plethora of constitutive and induced basal defenses to defend against pathogens. The phytohormones salicylic acid (SA), jasmonic acid, and ethylene are known to participate in regulating defenses in plants (Pieterse and Van Loon, 1999; van Wees et al 2000; Glazebrook 2001; Spoel et al 2003; Thaler et al 2004). SA is predominantly associated with resistance against biotrophic and hemibiotrophic pathogens, and triggering systemic acquired resistance (SAR; Grant and Lamb 2006). Although the complete mechanism of SA-mediated plant defense is not understood, the central role of SA in plant defense is universally accepted (for review, see Pieterse and Van Loon 1999; Shah 2003; Grant and Lamb 2006). Exogenous application of SA or its functional analogs, such as 2,6-dichloroisonicotonic acid and benzo-(1,2,3)-thiadiazole-7-carbothioic acid *S*-methyl ester (BTH), induces SAR in plants, resulting in resistance to certain pathogens (Ryals et al 1996; Achuo et al 2004; Wang et al 2005). Conversely, plants expressing the bacterial *NahG* gene (encoding salicylate hydroxylase, which converts SA to catechol) are more susceptible to several pathogens (Gaffney et al 1993). Direct evidence for the role of SA in plant defense comes from the identification and characterization of an Arabidopsis (*Arabidopsis thaliana*) isochorismate synthase (*ICS*) mutant (*sid2-2*) that is defective in SA biosynthesis (Wildermuth et al 2001). Endogenous SA levels in plants can also affect their interaction with symbiotic microorganisms, as demonstrated by increased root nodulation and infection upon inoculation with *Mesorhizobium loti*, in transgenic *Lotus japonicus* and *Medicago truncatula* expressing *NahG* (Stacey et al 2006). Only a limited number of studies has demonstrated the direct effects of SA on microbes including *Pseudomonas aeruginosa* (Prithiviraj et al 2005b), *Staphylococcus aureus* (Prithiviraj et al 2005a), *Sinorhizobium meliloti* (Martínez-Abarca et al 1998), and, more recently, *A. tumefaciens* (Yuan et al 2007).

The most significant criterion for Induced Resistance of the SAR-type is its dependency on salicylic acid (SA) as an essential component of the pathway for induction of SAR genes (Volt et al 2009). However, it is still questionable as to whether this well accepted requirement applies to monocots similar as to dicot plants. A single report has been published on biological induced rice plants, where an inducer infection by *Pseudomonas syringe* pv. *syringae* conferred systemic resistance against *Magnaporthe grisea*, the causative agent of rice blast (Smith and Métraux 1991). Importantly, phenylalanine ammonium-lyase (PAL), the key enzyme in SA synthesis, was not enhanced in activity indicating that protection did not rely on SA accumulation. In wheat and barley, prior inoculation by an avirulent race of *Blumeria graminis* resulted in enhanced protection against virulent races of the fungus (Cho and Smedegaard-Petersen 1986; Thordal-Christensen and Smedegaard-Petersen 1988; Hwang and Heitefuss 1992). Protection of barley against *Drechslera teres* was obtained by pre-inoculation with the barley non-pathogens *Bipolaris maydis* and *Septoria nodorum* (Jorgensen et al 1998). However, because a systemic resistance response has not been reproducibly demonstrated, implication of the SAR pathway has not been unequivocally proved. Alternatively, induced protection in cereals might rather be a consequence of a locally confined interference of established infections with the challenger inoculum (Ouchi et al 1974).

To gain further insight in the possible involvement of SA in resistance response to the powdery mildew fungus, we have kinetically analysed endogenous SA concentration during the development of the fungus in primary leaves of near-isogenic barley genotypes bearing different resistance genes. We distinguished between defence based on the hypersensitive cell death (HR, conferred by gene *Mlg*), on papillae formation (conferred by gene *mlo5*), and multi-cell necrosis (conferred by gene *Mla12*. Most unexpectedly, accumulation of SA was not observed regardless of the resistance phenotype. This result is not consistent with the SAR phenotype found in dicots (Hückelhoven *et al.* 1999). Instead, constitutive concentration of SA (non-inoculated leaves) varied significantly within independent plant batches covering a range from 101 to 1122 ng/g fresh weight indicating that barley compared to many dicot plants contains high levels of SA. In contrast, however, maize plants showed enhanced expression of defense-related genes along with increased levels of SA after infection with the pathogens *Colletotrichum graminicola*, the causative agent of anthracnose leaf blight and *Bipolaris maydis*, the causative agent of southern corn leaf blight (Morris et al 1998). PR gene expression and SA accumulation suggests that maize is similar to dicot plants in many aspects of Induced Resistance. This reinforces the notion that an ancient plant-inducible defense pathway against pathogen attack is shared between mono- and dicotyledonous plants.

Chemical induced resistance (CIR) has been demonstrated in cereals using biological extracts (B50, Milsana) and synthetic compounds. Most effective chemicals like 2,6-dichloroisonicotinic acid (DCINA) and benzo (1,2,3) thiadiazole-7-carbothioic acid S-methyl ester (BTH) are thought to mimic SA and therefore this type of resistance may rely on the natural SAR pathway (Kogel et al1994; Görlach et al 1996; Schweizer et al 1999). SA application to leaves of barley (Kogel et al 1994) or wheat (Görlach et al 1996) induced the same set of pathogenesis-related (PR) genes and the same type of cellular responses as the synthetic chemicals further supporting involvement of the SAR pathway. Cytological analysis of the CIR phenotype in the interaction of barley with powdery mildew fungus demonstrated that fungal growth restriction is always linked to formation of effective papillae at the site of fungal attack (Kogel et al 1994). SAR can be activated in corn, however, widely adapted commercial like Benzothiadiazole may not be efficient against *P. sorghi* causing common rust (Gautam and Stein, 2011).

Involvement of Reactive Oxygen Intermediates

In the interaction of barley with powdery mildew fungus "active defence" e.g. formation of effective papillae, expression of HR, and multi-cell necrosis is closely associated with a localised H_2O_2 burst (Hückelhoven et al 1999). Cytological inspection of the burst using diaminobenzidine as a histochemical tool for H_2O_2 verification showed a distinct stain starting off from cell walls. Whole cell H_2O_2 accumulation always resulted in cell death. Previous data had suggested that H_2O_2 is involved in cross-linking reactions of soluble cell wall components as proteins and phenols resulting in cell wall toughening. The question beyond that as to whether H_2O_2 is a signal for defense gene activation is still open. However, the finding that

H_2O_2 exogenously applied (injection) into primary leaves failed to activate expression of lipoxygenase or PR1-type genes known to be markers for monocot CIR, argues for H_2O_2 not being a signal.

SAR – a novel strategy of plant protection

SAR is associated with the accumulation of elevated levels of salicylic acid (SA) and expression of the pathogenesis-related (PR) group of genes, some of which encode antimicrobial proteins (Ryals et al 1996). The role of SA in SAR has been well characterized (Ryals et al 1996; Shah 2003; Shah et al 1999). The *Arabidopsis NPR 1* (At*NPR 1*) gene is a key regulator of SAR. Loss-of-function mutations in the At*NPR1* gene compromised the activation of SAR and enhanced susceptibility to a variety of pathogens (Durrant and Dong 2004). In contrast, over-expression of At*NPR1* in transgenic *Arabidopsis* enhanced resistance to bacterial and oomycete pathogens (Friedrich et al 1996; Cao et al 1998). The enhanced disease resistance in *Arabidopsis* conferred by over-expression of At*NPR1* was associated with the faster response of these plants to SA and its functional analog, benzothiadiazole (BTH) (Friedrich et al 1996; Cao et al 1997). At *NPR 1* expression also enhanced disease resistance in transgenic rice (Chern et al 2001). Considerable progress has been made in understanding the role of the NPR 1 protein in the induction of SAR in *Arabidopsis* (Durrant and Dong, 2004). SA application and exposure of the plant to the pathogen triggers a change in the molecular state of NPR 1, resulting in its translocation into the nucleus where it interacts with TGA family of DNA – binding proteins (Mou et al 2003). The translocation of NPR 1 into the nucleus and its interaction with the TGA proteins is essential for the expression of the *PR 1* gene and in governing disease resistance (Durant and Dong, 2004). In wheat, a SAR like mechanism has been documented (Schweizer et al 1989) and BTH application shown to activate resistance to fungal pathogens (Gorlach et al 1996). Expression of the wheat *PR* genes is induced in response to *Fusarium graminearum* inoculation and SA application (Anand et al 2003; Pritsch et al 2000), and the developmental resistance to FHB correlated with an increase in the endogenous level of SA (Anand et al 2003). Moreover, an NPR 1 homolog in present in wheat (Makandar et al 2006).

Commercialized SAR compounds

Harpin. This protein is marketed by Eden Biosciences Corp. as Messenger. The harpin protein in this formulation is identical to the protein that was initially discovered and naturally produced by *Erwinia amylovora* (the fire blight pathogen). The label states that this is "A biochemical pesticide used for plant disease management, insect suppression, and plant growth enhancement." Although the label states generally that Messenger will "aid in the management of diseases," several viral, bacterial, and fungal pathogens are specifically mentioned. Messenger is labeled for application on a wide range of vegetable, fruit, field, tree, and vine crops. However, it is not labeled for turfgrass (except seed production) or ornamentals. A cursory review of several recent experimental turf-disease trials revealed that Messenger did not protect against infection by *Pythium* or *Rhizoctonia.* However, I encourage researchers and practitioners alike to recognize the importance of the following statement found on the Messenger label: "Plants require 5-7 days to fully induce resistance." Discoveries about antimicrobial peptides and plant defence activators have made possible the de novo and rational design of novel peptides for use in crop protection. Here we report a novel chimeric protein, Hcml, which was made by linking the active domains of cecropin A and melittin to the hypersensitive response (HR)-elicitor Hpal of *Xanthomonas oryzae* pv. *oryzicola*, the causal agent of rice bacterial leaf streak. The resulting chimeric protein maintained not only the HR-inducing property of the harpin, but also the antimicrobial activity of the cecropin A-melittin hybrid. Hcml was purified from engineered Escherichia coli and evaluated in terms of the minimal inhibitory concentration (MIC) and the 50% effective dose (ED 50) against important plant pathogenic bacteria and fungi. Importantly, the protein acted as a potential pesticide by inducing disease resistance for viral, bacterial and fungal pathogens. This designed drug can be considered as a lead compound for use in plant protection, either for the development of new broad-spectrum pesticides or for expression in transgenic plants (Che *et al*, 2011).

Acibenzolar-S-methyl (ASM). This compound is marketed by Syngenta as Actigard. The proposed common name of acibenzolar-S-methyl has many variants, most notably benzothiadiazole (BTH). The label states that Actigard "is a selective, systemic compound used for the control of downy mildew of cole crops and leafy vegetables, bacterial leaf spots of tomato, and blue mold of tobacco." This product is not labeled for turfgrass or ornamentals. A cursory review of several recent experimental turf and ornamental disease trials revealed that Actigard offered no protection against Entomosporium leaf spot of photinia or Rhizoctonia brown patch of turfgrass and less than acceptable protection against Phytophthora root rot of snapdragon and dollar spot of turfgrass. Once again, I encourage researchers and practitioners to recognize the importance of the following statement found on the Actigard label: "Maximum disease control is normally obtained 4 days after an Actigard application."

Phosphonates. Following the Fungicide Resistance Action Committee's (FRAC 2004) lead to the combination of fosetyl-aluminum (marketed by Bayer as Aliette and Signature and by Lesco as Prodigy) and phosphoric acid (marketed by Cleary's as Alude; Arbor Systems as Whippet; and JH Biotech as Fosphite) compounds into the phosphonate group. From the literature, it appears that the debate continues as to whether or not phosphonates technically fit the SAR definition (they are plant nutrients, and they are slightly but directly toxic to certain fungi). Nevertheless, and for practical purposes, it is important to note that fosetyl-aluminum is commonly listed as "aluminum tris (ethyl phosphonate)," which is the chemical name. Similarly, phosphorous acid has several common synonyms, such as potassium phosphite and mono- and di-potassium salts of phosphorous acid. The phosphonate compounds are widely available in many markets, including turf and ornamentals. Their protection against *Pythium, Phytophthora,* and downy mildew diseases is well documented.

Although the SAR phenomenon has been studied for quite some time, many basic questions remain unanswered. As research explains more about molecular plant--microbe interactions, we will learn how to manipulate these and future SAR compounds like the relative precision instruments that they are.

References

Achuo EA, Audenaert K, Meziane H and Hofte M 2004 The salicylic acid-dependent defence pathway is effective against different pathogens in tomato and tobacco, *Plant Pathology* **53** 65-72

Anand A, Schmelz EA and Muthukrishnan S 2003 Development of a lesion-mimic phenotype in a transgenic wheat line overexpressing genes for pathogenesis-related (PR) proteins is dependent on salicylic acid concentration, *Molecular Plant-Microbe Interactions* **16** 916-925

Cao H, Bowling SA, Gordon AS and Dong X 1998 Characterization of an *Arabidopsis* mutant that is nonresponsive to inducers of systemic acquired resistance, *Plant Cell* **6** 1583-1592

Cao H, Glazebrook J, Clarke JD, Volko S and Dong X 1997 The Arabidopsis *NPR1* gene that controls systemic acquired resistance encodes a novel protein containing ankyrin repeats, *Cell* **88** 57-63

Che YZ, Li YR, Zou HS, Zou LF, Zhang B and Chen GY 2011 A novel antimicrobial protein for plant protection consisting of a *Xanthomonas oryzae* harpin and active domains of cecropin A and melittin, *Microbial Biotechnology* 4(6) 777-793

Chern MS, Fitzgerald HA, Yadav RC, Canlas PE and Dong X 2001 Evidence for a disease-resistance pathway in rice similar to the NPR1-mediated signaling pathway in *Arabidopsis*, *Plant Journal* **27** 101-113

Cho BH and Smedegaard-Petersen V 1986 Induction of resistance to *Erysiphe graminis* f. sp. *hordei* in near-isogenic barley lines, *Phytopathology* **76** 301-305

Durrant WE and Dong X 2004 Systemic acquired resistance, *Annual Review of Phytopathology* **42** 185-209

FRAC 2004 Fungicide List 1 (arranged by FRAC Code), *Fungicide Resistance Action Committee* <http://www.frac.info/publications.html> (Accessed July 15, 2004)

Friedrich L, Lawton K, Ruess W, Masner P, Specker N, Rella MG, Meier B, Dincher S, Staub T and Uknes S 1996 A benzothiadiazole derivative induces systemic acquired resistance in tobacco, *Plant Journal* **10** 61-70

Gaffney T, Friedrich L, Vernooij B, Negrotto D, Nye G, Uknes S, Ward E, Kessmann H and Ryals J 1993 Requirement of salicylic acid for the induction of systemic acquired resistance, *Science* **261** 754-756

Gautam P and Stein J 2011 Induction of Systemic Acquired Resistance to *Puccinia sorghi* in Corn, *International Journal of Plant Pathology* **2** 43-50

Glazebrook J 2001 Genes controlling expression of defense responses in Arabidopsis-2001 status, *Current Opinion on Plant Biology* **4** 301-308

Gorlach J, Volrath S, Knauf-Beiter G, Hengy G, Beckhove U, Kogel KH, Oostendorp M, Staub T, Ward E and Kessmann H 1996 Benzothiadiazole, a novel class of inducers of systemic acquired resistance, activates gene expression and disease resistance in wheat, *Plant Cell* **8** 629-643

Grant M and Lamb C 2006 Systemic immunity, *Current Opinion on Plant Biology* **9** 414-420

Hückelhoven R, Fodor J, Preis C, and Kogel KH 1999 Hypersensitive Cell Death and Papilla Formation in Barley Attacked by the Powdery Mildew Fungus are Associated with H_2O_2 Accumulation but are not Accompanied by Enhanced Concentrations of Salicylic Acid, *Plant Physiology* **119** 1251-1260

Hwang BK and Heitefuss R 1982 Induced resistance of spring barley to *Erysiphe graminis* f.sp. *hordei*, *Phytopathologische Zeitschrift* **103** 41-47

Jørgensen HJL, Lübeck PS, Thordal-Christensen H, de Neergaard E, and Smedegaard-Petersen V 1998 Mechanisms of induced resistance in barley against *Drechslera teres*, *Phytopathology* **88** 698-707

Kogel KH, Beckhove U, Dreschers J, Münch S, and Rommé Y 1994 Acquired resistance in barley: Analysis of the interaction with powdery mildew reveals the mechanism induced by 2,6-dichloroisonicotinic acid as a phenocopy of a genetically based mechanism governing race-specific resistance, *Plant Physiology* **106** 1269-1277

Makandar R, Essig JS, Schapaugh MA, Trick HN, and Shah J 2006 Genetically engineered resistance to fusarium head blight in wheat by expression of *Arabidopsis NPR1*, *Molecular Plant-Microbe Interaction* **19** 123-129

Martínez-Abarca F, Herrera J, Bueno P, Sanjuan J, Bisseling T, and Olivares J 1998 Involvement of salicylic acid in the establishment of the *Rhizobium meliloti*-alfalfa symbiosis, *Molecular Plant Microbe Interaction* **11** 153-155

Morris SW, Vernooij B, Titatarn S, Starrett M, and Thomas S 1998 Induced resistance responses in maize, *Molecular Plant Microbe Interaction* **11** 643-658

Mou Z, Fan W, and Dong X 2003 Inducers of plant systemic acquired resistance regulate NPR1 function through redox changes, *Cell* **113** 935-44

Ouchi SH, Oku Ch, and Akiyama I 1974 Induction of accessibility to a nonpathogen by preliminary inoculation with a pathogen, *Phytopathology Z* **79** 142-154

Pieterse CMJ and Van Loon LC 1999 Salicylic acid-independent plant defence pathways, *Trends Plant Science* **4** 52-57

Prithiviraj B, Bais HP, Jha AK, and Vivanco JM 2005a *Staphylococcus aureus* pathogenicity on *Arabidopsis thaliana* is mediated either by a direct effect of salicylic acid on the pathogen or by SA-dependent, NPR1-independent host responses, *Plant Journal* **42** 417-432

Prithiviraj B, Bais HP, Weir T, Suresh B, Najarro EH, Dayakar BV, Schweizer HP, and Vivanco JM 2005b Down regulation of virulence factors of *Pseudomonas aeruginosa* by salicylic acid attenuates its virulence on *Arabidopsis thaliana* and *Caenorhabditis elegans*. *Infection Immunity* **73** 5319-5328

Pritsch C, Muehlbauer GJ, Bushnell WR, Somers DA, and Vance CP 2000 Fungal development and induction of defense response genes during early infection of wheat spikes by *Fusarium graminearum*, *Molecular Plant-Microbe Interactions* **13** 159-169

Ross AF 1961 Systemic acquired resistance induced by localized virus infections in plants, *Virology* **14** 340-358

Ryals JA, Neuenschwander UH, Willits MG, Molina A, Steiner HY, and Hunt MD 1996 Systemic acquired resistance, *Plant Cell* **8** 1809-1819

Schweizer P, Hunziker W, and Mösinger E 1989 cDNA cloning, in vitro transcription and partial sequence analysis of mRNAs from winter wheat (*Triticum aestivum* L.) with induced resistance to *Erysiphe graminis* f. sp. *tritici*, *Plant Molecular Biology* **12** 643-654

Schweizer P, Schlagenhauf E, Schaffrath U, and Dudler R 1999 Different patterns of host genes are induced in rice by *Pseudomonas syringae*, a biological inducer of resistance, and the chemical inducer benzothiadiazole (BTH), *European Journal of Plant Pathology* **105** 659-665

Shah J 2003 The salicylic acid loop in plant defense, *Current Opinion on Plant Biology* **6** 365-371

Shah J, Kachroo P, and Klessig DF 1999 The *Arabidopsis ssi1* mutation restores pathogenesis-related gene expression in *npr1* plants and renders defensin gene expression salicylic acid dependent, *Plant Cell* **11** 191-206

Smith JA and Metraux JP 1991 Pseudomonas syringe pv. syringae induces systemic resistance to Pyricularia oryzae in rice, Physiological and Molecular Plant Pathology **39** 451-461

Spoel SH, Koornneef A, Claessens SMC, Korzelius JP, Van Pelt JA, Mueller MJ, Buchala AJ, Metraux JP, Brown R, and Kazan K 2003 NPR1 modulates cross-talk between salicylate- and jasmonate-dependent defense pathways through a novel function in the cytosol, *Plant Cell* **15** 760-770

Stacey G, McAlvin CB, Kim SY, Olivares J, and Soto MJ 2006 Effects of endogenous salicylic acid on nodulation in the model legumes *Lotus japonicus* and *Medicago truncatula, Plant Physiology* **141** 1473-1481

Thaler JS, Owen B, and Higgins VJ 2004 The role of the jasmonate response in plant susceptibility to diverse pathogens with a range of lifestyles, *Plant Physiology* **135** 530-538

Thordal-Christensen H and Smedegaard-Petersen V 1988 Comparison of resistance-inducing abilities of virulent and avirulent races of *Erysiphe graminis* f.sp. *hordei* and a race of *Erysiphe graminis f.sp. tritici* in barley, *Plant Pathology* **37** 20- 27

van Wees SCM, de Swart EAM, van Pelt JA, van Loon LC, and Pieterse CMJ 2000 Enhancement of induced disease resistance by simultaneous activation of salicylate- and jasmonate-dependent defense pathways in *Arabidopsis thaliana, Proceedings of National Academy of Science of USA* **97** 8711-8716

Vlot AC, Dempsey DA, and Klessig DF 2009 Salicylic Acid, a multifaceted hormone to combat disease, *Annual Review of Phytopathology* **47** 177-206

Wang D, Weaver ND, Kesarwani M, and Dong X 2005 Induction of protein secretory pathway is required for systemic acquired resistance, *Science* **308** 1036-1040

Wildermuth MC, Dewdney J, Wu G, and Ausubel FM 2001 Isochorismate synthase is required to synthesize salicylic acid for plant defence, *Nature* **414** 562-565

Yuan ZC, Edlind MP, Liu P, Saenkham P, Banta LM, Wise AA, Ronzone E, Binns AN, Kerr K, and Nester EW 2007 The plant signal salicylic acid shuts down expression of the vir regulon and activates quormone-quenching genes in *Agrobacterium, Proceedings of National Academy of Science of USA* **104** 11790-11795

Zhou C and Zhou Y 2012 Antiviral resistance in plants, *Methods in Molecular Biology* **894** 69-81.

10

Genetics of Resistance to Apple Scab: New Insights

K. Kumar[1], R. Kaur[2], Nirmal Sharma[1], Dinesh Singh[1] and Era Vaidya[2]

[1]Department of Fruit Science, [2]Department of Biotechnology, University of Horticulture and Forestry, Nauni - Solan (H.P.) India – 173 230

Breeding for pest and disease resistance is one of the major breeding objectives after higher yield and eating quality in any fruit breeding programme. Apple is host of a wide range of pests and diseases (Way et al 1989), many of which need to be controlled so that commercial production could be profitable. Apple scab, caused by fungus *Venturia inaequalis* (Cke.) Wint. (anamorph *Spilocea pomi* Fr.) is the most serious and widespread disease of this crop in the apple growing regions of the world. The exact origin of the scab fungus is not known is but hypothecated to have originated in the regions where apple has originated. Though scab has wide geographical distribution but it is serious in the temperate regions with cool, moist weather in early spring. Scab was first reported from Sweden (Fries 1819) and 15 years later from Germany (Wallroth 1833). Apple orchardists depend heavily on chemical control to produce disease free fruits. The use of plant resistance is widely regarded as the preferred means of controlling pests and diseases. There are many socio- economic advantages in using resistant cultivars, because they help reduce production costs and diminish the effects on the user and environment due to reduced requirements for equipment, labour and fossil fuels (Way et al 1989, Hogenboom 1993). The scab resistance apple breeding programmes have been conceived to increase grower's efficiency by reducing fungicidal sprays. The impact of the environmental movement and ever rising costs of energy have greatly increased the relevance of scab resistant cultivars the world-over. Widespread adoption of disease resistant cultivars could impact food safety, ground water purity, protection of non-target organisms, applicator's exposure to pesticide and also lower the cost of production. However, while the potential benefits of resistance breeding are large with regard to the wider impact of pesticide use, the savings to the grower in the direct costs of disease protection are only about 4 per cent of annual crop value (Merwin et al 1994).

Early History

Scab fungus was first reported from Sweden in 1819 and later on reported from other apple growing countries (Gupta 1993). Rudolf and Schmidt (1934, 1935) were first to initiate apple breeding programme for scab resistance. They transferred scab resistance from wild *Malus* spp. namely *Malus atropurpurea*, *Malus micromalus* and *Malus spectabilis* Kaido, but the dominant nature of small fruited character of wild *Malus* spp. was the problem. Therefore, Schmidt (1936-42) used large fruited Antonovka and Ernst Bosch as female parent. Selection evolved by him carrying Antonovka gene were later used by several workers (Granhall 1953, Zwintscher 1954, Spangelo et al 1956, Lamb and Hamilton 1956 and Shay 1960). In 1944 L. F. Hough discovered field immunity to scab in hybrid progenies obtained from inter-specific crosses [(Rome Beauty × *Malus floribunda* 821) × (Rome Beauty × *Malus floribunda* 821)] made by Crandall at University of Illinios, USA (Crandall 1926). Studying the inheritance pattern, suggested multiple factor inheritance and evidence of cytoplasmic influence on scab resistance was also indicated at the time. Absolute immunity to scab was noted in six Asiatic *Malus spp.* viz. *Malus atrosanguinea* 804, *M. floribunda* 821, *Malus prunifolia* 19651, *Malus ringo*, *Malus toringo* and *Malus zumi* originally collected by C.S. Crandall. On the basis of these observations a cooperative breeding programme was started in USA by Hough of the University of Illinois and Shay of Purdue University in 1945 and later joined by Rutgers university in 1948 named as PRI programme. Subsequently separate scab resistant programmes were initiated by fruit breeders in USA, France, Germany, UK, Canada, Brazil, Czech Republic, Sweden, Hungry, Italy, Romania and several other countries (Crosby et al 1992). The most promising selections from PRI programme were given Co-op numbers. By the year 1988, nine scab resistant cultivars having *Malus floribunda* 821 (V_f gene) in their pedigree were released through PRI programme viz. Prima, Priscilla, Priam, Sir Prize, Jonafree, Red Free, Williams Pride, Dayton and McShay. Later on Liberty and Freedom were released in 1983 by New York State Agriculture Experiment Station (Lamb et al 1978 and Lamb et al 1985). The parentage of both Liberty and Freedom included male parent carrying *Malus floribunda* 821 (V_f) type resistance and female Antonovka type (Polygenic). Nova Easygrow (Spangelo et al 1974) and MacFree (Crowe 1975) were released by Canadian apple breeding programme at Kentville and Ottawa, respectively. Apple cv. Nova Easygrow carries scab resistance V_r from Russian seedling R12740-7A and was the first scab resistant variety derived from a source other than *Malus floribunda* 821. Murray and Rouvilla carrying V_m gene from *Malus micromalus* were also released from Canadian breeding programmes but were susceptible to race 5 of *Venturia inaequalis*. A cultivar carrying polygenic resistance from *Malus spectabilis* Kaido named Genros was bred in Romania and released in 1972 (Lespinasse 1989). Brazilian breeding programme released Primicia (V_f resistance gene) in 1988 (Denardi et al 1988) and breeders in England released Gavin (V_f resistance) in 1977. More than 70 scab resistant apple cultivars have been bred till 1996 worldover including Enterprise, Gold Rush, Jonafree, Liberty, Pristine and Judeline (Janick et al 1996) showing promise for commercial cultivation and some of them viz. Enterprise, Freedom and Liberty have shown multiple disease resistance.

The breeding strategy followed all over the world has been through a series of modified backcrosses. The best selections are selected and given a name of scab resistant cultivar. The most important V_f progenators by order of importance are Prima, Priscilla, Florina, Freedom and to small extent Redfree, Coop 17, Liberty and more recently Gold Rush, Enterprise and BM 1497 (Laurens 1999). But often the discovery of new races of scab fungus, forced breeders to change the breeding strategy. The aim is now to combine several type of resistance in one genotype referred to as durable resistance. However, a major problem in breeding resistant cultivars of perennial crops is the requirement of durable resistance. The disease resistance, lasting for the period and the cultivar grown and preferred by growers is termed as durable (Johnston 1981, Hogenboom 1993). Until recently the apple cultivars carrying V_f gene from *Malus floribunda 821* have remained free from scab over 50 years. But identification of race 6 and 7 which infects the cultivars/selections carrying V_f gene warrants the diversification of source(s) of resistance to apple scab.

Genetics and Inheritance of Apple Scab Resistance

Delineation of genes controlling specific traits is an important part of any breeding programme. Through the logical choice of parents, breeder is able to incorporate useful characteristics into breeding lines (Alston 2000). From the studies it has been revealed that the resistance conferred by V_f gene is a single qualitative dominant gene or by a block of closely linked quantitative genes and could be transferred to the high quality commercially acceptable apple cultivars through modified backcross method. Fourteen resistance genes have been identified and characterized till date and V_f gene has been the most popular in apple breeding programmes (Baniulis et al 2008). Six loci for qualitative resistances have been identified till date viz. V_f (*Malus floribunda* 821), V_m (*Malus micromalus*), V_r (*Malus domestica* R12740-7A), V_{bj} (*Malus baccata* Jackii), V_b (Hansen's baccata #2) and V_a (Antonovka PI 172623) and 6 different virulent races of *V. inaequalis* have been identified. V_f is a simply inherited gene that confers a class 3 reaction type and additional genes enhance or augment the affect of V_f to give class 1, 2 and M type of resistance. It is very likely that these additional or modifier genes are complexly inherited quantitatively. As breeders cannot afford the time to develop test crosses to assess the ability of crossing combinations to achieve the breeding goals (Bringhurst 1983), there will be an aspect of chance in the parent selection for a high specific combining ability (SCA) with regard to quantitatively inherited traits. The effect of parents with poor fruit quality is illustrated by the breeding of scab resistant cultivars carrying the V_f gene from *Malus floribunda* 821, a crab apple with small fruits of low quality. Bus et al (2000a) recorded moderate heritability for scab resistance (V_f) which is sufficient for the successful development of new apple cultivars with desirable resistance in a breeding strategy based on recurrent selection. Two types of scab resistance exists in *Malus* i.e. monogenic and polygenic (Table 1) but more than 80% scab resistant cultivars released till date and the parents currently included in breeding programme carry V_f gene from *Malus floribunda* 821. Other monogenic sources such as V_r, V_b, V_{bj}, V_m or the polygenic resistance have been used only sporadically (Kumar and Sharma 1999).

Table 1. Sources of Scab Resistance

Source	Gene effect	Reference	Symbol
Antonovka PI 172612	Monogenic	Hough *et. al.* (1970)	V_a
Hansen's baccata #2	Monogenic	Dayton and William (1968)	V_b
M. baccata jackii Dg27T1	Monogenic	Dayton and William (1968)	V_{bj}
Cathay	Monogenic	Korban and Chen, 1992	V_{bc}
Durello di Forl	Monogenic	Tartarini *et. al.* 2004	V_d
Prima	Monogenic	Dayton and William (1968)	V_f
Malus floribunda 821	Monogenic	Gianfranceschi *et. al.* (1998)	V_{fh}
Golden Delicious	Monogenic	Benaouf et al 1997	V_g
Russian apple R12740-7A	Monogenic	Bus *et al* 2005b	V_{h2}
Russian apple R12740-7A	Monogenic	Bus *et al* 2005b	V_{h4}
Malus severii W193B	Monogenic	Bus *et al* 2005a	V_{h8}
Jonsib	Monogenic	Korban and Chen, 1992	V_j
Malus micromalus 245-38	Monogenic	Dayton and William (1968)	V_m
Malus atrosanguinea 804	Monogenic	Lespinase (1989); Lespinasse and Durel (1999)	V_m
Russian seedling R 12740-7A	Monogenic	Dayton and William (1968)	V_r
Malus baccata (selected seedlings)	Polygenic	Lespinasse (1989); Lespinasse and Durel (1999)	
Malus sargentii 843	Polygenic	Lespinasse (1989); Lespinasse and Durel (1999)	
Malus sieboldii 2982-22	Polygenic	Lespinasse (1989); Lespinasse and Durel (1999)	
Malus zumi calocarpa	Polygenic	Lespinasse (1989); Lespinasse and Durel (1999)	
Malus toringo 852	Polygenic	Lespinasse (1989); Lespinasse and Durel (1999)	
Old European cultivars	Polygenic	Lespinasse (1989); Lespinasse and Durel (1999)	

Due to the dependence on the monogenic source of scab resistance, the breakdown has been reported in *Malus micromalus* 245-38 and *M. atrosanguinea* 804 (Williams and Brown 1968; Williams and Kuc 1969), Liberty (Fischer et al 1983), Coop 2, Coop 7, Coop 9 and Coop 10 (Kruger 1988, 1989, 1991), Baujade, Coop 28, Florina, Liberty and Priscilla (Parisi et al 1993). Thus till date as far as seven different virulent races have been identified (Table 2).

Gene Pyramiding

The genetic adaptivity of *V. inaequalis* has resulted in the emergence of pathogen strains that overcome the monogenic resistance. This fact stresses the importance of development of breeding strategy based on combination of multiple resistance genes. An effective "pyramidization" strategy requires new apple scab resistance gene identification and integration into breeding programmes. The genetic locus for resistance to apple scab most frequently used in apple breeding is V_f, derived from *Malus floribunda* 821 (Boudichevskaia et al 2009), but this type of monogenic resistance has been shown to be subject to breakdown under severe infection conditions repeatedly (Parisi et al 1993, Roberts and Crute 1994; Fischer et al 1999). The only safeguard against the loss of a resistant variety from the newly arising or mutated races of the fungus is the incorporation of both monogenic and polygenic resistant genes in the breeding programmes. Therefore, durable resistant varieties through pyramiding resistance genes and resistance gene types should be developed. Polygenic resistance based on quantitative gene support is considered to be durable (Parlevliet 1993). A combination of several resistant genes in one genotype might be the solution e.g. cultivar Freedom carrying monogenic resistance genes V_f from *Malus floribunda* and polygene from Antonovka. It is vital to select varieties with both parents contributing genes for resistance to scab and not to depend on the backcrossing single gene in to susceptible types (Fischer, 1993). Fischer et al (1994) reported durable resistance in a number of new selections at Pillnitz resulting from combinations of alleles from several crosses i.e. $V_f \times V_r$, $V_f \times V_a$ and $V_r \times V_a$. Several workers have suggested to work with various independent resistance mechanisms (Lespinasse 1989; Kellerhals et al 2009) and incorporate genes which favours dissociation of virulent genes thereby resulting in stable resistance (Gessler 1989). In particular polygenic resistance and major resistance gene should be combined.

Table 2. Virulent Races of *Venturia inaequalis* and Susceptible Genotypes

Race source/distribution	Susceptible material
World wide	Most of world's cultivars
South Dakota, USA	*M. baccata*, Dolgo, Alexis, Biitercrab, segregates of Russian seedling R12740-7A, Genva (*M. pumila* var. Niedzwetz)
Nova Scotia, Canada	Geneva
Lafayette, Indiana, USA	Segregates of Russian seedling R12740-7A
Norwich, England	*Micromalus* pit types resistance, *M. atrosanguinea* 804
Ahrensburg, Germany	Prima(V_f) but not Evereste, *Malus*×Perpetu and *M. floribunda* 821
England - Europe	*Malus floribunda* 821

Source : Janick et al 1996

The first scab resistant apple cultivar, Prima (Dayton 1970) is a F_4 descendent of *Malus floribunda* 821 and was introduced about 30 years after the PRI breeding programme started with specific objective of developing pest and disease resistant cultivars (Crosby et al 1992). Inspite of an unceasing, single minded emphasis on moving the V_f gene into an adapted type (Janick et al 1996), 40 years later there is still no cultivar that has had a considerable impact on fruit production by replacing major susceptible cultivars. Breeders have not been able to make the scab resistant cultivars "catchup" with the eating quality expected of new cultivars today. Nevertheless, the programme might have made still less progress if the breeders had been aiming to achieve too many breeding objectives at the same time, which creates inefficiencies as large number of seedlings are required to improve the chances of meeting all selection criteria (Brown 1975 and Oraguzie et al 2004). Introduction of multiple genes into single apple tree cultivar is a complicated task to achieve using traditional breeding methods. Application of modern molecular biology methods provides tools for resistance gene localization on genetic maps and development of molecular markers used for an efficient gene donor selection and analysis of inheritance of resistance traits.

Molecular Markers

Identification of apple scab resistant genes has attracted the most attention of genetists because of its economic importance. Identification of tightly linked markers with the resistance genes helps in identifying plants in early stage of their seedling stage without subjecting them to artificial inoculation of the pathogen. Development of saturated genetic maps and tightly linked molecular markers to the resistant genes have opened the ways for marker assisted selection in breeding programmes. As apple has proved to be a rich source of simply inherited resistance genes with major effects, different types of markers such as microsatellite (SSR) and Single nucleotide Polymorphism (SNP) markers are required to identify resistance genes that are linked or residing in a cluster (Bus et al 2005b). Besides the use of isoenzymes (Weeden and Lamb 1985, 1987; Manganaris and Alston 1992), molecular markers (Nybom and Schaal 1990; Mulcahy et al 1993) have been successfully demonstrated for the apple genotype identification. In the development of molecular markers most work in apple has been done on the tagging of V_f gene and over 40 markers have been identified associated with V_f. Markers for other scab resistance genes have also been identified by many groups and include V_h from Russian seedling R12740-7A of *Malus sieversi* (Hemmat et al 2002; Bus et al 2005a,b; Boudichevskaia et al 2006), V_m (Cheng et al 1998; Patocchi et al 2005); V_a and V_b (Hemmat et al 2003; Erdin et al 2006), V_d (Tartarini et al 2004), Vbj (Gygax et al 2004) and V_g (Durel et al 2000; Calenge et al 2005). SCAR markers have been identified for V*bj* from *Malus buccata* Jackii (Gygax et al 2004) and V_m from *Malus atrosanguinea* 804 (Cheng et al 1998). By using BSA (bulked segragant analysis) number of RAPD markers linked to V_f have been identified (Table 3). Patocchi et al (2005) used genome scanning to identify microsatellites tightly linked to V_m.

Table 3. List of RAPD Markers linked to V_f Locus

Marker name	Reference
OPA 15	Durham and Korban (1994)
OPU 01, OPM 18	Koller *et al.* (1994)
OPC 09	Tartarini *et al.* (1996)
OPD 20	Yang and Krugar (1994)

Later on markers linked to V_f gene were identified with BSA and mapped by RAPD or after conversion to SCAR or CAPS markers (Table 4).

Table 4. List of Markers other than RAPDs linked different Resistance Loci

Marker name	**Gene**	**Reference**
OPH 01, OPR 16, OPU 01, OPM 18, OPAM 19, OPAL 07, OPAR 4, OPK 16, SS, B505, S 29, P 198, B 398, VfC	V_f	Gardiner et al 1996, Gianfranceschi et al (1996), Tartarini et al 1996, Yang et al (1997a), Yang et al (1997b), Hemmat et al (1998), Urbanovich and Kazlovskaya (2008)
OPB12, OPB12687, Hi07h02	V_m	Cheng et al 1998; Urbanovich and Kazlovskaya (2008), Patocchi et al (2005) Cheng et al unpublished
CH02b10, OPL19, OPZ13, OPL19	V_{h2}	Gygax et al 2004; Bus et al 2005b, Urbanovich and Kazlovskaya (2008)
B220	V_b	Hemmat et al 2003
P136	V_a	
OPAS07, OPAS11	V_{mis}	Gardiner et al 2001
AD13	V_{r1}	Urbanovich and Kazlovskaya (2008)

Genetic Maps

Construction of genetic maps in apple started during early nineties. Besides readily available RAPD markers, researchers also included a small numbers of isozyme markers (Hemmat et al 1994; Conner et al 1997). With time, RAPDs due to their poor transferability were replaced by number of codominant transportable markers e.g. AFLPs and few microsatellite markers (Maliepaard et al 1998). A number of apple linkage maps have been published using several different sets of parents: White Angel × Rome Beauty (Hemmat et al 1994), Wijcik MacIntosh × NY 75441-58 (Conner et al 1997), Prima×Fiesta (Maliepaard et al 1998), Iduma × A679-2 (Gianfranceschi et al 1998), Fiesta × Discovery (Liebhard et al 2003a; Silfverberg – Dilworth et al 2006) and Telamon × Braeburn (Kenis and Koulemans 2005). The one with the greatest genome coverage and marker density is that of Liebhard et al 2003_b) with 475 AFLPs, 235 RAPDs, 129 SSRs and 1 SCAR markers. Two parental maps were also constructed that spanned 1140 and 1450 cM. Prior to this Maliepaard et al (1998) with the help of microsatellites, codominantly segregating isoenzymes and RFLPs generated first integrated apple map comprising 17 linkage groups from the population arising from Prima×Fiesta. In order to orientate the existing genetic maps Calenge et al (2004, 2005a) employed two microsatellites per linkage group. They then used NBS (Nucleotide binding site) profiling (Van der Linden et al 2004) to simultaneously amplify and map 23 polymorphic markers with sequence homology to Malus NBS-LRRs as well as 20 markers with unrelated homologies (Calenge et al 2005_a). NBS markers mapped close to major scab resistant genes on LG2, LG10 and LG12 (V_r, V_d and V_g, respectively). In Europe under the project DARE (Durable Apple Resistance in Europe) Prima × Fiesta and other related progenies have been used to identify major genes associated with resistance (Durel et al 2002, 2003). The major scab resistance gene was found on LG12 and several different NBS type resistance gene analogues were clustered at the bottom of LG5 and at the top of LG10. On the basis of different studies the major apple scab resistance genes which have been assigned to different linkage groups (Table 5).

Table 5. Distribution of different Qualitative Resistance Gene Loci on different Linkage Groups.

Gene	Linkage Group	Reference
Vf	LG1	Maliepaard et al (1998)
Vm	LG17	Patocchi et al (2005)
Vr2, Vh2, Vh4, Vt57, Vbj, Vh8	LG2	Bus et al (2004); Gygax et al (2004); Patocchit et al (2004); Bus et al (2005a, b)
Vd	LG10	Tartarini et al (2004)
Vg, Vb	LG12	Durel et al (1999); James et al (2004); Erdin et al (2006)

QTL Mapping for Apple Scab

Quantitative trait is the trait expressed phenotypically which is the result of many genes and each gene has small effect on the trait. Genetic mapping of QTL refers to determination of association between the trait and a set of genetic markers. The development of QTL markers for polygenic resistance and for different major resistance gene can allow a broad application of marker assisted selection (Kellerhals et al 2000). The accurate QTL identification is only possible if the saturated genetic map of the entire genome is available. Several workers (Hemmat et al 1994, Conner et al 1997, Seglias and Gesselar 1997, Maliepaard et al 1998, Liebhard et al 2002, Liebhard et al 2003a, Gianfranceschi et al 1998) have given the Malus genome maps in the past. Subsequently aligning the genetic maps and integrating with each other, saturated maps have been developed which provide a valuable tool for QTL detection and analysis in apple. In addition to the map saturation technique to identify QTLs in apple, genetic markers based on the sequence homology between the NBS domain in plant resistance genes identified loci which co-segregate with apple scab have been detected in Discovery × TN 10-8 progeny (Calenge et al 2005a). From the previous works it is clear that apple scab resistance is conferred by single major gene and complies character for gene hypothesis while conferring resistance to the pathogen. Thus is easily overcome by the pathogens ability to mutate to virulent strains (Crute and Pink 1996). Thus an effective durable resistance to apple scab pathogen will be established by gene pyramiding in a cultivar of different resistances. Inclusion of QTLs along with the major gene pyramiding will be more effective than combining major genes alone (Parlevliet and Zadoks 1977). Using the reference genetic maps (Durel et al 2003, Liebhard et al 2003c, Calenge et al 2004) QTLs for apple scab have been identified. Durel et al (2003) identified 4 genomic regions that were significantly involved in partial resistance in which one region was lying close to original V_f gene and may be result of the closely linked gene or residual effect of the overcome V_f gene (Durel et al 2003). The remaining 3 genomic regions were on LG 15, LG 11 and LG 17. Liebhard et al (2003b) identified eight QTLs responsible for apple scab (six for leaf scab and two for fruit scab). The strongest scab resistance QTL have been mapped on LG17 in Prima×Fiesta population (Liebhard et al 2003b) coinciding with the QTLs which was given by Durel et al 2003. Similarly, regions showing resistance were located on LG11 in both the studies. Another QTLs detected by Liebhard et al (2003b) was on LG12 on a position comparable to V_g (Van de Weg unpublished). Calenge et al (2004) identified three major QTLs on LG1, LG2 and LG17. The QTLs identified on LG1 corresponds to the region around V_f (Durel et al 2003). Subsequently other QTLs were also detected on different linkage groups by Calenge et al (2004) and one of them of interest is the QTL which spans around the major scab resistance V_{bj}, V_{h2} and V_{h8} (Calenge et al 2004; Durel et al 2004).

Marker Assisted Breeding

The major advantage in using markers in breeding programmes is that they enhance the breeding efficiency by enabling accurate early selection, simultaneous selection for multiple traits, including resistance gene pyramiding and selection of traits that are expensive to phenotype. The use of MAS in fruit breeding programmes is considerably limited with the extent of cost reduction with their use and application of MAS in apple scab resistance gene pyramiding helps to lower down the internal cost by not having to make test crosses and not having to evaluate the seedlings while external benefits in form of developing a new cultivar with durable resistance(s) (Bus et al 2000b). MAS is only beneficial if the linkage is tight between the marker and the gene in question or to use flanking markers. At an average recombination rate of 1% between marker and a gene 3.0% in case of 3 pyramided genes and 5.9% in case of 6 pyramided genes, progeny would be not be having all the pyramided genes (Bus et al 2000b) which could be acceptable to breeders. After the greenhouse inoculation of the seedlings and the reproducibility of results under field conditions have been strongly advisable for accuracy and MAS has been strongly advisable for accuracy and its availability to identify homozygous plants (homozygous V_f confers higher degree of resistance) (Tartarini et al 2000).

Pyramiding various resistances into commercial cultivars comes closest to a functional diversity. Owing to availability of V_f resistant cultivars and selections, V_f can easily be exploited as first source of resistance (being effective in several regions), to which others should be added to gain momentarily in this arm race between the pathogen and host. As breeders are receiving more and better tools new cultivars with several effective resistances can be released in shorter intervals. Currently sufficient resistances with closely associated molecular markers are available, so to pyramid resistances with ''marker assisted selection'' (Kellerhals et al 2009; Patocchi et al 2009). Faster generation cycles will also help by reducing the juvenility. Recently, a gene has been inserted from silver birch into Pinova by Flachowsky et al (2011) who were able to reduce the juvenility phase drastically to a generation turnover of 1 year.

Map Based Cloning

Since the most detailed maps in apple are given around the V_f locus. Therefore number of map based cloning to isolate V_f gene have been initiated by many research organization world-wide. Apple BAC libraries have been developed from atleast 4 different sources containing V_f gene (Gardiner et al 2007). First BAC library of apple cultivar containing V_f gene was developed by Vinatzer et al. (1998). Later on Xu et al (2001), Xu and Korban (2002 a,b) and Rikkerink et al 2006 (Unpublished) also developed apple BAC libraries containing scab resistant genes. For the V_f locus a cluster of four resistance gene paralogs (called as HcrVf genes) encoding receptor-like proteins (RLP) with similarity to the tomato Cf resistance genes is known. A series of 13 HcrV$_f$ candidate genes have been partly cloned from the PCR fragments spanning N-terminal LRRs 20-30 by (Boudichevskaia et al 2009). While screening Malus x domestica 'Florina' (V$_f$) bacterial artificial chromosome (BAC) genomic library by hybridization using HcrV$_{f2}$ as a probe Broggini et al (2009) obtained contained HcrV$_{f2}$ paralogs only on linkage groups 1 and 6.

Genetic Transformation

Introducing a gene by transformation is not only a useful analytical tool but also is a way to circumvent the difficulty of introgressing useful single genes into new varieties by pseudo-backcrossing. The introgression of useful simply inherited characters such as disease resistance is a relatively standard breeding strategy for autogamous crops. But introgression is not simple in allogamous crops like apple. The combination of various gene cloning/identification methods and gene transformation has made possible the gene introgression in allogamous crops like apple. First apple transformation was done by James et al (1989) using *Agrobacterium tumefaciens*. Since then much work has been done to improve the efficiency of gene transfer and regeneration and a wide array of cultivars have been transformed and regenerated (Hammerschlag and Liu, 2000; Brown and Maloney 2003). The HcrV$_{f2}$ gene, derived from the V$_f$-region of the wild apple *Malus floribunda* 821 and encoding a receptor-like protein, has proved to confer scab resistance in a transgenic susceptible cultivar (Szankowski 2009). Belfanti et al (2004) isolated the Hcr V$_{f2}$ gene from *Malus floribunda* and found it confers resistance in transgenic Gala. The expression of endochitinase from *Trichoderma atroviride* have been found to increase the resistance in transgenic Marshall MacIntosh (Bolar et al 2001) and the expression of endochitinase and exochitinase from same source had synergistic effect against scab fungus in transgenic apple (Bolar et al 2000). Hanke et al (2000) also successfully produced transgenic lines for resistance to scab by introduction of chitinolytic protein genes originated from *Trichoderma harzianum* and integration of transgenes was also confirmed. Malnoy et al (2008) have recently demonstrated that two receptor-like genes, *Vfa* 1 and *Vfa* 2, conferred resistance to apple scab. Szankowski et al (2009) obtained highly scab resistant 'Elstar' and 'Gala' apple plant transformed with HcrV$_{f2}$ gene controlled by its native promoter which is effective in conferring resistance to *Venturia inaequalis* similarly as V$_f$ introgressed in apple cultivars through classical breeding. They found that varying the length of the native promoter of HcrV$_{f2}$ influenced the degree of resistance expressed.

The transformation of apple with the cloned scab resistance gene (HcrV$_{f2}$) from *Malus floribunda* by Belfanti et al (2004) provided the first functional confirmation of a cloned apple gene. This research has progressed to expression profiling in Hcr-V$_f$-2-transformed apple plants in response to *Venturia inaequalis,* with 523 unigenes identified (Paris et al 2009). Recently, apple cv. Gala was genetically modified by inserting apple scab resistance gene Hcr V$_{f2}$ (now known as R$_{vi}$ E) under the control of its own regulatory sequence (Vanblaere et al 2013). The gene was deviced from cv. Florina and the vector used for genetic modification allowed a post-selection marker gene eliminator to achieve cis genesis.

Future Thrust

Apple scab has remained a serious problem with the orchardists adding to input costs and management efforts. The widespread use of chemicals to control the disease is being discouraged the worldover for health and environmental concerns. The advent of scab-resistant cultivars would always be preferred but given the cumbersome perennial crop breeding coupled with lack of precise information on genetics of scab resistance have been major hurdles. Though scab-resistant cultivars have been evolved, especially PRI series, still these have not been accepted widely owing to inferior fruit quality or other factors. Furthermore, apple scab resistance is not durable because of several reports of its breakdown. It necessitates the evolution of apple cultivars with durable resistance, which can be achieved through pyramiding of genes. Thus desirable resistant varieties through pyramiding resistance genes are required to be developed. Since these strategies look cumbersome and complicated through traditional breeding techniques, new insights into modern molecular biology provides tools for resistance gene localization on genetic map, gene donor selection and analysis of inheritance. The correct identification of gene sources and information on inheritance pattern would be pre-requisite. Over the years, marker technologies have been developed which has presented deep insight into the genetics of apple scab resistance but still it is a long way to go. There is a continuous need to elucidate further into the genetics especially identification of QTL / markers linked to apple scab. Genetic engineering may render a popular cultivar disease resistant by creating functional resistance diversity in an otherwise homogeneous genetic cultivar. Once better technologies are available, such as replacing non-functional resistance alleles with still effective resistance alleles such genetically modified cultivars may become more acceptable. The main constraint is currently the lack of cloned resistance genes, being $HcrV_{f2}$ cloned in 2004, the only one available. In a longer time span new approaches could derive from the identification and sequencing of avirulence genes of *V. inaequalis*, for example, the product of the avir V_f which is recognized by $HcrV_{f2}$, as well as the mutated gene whose product is no longer recognized, rendering the host susceptible once again. Once we understand the binding patterns and changes that lead to non-recognition, it may be possible to create artificial resistance alleles by making small changes in the recognition zone (LRR) and selecting artificial mutations of the $HcrV_{f2}$ so that the pathogen product may once again be recognized. On the whole, mechanism of action of scab resistance gene is still largely unknown, and there is absolute need to continue developing insights form,the development and application of new markers in breeding programmes. This would pave the way for evolving new scab resistant apple cultivars with desirable resistance (Daniels 2013).

References

Alston FH, Phillips KL, and Evans KM 2000 A *Malus* list, *Acta Horticulturae* **538** 570

Baniulis D, Gelvonauskienė D, Gelvonauskis B, Rugienius R, Bendokas V, and Stanys V 2008 Genetic studies of apple scab resistance and application of molecular biology methods in apple tree breeding, *Sodininkystė ir Daržininkystė,* **27** 61-75

Belfanti E, Silfverberg-Dilworth E, Tartarini S, Patocchi A, Barbieri E, Zhu J, Vinatzer BA, Gianfranceschi L, Gessler C, and Sanasavini S 2004 The HcrVf2 gene from a wild apple confers scab resistance to a transgenic cultivated variety, *Proceedings of National Academy of Sciences USA* **101** 886-890

Benaouf G, Parisi L, and Laurens F 1997 Inheritance of *Malus floribunda* clone 821 resistance to *Venturia inaequalis*, *IOBC/WPRS Bulletin* **20** 1-7

Bolar JP, Norelli JL, Harman GE, Brown SK, and Aldwinkle HS 2001 Synergistic activity of endochitinase and exochitinase from *Trichoderma atroviride* (*Trichoderma harzianum*) against the pathogenic fungus (*Venturia inaequalis*) in transgenic apple plants, *Transgenic Research* **10** 533-543

Bolar JP, Norelli JL, Wong KW, Hayes CK, Harman GE, and Aldwinkle HS 2000 Expression of endochitinase from *Trichoderma harzianum* in transgenic apple increases resistance to apple scab and reduce vigor, *Phytopathology* **90** 72-77

Boudichevskaia A, Flachowsky H, and Dunemann F 2009 Identification and molecular analysis of candidate genes homologous to HcrVf genes for scab resistance in apple, *Plant Breeding* **128** 84-91

Boudichevskaia A, Flachowsky H, Peil A, Fischer C, and Dunemann F 2006 Development of a multiallelic SCAR marker for the scab resistance gene Vr1/Vh4/Vx from R12740-7A apple and its utility for molecular breeding, *Tree Genetics and Genomes* **2** 186-195

Bringhurst RS 1983 Breeding Strategy, in *Methods in Fruit Breeding* (eds) JN Moore and J Janick, West Lafayette, IN, USA: Purdue University Press

Broggini GAL, Galli P, Parravicini G, Gianfranceschi L, Gessler C, and Patocchi A 2009 HcrVf paralogs are present on linkage groups 1 and 6 of Malus, *Genome* **52** 129-138

Brown AG 1975 Apples, in *Advances in Fruit Breeding,* (eds) JN Moore and J Janick, West Lafayette, IN, USA: Purdue University Press

Brown SK and Malooney KE 2003 Genetic improvement of apple: breeding, markers, mapping and biotechnology, in *Apples: Botany, Production and Uses* (eds) DC Ferree and I Warrington, Cambridge, MA: CABI Publishing, 31-59 pages

Bus VGM, Bradley S, Hofstee M, Alspach P, Brewer L and Luby I 2000a. Increasing genetic diversity in apple to improve the durability of pests and disease resistance. *Acta Horticulturae* **538** 185-190

Bus VGM, Ranatunga C, Gardiner S, Basset H, and Rikkerink E 2000b Marker assisted selection for pest and disease resistance in the Newzealand apple breeding programme, *Acta Horticulturae* **538** 541-545

Bus VGM, Laurens FND, van de Weg WE, Rusholme RL, Rikkierink EHA, Gardiner SE, Bassett HCM, Kodde LP and Plummer KM 2005a The Vh8 locus of a new gene for gene interaction between *Venturia inaequalis* and the wild apple *Malus sieversii* is closely linked to the Vh2 locus in *Malus pumila* R 12740-7A, *New Phytologist* **166** 1035-1049

Bus VGM, Rikkerink EHA, van de Weg WE, Rusholme RL, Gardiner SE, Bassett HCM, Kodde LP, Parisi L, Laurens FND, Meulenbroek EJ and Plummer KM 2005b The Vh2 and Vh4 scab resistance genes in two differential hosts derived from Russian apple R 12740-7A map to the same linkage group of apple, *Molecular Breeding* **15** 103-116

Bus VGM, van de Weg WE, Durel CE, Gessler C, Calenge F, Parisi L, Rikkerink E, Gardiner S, Patocchi A, Meulenbroek M, Schouten H and Laurens F 2004 Delineation of a scab resistance gene cluster on linkage group 2 of apple, *Acta Horticulturae* **663** 57-62

Calenge F, Faure A, Goerre M, Gebhardt C, Van de Weg WE, Parisi L and Durel CE 2004 Quantitative trait loci (QTL) analysis reveals both broad spectrum and isolate-specific QTL for scab resistance in an apple progeny challenged with eight isolates of *Venturia inaequalis, Phytopathology* **94** 370-379

Calenge F, Van der linden CG, Van de Weg E, Schouten HJ, Van Arkel G, Denance C and Durel CE 2005a Resistance gene analogues identified through the NBS profiling method map close to major gene and QTL for disease resistance in apple. *Theoretical and Applied Genetics* **110** 660-668

Cheng FS, Weeden NF, Brown SK, Aldwinkle HS, Gardiner SE, and Bus VG 1998 Development of a DNA marker for Vm, a gene conferring resistance to apple scab, *Genome* **41** 208-214

Conner PJ, Brown SK, and Weeden NF 1997 Randomly amplified polymorphic DNA based genetic linkage map of three apple cultivars, *Journal of American Society for Horticultural Science* **122** 350-359

Crandall CS 1926 Apple breeding at the University of Illinois, *Illinois Agr. Expt. Station Bulletin* **275** 341-600

Crosby JA, Janik J, Pecknold PC, Korban SS, O'Conner PA, Ries SM, Goffreda S, and Voordeckers A 1992 Breeding apples for scab resistance:1945-1990, *Acta Horticulturae* **317** 43-90

Crowe AD 1975 Nova Easygrow apple, *Fruit Varieties Journal* **29** 37

Crute IR and Pink DAC 1996 Genetics and utilization of pathogen resistance in plants, *Plant Cell* **8** 1747-1755

Daniels B 2013 Response of apple (*Malus* x *domestica*) to *Venturia inaequalis* the caused agent of apple scab : A real-time PGR and proteomics study, Ph.D. thesis, University of Leuven, Belgium.

Dayton DF and Williams EB 1968 Independent genes in *Malus* for resistance to *Venturia inaequalis, Proceeding of American Society for Horticultural Science* **92** 89-94

Dayton DF, Mowry JB, Hough LF, Bailey CH, William EB, Janick J, and Emerson FH 1970 Prima-an early fall red apple with resistance to apple scab, *Fruit Varieties Journal* **24** 20-22

Denerdi F, Hough LF, and Camilo, AP 1988 'Primicia' apple. *Hort Science* **23** 632

Durel CE, Lespinasse Y, Calenge F, van der Weg WE, Koller B, Dunemann F, Thiermann M, Evans K, James C, and Tartarini S 2002 Four years of the European D.A.R.E. Project: numerous results on apple scab and powdery mildew resistance gene mapping, in *Plant Animal and Microbe Genomes X Conference*, San Diego, California. http://www.intl-pag.org

Durel CE, Parisi L, Laurens F, Van de Weg WE, Liebhard R and Jourjon MF 2003 Genetic dissection of partial resistance to race 6 of *Venturia inaequalis* in apple, *Genome* **46** 224-234

Durel CE, Van de Weg WE, Venisse JS, and Pariori L 2000 Localization of a major gene for scab resistance on the European genetic map of the Prima x Fiesta cross. 1OBC/WPRS Bulletin **23** 245-248

Durel CE, van der Weg WE, Venisse JS, and Parisi L 1999 Localisation of a major gene for apple scab resistance on the European genetic map of the Prima x Fiesta cross, 5th Workshop on Integrated Control of Pome Fruit Diseases, Fontvraud, Abbaye, France

Durham RE and Korban SS 1994 Evidence of gene introgression in apple using RAPD markers, *Euphytica* **79** 109-114

Erdin N, Tartarini S, Broggini GAL, Gennari F, Sansavini S, Gessler C, Patocchi A 2006 Mapping of the apple scab resistance gene Vb, *Genome* **49** 1238-45

Fischer C 1993 Results in the resistance breeding of apple in Dresdan-Pillnitz, *Acta Horticulturae* **347** 163-168

Fischer C, Bukartschuk V, Bondarenko A, and Artamonova E 1983 Erste Ergebnisse zur Stabilitat der Scorfresistenz beim Apfel unter Verschiedenen Okologioshen Bedingungen in der UdSSR and DDR-vorlaufige Mitteilung, *Arch. Gartenbau* **31** 263-264

Fischer C, Schreiber H, Buttner R, and Fischer M 1999 Testing scab resistance stability of new resistant cultivars within the apple breeding programme, *Acta Horticulturae* **484** 449-454

Fischer C, Bonderenko A, and Artomonova E 1994 Results on the stability of scab resistance in apple breeding. *In Proceedings of EUCARPIA Symposium on Progress in Temperate Fruit Breeding*, Wademwl/Einsiedder, Switzerland

Fisher 2000 Multiple resistant apple cultivars and consequences for apple breeding in the future, *Acta Horticulturae* **538** 229-234

Flachowsky H, Le Roux PM, Peil A, Patocchi A, Richter K, and Hanke MV 2011 Application of a high-speed breeding technology to *apple* (*Malus domestica*) based on transgenic early flowering plants and marker-assisted selection, *New Phytologist* **192** 364-377

Fries E 1819 Spilocaea pomi. Fr. Nov. Fl. Suec. **5** 79

Gardiner S, Basset H, Murdoch J, Meech S, Cook M, Bus V, Ranatunga C, and Rikkerink E 2001 Major pest and disease resistance loci in apple available to breeders, *In Plant and Animal Genome IX Conference San Diego, CA, USA*

Gardiner S, Bus V, White A, Noiton D, Basset, H, Rikkerink E, Ball R, Guilford P, and Foriter R 1996 A genetic map around the Vf gene for resistance to apple scab. Ibid. pp. 9 (Abstr. No. 33)

Gardiner S, Bus VGM, Rushmore RL, Chagne D, and Rikkerink EHA 2007 Apple, in *Genome Mapping and Molecular Breeding in Plants, Vol 4: Fruits and Nuts* (ed) C Kole, Berlin, Heidelberg, Germany: Springer-Verlag, 1-62 pages

Gessler C 1989 Genetics of the interaction *Venturia inaequalis*-Malus: the conflict between theory and reality, in *Integrated Control of Pome Fruit Diseases*, (eds) C Gessler, DJ Butt and B Koler, IOBC Bulletin Vol. II. 346

Gianfranceschi L, Kellerhals M, and Gessler C 1996 Molecular markers for scab resistance in apple, 9, *In Eucarpia Symposium on Fruit Breeding and Genetics*, Oxford, UK

Gianfranceschi L, Seglia N, Kellerhals M and Gessler C. 1998. Molecular markers applied to apple breeding, analysis of oligogenic and single gene resistances, *Acta Horticulturae* **484** 417-428

Granhall I 1953 Kan apfelskorven bemastras genom vaxforadding. *Svrig. Pomol. Foren Arsskr* **54** 17-28

Gupta G K 1993 *Apple Scab* (*Venturia inaequalis*), Bombay: Sandoz (India) Limited

Gygax M, Gianfranceschi L, Liebhard R, Kellerhals M, Gessler C, and Patocchi A 2004 Molecular markers linked to the apple scab resistance gene Vbj from *Malus baccata* Jackii, *Theoretical and Applied Genetics* **109** 1702-1709

Hammerschlag FA and Liu Q 2000 Generating apple transformants free of *Agrobacterium tumefaciens* by vacuum infiltrating explants with an acidified medium and with antibiotic, *Acta Horticulturae* **530** 103-111

Hanke V, Hiller I, Klotzsche G, Norelli JL, and Aldwinkle HS 2000 Transformation in apple for increased disease resistance, *Acta Horticulturae* **538** 611-616

Hemmat M, Brown SK, and Weeden NF 2002 Tagging and mapping scab resistance gene from R 12740-7A apple, *Journal of American Society for Horticultural Sciences* **127** 365-370

Hemmat M, Brown SK, Aldwinkle HS, Mahalenbacher SA, and Weeden NF 2003 Identification and mapping of markers for resistance to apple scab from Antonovka and Hensen's baccat #2, *Acta Horticulturae* **622** 153-161

Hemmat M, Weeden NF, Aldwinkle HS, and Brown SK 1998 Molecular markers for scab resiatnce (Vf) region in apple, *Journal of American Society for Horticultural Science* **123** 992-996

Hemmat M, Weeden N, Manganaris A, and Lowson D 1994 Molecular marker linkage map for apple, *Journal of Heredity* **85** 4-11

Hogenboom NG 1993 Economic importance of breeding for disease resiatnce, in *Durability of disease resistance* (eds) TH Jacobs and JE Parlevliet, Kluwer Academic Publishers, 5-9 pages.

Hough LF 1944 A survey of the scab resistance of the foliage on seedling in selected apple programmes, *Proceedings of. American Society of Horticulture Sciences* **44** 260-272

Hough LF, Williams EB, Dayton DF, Shay JR, Bailey CH, Mowry JB, Janick J, and Emerson FH 1970 Progress and problems in breeding apples for scab resistance, in *Proc. Angers Fruit Breed. Symp. Editions S.E.I. Versailles, France*

James CM, Clarke JB, and Evans KM 2004 Identification of molecular markers linked the mildew resistance gene *Pl-d* in apple, *Theoretical and Applied Genetics* **110** 175-181

James DJ, Passey AJ, Barbaro DJ, and Bevan MW 1989 Genetic transformation of apple (*Malus pumila* Mill.) using a disarmed Ti-binary vector, *Plant Cell Report* **7** 658-661

Janick J, Cummins JN, Brown SK, and Hemmat M 1996 Apples, in *Fruit Breeding Vol. I: Tree and Tropical Fruits* (eds) J Janick and JN Moore, New York, USA: John Wiley and Sons, 1-77 pages

Johnston R 1981 Durable resistance: definition of genetic control and attainment in plant breeding, *Phytopathology* **71** 567-568

Kellerhals M, Dolega E, Koller B, and Gessler C 2000 Advances in marker assisted apple breeding, *Acta Horticulturae* **538** 535-540

Kellerhals M, Székely T, Sauer C, Frey JE, and Patocchi A 2009 Pyramiding scab resistances in apple breeding, *Erwerbsobstbau* **51** 21-28

Kenis K and Keuemans V 2005 Genetic linkage maps of two apple cultivars (*Malus* × *domestica* Borkh.) based on AFLP and microsatellite markers, *Molecular Breeding* **15** 205-219

Koller B, Gianfranceschi L, Seglias N, McDermott J, and Gessler C 1994 DNA markers limked to Malus floribunda 821 scab resistance, *Plant Molecular Biology* **26** 597-602

Korban SS and Chen H 1992 Biotechnology of apple, in *Biotechnology of Perennial Fruit Crops* (eds) R Litz and FA Hammerschleg, Wallingford, UK: CAB International 203-227 pages

Kruger J 1988 Bestandigkeit der Schorfresistenz aus Malus floribunda 821 aufdem Versuchsfeld der Bundesforschungsanstalt fiir gartenbauliche Pflanzenzuchtung in Ahrensberg, *Erwerbsobstbau* **30** 52

Kruger J 1989 Scab resistance of apple cultivars, selection and progenies with the V_f gene, *OILB/WPRS Bulletin*, **XII/6** 161-167

Kruger J 1991 Schorfbefall von Nachommen aus Kreuzungen mit der schorfresistenten Apfelsorte Prima, *Journal of Plant Disease Protectection* **98** 73-76

Kumar K and Sharma SD 1999 Breeding scab resistant apples: new directions, *Journal of Genetics and Breeding* **53** 155-164

Lamb RC and Hamilton JM 1956 Breeding scab resistance apples, New York State Agriculture Experimental Station, Geneva Farm Reporter **22**15

Lamb RC, Aldwinkle HS, and Terry DD 1985 Freedom, a disease resistant apple, *Hort Science* **20** 774-775

Lamb RC, Aldwinkle HS, Way RD, and Terry DD 1978 Liberty, a new disease resistant apple, *New York's Food and Life Science Bulletin* **73**

Lespinasse Y 1989 Breeding pome fruits with stable resistance to diseases, Genes, resistance mechanism, present work and prospects in *Integrated control of pome fruit diseases, Vol. II.* (eds) C Gessler, DJ Butt, and B Koller, 100-115 pages

Lespinasse Y and Durrel CE 1999 DARE- Durable apple resistance in Europe: a new European project on apple scab and powdery-mildew resistance based on a close collaboration between geneticists, pathologists, breeders and nurserymen, In *Plant and Animal Genome VII Conference, San Diego, CA, USA*

Laurens F 1999 Review of the current apple breeding programmes in the world: Objectives for scion cultivar improvement, *Acta Horticulturae* **484** 163-170

Liebhard R, Gianfranceschi L, Koller B, Ryder CD, Tarchini R, Van de Weg WE, and Gesler C 2002 Development and characterization of 140 new microsatellites in apple (*Malus* × *domestica* Borkh.), *Molecular Breeding* **10** 217-241

Liebhard R, Koller B, Gianfranceschi L, and Gessler C 2003a Creating a saturated reference map for the apple (*Malus* × *domestica* Borkh.) genome, *Theoretical and Applied Genetics* **106** 1497-1508

Liebhard R, Koller B, Patocchi A, Kellerhals M, Pfammatter W, Jermini M, and Gessler C 2003b Mapping quantitative field resistance against apple scab in a Fiesta × Discovery progeny, *Phytopathology* **93** 493-501

Maliepaard C, Alston F, van Arkel G, Brown L, Cheve AUE, Dunemann F, Evans KM, Gardiner S, Guilford P, van Heusden AW, and Janse J 1998 Aligning male and female linkage maps of apple (*Malus pumila* Mill.) using multi-allelic markers, *Theoretical and Applied Genetics* **97** 60-73

Malnoy M, Xu M, Borejsza-Wysocka EE, Korban S, and Aldwinckle HS 2008 Two receptor-like genes, *Vfa* 1 and *Vfa* 2, confer resistance to the fungal pathogen *Venturia inaequalis* inciting apple scab disease, *Molecular Plant-Microbe Interactions* **21** 448-458

Manganaris AG and Alston FH 1992 Genetics of leucine aminopeptidase in apple, *Theoretical and Applied Genetics* **83** 345-352

Merwin IA, Brown SK, Rosenberger DA, Cooley DR, and Berkett LP 1994 Scab resistant apples for the northeastern United States: New prospects and old problems, *Plant Disease* **78** 4-10

Mulcahy DL, Cresti M, Sansavini S, Douglas GC, Linskens HF, Mulcahy GB, Vignani R, and Pancaldi M 1993 The use of random amplified polymorphic DNAs to fingerprint apple genotypes, *Scientia Horticulturae* **54** 89-96

Nybom H and Schaal B A. 1990. DNA fingerprinting applied to paternity analysis in apples (Malus × domestica), *Theoretical and Applied Genetics* **79** 763-768

Nybom H and Schaal BA 1990 DNA fingerprinting applied to paternity analysis in apples (*Malus* X *domestica*), *Theoretical and Applied Genetics* **79** 763-768

Oraguzie NC, Rikkerink E, Gardiner S, Bus V, Currie A, Rushmoe R, and Volz R 2004 A review of developments in breeding techniques and gene technology tools for new cultivar development in apple, in *Recent Research Developments in Genetics and Breeding Vol I* (ed) SG Pandalai, UK: CAB International, 223-257 pages

Paris R, Cova V, Pagliani G, Tartarini S, Komjanc M, and Sansavini S 2009 Expression profi ling in Hcr-Vf-2-transformed apple plants with in response to *Venturia inaequalis, Tree Genetics and Genome* **5** 81-91

Parisi L, Lespinasse Y, Guillaumes J, Kruger J 1993 A new race of *Venturia inaequalis* virulent to apple with resistance due to the V_f gene, *Phytopathology* **83** 533-537

Parlevliet JE 1993 What is durable resistance, a general outline, in *Durability of disease resistance* (eds) TH Jacobs and JE Parlevliet, Kluwer Academic Publishers, 23-39 pages

Parlevliet JE and Zadoks JC 1977 The integrated concept of disease resistance: A new view including horizontal and vertical resistance in plants, *Euphytica* **26** 5-21

Patocchi A, Bigler B, Koller B, Kellerhals M and Gessler C 2004 Vr2: a new apple scab resistance gene. *Theoretical and Applied Genetics* **109** 1087-1092

Patocchi A, Frei A, Frey JE, and Kellerhals M 2009 Towards improvement of marker assisted selection of apple scab resistant cultivars: *Venturia inaequalis* virulence surveys and standardization of molecular marker alleles associated with resistance genes, *Molecular Breeding* **24** 337-347

Patocchi A, Walser M, Tartarini S, Broggini GAL, Gennari F, Sansavini S and Gessler C 2005 Identification by genome scanning approach (GSA) of a microsatellite tightly associated to the apple scab resistance gene V_m, *Genome,* **48** 630-636

Rikkerink EHA, Hilario E, Thrush A, Gardiner SE, Bassett HC, Jordan M, Bus VGM, Forster RLS, and Crowhurst RN 2006 PCR based cloning of fragments from the apple NBS LRR gene class and phylogenetic analysis identifies clades widely conserved in the Rosids (Submitted-May, 2006 to the EMBL/GenBank/DDBJ databases)

Roberts AL and Crute IR 1994 Apple scab ressitance from Malus floribunda 821 (V_f) is rendered ineffective by isolates of *Venturia inaequalis* from *Malus floribunda. Norwegian Journal of Agricultural Sciences* **17** 403-406

Rudolf CF and Schmidt M 1934 *Venturia inaequalis* (Cooke.) Aderh. II. Zur schorfwiderstandsfahiger Apfelsorten Den Zuchter **6** 288-294

Rudolf CF and Schmidt M 1935 Der Erreger des Apfelschorfes Venturia inaequalis (Cooke) Aderh. Grundlagen and Moglichkeiten fur senie Bekampfung auf Zuchterischen Wage II Der Zuchter **7** 65-74

Schmidt M 1936a. *Venturia inaequalis* (Cooke) Aderhold. V. Weitere Untersuchungen uber die auf verschiedenen *Baumen lebenden* Populationen des Apfelschorpilzes, *Gartenbauwissenschaft* **10** 422-427

Schmidt M 1936b *Venturia inaequalis* (Cooke) Aderhold. VI. Zur Frage nach dem Vakommen physiologisch speialisierter Rassen beim Erreger des Apfelschorfes, Erste Mitteilung. Afelschorfpilzs **10** 478-499

Schmidt M 1942 Untersuchungen uber den Biologie von Venturia inaequalis in Zusammendang mit der Zuchtung Schorfwiderstandfahiger Apfelsorten, *Gartenbauwissenschaft* **18** 19-20

Seglias NP and Gessler C 1997 Genetics of apple powdery mildew resistance from *Malus zumi* (Pl2), *IOBC/WPRS Bulletin* **20** 195-208

Shay JR 1960 Breeding vegetable and fruit crops for resistance to disease. Biological and chemical control of plant and animal pests. *American Association for the Advancement of Science* **61** 229-244

Silfverberg-Dilworth E, Matasci CL, Van de Weg WE, Van Kaauwen MPW, Walser M, Kodde LP, Soglio V, Gianfranceschi L, Durel CE, Costa F, Yamamoto T, Koller B, Gessler C, and Patocchi A 2006 Microsatellite markers spanning the apple (*Malus* × *domestica* Borkh.) genome, *Tree Genetics and Genomes,* **2** 202-224

Spangelo LPS, Julien JB, Racicot HN, and Blair DS 1956 Breeding apples for resistance to scab, *Canadian Journal of Agricultural Science* **36** 329-338

Spangelo LPS, Lenty SJ, Heeney HB, and Modderman LL 1974 'MacFree' apple. *Canadian Journal of Plant Science* **54** 847

Szankowski I, Waldmann S, Degenhardt J, Patocchi A, Paris R, Silfverberg-Dillworth E, Broggini G, and Gessler C 2009 Highly scab-resistant transgenic apple lines achieved by introgression of HcrVF2 controlled by different native promoter lengths, *Tree Genetics & Genomes* **5** 349-358

Tartarini S 1996 RAPD markers linked to the Vf gene for scab resistance in apple, *Theoretical and Applied Genetics* **92** 803-810

Tartarini S, Gennari F, Pratesi D, Palazzetti C, Sansavini S, Parisi L, Fouillet V, and Durel C E 2004 Characterization and genetic mapping of a major scab resistance gene from the old Italian apple cultivar "Durello di Forli", *Acta Horticulturae* **663** 129-133

Tartarini S, Sansavini S, Vinatzer B, Gennari F, and Domizi C 2000 Efficiency of marker assisted selection (MAS) for the V_f scab resistance gene, *Acta Horticulturae* **538** 549-552

Urbanovich O and Kazlovskaya Z 2008 Identification of scab resistance genes in apple trees by molecular markers, *Sodininkystė ir Daržininkystė* **27** 347-357

Van der Linden CG, Wouters DCAE, Mihalka V, Kochieva EZ, Smulders MJM, and Vosman B 2004 Efficient targeting of plant disease resistance loci using NBS profiling, *Theoretical and Applied Genetics* **109** 384-393

Vanblaere T, Flachowsky H, Gess ler C and Broggini GAL 2013 Molecular characterization of cisgenic lines of apple `Gala' graying the Rui6 scab resistance gene, *Plant Biotechnology Journal* DoI: 10.1111/pp. 12110 (published online 3 September 2013)

Vinatzer BA, Zhang HB and Sansavini S 1998 Construction and characterization of a bacterial artificial chromosome library of apple, *Theoretical and Applied Genetics* **97** 1183-1190

Wallroth FG 1833 *Cladosporium denedriticum, Flora Cryptogamic Germaniae* **1** 169

Way RD, Aldwinkle HS, Lamb RC, Rejman A, Sansavini S, Shen T, Watkins R, Westwood MN, and Yoshida Y 1989 Apples (*Malus*), *Acta Horticulturae* **290** 1-62

Weeden N F and Lamb R C. 1987. Genetics and linkage analysis of 19 isozyme loci in apple scab. J. Amer. Soc. Hort. Sci. 112:865-872.

Weeden NF and Lamb RC 1985 Identification of apple cultivars by isozyme phenotypes, *Journal of the American Society for Horticultural Science* **110** 509-515

Williams EB and Brown AG 1968 A new physiological race of *Venturia inaequalis*, incitant of apple scab, *Plant Disease Reports* **52** 799-801

Williams EB and Kuc J 1969 Resistance in *Malus* to *Venturia inaequalis, Annual Review of Phytopathology* **1** 223-246

Xu M and Korban SS 2002a AFLP-derived SCARs facilitate construction of a 1.1.Mb sequence-ready map of a region that spans the Vf locus in the apple genome, *Plant Molecular Biology* **50** 803-818

Xu M and Korban SS 2002b A cluster of four receptor like genes resides in the Vf locus that confers resistance to apple scab disease, *Genetics* **162** 1995-2006

Xu M, Song J Q, Cheng Z K, Jiang J M and Korban S S. 2001_b. A bacterial artificial chromosome library (BAC) of *Malus floribunda* 821 and contig construction for positional cloning of the apple scab resistance gene Vf. *Genome* 44:1104-1113.

Yang H Y, Korban S S, Kruger J, and Schmidt H 1997b A randomly amplified polymorphic DNA (RAPD) marker tightly linked to the scab resistance gene V_f in apple, *Journal of the American Society for Horticultural Science* **122** 47-52

Yang HY and Kruger J 1994 Identification of a RAPD marker linked to the V_f gene for scab resistance in apple, *Euphytica* **77** 83-87

Yang HY, Korban SS, Kruger J, and Schmidt H 1997a The use of a modified bulk segregant analysis to identify a molecular marker linked to scab resistance gene in apple, *Euphytica* **94** 175-182

Zwintscher M 1954 Uber die Widerstandrfahigkeit van F_2-Barterden den Apfels gegenuber dem Scorferreger *Venturia inaequalis* (Cooke) Aderhold, *Gartenbauwissenschaft* **19** 22-25.

11

Biotechnological Approaches for the Plant Disease Management

Monica Sharma and Sapna Devi

Department of Plant Pathology, Dr. Y.S. Parmar University of Horticulture & Forestry, Nauni, Solan, HP-173230

Since the beginning of agriculture, farmers have had to develop means for managing weeds, insect pests, and diseases. Even today, with all the scientific research on crop protection, it is estimated that insects, diseases, weeds, and animal pests eliminate half the food produced in the world during the growing, transporting, and storing of crops (Ellis and Boehm, 2008). Emerging, re-emerging and endemic plant pathogens continue to challenge our ability to safeguard plant health world-wide. Plants can not move so they have to defend themselves against the diseases by building up their resistance. They showed to be equipped with the arms to fight against the diseases causing organisms. Biotechnological tools provide a great opportunity to build up resistance in plants against diseases and microorganisms causing diseases. Exploitation of biotechnological approaches is a faster/quicker method compared to conventional methods of detection and management of plant diseases.

A. Detection of phytopathogens

Accurate detection of phytopathogenic organisms is crucial for virtually all aspects of plant pathology, from basic research on the biology of pathogens to the control of the diseases they cause. Early and accurate diagnoses and pathogen surveillance on local, regional, and global scales are necessary to predict outbreaks and allow time for development and application of mitigation strategies. Disease diagnosis and pathogen detection are central to our ability to protect crops and natural plant systems, and are the crucial prelude to undertaking prevention and management measures. The concept of integrated pest management (IPM) is predicated on the ability to detect and diagnose diseases and pests, which then informs management decisions. Failures in pathogen detection and disease diagnosis lead directly to inadequate disease control and reductions in crop production and quality, and hence trade. Trade is further negatively impacted by noncompliance with sanitary and phytosanitary (SPS) requirements of importing countries. Traditionally, the most predominant techniques used to identify plant pathogens relied upon culture-based morphological approaches. The major limitations of these methods, however, are the reliance on the ability of the organism to be cultured, the time consuming and laborious nature, and the requirement for extensive taxonomical knowledge, all together often complicating timely disease management decisions. After the entrance and establishment of phytopathogens like bacteria, phytoplasmas, viruses and viroids in the host plant, these pathogens are difficult to control. Thus preventive measures are essential to minimize the losses they cause each year in different crops. Rapid and accurate methods for detection and diagnosis of these plant pathogens are essential to apply preventive measures. Precise and quick detection will help to apply treatments, undertake agronomic measures or proceed with eradication practices, particularly for quarantine pathogens. With the opening of the borders of many countries and increased free-trade agreements, rapid testing for possible contamination with quarantine organisms is in high question. There is a need for techniques which are rapid and of high accuracy for quarantine phytopathogens, because the risk of the disease and the spread of the inoculum must be reduced to nearly zero (López et al 2003).

Molecular techniques can circumvent many of the drawbacks of traditional detection methods. Molecular methods of detection have been revolutionized with the use of the polymerase chain reaction (PCR) (Mullis and Faloona 1987). In general, these methods are much faster, more specific, more sensitive, and more accurate, and can be performed and interpreted by personnel with no specialized taxonomical expertise. Perhaps even more important, these techniques allow detection of non-culturable microorganisms. One of the most important advantages that molecular-based detection techniques have over conventional diagnostic methods is the power to, in principle, distinguish closely related organisms. PCR testing for many pathogens is now routine and affordable in the developed world.

Detection of fungal pathogens

For the detection of fungal pathogens, nuclear ribosomal DNA (rDNA) has been used most extensively in the current era. rDNA is being used frequently in fungal molecular diagnostics and fungal molecular phylogenetic studies (White et al 1990). A large amount of rDNA sequence data is available in public databases, which helps in a diagnostic assay. The availability of extensive sequence data allow comparison of sequences in question and helps in specific identification of the fungal pathogens. Moreover, rDNA genes are present in high copy number in any genome and this permits a highly sensitive detection of fungal pathogens. rDNA gene occurs as a structured unit consisting of three ribosomal RNA subunit genes: Large ribosomal unit, small ribosomal subunit and internal transcribed spacers (ITS) (Fig 1). ITS region contains alternating areas of high conservation and high variability and is therefore area of particular importance to fungal diagnostics. The variability in ITS region allows the detection and discrimination below the species level (Atkins et al 2003). In addition to the ITS region, other house-keeping genes such as beta-tubulin, actin, and elongation factor 1-alpha (O'Donnell et al 1998; Fraaije et al 1999; Weiland and Sundsbak 2000) are also being studied more intensively during these days for the fungal phylogenetic analyses.

Another strategy for the detection and characterization of fungal pathogens involves the screening of random oligonucleotides in the genome. This can be achieved by several techniques, including random amplified polymorphic DNA (RAPD) and amplified fragment length polymorphism (AFLP) technology. Random primers are used to amplify the regions containing the corresponding sequences. PCR based methods have very high specificity, so precautions must be taken to avoid the "false positive" results which may arise due to the slightest carry-over contamination of DNA. Therefore, stringent conditions and controls are necessary, such as guarding the reagents and samples for accidental DNA contamination via aerosols, running negative controls simultaneously with the test samples and having separate dedicated areas for pre- and post-PCR handling (Kwok and Higushi, 1989). The efficiency of PCR assays could be drastically reduced or even inhibited due to the presence of many naturally occurring compounds in the extracted DNA. Mainly these compounds include phenolic compounds and heavy metals and in addition humic acids and fulvic acids also cause problems in DNA amplification during PCR. To avoid these problematic compounds in extracted DNA, improved extraction methods or optimized extraction kits should be used for the extraction highly purified DNA from samples (McCartney et al 2003). Before in invention and popularization of PCR based techniques, specific probes based techniques such as restriction fragment length polymorphism (RFLP) were used for the detection of fungal DNA. These probe based techniques had the limitation of specificity and often led to "false negatives". PCR assays are limited in the number of pathogens that can be tested at once, and thus a preliminary differential diagnosis must be done to narrow down the possible causes and reduce costs due to unnecessary testing. Crops are often infected by numerous pathogens and in case of complex diseases where more than one pathogen is involved in the disease development, it is desirable to develop assays that can detect multiple pathogens simultaneously. One possibility to achieve this is the use of multiple primer sets in the same PCR reaction. Nevertheless, the development of a reliable multiplex PCR, in order to resolve at least a few amplicons by gel electrophoresis, is a significant technical challenge (Elnifro et al 2000). In real-time / quantitative PCR (qPCR), using the fluorescence dyes which binds to the double stranded DNA, the amplification process is monitored on computer screen and hence no gel electrophoresis is required (Heid et al 1996). The amplicons are separated based on their melting points. DNA-based assays, particularly PCR and real-time/quantitative PCR (qPCR), are being adopted in diagnostic laboratories. Quantitative PCR is more expensive than PCR as a result of high start-up costs and expensive materials, but it has been adopted for high-consequence pathogens, especially in regulatory circumstances, because of their high sensitivity and specificity (Frederick et al 2002; Li et al 2007). Currently, in plant pathology, real-time/q PCR is being used to quantify the pathogens in plants and for the quantification of disease progress (Brouwer et al 2003, Gachon et al 2004). However, for the detection of more than one pathogen in a given sample through real-time PCR, there is limitation of the availability of dyes emitting fluorescence at different wavelengths. In addition, there is limitation of the monochromatic character of the energizing light source in real-time PCR instruments (Mackay et al 2002). As a result, detection of more than a few pathogens per assay is currently not possible using these strategies. Considerable emphasis is now being placed on generic platforms such as DNA microarrays and sequencing (DNA barcoding), which will facilitate diagnosis of many unknown pathogens. Hybridization technology is combined with PCR-based detection. This combination of DNA hybridization with PCR amplification results in high levels of sensitivity, specificity, and throughput capacity (Lievens et al 2003). A pathogen detection array typically consists of many discretely located pathogen-specific detector sequences that are immobilized on a solid support, such as a nylon filter or a glass slide, to create a macroarray or a microarray, respectively. For signal amplification, in general the target DNA to be tested is amplified using consensus primers that target a genomic region containing the pathogen-specific sequences, and is labeled simultaneously or subsequently. In this way, it may thus be possible to differentiate a large number of organisms using a single PCR, provided that sufficient discriminatory potential exists within the region that is used. Subsequently, labeled amplicons are hybridized to the array under stringent conditions. This kind of approach could be used to develop diagnostic kits that can detect all the relevant pathogens of a specific crop.

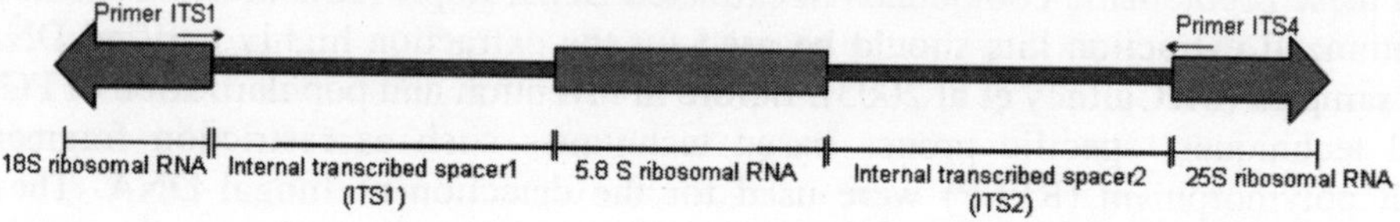

Fig 1. Schematic representation of internal transcribed spacer (ITS1-5.8S-ITS2) region

Detection of bacterial pathogens

For the detection of bacterial pathogens using PCR, primers are designed for any of the three main regions of bacteria genome: ribosomal genes, pathogenicity/ virulence genes or plasmid genes. The ribosomal genes have been employed as a source of primers in many bacteria. The advantage of this target is the universality of the ribosomes in all bacteria, and a size (1,600-2,300 bp), which enables the whole sequence of ribosomal operon to be sequenced quickly and suitable primers to be selected for further in-detail analysis. Databases of ribosomal sequences are available and specific primers can thus be designed for detection purposes. Genus-specific primers for phytobacteria are now available to detect many phytopathogenic (Louws et al 1999; Botha et al 2001; Wang et al 2006). The drawback of choosing this region is the relatively low complexity of the sequences, mainly in bacteria belonging to the same genus or species. This can be solved using sequences from the Internal Transcribed Spacer (ITS), which is more variable in its nucleotide composition. Then, this strategy employs primers from conserved regions of the 16S and 23S ribosomal genes to amplify the ITS region, which can include several tRNA genes and noncoding regions. In this case, the identification of bacteria may be based on the differences in the number and length of the amplified PCR products (Jensen et al 1993). Despite this universality, several primer sets have been developed for the specific detection of some plant pathogenic bacteria from the DNA sequence comparison of the ITS region. The pathogenicity genes used as targets can be involved in any of the several steps leading to symptoms development and can be related to virulence factors (Sato et al 1997; Coletta-Filho et al 2006), virulence or avirulence genes (Haas et al 1995; Kousik and Ritchie, 1998), toxin products (Lydon and Patterson, 2001), other factors (Tan-Bian et al 2003; Schonfeld et al 2003) etc.

These genes can be situated in plasmids or in the chromosome and are specific to a pathogen or to a group of pathogens. Plasmid DNA is also widely employed in the design of primers for important bacterial diseases. The plasmid genes amplified may be associated to pathogenicity or be of unknown function. There is problem of stability of plasmid genes except in those bacteria where the plasmid provide special fitness or pathogenicity traits (Louws et al 1999). In some cases, primers targeting plasmid genes have not been found useful for the universal detection of a pathogen, due to the lack of the plasmid in some strains.

Detection of viral and other plant pathogens

Detection of latent infections of bacteria, viruses, viroids, phytoplasm etc. is the most limiting feature of traditional methods of detection. It is not possible to detect these infections through traditional conventional methods. The most commonly used methods for the detection of viruses, viroids or graft-transmissible virus-like pathogens associated with plant material are biological indexing and serological tests. Biological assays were developed first and due to their simplicity and requirement of minimal knowledge about the pathogen, these are used widespread. Furthermore, biological indexing is still the only method of choice to detect uncharacterized but graft-transmissible agents. The sensitivity of biological assays is considered to be very high because the viral multiplication in the host plant used as indicator. One of the major limitations of such assays is the escape in detection of pathogens which multiply in host but do not show any visible symptoms. Other limitations are requirement of long time (weeks to months) to obtain results, and the impossibility of large-scale use. The accurate detection of plant viruses could be achieved with the serological techniques, especially ELISA (Clark and Adams 1977). Clark and Adams (1977) demonstrated the application of ELISA for the detection of plant viruses, and since then ELISA has been utilized for the detection and quantitative estimation of different plant pathogens such as viruses, bacteria, fungi and phytoplasma. The most important regent of ELISA is a conjugate, which is prepared by covalently linking a protein molecule that is capable of binding antibody, such as protein A with an enzyme. The basic principle of ELISA is based on the two properties of the conjugate, immunospecificity of the antibody and catalytic activity of the enzyme. ELISA is simple to perform, involves low cost and analysis of large number of samples is possible. Polyclonal antibodies are frequently used in ELISA with which problems of specificity could arise. However, this problem could be resolved with the use of specific monoclonal and recombinant antibodies (Köhler and Milstein 1975; Terrada et al 2000) yet low sensitivity of this method remains major limitation of this method. Nonetheless, ELISA remains the most easy and readily available method for detection of viral diseases in plants. At present time, ELISA has become the most widely used serological tool all over the world. ELISA is an extremely sensitive serological technique, which can detect 1 to 10 ng of virus per ml. The beauty of ELISA is that the virus can be detected in crude extract of plant and insect vector with an extraordinary specificity and sensitivity. ELISA kits are being developed by India's Ministry of Science and Technology for virus disease diagnosis in high-value crops including ornamentals, fruits, and spices. ELISA kits are convenient and easy to use, and have already been adopted widely in many developed countries.

In the last decade or two, many molecular methods have been developed for the detection of many bacteria, Spiroplasma, phytoplasmas, viruses, and viroids in plant or environmental samples (Alvarez, 2004). These detection methods could be grouped into two broader categories: a) RNA level: RT-PCR, NASBA or AmpliDet RNA; and b) DNA level: hybridisation, FISH (Fluorescent *in situ* hybridization), and PCR variants (conventional PCR, nested PCR, cooperative PCR, multiplex PCR, qPCR) (Bonants et al 2005). The molecular diagnostics can provide the degree of discrimination needed to detect and monitor plant diseases, which is not always obtained by other types of analysis. The use of molecular techniques in bacterial taxonomy allows different taxa of etiologically significant bacteria to be separated (De Boer et al 2007). Furthermore, this is the age of genomics, in which entire DNA or RNA sequences of organisms and their genetic mapping are being determined, providing the data needed to generate microarrays for the detection and identification of plant pathogens (Lévesque, 2001). Some of the latest diagnostic techniques employed in the detection of fungi, including fluorescence in situ hybridisation (FISH), DNA array technology, Multiplex tandem PCR, and Padlock probe technology with rolling circle amplification, loop-mediated isothermal amplification (LAMP) and lateral flow assay. FISH is a powerful method for the in situ detection of active growing organisms in environmental samples (Amann et al 1995). This method discriminates the metabolic active pathogen and dormant pathogens. The fluorescent tagged oligonucleotide probes are hybridized to the DNA or RNA in biological materials and finally these sites of hybridization are detected and visualized through fluorescent or confocal laser scanning microscope. In Multiplex tandem PCR, a combination of primers is designed in such a way to detect more than one pathogenic genera/species in a single PCR and the amplicons are of different sizes for different pathogenic genera/species. It enables to detect and quantify a large number of pathogens in a single assay. The development of efficient and accurate multiplex formats is often difficult and typically requires extensive optimization of reaction conditions in order to properly discriminate at least a few amplicons per reaction. Padlock probes (Nilsson et al 1994) were originally developed as a new approach for molecular analysis of DNA samples, including analysis of alleles and point mutations in the human genome (Nilsson et al 1997). A padlock probe consists of a single stranded linear oligonucleotide of about 70-100 nucleotides in length with a target-complementary region at both ends and a linker segment in between. The 5' and 3' end regions are designed to hybridize next to each other on a target strand. When properly hybridized to the target sequence, the molecule can be circularized upon ligation. Because of the need for precise base pairing at the junction where ligation should take place and the simultaneous hybridization of two different fragments, padlock probes ensure high specificity (Nilsson et al 1997). For sensitive pathogen detection, however, signal amplification is a prerequisite. One approach for the amplification of padlock probes is a PCR reaction using primers that hybridize to sequences within the spacer region of the probe (Thomas et al 1999). Another method to amplify padlock probes is rolling circle amplification (RCA), analogous to replication mechanisms of several viruses with circular genomes (Fire and Xu, 1995; Baner et al 1998; Thomas et al 1999; Nilsson et al 2002). Two types of RCA have been described: linear and hyperbranched RCA. In the first procedure, a primer hybridized at some point on the circular DNA is extended continuously using a DNA polymerase that lacks exonuclease activity. As a result, a long linear fragment composed of many tandem repeats of the complement to the circularized molecule is generated. In addition, hyperbranched (or cascade) RCA uses a second primer that binds to each generated RCA repeat. During elongation, the exoncuclease deficient DNA polymerase displaces the polymerized strand in front of it. Next, the displaced strands which are tandem repeats with identical sequences to the original padlock probe, serve again as template for the first primer, resulting in a cascade of DNA amplification. LAMP and lateral flow assay are methods for rapid and on-site diagnosis. Lateral flow assay is a new formats using antibody-based detection for very rapid presumptive on-site diagnosis. These do not require specialized equipment or knowledge. Most of them use a membrane-based lateral flow assay, in which capillary forces generate a migration of the sample extract over specific antibodies. Loop-mediated isothermal amplification (LAMP) is a powerful and novel nucleic acid amplification method. No specific thermocycler is required to perform LAMP but it needs a simple instrument that can maintain the temperature at 60-65 C. A few copies of target DNA are amplified with high specificity, efficiency, and rapidity under isothermal conditions, using a set of four to six specially designed primers and a DNA polymerase with strand displacement activity (Tomita et al 2008). LAMP products can be directly observed by the naked eye by observing the turbidity or using a UV transilluminator in the reaction tube by adding SYBR Green stain to the reaction tube separately.

For the accurate detection of pathogens in plant material or environmental samples, efficient and reproducible methods of the isolation of nucleic acid (DNA or RNA) are required. The nucleic acid should be pure and in sufficient amounts and should undergo the processing without inhibition during molecular methods like amplification, restriction digestion etc. There are different methods for the isolation of nucleic acids from different plant tissues, organisms, soil, water, air etc. There are commercial kits available through which nucleic acids can be isolated more easily, quickly and of high quality.

Most of the detection procedures developed are used in specific research areas and are only evaluated to work under the experimental conditions used. The evaluation of such detection system for their direct use in practice is rarely taken although some progress has been made recently and some methods like LAMP and lateral flow devices have been developed to assist the on-site diagnosis of phytopathogens. The detection methods should be validated and standardized for worldwide conditions before being introduced and recommended. Factors involved in validation should include (i) specificity, (ii) sensitivity, (iii) reproducibility, (iv) accuracy of results and (v) consistency and reliability of detection. The evaluation process must be monitored by an internationally recognized organization that ensures suitable expertise for the crop(s) and pathogen(s) involved in evaluating the test (Martin et al 2000)

B. Plant Disease Management

The two basic techniques of plant biotechnology which have potential to be used in the plant disease management strategies are:

(i) Plant Tissue culture (ii) Plant Genetic Engineering

(i) Plant Tissue culture

Plant tissue culture, in a broad sense, is the cultivation of plant organs, tissues or cells in test tubes on artificial media (Singh, 2002). The tissue culture techniques could be broadly grouped into following categories:

(a) Meristem tip culture (b) Somaclonal variants

(c) Tissue or cell culture (d) Protoplast fusion and somatic hybridization

(e) Embryo culture (f) Anther or pollen culture

(a) Meristem Tip Culture: Apical meristem which generally remains virus free has been used to produce virus free plants since 1952 when first report or Dahlia appeared (Morel and Martin, 1952). Meristem tip are culture in vitro and induced to develop and complete plant. The plants are infixed using ELISA or other techniques. This technique has been used successfully in many crops (Table 1) including Apple, Papaya, Banana, Potato etc.

Table 1. Virus free plant species obtained by meristem tip culture

Plant species	Virus eliminated
Brassica oleracea (cauliflower)	C_bBRSV, TUMV, CaMV
Chrysanthemum spp.	CC MVd
	CVB
Citrus spp.	CEVd
	CTV, CVV, Vein enation, yellow vein
Dahlia spp.	Complex of viruses DMV, TAV, CVMV, CVB
Dianthus caryophyllus (carnation)	CERV
	CLV, CVMV, CRSV
	CarMv
Malus spp. (apple)	Latent viruses
	ASGV
Malus pumila	ACLSV
Pisum sativum	PSbMV
Prunus spp.	PDV, PNRSV
Solanum tuberosum (Potato)	PLRV
Manihot spp. (Cassava)	ACMV, CBSV
Fragaria spp. (Strawberry)	SCrV, SVBV, Latent A SLRSV
Lilium	CMV, HyMV, LSV, LMV, latent
Solanum tuberosum	PVS, PVX, PVY, PAMV,
Lilium spp.	Carlavirus

ACMV = African cassava mosaic virus, ACLV=Apple Chlorotic leaf spot virus, ASGV = Apple stem grooving virus, CbBRSV = Cabbage black ringspot virus, CBSV = Cassava brown streak virus, CERV = Carnation etched ring virus, CaMV = Cauliflower mosaic virus, CSvd = Chrysanthemum stunt viroid, CVB = Chrysanthemum virus B, $CCMV_2$ = Chrysanthemum chlorotic mottle viroid, CEVd = Citrus exocortis viroid, CTV = citrus tristeza virus, CVV = citrus variegation virus, CMV = cucumber mosaic virus, DMV = Dahlia mosaic virus, PsbMV = Pea seed borne mosaic virus, PAMV = Potato acuba Mosaic virus, PLRV = potato leaf roll virus, PSTVd = Potato spindle tuber viriod, PVA = Potato virus A, PVS = Potato virus S, PVX = Potato virus X, PVY = Potato virus Y, PNRSV = Prunus necrotic ringspit virus, PDV = Prune dwarf virus, SCrV = Strawberry crinkle virus, SLRSV = Strawberry patent ringspot virus, SVBV = Strawberry vein banding virus, TuMV = Turnip mosaic virus.

(b) Somaclonal variants: *In-vitro* screening and selection approach has shown great potential of regenerating disease free plants in fruits, vegetables and ornamentals (Hammerschaleg et al 1995). Screening of somaclonal variants and selection of resistant cells is exploited to generate disease-free plants (Table 2). It is now well established fact that some permanent changes do occur in genomes of cells under tissue culture conditions. This has been a good tool to screen out disease resistant cells (Hammerschleg et al 1995). Somaclones are the plants regenerated from somatic cells or tissue cultured *in vitro* on synthetic and semi synthetic media. Some of the somaclones may differ from donor plants in one or more characteristics a phenomenon called somaclonal variations (Larkin and Scoweroft, 1981). The alternation in genetic mark up was found to be due to changes in the structure and number of the chromosomes and point mutation.

Table 2. Disease resistant plants of different crops generated through somaclonal variation

Crop	Pathogen	Resistance observed
Barley	*Rhynchosporium recalis*	Increase field resistance
Maize	*Helminthosporium maydis*	Resistance to Hm toxin
Potato	*Steptomyces scabies*	Increased resistance
Sugarcane	*Ustilago scitaminea*	Increased resistance
Tobacco	*Phytopthora parasitica var. nicotianae pseudomonas solanacearum*	Increased resistance
Tomato	*Fusarium oxysporum* f.sp. lycopersici race 2	Resistance Monogenic
Tomato	*Tobacco mosaic* virus	Resistance
Tomato	*P. solanacearum*	Suppressed bacterial growth
Wheat	*Pseudomonas syringae* pv syringae	Increased resistance
Wheat	*Erysiphe graminis* f. sp. tritici	Increased resistance
Wheat	*N. indica Erysiphe graminis*	Increased resistance

(c) Tissue and cell culture: Another approach is to apply a selection pressure so that all sensitive cells die and only resistant survive from which plants are subsequently regenerated. Selection pressure is applied through toxins or compounds produced by their analogues (Daub 1986, Hammerschleg et al 1995). Some of the cells in developing callus are generally free of viruses as virus replication cannot complete with rapid multiplication of callus cells. Hence, plants can be regenerated from these virus free cells. Carlson (1973) first demonstrated that plant cell and protoplast could be related in culture for resistance to a pathogen toxin and the plants with an altered response to infection by the pathogen could be regenerated from these cultured cells. Since then a great many claims have been made about the vast potential of cell culture selection for the development of novel disease resistant plants (Table 3). Tissue culture techniques have already substantially contributed to plant disease control while some recent ones of these methods have been shown to have great promise in this connection. A part from inducing resistance in host's one more short term approach is production of disease free planting material.

Table 3. Disease resistant crops obtain by *in vitro* selection

Crop	Pathogen	Selection agents
Food crops		
Rape seed	*Phoma lingam, Alternaria brassicicola*	CF
Rice	*Helminthosporium oryzae*	CT
	Xanthomonas oryzae	Bacterial cells
Barley	*Helminthosporium sativum*	CT
	Fusarium spp.	Fusaric acid
Maize	*Helminthosporium maydis*	Hm T toxin
Oats	*Helminthosporium victoriae*	Victorin
Wheat	*Helminthosporium sativum, Fusarium graminearum*	CT
	Pseudomonas syringae	Syringomycin
Sugarcane	*Helminthosporium sacchari*	Toxin
Horticultural and other crops		
Tobacco	*Pseudomonas sringae* pv. tabaci	Methionine sulfox
	Alternaria alternate; *P. syringae* pv. *tabaci*	Toxin
	Tobacco mosaic virus	Virus
	Fusarium oxysporum f.sp. *nicotianae*	CF
Potato	*Phytophthora infestans, Fusarium oxysporum*	CF
	Erwinia carotovora	Pathogen
Pea	*Fusarium oxysporum* f.sp. *pisi*	CF
Alfalfa	*F. oxysporum* f.sp. *medicagnis*	CF
Tomato	Tobacco mosaic virus	Virus
	Pseudomonas solanacearum	CF
Egg plant	*Verticillium dahlie*	CF
	Little leaf disease	Pathogen
Peach	*Xanthomonas campestris* pv. *pruni*	CF
Hop	*Verticillium albo-atrum*	CF
Celery	*Septoria apiicola*	CF

CF: Culture Filtrate; CT: Crude Toxin (Source: Chawla, 2002)

(d) Protoplast fusion and somatic hybridization: Fusion of the protoplast is called protoplast fusion. Disease resistance is transferred from one species to another even if the two species are genetically distant where it is impossible to cross them under normal conditions. Somatic hybridization also follows to bring together cytoplasmicgenes of the two species to produce hybrid, which is close wise, not possible through sexual crossing. The first somatic hybrid between *Nicotiana glanca* and *N. largsdorffii* was reported by Carlson et al (1972) other examples of somatic hybrids include (*Arabididospis thaliana + Brassica compestris* (Gleba and Hoffman, 1980) and *B. hapus + Eruca sativa* (Fahleson et al 1988) (Resistance to late blight and potato leaf roll virus of Solanum brevidens is expressed in its somatic hybrid with *S. tuberosum* (Helgeson et al 1986). Host specific as well as host non specific pathogen metabolites such as toxins and crude culture filtrates, have been sued as selection agents to isolate plant cells which enhanced resistance to diseases (Table 4). Similarly, resistant lines were not obtained when cercosporin, a toxin produced by the members of the genus *Cercospora* was as selection agents (Daub, 1986).

Table 4. Genetic traits transferred via protoplast fusion

Somatic hybrids	Traits (Resistance)
Nicotiana tabacum + N. nesophila	Tobacco mosaic virus
Solanum tuberosum + S. chacoense	Potato virus x
N. tabacum + N. nesophila	Tobacco horn worm
S. tuberosum + S. brevidens	Potato leaf roll virus
S. eircalifolium + S. tuberosum	*Phytophthora*
S. melongena + S. sanitwongsei	*Pseudomonas solanacearum*
S. tuberosum + S. commersonii	Frost tolerance
Brassica napus + B. nigra	*Phoma lingam*
B. oleracea + Sinapis alba	*Alternaria brassiceae*
B. napus + Sinapis alba	*A. brassiceae*
Citrus sinensis + Poncirus trifoliata	*Phytophthora*
Lycopersicon esculentum + L. peruvianum	TMV. Spotted wilt virus, Cold tolerance
Solanum lycopersicoides + L. esculentum	Cold tolerance
Solanum ochranthum + L. esculentum	Tomato diseases and insect pest
B. oleracea + B. napus	Black rot
B. oleracea sp. capitata + B. oleracea (ogura cms line)	Cold tolerance
S. tuberosum + S. Cammersonii	Frost tolerance
R. sativa (Japanese radish)+B. aleracea var. botrytis	Club rot disease
B. oleracea var. botrytis + S. alba + B. carinata	*Alternaria brassicola* and *Phoma lingum*
Citrullus lanatus + Cucumis melo	Club rot resistance and salt tolerance
Hardeum vulgare + Daucus carota	Frost and salt tolerance
Salanum melongena + S. sisymbrifolium	Nematode
S. tuberosum + S. bulbocastanum	Nematode
S. tuberosum + S. bulbocastanum	Root knot nematode
Raphanus sativus + Brassica napus	Beet cyst nematode
Sinapis alba + R. sativus + B. napus	Beet cyst nematode

Source: Chawla, 2002

(ii) Genetic engineering

Genetic engineering offers plant breeders access to an infinitely wide array of novel genes and traits, which can be inserted through a single event into high-yielding and locally-adapted cultivars. This approach offers rapid introgression of novel genes and traits into elite agronomic backgrounds. It takes five to six generations to transfer a trait within a species into the high yielding locally adapted cultivars through the conventional breeding, and one has to plant a large number of progenies to be able to select the plants with appropriate combination of traits (Fig 2). The improved lines developed then have to go through a set of multi-location tests, before a variety could be identified for cultivation by the farmers. This process takes minimum of 7/10 years. However, genetic transformation provides access to genes from other species, which can be used for producing transgenic crops, ability to change the level of gene expression, and capability to change the spatial and temporal pattern of gene expression. The genes of interest can be transferred into the target crops/cultivars in a single event, and it takes 5/6 years to develop cultivars with stable gene expression. The lines thus produced can be released for cultivation by the farmers or used as donor parents in the conventional plant breeding and/or marker assisted selection.

With the help of genetic engineering, crop production for many beneficial plant traits could be improved, particularly enhanced ability to withstand or resist attack by plant pathogens. New approaches to plant disease control are particularly important for pathogens that are difficult to control by existing methods.

There are four major areas of research and application of plant genetic engineering for resistance to plant pathogens:-

1. Enhancing resistance with plant genes: Plants have their own networks of defense against plant pathogens that include a vast array of proteins and other organic molecules produced prior to infection or during pathogen attack. Not all pathogens can attack all plants and a single plant is not susceptible to the whole plethora of plant pathogenic fungi, viruses, bacteria or nematodes. Pathosystem-specific plant resistance genes, *e.g.* Pto (Tang et al 1999), Cf-9 (Hammond et al 1998) etc. have been used as transgenes to confer resistance in different plants. Briefly, a gene that confers resistance to a certain pathogen in plant species A is identified, cloned and transformed into plant species B. Another approach to enhance resistance in plants is Recombinant DNA technology which allows the enhancement of inherent plant responses against a pathogen by either using single dominant resistance genes not normally present in the susceptible plant (Keen 1999) or by choosing plant genes that intensify or trigger the expressions of existing defense mechanisms (Bent and Yu 1999, Rommens and Kishore 2000), e.g., Incorporation of genes coding for Pathogenesis Related (PR) proteins and Phytoalexins in plants could provide resistance against some specific group of pathogens. Furthermore, biotechnology facilitates the discovery and elucidation of the molecular interactions between plants and pathogens. Understanding the molecular basis of plant-pathogen interactions increases our ability to deploy selected plant resistance genes from virtually any plant e.g., production of active oxygen species (superoxide anions, hydroxy radicals and hydrogen peroxide) have been observed in many plant-pathogen interactions and are known to play an important role in plant defense (Wu et al 1997) and thus plants have been engineered to continuously produce active oxygen species.

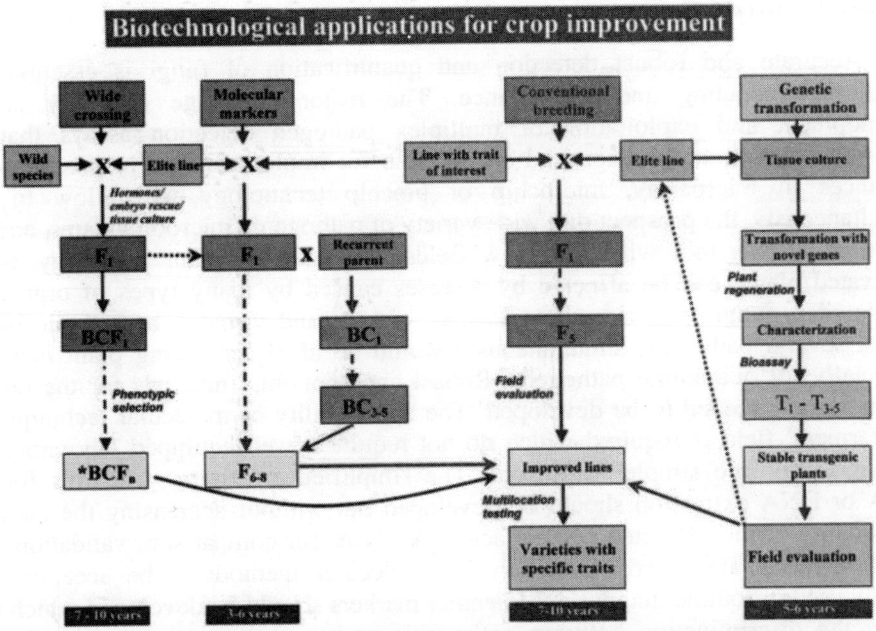

Fig. 2. A schematic outline of biotechnological approaches in crop improvement. Lines derived through genetic transformation can be released as varieties or used as a donor parent in the conventional breeding. The lines derived from wide crossing can take many generations (BCFn) to obtain homozygous and stable lines, and such material can either be used as improved lines or as a donor parent in conventional breeding or marker-assisted selection. (Source: Sharma et al 2002)

2. Pathogen derived resistance: Plants can be protected from diseases with transgenes that are derived from the pathogens themselves, a concept referred to as pathogen-derived resistance. For example, plant viral transgenes can protect plants from infection by the virus from which the transgene was derived. Genetic engineering brings new hope for the effective control of plant virus diseases. The concept of pathogen derived resistance has stimulated research on obtaining virus resistance through genetic engineering. Pathogen-derived resistance is mediated either by the protein encoded by the transgene (protein-mediated) or by the transcript produced from the transgene (RNA-mediated).

3. Antimicrobial proteins: By genetic engineering, a gene found in fungi, insect, animal, yeast or plant, which produce antimicrobial peptides or proteins, can be introduced in the plant to improve the plants' defense against a particular pathogen. Lytic peptides, lysozymes, chitinase, attacins etc. are few examples of antimicrobial peptides which are frequently used in plants to achieve resistance against pathogens.

4. Plantibodies: Although plants have mechanisms to protect themselves against pathogen attack, in contrast to animals, there is no "immune system" *per se* in plants. With the advent of genetic engineering, plants can be engineered to express an antibody against a protein crucial for pathogenesis resulting in a level of immunity or resistance to the pathogen.

Future Thrusts

Accurate and robust detection and quantification of fungi is essential for diagnosis, modeling and surveillance. The major challenge currently is the development and exploitation of multiplex pathogen detection assays that, in addition, allow quantification and are suitable for implementation in practice. The advances in microarray, microchip or biochip technology will allow to test simultaneously, the prospect of a wide variety of pathogenic microorganisms, and the potential of this tool will open new fields of studies in plant pathology. Since cultivated plants can be affected by diseases caused by many types of organisms (nematodes, fungi, bacteria, phytoplasmas, viruses and viroids), a method able to detect several pathogens simultaneously would be ideal for testing plant material, especially for quarantine pathogens. Robust and economic methods for the on-site diagnosis are needed to be developed. The applicability of molecular techniques at the farmers' field is required which do not require a well equipped laboratory but required only one simple instrument. The simplified extraction protocols for the RNA or DNA extraction should be developed but without decreasing the quantity and quality of the extracted nucleic acids. Analyses for comparison, validation, and standardization are strictly necessary for molecular methods to be accepted and widely used in routine diagnosis. Molecular markers should be developed which will allow the discrimination between pathogenic and nonpathogenic strains of plant pathogens especially in complex species such as *Fusarium oxysporum*, *F. solani*, and *Rhizoctonia solani*. New crop cultivars with resistance to insect pests and diseases combined with bio-control agents should lead to a reduced reliance on pesticides, and thereby reduce farmers' crop protection costs, while benefiting both the environment and public health.

References

Alvarez AM 2004 Integrated approaches for detection of plant pathogenic bacteria and diagnosis of bacterial diseases, *Annual Review of Phytopathology* **42** 339-366

Amann RI, Ludwig W, and Schleifer KH 1995 Phylogenetic identification and in situ detection of individual microbial cells without cultivation, *Microbiology Reviews* **59** 143-169

Atkins SD, Hidalgo-Diaz L, Clark IM, Morton CO, Montes De Oca N, Gray PA, and Kerry BR 2003 Approaches for monitoring the release of *Pochonia chlamydosporia* var. *catenulata*, a biological control agent of root-knot nematodes, *Mycological Research* **107** 206-212

Baner J, Nilsson M, Mendel-Hartvig M, and Landegren U 1998 Signal amplification of padlock probes by rolling circle replication, *Nucleic Acids Research* **26** 5073-5078

Bent AF and Yu IC 1999 Applications of molecular biology to plant disease and insect resistance, *Advances in Agronomy* **66** 251-298

Bonants PJM, Schoen CD, Szemes M, Speksnijder A, Klerks MM, van den Boogert PHJF, Waalwijk C, van der Wolf JM, and Zijlstra C 2005 From single to multiplex detection of plant pathogens: pUMA, a new concept of multiplex detection using microarrays, *Phytopathology Polonica* **35** 29-47

Botha WJ, Serfontein S, Greyling MM, and Berger DK 2001 Detection of Xylophilus ampelinus in grapevine cuttings using a nested polymerase chain reaction, *Plant Pathology* **50** 515- 526

Brouwer M, Lievens, B, Van Hemelrijck W, Van den Ackerveken G, Cammue, BPA, and Thomma BPHJ 2003 Quantification of disease progression of several microbial pathogens on *Arabidopsis thaliana* using real-time fluorescence PCR, *FEMS Microbiology Letter* **228** 241-248

Carlson PS, 1973 Methionine sulfoxinine-resistant mutants of tobacco, *Science* **180** 1366-68

Carlson PS, Smith HH, and Dearing RD 1972 Parasexual Interspecific Plant Hybridization, *Proceddings of National Academy of Sciences of USA* **69** 2292-2294

Chawla HC 2002 *Introduction to plant Biotechnolog, Second edition*, New Delhi: Oxford and IBH publishing co. Private Limited

Clark MF and Adams AN 1977 Characteristics of the microplate method of enzyme- linked immunosorbent assay for the detection of plant viruses, *Journal of General Virology* **34** 475-483

Coletta-Filho HD, Takita MA, de Souza AA, Neto JR, Destefano SAL, Hartung JS, and Machado MA 2006 Primers based on the rpf gene region provide improved detection of *Xanthomonas axonopodis* pv. *citri* in naturally and artificially infected citrus plants, *Journal of Applied Microbiology* **100** 279-285

Daub MF 1986 Tissue culture and the selection of resistance to pathogens pathogens, *Annual Review of Phytopathology* **24** 159-186

De Boer SH, Elphinstone JG, and Saddler G 2007 Molecular detection strategies for phytopathogenic bacteria, in *Biotechnology and Plant Disease Management* (eds) ZK Punja, SH De Boer, and H Sanfançon, UK: CAB International 165-194 pages

Ellis SD and Boehm MJ 2008 Plants Get Sick Too! An introduction to plant diseases, Ohio State University, *Agriculture and Natural Resources*, 401-01 page

Elnifro EM, Ashshi AM, Cooper RJ and Klapper PE 2000 Multiplex PCR: Optimization and application in diagnostic virology, *Clinical Microbiology Review* **13** 559-570

Fahleson J, Rahlen L and Glimelinus K 1988 Analysis of plants regenerated from protoplast fusion between *Brassica napus* and *Eruca sativa, Theoretical and Applied Genetics* **76** 507-521

Fire A and Xu SQ 1995 Rolling replication of short DNA circles, *Proceedings of National Academy of Science of USA* **92** 4641-4645

Fraaije B, Lovell DJ, Rohel EA, and Hollomon DW 1999 Rapid detection and diagnosis of *Septoria tritici* epidemics in wheat using a polymerase chain reaction/PicoGreen assay, *Journal of Applied Microbiology* **86** 701- 708

Frederick RD, Snyder CL, Peterson GL, and Bonde MR 2002 Polymerase chain reaction assays for the detection and discrimination of the soybean rust pathogens *Phakopsora pachyrhizi* and *P. meibomiae*, *Phytopathology* **92** 217–27

Gachon C, Mingam A and Charrier, B 2004 Real-time PCR: What relevance to plant studies? *Journal of Experimental Botany* **55** 1445-1454

Gleba YY and Hoffman F 1978 Hybrid cell lines *Arabidopsis thaliana* + *Brassica campestris*: no evidence for specific chromosome elimination, *Molecular and General Genetics* **165** 257-264

Haas JH, Moore LW, Ream W and Manulis S 1995 Universal PCR primers for detection of phytopathogenic *Agrobacterium* strains, *Applied and Environmental Microbiology* **61** 2879-2884

Hammerschlag FA, Ritchie D, Werner D, Hashmi G, Krusberg L, Meyer R and Huettel R 1995 *In vitro* selection of disease resistance in fruit trees, *Acta Horticultuare* **392** 19-26

Hammond KKE, Tang S, Harrison K and Jones JD 1998 The tomato Cf-9 disease resistance gene functions in tobacco and potato to confer responsiveness to the fungal avirulence gene product Avr9, *Plant Cell* **10** 1251-1266

Heid CA, Stevens J, Livak KJ and Williams PM 1996 Real time quantitative PCR, *Genome Research* **6** 986-994

Helgeson JP, Hunt GJ, Geraldine TH and Austin S 1986 Somatic hybrids between *Solanum brevidens* and *Solanum tuberosum*: Expression of a late blight resistance gene and potato leaf roll resistance, *Plant Cell Reports* **3** 212-214

Jensen MA, Webster JA and Straus N 1993 Rapid identification of bacteria on the basis of polymerase chain reaction-amplified ribosomal DNA spacer polymorphisms, *Applied and Environmental Microbiology* **59** 945-952

Keen NT 1999 Plant disease resistance: progress in basic understanding and practical application, *Advances in Botanical Research* **30** 292-328

Köhler G and Milstein C 1975 Continuous cultures of fused cells secreting antibodies of predefined specificity, *Nature* **257** 495-497

Kousik CS and Ritchie DF 1998 Response of bell pepper cultivars to bacterial spot pathogen races that individually overcome major resistance genes, *Plant Disease* **82** 181-186

Kwok S and Higushi R 1989 Avoiding false positives with PCR, *Nature* **339** 237-238

Larkin PJ and Scowcroft WR 1983 Somaclonal variation and eye spot toxin tolerance in sugarcane, *Plant Cell Tissue and Organ Culture* **2** 111-121

Lévesque CA 2001 Molecular methods for detection of plant pathogens-What is the future? *Canadian Journal of Plant Pathology* **24** 333-336

Li W, Hartung JS and Levy L 2007 Evaluation of DNA amplification methods for improved detection of "*Candidatus Liberibacter* species" associated with *citrus huanglongbing, Plant Disease* **91** 51-58

Lievens B, Brouwer M, Vanachter ACRC, Lévesque CA, Cammue BPA and Thomma BPHJ 2003 Design and development of a DNA array for rapid detection and identification of multiple tomato vascular wilt pathogens, *FEMS Microbiology Letter* **223** 113-122

López MM, Bertolini E, Olmos A, Caruso P, Gorris MT, Llop P, Penyalver R, and Cambra M 2003 Innovative tools for detection of plant pathogenic viruses and bacteria, *International Microbiology* **6** 233-243

Louws FJ, Rademaker JLW, and Brujin FJ 1999 The three Ds of PCR-based genomic analysis of phytobacteria: diversity, detection, and disease diagnosis, *Annual Review of Phytopathology* **37** 81-125

Lydon J and Patterson CD 2001 Detection of tabtoxin-producing strains of *Pseudomonas syringae* by PCR, *Letters in Applied Microbiology* **32** 166-170

Mackay IM, Arden KE, and Nitsche A 2002 Real-time PCR in virology, *Nucleic Acids Res*earch **30** 1292-1305

Martin RR, James D, and Lévesque CA 2000 Impacts of molecular diagnostic technologies on plant disease management, *Annual Review of Phytopathology* **38** 207-239

McCartney HA, Foster SJ, Fraaije BA, and Ward E 2003 Molecular diagnostics for fungal plant pathogens, *Pest Management Science* **59** 129- 142

Morel G and Martin C 1952 Guerison de dahlias atteints d'ume Maladie a Virus, CR Academy of Sciences of Paris **235** 1324-1325

Mullis KB and Faloona FA 1987 Specific synthesis of DNA *in vitro* via a polymerase-catalyzed chain reaction, *Methods in Enzymology* **155** 335- 350

Nilsson M, Gullberg M, Dahl F, Szuhai K, and Raap AK 2002 Real-time monitoring of rolling-circle amplification using a modified molecular beacon design, *Nucleic Acids Research* **30** e66

Nilsson M, Krejci K, Koch J, Kwiatkowski M, Gustavsson P, Landegren U 1997 Padlock probes reveal single-nucleotide differences, parent of origin and in situ distribution of centromeric sequences in human chromosomes 13 and 21, *Nature Genetics* **16** 252-255

Nilsson M, Malmgren H, Samiotaki M, Kwiatkowski M, Chowdhary BP, Landegren U 1994 Padlock probes: circularizing oligonucleotides for localized DNA detection, *Science* **265** 2085-2088

O'Donnell K, Kistler HC, Cigelnik E, and Ploetz RC 1998 Multiple evolutionary origins of the fungus causing Panama disease of banana: Concordant evidence from nuclear and mitochondrial gene genealogies, *Proceedings in National Academy of Science of USA* **95** 2044-2049

Rommens CM and Kishmore GM 2000 Exploiting the full potential of disease-resistance genes for agricutural use, *Current Opinion in Biotechnology*, **11** 120-125

Sato M, Watanabe K, Yazawa M, Takikawa Y, and Nishiyama K 1997 Detection of new ethylene-producing bacteria, *Pseudomonas syringae* pvs. *cannabina* and *sesami*, by PCR amplification of genes for the ethylene-forming enzyme, *Phytopathology* **87** 1192-1196

Schonfeld J, Heuer H, van Elsas JD, and Smalla K 2003 Specific and sensitive detection of *Ralstonia solanacearum* in soil on the basis of PCR amplification of fliC fragments, *Applied and Environmental Microbiology* **69** 7248-7256

Sharma HC, Crouch JH, Sharma KK, Seetharama N, and Hash CT 2002 Applications of biotechnology for crop improvement: prospects and constraints, *Plant Science* **163** 381-395

Singh BD 2002 Plant breeding: Principles and methods, Kalyani Publishers

Tan-Bian S, Yabuki J, Matsumoto S, Kageyama K, and Fukui H 2003 PCR primers for identification of opine types of *Agrobacterium tumefaciens* in Japan, *Journal of General Plant Pathology* **69** 258-266

Tang X, Xie M, Kim YJ, Zhou J, Klessig DF, and Martin GB 1999 Overexpression of Pto activates defense responses and confers broad resistance, *Plant Cell*, **11** 15-29

Terrada E, Kerschbaumer RJ, Giunta G, Galeffi P, Himmler G, and Cambra M 2000 Fully "Recombinant Enzyme-Linked Immunosorbent Assays" using genetically engineered single-chain antibody fusion proteins for detection of *Citrus tristeza virus*, *Phytopathology* **90** 1337-1344

Thomas DC, Nardone GA, and Randall SK 1999 Amplification of padlock probes for DNA diagnostics by cascade rolling circle amplification or the polymerase chain reaction, *Archives of Pathology & Laboratory Medicine* **123** 1170-1176

Tomita N, Mori Y, Kanda H, and Notomi T 2008 Loop-mediated isothermal amplification (LAMP) of gene sequences and simple visual detection of products, *Nature Protocols* **3** 877-882

Wang Z, Yin Y, Hu H, Yuan Q, Peng G, and Xia Y 2006 Development and application of molecular-based diagnosis for *'Candidatus Liberibacter asiaticus'*, the causal pathogen of citrus huanglongbing, *Plant Pathology* **55** 630-638

Weiland JJ and Sundsbak JL 2000 Differentiation and detection of sugar beet fungal pathogens using PCR amplification of actin coding sequences and the ITS region of the rRNA gene, *Plant Disease* **84** 475-482

White TJ, Bruns T, Lee S, and Taylor J 1990 Amplification and direct sequencing of fungal ribosomal RNA genes for phylogenetics, in *PCR Protocols: A Guide to Methods and Applications* (eds) MA Innis, DH Gelfand, JJ Sninsky and TJ White, San Diego, CA: Academic Press, 315-322 pages

Wu GS, Shortt BJ, Lawrence EB, Leon J, Fitzsimmons KC, Levine EB, Raskin I, Shah DM 1997 Activation of host defense mechanisms by elevated production of H_2O_2 in transgenic plants, *Plant Physiology* **115** 427-435.

12

Molecular Markers in Plant Disease Management

Monica Sharma

Department of Plant Pathology, Dr. Y.S. Parmar University of Horticulture & Forestry, Nauni, Solan, HP-173 230

The evolution of new races of the plant pathogens make the researchers to find new techniques and methods which would be helpful to develop effective disease management strategies including the development of new resistant varieties in a short time. The use of molecular markers has opened new avenues for efficient identification of resistance genes and screening of resistant varieties in a short span of time compared to conventional breeding. The molecular markers detected at genetic level are less affected by the environment and can be able to distinguish the varietal difference in crops. Molecular marker-assisted selection, often simply referred to as marker-assisted selection (MAS) involves selection of plants carrying genomic regions that are involved in the expression of traits of interest through molecular markers. With the development and availability of an array of molecular markers and dense molecular genetic maps in crop plants, MAS has become possible for traits both governed by major genes as well as quantitative trait loci (QTLs).

A marker may be biological, morphological, biochemical, cytological, or DNA-based molecular marker. The various molecular markers discovered so far ultimately depend upon the genomic DNA of an individual. Molecular markers should not be considered as normal genes as they usually do not have any biological effect. Instead, they can be thought of as constant landmarks in the genome. They are identifiable DNA sequences, found at specific locations of the genome and transmitted by the standard laws of inheritance from one generation to the next. They rely on a DNA assay, in contrast to morphological markers that are based on visible traits and biochemical markers that are based on proteins produced by genes. Molecular or DNA markers are either non-PCR (eg., RFLP, Restriction fragment length polymorphism) or PCR (Polymerase chain reaction) based.

(I) Non-PCR or hybridization based molecular markers

1. Restriction fragment length polymorphism (RFLP): It is the first molecular marker system to be conceived and developed which is based on DNA/

Southern hybridization technique. RFLP is technically complex, laborious and difficult to automate. The two main advantages of RFLP markers are co-dominance and high reproducibility. Disadvantages are the requirement of relatively large amounts of pure and intact DNA and the tedious experimental procedure. The technique centers around the digestion of genomic DNA digested with restriction enzymes (Fig 1). These enzymes are isolated from bacteria and consistently cut DNA at specific base pair sequences which are called restriction sites. These restriction sites are not associated with any type of gene and are distributed randomly throughout the genome. When genomic DNA is digested with one of these restriction enzymes, (of which there are thousands, each cutting at a specific sequence), a series of fragment are produced of varying length. Changes within the restriction sequences which can be caused by point mutations, insertions or deletions, result in DNA fragments of differing length and molecular weights. These fragments are separated using agarose or polyacrylamide gel electrophoresis (PAGE) and yield a characteristic pattern. DNA has a uniform charge per unit length when run under electrophoresis conditions which arises from the phosphates groups in its backbone. So when DNA fragments are separated via electrophoresis, the distance they travel is dependent only on their molecular weight. This allows their molecular weight to be determined with simple standard called DNA ladders which are run along side the DNA in the gel. When restriction fragments are separated on agarose gels a series of bands results. Each band corresponds to a restriction fragment of different length. The lighter they are the farther they have traveled. Variations in the characteristic pattern of a RFLP digest can be caused by base pair deletions, mutations, inversions, translocations and transpositions which result in the loss or gain of a recognition site resulting in a fragment of different length and polymorphism.

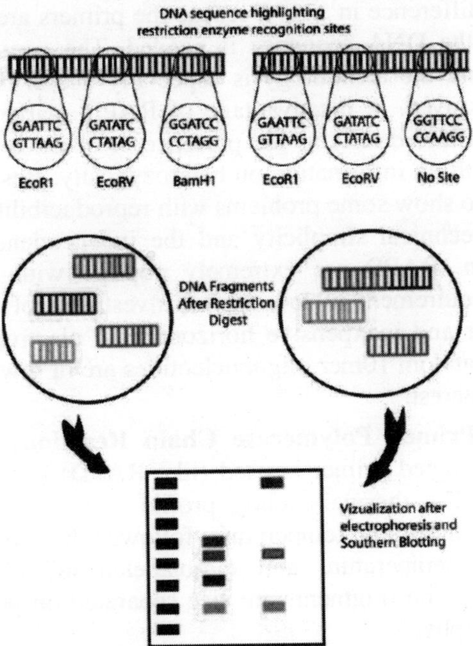

Fig. 1. Diagram of major events of RFLP

(II) PCR-based Molecular Markers

The polymerase chain reaction (PCR) has proven to be a revolutionary technique in molecular biology in which DNA is amplified *in vitro*. Two oligonucleotide primers complementary to opposing DNA strands are used. These oligonucleotides act as sequence-dependent priming sites for simultaneous primer extension on the opposing DNA strands. By appropriate selection of the primer sites and reaction conditions, the primer extension reactions with DNA polymerase will result in the synthesis of new DNA template. By combining a thermal - stable DNA polymerase with repeated cycling through reaction conditions favoring DNA template denaturation, followed by template-primer annealing and finally primer extension, exponential DNA amplification occurs. In a couple of hours, amplifies amount of DNA i.e., micrograms of DNA could be produces from little amount, as little as one molecule of template DNA in the starting. PCR is rapid, inexpensive, technically simple and extremely flexible. The PCR-based molecular markers can be divided into two categories:

(a) Sequence arbitrary methods (RAPD, AP-PCR, DAF, ASAP, ISSR, STS, RAMPs, AFLP, SAMPL)

(b) Sequence dependent methods (SSRs, SNPs, SACR, CAPs, SSCP, DGGE, MALDI-TOF MS)

(a) Sequence arbitrary methods

(i) Random Amplified Polymorphic DNA (RAPD): It employs a single, short decamer primer of arbitrary sequence which is annealed to the template DNA at low temperature. The resulting amplification products are usually separated on agarose gels on the basis of difference in size. Because the primers are arbitrarily chosen, no prior knowledge of the DNA sequence is needed. The genome is expected to be sampled randomly and the technology is especially useful if loci across an entire genome are to be assayed. A disadvantage of RAPD markers is the fact that the polymorphisms are detected only as the presence or absence of a band of a certain molecular weight, with no information on heterozygosity. Besides being dominantly inherited, RAPDs also show some problems with reproducibility of data. Their major advantages are the technical simplicity and the independence of any prior DNA sequence information. RAPD are extremely popular with scientists due to the simplicity and the requirement of low capital investment of the method. No more than a thermal cycler and inexpensive horizontal gel electrophoresis equipment is needed. In addition, random 10mer oligonucleotides are of low cost and may be used with any species of interest.

(ii) Arbitrary Primed Polymerase Chain Reaction (AP-PCR): A single, sequence-arbitrary designed primer is used (like RAPD) but primer concentrations are ten fold higher. The thermal-cycling profile begins with one or two cycles incorporating a low annealing temperature followed by 30 cycles using a high stringency annealing temperature and radiolabeled dCTP to label the newly synthesized fragments. The fragments are size separated on polyacrylamide gel and visualized via radiography.

(iii) DNA Amplification Fingerprinting (DAF): DAF is another PCR based marker which employs shorter (<10 nucleotides), usually 5-8 nucleotides (nt.) primer of arbitrary sequence. The primer annealing temperature is also low. The amplified products are separated on polyacrylamide gel electrophoresis (PAGE) and are detected by silver staining. Due to short primer length and low primer annealing temperature, there is problem in non-reproducibility of the above three marker systems and hence are generally less preferred in marker assisted selection (MAS).

(iv) Arbitrary Signatures from Amplification Profiles (ASAP): An approach to enhance the level of informativeness from DAF is called as. In this, a DAF amplification is first performed and then this amplification reaction is used as template in a second amplification in which either a mini-hairpin based primer or alternatively a primer with simple sequence repeat (SSR) at its 5'end is used.

(v) Inter Simple Sequence Repeats (ISSR): This method require only limited sequence information and hence not completely sequence arbitrary. In this marker system, primers based on a repeat of simple sequence or simple sequence repeats (SSR) e.g., ATGATGATGATG i.e., repeat of ATG, CACCACCACCAC i.e., repeat of CAC etc. Simple sequence repeats are present in the genomes of all eukaryotes and consists of several to over a hundred repeats of a 1-4 nucleotide motif. The resultant PCR amplifies the sequence between two SSR regions of the genome which are separated on agarose gel or polyacrylamide gel.

(vi) Randomly Amplified Microsatellite Polymorphisms (RAMPs): A radiolabeled SSR primer is combined with a 10mer RAPD primer. Only those amplification products containing the target SSR repeat motifs are visualized.

(vii) Amplified Fragment Length Polymorphism (AFLP): AFLP is a highly sensitive method based on the combined concepts of RFLP and RAPD. This technique is applicable to all species giving very reproducible results. The basis of AFLP is the PCR amplification of restriction enzyme fragments of genomic DNA i.e. selective PCR amplification of restriction fragments. The genomic DNA is first digested by two restriction endonucleases. Adapter molecules of known sequence are ligated to the ends of DNA fragments (Fig. 2). DNA amplification is carried out using primers with sequence complementary to the adapters as well as extra random nucleotides at their 3'end for selective amplification of fragments. Only those fragments are amplified, whose ends are perfectly complementary to the 3′ends of the selective primers. Thus, only a certain portion of the restriction fragments is amplified. The resulting amplification products are are resolved by gel electrophoresis on agarose gel or polyacrylamide gel and are either visualized by radiolabeling of one of the primer followed by fragment separation on polyacrylamide gel or by autoradiography, silver staining or fluorescence, resulting in a unique reproducible fingerprint for each individual. The possible reasons for AFLP-Polymorphisms are (i) sequence variations in a restriction site (the same as in RFLPs), (ii) insertions or deletions within an amplified fragment (also the same as in RFLPs) and (iii) differences in the nucleotide sequence immediately adjoining the restriction site (not detected with RFLPs). Thus, the usage of AFLP technologies results in the detection of higher levels of polymorphisms compared with RFLPs. AFLPs also have a much higher multiplex ratio (more markers per experiment) and better reproducibility than RAPDs. However, this technique is requires lengthier and costlier than other PCR based techniques and it requires larger investments in equipment. In addition, this technique also requires greater technical skill and as they require the use of polyacrylamide gels for detection and good quality of DNA is required for ensuring complete digestion by enzymes as partial digestion of DNA may result in non-reproducible variation in DNA profiles. However, this technique has proven to be quite popular. A number of high density genetic maps have been constructed in numerous plant species using AFLPs. In addition, a number of quantitative trait loci (QTL) mapping and fingerprinting studies have been carried out using AFLPs.

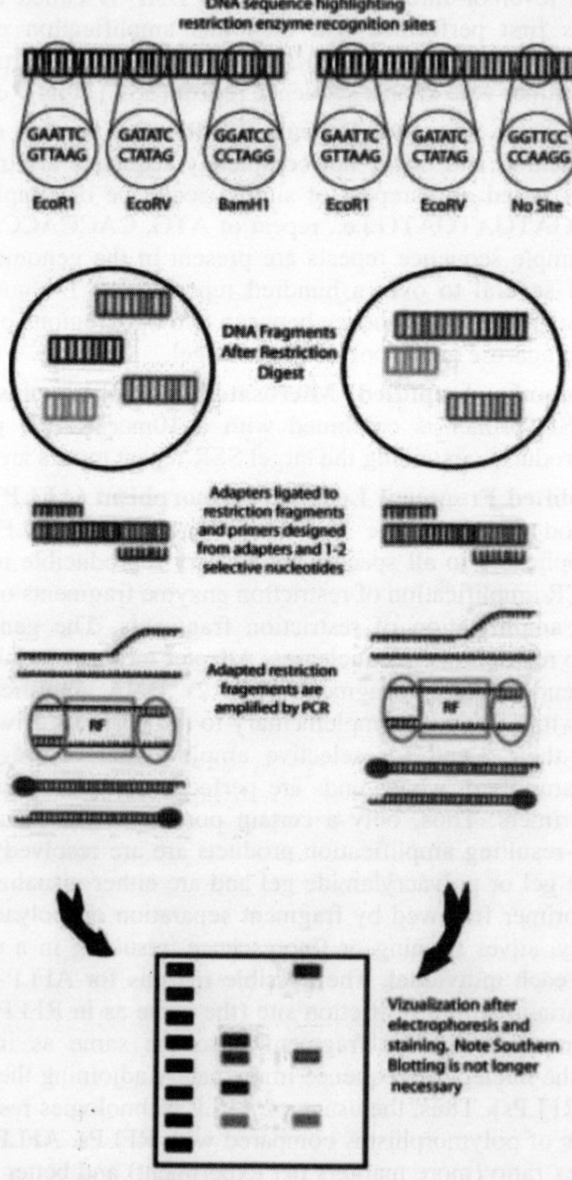

Fig. 2. Diagram of major events of AFLP

(viii) Selective Amplification of Microsatellite Polymorphic Loci (SAMPL): A hybrid/ combination of SSR sequence based method and AFLP method. As in ADLP, the genomic DNA is first digested with one or more restriction endonucleases and the adapters are ligated at the digested ends of the DNA. Then using a single oligonucleotide primer consisting of sequence complementary to SSR motif with additional anchoring bases and sequence complementary to adapter sequence. The amplification products are separated on polyacrylamide gels.

(b) Sequence dependent methods: These sequence-dependent methods are especially attractive for marker-assisted plant breeding.

(i) Simple sequence repeats (SSR): SSR markers are also known as microsatellite markers or variable number of tandom repeats (VNTR) or simple tandom repeats. SSRs are repeats of nucleotides usually 2-4 and sometimes longer. Compound repeats composed of two or more repeat motifs are also found. Sequence repeats are distributed throughout the genomes of all eukaryotic species. These markers are generally highly polymorphic which may be due to increased rates of sequence mutation affecting the number of repeat motifs present in an SSR locus. There are two approaches which could be used to develop SSR markers. In the first approach, a short probe consisting of repeat motifs/units is designed which is hybridized as in RFLP method. Other approach is amplification based in which a primer is designed corresponding to the flanking region of a specific SSR locus. Polymorphism will be based on either the flanking sequence, or more usually, the number of repeats. The technique reveals allele size differences even of a single base pair. A further advantage is the fact, that microsatellite markers can easily be distributed between labs by sharing primer sequences.

(ii) Single Nucleotide Polymorphisms (SNPs): Another marker system exploits Single Nucleotide Polymorphisms (SNPs) in the genome. SNPs are variations in DNA sequence due to the change in a single nucleotide in the sequence. SNPs are abundant in the genome of organisms e.g., in Human genome, SNPs have been found every 100 to 300 bases. SNPs may be found in both transcribed and non-transcribed regions. In some instances SNPs are the direct cause for observed phenotypic variation. SNPs do not change much from generation to generation i.e., they are evolutionary stable and this makes them easy to follow in population studies. Single nucleotide exchanges in genomes are numerous; therefore SNP markers provide a great marker density. Another important advantage of SNP is that it is not a gel-based technology. For the large-scale genotyping required in marker-assisted breeding programs, technologies based on gel electrophoresis are often too labor intensive and time consuming. The high density of SNP markers increases the probability to find polymorphisms in a target gene, which provides a huge advantage over previous markers that are at best closely linked to a locus of interest and not within. In the case of linkage it can easily happen that a linkage is lost when a marker is applied to other populations with different recombination patterns (Xu and Crouch 2008). Although the use of SNP markers in plants is still in its infancy, SNP markers are expected to become the marker system of choice in the near future, especially as the full sequences of more plant genomes will become available (Ganal et al 2009). There are several methods of SNP discovery, either following the database approach, where SNPs are detected by mining sequence databases, or following the experimental approach, where genome regions of interest are screened for SNPs with one of numerous techniques developed for the detection of SNPs. In principle, the SNP methods reveal differences between a probe of known sequence and a target DNA containing the SNP site. The target DNA sections are mostly PCR products and mismatches with the probe reveal SNPs within the amplified target DNA segment. The mismatching DNA segments can be sequenced then as the most direct way to identify SNP polymorphisms (Gupta et al 2001; Kwok 2001; Rafalski 2002; Weising et al 2005).

(iii) Sequence Characterized Amplified Region (SCAR): It is advancement of RAPD marker. The amplified products of RAPD are cloned and sequenced and primers corresponding to these regions are designed and used to observe polymorphism.

(iv) Cleaved Amplified Polymorphic Sequence (CAPs): It employs restriction endonucleases to cut PCR amplified products. These digested amplified products are separated on the gel to reveal polymorphism.

(v) Single-Strand Conformation Polymorphism (SSCPs): It is based on the differences in three-dimensional conformation of single stranded DNA which in turn depend upon the DNA sequence. In SSCP assay, the amplified products are first denatured to generate single strands of DNA. These strands are then electrophoresed under non-denaturing conditions. The difference in even a single base results in changes in three-dimensional conformation which result in difference in the rate of migration during electrophoresis.

Some other techniques used for molecular analysis

(i) Denaturing Gradient Gel Electrophoresis (DGGE): This method differentiates PCR amplification products depending upon the differences in their melting points under denaturing conditions due to differences in the sequence. The difference in the melting points ultimately results in differences in the rate of migration of DNA sequences. By using either a heat denaturation or chemical denaturant like urea or formaldehyde during electrophoresis, the sequences are denatured which results in difference in the migration of DNA molecules/sequences according to the difference in their melting points. Even single base pair difference in the sequence can be detected.

(ii) Genetic Bit Analysis (GBA): To detect SNPs, Genetic Bit Analysis (GBA) has been developed. In this, first amplification of the DNA region containing the base in question was amplified using specific primers. Then single stranded DNA is produced using exonuclease to digest the PCR products. One of the strands of DNA is protected from exonuclease digestion by adding phosphorothioate at the 5'end of the primer. The single stranded DNA is then hybridized with an oligonucleotide which immediately flanks the base in question at 3/end. Then labeled dideoxynucleotide, either biotin- or fluorescein- labeled, is used to extend the 3'end of the base in question. Labeled dideoxynucleotides are then detected.

(iii) Microarray assay: One of the most exciting and new technology for genetic analysis is 'Oligonocleotide Chips' or 'Silicon Chips' or 'Microarray Chips'. The oligonuclotides on microarray chips are arranged on glass slides using either

PCR-amplified cDNA or long oligonucleotides complementary to the transcribed part of genes. PCR amplified and labeled DNA or labeled RNA is hybridized with oligonucleotides on the microarray chip and then hybridization is conveniently detected. Thousands of nucleotides arranged on silicon chips have been used to scan DNA for differences and detect hundred of SNPs simultaneously. This technology is very expensive which limits its frequent use.

(iv) Matrix Assisted Laser Desorption Ionization-Time of Flight Mass Spectrometry (MALDI-TOF MS): It is another new and exciting technology for DNA analysis. This technique is extremely rapid and allows determination of accurate fragment size of DNA molecule. A sample is processed in few seconds or even less as it takes milliseconds to separate individual fragments of DNA and ionization from multiple laser pulses are measured to determine the accurate fragment size. MALDI-TOF MS can be used for the analysis of SNPs as well as SSR alleles.

Molecular markers and disease resistance

Molecular markers are used for germplasm screening for sources of resistance, marker-assisted selection for pyramiding resistance genes in a variety, map-based cloning of resistance genes and in host-pathogen interaction research. Molecular markers are used to characterize both vertical/monogenic as well as horizontal resistance/ polygenic resistance. Highly informative markers like SSR, SNPs, SCAR, AFLP etc. are required to efficiently identify resistance genes that are linked to or residing in clusters. Polygenic resistance generally describing quantitative trait loci (QTL) traits can be evaluated easily by using molecular markers. Molecular markers enable the selection of combinations of both specific genes and QTLs Collard et al 2005).

In general the term 'molecular breeding' or 'marker-assisted breeding' is used to describe the use of molecular markers in breeding. A major advantage of the use of molecular marker over conventional breeding is that they increase breeding efficiency by enabling early selection for adult traits, simultaneous selection for multiple traits including resistance gene pyramids and selection for traits that are expensive to phenotype. The developments of molecular marker technology have revolutionized the breeding programmes in plants. The molecular marker systems allow high-density DNA marker maps (i.e. with many markers of known location, interspersed at relatively short intervals throughout the genome) to be constructed for a range of economically important agricultural species, thus providing the framework needed for eventual applications of MAS (Chaudhary et al 2008). Using the marker map, putative genes affecting traits of interest can then be detected by testing for statistical associations between marker variants and any trait of interest. These traits might be genetically simple-for example, many traits for disease resistance in plants are controlled by one or a few genes (Young 1999). Alternatively, they could be genetically complex quantitative traits, involving many genes (i.e. so-called quantitative trait loci (QTL) and environmental effects. Having identified markers physically located near or even within genes of interest, in the next step it is now possible to carry out MAS, i.e. to select identifiable marker variants (alleles) in order to select for non-identifiable favorable variants of the genes of interest.

The success of a marker-based breeding or molecular breeding depend mainly on three important factors:

(1) A genetic map with an adequate number of uniformly spaced polymorphic markers to accurately locate desired quantitative trait loci (QTLs) or major genes.

(2) Close linkage between the QTL or a major gene of interest and adjacent markers.

(3) Adequate recombination between the markers and the rest of the genome.

The success of MAS is influenced by the relationship between the markers and the genes of interest. Dekkers distinguished three kinds of relationship:

- The molecular marker is located within the gene of interest. In this situation, one can refer to gene- assisted selection (GAS). This is the most favourable situation for MAS since, by following inheritance of marker allele inheritance of allele of gene of interest is followed. On the other hand, these kinds of markers are the most uncommon and are thus the most difficult to find.
- The marker is in linkage disequilibrium (LD) with gene of interest throughout the whole population. LD is the tendency of certain combinations of alleles to be inherited together. Population wide LD can be found when markers and genes of interest are physically very close to each other and/or when lines or breeds have been crossed in recent generations. Selection using these markers can be called LD-MAS.
- The marker is not in linkage disequilibrium (i.e. it is in linkage equilibrium [LE]) with gene of interest throughout the whole population. Selection using these markers can be called LE-MAS. This is the most difficult situation for applying MAS.

In the real context of MAS, DNA- based markers can be effectively utilized for two basic purposes: (i) Tracing favourable allele(s) (dominant or recessive) across generations and (ii) identifying the most suitable individual(s) among the segregating progeny, based on allelic composition across a part or the entire genome.

Foreground selection and background selection

The use of molecular markers to trace the presence of target genes is referred to as foreground selection while their use for accelerating the recovery of the recurrent parent genome is referred to as background selection. Marker-assisted backcrossing (MAB) improves the efficiency of backcross breeding in three ways: (i) If the phenotype of the desired gene cannot be easily assayed, backcross (BC) progeny possessing a marker allele from the donor parent at a locus near/within the target gene can be selected with a good probability of carrying the gene, (ii) markers can be used to select BC progeny with least amounts of donor parent germplasm in the genome outside the target region and (iii) markers can be used to select rare progeny that are the result of recombination near the target gene, thus minimizing the effects of linkage drag. Transfer of recessive genes through conventional breeding requires additional selfing generations after every backcross, a procedure that is prohibitively slow for most commercial breeding purposes.

Procedure of Marker Aided Selection

1. Selection of parents: The parents are selected by screening the germplasm. The parents should show a high level of polymorphism (variation) for the trait of interest i.e. the parents should have contrasting characters. The parents should be homozygous (in self-pollinated species) or inbred lines (in cross-pollinated species). The elite selected parent is crossed with another parent having trait(s) of interest.

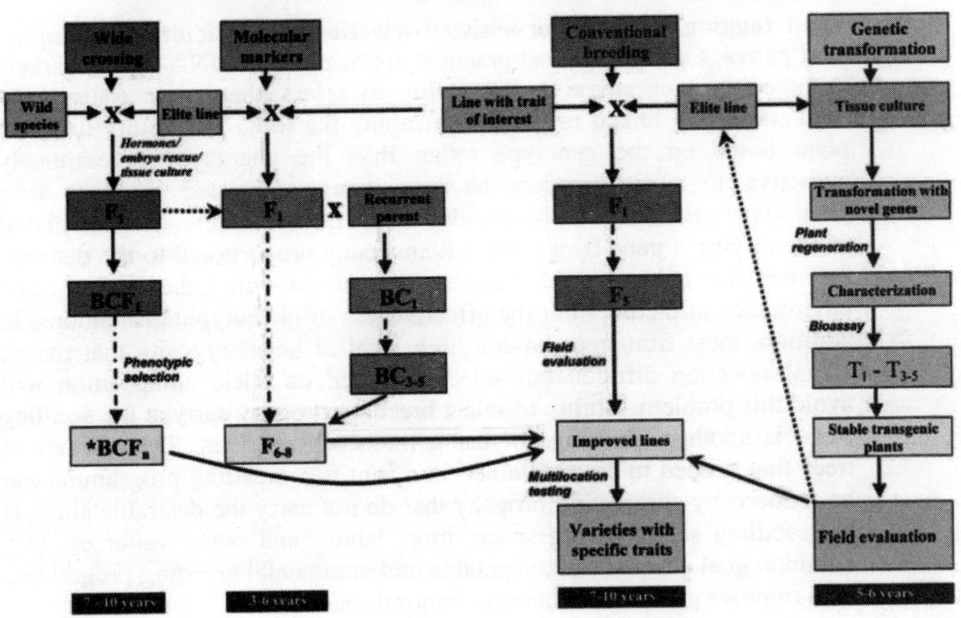

(Sharma et al 2002)

2. Development of breeding population: The selected parents are crossed and F1 plants are obtained. F1 plants are phenotypically homogenous but are heterozygous at the DNA level. The F1 hybrid is crossed with the recurrent parent (invariably the elite parent) (BC1), and the gene transfer is monitored through marker-assisted selection until BC3/5 [until the quantitative trait loci (QTL) or the gene of interest is transferred into the elite line]. In case of wild relatives that are not easily crossable with the cultivated types, the F1 hybrids may have to be produced through embryo rescue and tissue culture, and the progenies advanced as in the conventional backcross breeding approach (phenotypic selection) or through-marker assisted selection, using cultivated species as the recurrent parent. Progenies from F2 to F6 generations can also be advanced as per conventional pedigree breeding, and plants with appropriate combination of traits can be used as improved varieties or as donor parents in conventional breeding. The F6 progenies can also be used as random inbred lines (RILs) for mapping the trait(s) of interest if 250/300 plants are selected at random in F2s, and advanced by selfing the plants at random in each line in each generation. The plants obtained in BC5 can be used as isogenic lines to study the inheritance or role of traits of interest. The marker assisted selection takes 3/6 years, and thus speeding up the pace of transferring the traits of interest into the improved varieties, and it does not require large scale planting of the progenies up to crop harvest, as the plants showing the presence of the trait or QTL only need to be maintained up to maturity. Wide hybridization may take 7/10 years or longer (BCFn), depending on the success in transferring the trait(s) of interest into the elite line without other wild traits that influence the quality of the produce and productivity potential of the crop.

Applications of molecular markers in crop breeding

1. **Trait tagging and marker-assisted-selection of horticulturally important genes:** One of the most practical applications of DNA-based markers in breeding programme is the ability to select phenotypic traits using markers tightly linked to genes controlling the trait. The ability to select plant based on the genotype rather than the phenotype is extremely attractive to plant breeders because many associated problem with phenotypic selection will be avoided using DNA-markers. The likelihood of identifying a gene by a marker is inversely proportional to the distance between the genc and the marker. Interactions with other genetic and environmental factors limit the effectiveness of phenotypic evaluations. In addition, most fruit trees have a high level of heterozygosity that makes visual selection difficult but selection based on allele composition will avoid this problem. Ability to select breeding progeny early at the seedling stage is another advantage of using molecular markers. The number of trees that needed to be maintained in a fruit tree breeding programme can be reduced by eliminating progeny that do not carry the desirable allele at the seedling stage, saving space, time, labour and other resources. One common goal of most fruit, vegetable and ornamental breeding programme is to improve genetic resistance to major diseases.
2. **Gene introgression from wild germplasm:** Markers can be employed for crop improvement in introgressing beneficial traits from wild germplasm into crop cultivars. Markers linked to the genes from the wild parent (donor) parent as well as marker distributed throughout the genome of the improved cultivar (recurrent) parent in the form of genetic map, are used in selection of breeding progeny. Markers will be used in tracking desirable alleles from the donor parent and also, it will help reduce the genetic background of the donor parent in the progeny.
3. **Germplasm characterization:** Molecular markers are used to evaluate variation in existing gemplasm. Multiloci markers like random amplified polymorphic DNA (RAPD) and amplified fragment length polymorphism (AFLP) markers that can scan the entire genome quickly are efficient for this purpose. Molecular markers can help dentify the genetic diversity or lack of it in the material available to breeders because understanding the genetic relationships among the germplasm helps to select appropriate parental plants for crossing and make informed decisions on breeding strategies too. Many horticultural crops might have very narrow genetic base which need to infuse genetic donors in the breeding programs

(Sosinki et al 2000). Wild relatives of crop plants are source of beneficial traits for crop improvement. The use of molecular markers to study genetic relatedness between wild and cultivated species provides information on selecting closest wild relatives to use in breeding programmes especially when crossing between the wild species and the cultivated species is difficult to perform. Molecular markers can be used to identify core collections at germplasm repositories of collection centers in order to eliminate duplicate and unidentified materials but only represent the diversity available in all accessions present in the entire collection. This narrowing of the genetic materials allow breeders to use them more efficiently.

4. **Construction of genetic linkage maps:** Prior to the invention of molecular markers, map construction was based on phenotypic mutations which are rare and come from different genetic background and they are difficult to assemble into a single population. The advent of methods to generate DNA markers has greatly empowered genetic mapping of horticulturally important species and this discovery allowed map construction using only a single progeny set. In genetic mapping, molecular markers consist of short segments of DNA that provide landmarks along the chromosomes which now provides a scaffold of the entire genome. Genetic linkage map derived from the use of molecular markers provide various levels of genomic coverage and marker saturation.

Mapping of disease resistance gene

For mapping the genes for monogenic resistance in plants, three different approaches can be used: a) Use of a complete linkage map; b) Use of near isogenic lines (NILs); c) Bulk segregation analysis of F_2 population. Out of these, use of complete linkage map is most directed method of gene mapping. With the availability of complete genetic linkage map constructed on the basis of segregation and linkage analysis of a large number of molecular markers, two markers whose distance is less than 20 centimorgan (20 cM) are selected for the mapping of resistance gene. Co-segregation of any of these markers with resistance phenotype reveals linkage which eventually facilitates quick localization of the resistance gene in the linkage map as the marker position is already known. For using near isogenic line (NILs) approach, first near isogenic lines are developed and then molecular markers which distinguish resistant and susceptible cultivars are identified. Bulk segregation analysis (BSA) of F_2 population approach is most rapid way of mapping resistant genes. A resistant bulk and a susceptible bulk is constituted by pooling DNA from ten homozygous resistant and ten homozygous susceptible respectively F_2 plants. DNA markers which distinguish these two bulks are selected.

For mapping the genes for polygenic resistance in plants, three different methods can be used: a) Single marker analysis; b) Interval mapping; c) Composite interval mapping. All three methods require the generation of permanent mapping population such as recombinant inbred lines or double haploids which can be replicated over seasons and locations. In addition, information of a complete linkage map is required except for single marker analysis.

Table 1. Mapping of disease resistance genes in crops using molecular markers

Crop	Disease/ Pathogen	Method	Marker system
Apple	Scab	BSA	RAPD
		BSA	RAPD
		Genetic mapping	SCAR
		Genetic mapping	SCAR
		Genetic mapping	SCAR
		Comparative mapping	Microsatellite, SCAR
	Powdery mildew	Population screen	RAPD
		Whole genome scan	Microsatellite
		BSA	AFLP, SCAR
		Genetic mapping	SCAR
Grapes	Powdery mildew	BSA	AFLP
		RGA	-
	Downy mildew	BSA	RAPD, SCAR
	Pierce's disease	-	AFLP, ISSR, RAPD
Plum	Plum pox virus	-	SSR
Banana	Banana stunty virus	-	AFLP
Strawberry	*Phytophthora fragariae* (Red stele)	-	RAPD, SCAR
Tomato	*Alternaria solani* (Early blight)	-	RFLP
Potato	*Phytophthora* (Late blight)	-	RFLP, RAPD
	Potato virus X	-	AFLP
Soybean	*Phytophthora rojae*	-	RFLP
	Phialophora gregota	-	RFLP
Vicia faba	*Uromyces viciae-faba*	-	RAPD
Wheat	Fusarium head blight (*Fusarium graminearum*)	RIL	SSR
	Loose smut		SCAR
Maize	Leaf rust		RFLP
	Powdery mildew		RFLP

Using molecular markers, several disease resistance genes have been identified in many crops (Table 1). Recently, for tagging resistance genes, a new PCR-based approach for rapid generation of genetic markers has been developed and used in many crops. This approach exploits the fact of presence of upto seven conserved domains in the nucleotide binding sites (NBS) of the resistance (R) genes by using degenerate primers, one or more conserved domains are amplified and which ultimately facilitate the cloning of resistance gene analogues (RGA) from diverse species. NBS encoding sequences tend to be clustered in the genome at, or near, previously identified resistance loci. Therefore, the identification of RGAs represents a potential powerful strategy to develop new markers around the resistance genes and a good aid for map-based positional cloning of genes.

Future Thrust

Molecular marker maps, the necessary framework for any MAS programme, have been constructed for the majority of agriculturally important species but the density of the maps varies considerably among species. Marker-assisted background selection will become a standard breeding technique in as much as transgenic crop cultivars become more popular in the future. The marker-assisted transfer (e.g. pyramiding) of major disease resistance genes will remain a limited application as long as the costs for DNA marker analyses are high. Even with MAS, the conversion of elite inbred lines still takes about 2 years. This means the conversion breeder is always 2 years behind the "pipeline", i.e. the best new lines emerging from the basic breeding program. In any case, the utility of MAS will therefore remain restricted to true major genes and exclude QTL with only small effects. The transfer of single, important QTL alleles from otherwise inferior genetic resources is a special case deserving more research. Marker-assisted foreground selection will certainly become more popular as allele-specific markers become available through genomics research programs. Bioinformatic tools are now available to optimize MAS breeding plans. These will help to predict costs and time lines of conversion programs more accurately.

References

Choudhary K, Choudhary, OP, and Shekhawat NS 2008 Aarker assisted selection: A novel approach for crop improvement, *American-Eurasian Journal of Agronomy* **1 (2)** 26-30

Collard BCY, Jahufer MZZ, Brouwer JB, Pang ECK 2005 An introduction to markers, quantitative trait loci (QTL) mapping and markerassisted selection for crop improvement: The basic concepts, *Euphytica* **142** 169-196

Dekkers JCM 2004 Commercial application of marker-and gene-assisted selection in livestock: strategies and lessons, *Journal of Animal Sciences* **82** E313-E328

Ganal MW, Altmann T, and Röder MS 2009 SNP identification in crop plants, *Current Opinion in Plant Biology* **12** 211-217

Gupta PK, Roy JK, and Prasad M 2001 Single Nucleotide Polymorphisms: A new paradigm for molecular marker technology and DNA polymorphism detection with emphasis on their use in plants, *Current Science* **80** 524-535

Kwok PY 2001 Methods for Genotyping Single Nucleotide Polymorphisms, *Annual Review of Genomics and Human Genetics* **2** 235-258

Rafalski A 2002 Applications of Single Nucleotide Polymorphisms in crop genetics, *Current Opinion in Plant Biology* **5** 94-100

Sharma HC, Crouch JH, Sharma KK, Seetharama N, and Hash CT 2002 Applications of biotechnology for crop improvement: prospects and constraints, *Plant Science* **163** 381-395

Sosinki B, Gannavarapu M, Hager LD, Georgi L, Beck E, Rajapakse S, Baird WV, Ballard RE, and Abbott AG 2000 Characterization of microsatellite markers in peach (*Prunus persica* (L.) Batsch), *Theoretical and Applied Genetics* **101** 421-428

Weising K, Nybom H, Wolff K, and Kahl G 2005 *DNA Fingerprinting in Plants - Principles, Methods, and Applications*, Boca Raton: CRC Press

Xu Y and Crouch JH 2008 Marker-assisted selection in plant breeding: From publications to practice, *Crop Science* **48** 391-407

Young ND 1999 A cautiously optimistic vision for marker-assisted breeding, *Molecular Breeding* **5** 505-510.

13

Fungicides in Plant Disease Management

T.S. Thind
Department of Plant Pathology, Punjab Agricultural University, Ludhiana 141 004

Plant diseases are estimated to cause yield reductions of almost 20 per cent in the principal food and cash crops worldwide (Thind 2012). Of the over 1,00,000 described species of fungi in the work, approximately 20,000 produce one or more diseases in various plants. Over the past 150 years, much has been learnt about the control of plant diseases and several complementary approaches such as cultural methods, biological methods, chemical control and regulatory measures were developed for their control. Depending on the crop, the disease and the availability of control methods, a different set of approach is employed.

Although losses due to plant diseases may be reduced by the use of disease resistance cultivars, crop rotation or sanitation practices, fungicides are often essential to maximize crop yields. The predicted growth of the world's population from 7 billion at present (1.2 billion in India alone) to 8.3 billion by 2030 presents a major global challenge to meet necessary food demand. At the same time, per capita farmland is decreasing drastically. This will result into increased demand of food, feed and fiber. India has produced 241 MT of food grains during 2010-11 and with the increase in population growth rate, it would require about 400 MT by 2050. Yield improvements are attainable only if farm inputs are optimized, including fungicides. Fungicides also play important role in improving food quality. They contribute to food safety by controlling many fungi that produce mycotoxins. About one quarter of food crops worldwide are affected by mycotoxins such as aflatoxins, ergot toxins, Fusarium toxins, patulin and tenuazonic acid (Knight et al 1997). In U.S.A. alone, growers apply fungicides for controlling more than 200 diseases of 50 crops under field conditions. Fungicides are now well considered to be the second line of defense in plant disease control programmes after disease resistance.

The benefits and risks of fungicides

The application of any chemical to a crop or food raises the question of risks and benefits. This discussion of risk has shifted from dealing with the toxicity to the uses in the field and the consumer (Frazer 1963) to a much wider focus that includes the whole environment and the ecosystem in which the crops are growing (Schlunt 2002). As a consequence more and more studies are required before a fungicide can be used, leading to enormous development costs. In the U.S.A., the IR-4 programme has been established to provide safe and effective pest management solutions for speciality crops growers.

Overall, most analyses come to the conclusion that the benefits of fungicides far out weight the risks, if they are used carefully and according to the label recommendations. Currently, more than 80 per cent of the fruit and vegetable crops grown in the U.S.A. receive a fungicide every season. The benefit of fungicide use in the U.S. agriculture is reported to boost farm income by nearly $13 billion annually (Gianessi and Reigner 2006). In general, the cost-benefit ratio of fungicide spray application in managing foliar diseases comes out to be 1:3 in most of the cases, which is still much higher in case of seed treatment. The alternatives proposed by organic farmers, who are opposed to intensive farming altogether, exclude the use of synthetic fungicides, but allow the use of copper and sulfur-based inorganic fungicides. These are still an ongoing debate as to whether traditionally or organically grown produces are safer for the consumer. For example, a growing number of studies are being conducted to evaluate the risk of mycotoxins in the two farming systems.

Fungicide usage

Out of total pesticides sales of $32.5 billion, total sale value of fungicides is $6.0 billion. Europe leads the world in usage of fungicide (40%) followed by Japan (28%) and USA (21%). Major share of fungicide consumption in Europe is on wheat, vines and barley (87% of the total) while Japan has maximum usage of fungicides in rice and top fruits. Based on sales, mancozeb, copper compounds, sulphur, chlorothalonil, propiconazole, tebuconazole, azoxystrobin, metalaxyl, carbendazim and iprodione are the top ten fungicides used in the world.

Overall use of pesticides in India is quite low as compared to developed countries. In India, consumption of pesticides (calculated on the basis of gross cultivated area and total consumption of technical grade pesticides) is about 500 g/ha while it is 12000 g/ha in Japan, 3000 g/ha in Europe and 2500 g/ha in USA (Singhal, 2000). The use of fungicides in India is less than insecticides and herbicides. The consumption of fungicides was 10563 Mt compared to 38788 MT of insecticides and 6040 MT of herbicides in 1995-96 that decreased to 8307 Mt in 2000-01 compared to 26756 MT of insecticides and 6040 MT of herbicides. In the year 2003-04, sale of fungicides accounted for 12 per cent compared to 70 per cent of insecticides and 16 per cent of herbicides among total crop protection sales. According to another estimate, the current fungicide use in India stands at 19 per cent compared to 61 per cent of insecticides and 17 per cent weedicides (Krishna et al 2004). The current fungicide market in India is worth Rs.430 crores.

Historical developments

The development of fungicides for controlling plant diseases has passed through several stages. Generally, the type of materials developed corresponded with the knowledge of disease etiology and also the chemical and biological properties of the compounds. Initially, simple inorganic chemicals were put to use, but as our understanding of pathogenic processes increased, more complex organic compounds were developed with enhanced efficacy. Several incidents of devastating effects of diseases during 18th and 19th century prompted the growers to try various rudimentary chemicals like sulphur, lime sulphur, sodium chloride, copper sulphate, copper carbonate, ammonium sulphate, ammonium carbonate etc. (Hewitt, 1998). The real interest in the development and use of fungicides started with the discovery of Bordeaux mixture by Millardet in 1885 for the control of grape downy mildew in France (Millardet and Gayon, 1887). This period witnessed the beginning of commercial use of fungicides on a large scale. Most of the earlier organic fungicides were developed from dyestuff intermediates leading to synthesis of several phenolic compounds containing metallic elements such as mercury, copper and tin. Since then many fungicides belonging to a number of different chemical classes have been developed.

The agrochemical companies became more active after the second world war and realize the potential of fungicide use in crop protection, developed several classes of versatile non-systemic, protectant fungicides like sulphur, copper-based, dithiocarbamates, phthalimides, crotonates, organotins, dodine, quintozene etc. between 1930s and 1950s. These were unable to control pathogens already established within the plant tissues. The introduction of systemic fungicides in late 1960s caused a revolution in plant disease control and several damaging diseases, which were difficult to be controlled by earlier non-systemic protectant fungicides, could be effectively managed by the use of these systemic compounds. In late 1960s and 1970s, several new classes of fungicides, mostly systemic in nature, such as oxathiins, 2-aminopyrimidines, benzimidazoles, morpholines, dicarboximides, phenylamides, phosphorothiolates, phenylamides, alkyl-phosphonates were introduced and development of their related compounds with improved properties continued thereafter in 1980s. These fungicides were more potent and specific in their activity and were effective at much lower doses than the earlier surface protectants.

While the development of non-systemic multisite action protectant fungicides became almost static after 1970, new groups of site specific systemic fungicides continued to emerge and still new such molecules are being discovered every year.

During the past decade, more novel compounds with different modes of action notably phenylpyrroles, probenazole, anilinopyrimidines, quinolines and more important strobilurins have been developed by different companies. These are generally effective at low doses and possess novel target sites (Gullino et al 2000). Presently, more than 150 fungicidal compounds are used the world over to manage different diseases (Thind 2007, Thind 2012).

First generation fungicides

Traditional fungicides that were protectant in nature have contributed much in the control of seed-borne, foliar and fruit diseases. Inorganic and organic sulphur compounds have a major share among these fungicides and are still being widely used to protect agricultural and horticultural crops against a number of diseases. Elemental sulphur is still a common choice for powdery mildew control in fruit trees and other crops because of low cost, higher efficacy and environmental safety. However, in some crop like cucurbits, its use is restricted due to phytotoxic effects.

The dithiocarbarmates became commercially available in USA in 1940s (Tisdale and Flenner, 1942) and called organic sulphur compounds, these represented the most widely used class of broad spectrum fungicides followed by copper-based fungicides (fixed copper compounds such as copper oxychloride) developed in the same period during 1930-34 as a substitute to Bordeaux mixture that caused phytotoxic effects in some crops. Kittleson (1952) introduced phthalimides class of fungicide with captan as the representative fungicide. Various other compounds having protectant and broad spectrum activity were developed till 1965. As protective applications, dithiocarbamates (maneb, mancozeb,propineb, ziram, zineb), copper compounds (copper oxycloride, Bordeaux mixture, copper hydroxide), phthalimides (captan, captafol, folpet), chlorophenyl compounds such as chlorothalonil etc. were available to provide effective control of various foliar and fruit diseases e.g. early and late blight of potato and tomato, Alternaria blight, white rust and downy mildew of crucifers, anthracnose, Alternaria blight and downy mildew of cucurbits, purple blotch of onion, fruit rot of chillies, Cercospora leaf spot of ground nut and other crops, Ascochyta blight of chickpea and peas, rusts and blights in several field and vegetable crops, apple and pear scab and leaf spots, peach leaf curl, anthracnose, brown leaf spot and downy mildew of grapevine, downy mildew or black shank in tobacco and hops, blister blight in tea and many other diseases on various agricultural, horticultural, ornamental and plantation crops. The use of captafol is now restricted to seed treatment only because of toxicological considerations. Dichlofluanid, a sulphamide compound, is close to phthlamide fungicide chemistry and has a similar mode of action and spectrum of activity. Similarly, dithianon, a quinine fungicide, is effective against foliar diseases of pome and stone fruits, grapevine hops and coffee. Being subjected to weathering effects, these fungicides had to be applied frequently at regular and shorter intervals starting before disease appearance to maintain acceptable level of disease control.

Thiram and captan and organomercurial fungicides have proved to be highly effective in controlling externally seed-borne pathogens (bunts, external smuts, seed rots, damping off of seedlings due to *Pythium, Fusarium, Rhizoctonia* spp. etc.) in different crops resulting in improved seed germination and crop stand. The use of most of the mercurial fungicides has been withdrawn because of wide spread toxicological problems and potential environmental danger. In India, the use of phenyl mercury acetate has been banned while that of methoxy ethyl mercury chloride has been restricted (Anonymous 2006).

The use of organotin compounds, developed in 1954 by Van der Kerk and Luijten, remained limited in agriculture because of their general biocidal activity. Fentin acetate and fentin hydroxide were earlier used in crops like potatoes, sugarbeet, coffee, rice and vegetables to control pathogens such as *Phytophthora infestans*, *Alternaria* spp., *Helminthosporium* spp., *Cercospora* spp., *Ramularia* spp., *Glomerella cingulata, Pyricularia grisea* etc., but were found potentially phytotoxic.

Copper, sulphur, mercury-based, dithianon and tin-based fungicides disrupt cell function in a non-specific manner and are also known as general cell toxicants. They disrupt several biochemical processes through their ability to bind with chemical groups such as thiol moiety (-SH) common to many enzymes and form chelates with heavy metals in fungal cells. Captan, captafol and folpet preferably react with –SH enzymes but may also attach amino groups. Broad spectrum activity and multisite mode of action of these compounds are useful features in strategies to manage resistance to systemic fungicides. Tin compounds are reported to inhibit mitochondrial oxidative phosphorylation in target fungi.

Dinocap, a nitrophenol derivative, developed in late 1946 by Rhom and Hass Co. of the USA as the first organic fungicide to have commercial potential in the control of powdery mildews. It was developed as a substitute to sulphur specially for sulphur sensitive crops such as certain curcurbits and apple varieties. Dinocap showed better activity than inorganic sulphur in controlling powdery mildews in various crops. Binapacryl, now a superseded compound, was developed in the same period for powdery mildew control and showed structural similarity to dinocap. These compounds act as uncouplers of mitochondrial oxidative phosphorylation. Dodine, a guanidine compound, introduced in late 1950s, had some eradicant action and was recommended for use in fruit, vegetable and ornamental crops for the control of *Venturia* spp., *Taphrina deformans, Monilinia laxa, Mycosphaerella* and *Cladosporium* spp. Another guanidine fungicide, guazatine was developed in this period for use as seed treatment in cereals against *Septoria nodorum* and foliar applications against *Cercospora* spp. in peanuts and rice blast. These compounds disrupt the fungal membranes in a non-specific way thus interrupting permeability.

Few aromatic hydrocarobons e.g. pentachloronitrobenzene (PCNB), chloroneb, dicloran and phenyl phenol were discovered during 1940s and 1950s because of their mutagenic properties in fungi. They showed commercial potential in the control of several pathogens including soil-borne and seed-borne diseases of cotton, vegetables, sugarbeet and horticultural crops. They induce gross changes in membrane function. With the ban on use of PCNB (quintozene), presently there is no good fungicide available for the control of *Rhizoctonia* infections and other soil-borne pathogens. However, with the introduction of pencycuron, a phenylurea based fungicide; the control of seed-borne *Rhizoctonia* infections would become easy in many crops (Thind et al 2002a).

Second generation fungicides

Development of systemic fungicides in the late 1970s caused a revolution in the control of many serious plant diseases which were not effectively controlled by non-systemic conventional fungicides as the earlier fungicides were unable to check the disease once infection had occurred. The need for internal curative action to eradicate established infection had led to the development of more specific compounds with precise and distinct anti-fungal spectrum. Commercial use of such fungicides brought in a new era of practical disease control with considerable benefits to the farmers. Being systemic in nature, these were absorbed by plant tissues after application and were translocated upwards (with few exceptions) to newer plant parts away from site of application.

Carboxin and oxycarboxin to oxathins (carboxamides) group were the first successful systemic fungicides developed in 1966 on commercial scale for the control of internally seed-borne smuts and rusts in cereals (Von Schmeling and Kulka, 1966). Carboxin is still used widely for control of smuts caused by *Ustilago* and *Urocystis* species in cereals, *Sphacelotheca sorghi* (sorghum), *Exobasidium vexans* (tea) *Drechslera* species (barely), *Rhizoctonia solani* (rice). Major fungal pathogens controlled by oxycarboxin are *Puccinia* spp. in different cereals and other crops, *Hemileia vastatrix* (coffee), *Melampsora lini* (flax). Caoboxamide chemistry has been well explored leading to more active compounds like fenuram and methfuroxam which are used for the control of seed-borne pathogens in cereals. Similarly, mapronil was developed to control *R. solani* in rice and *P. recondita* in wheat. Mebenil and banodanil were also introduced to control rusts in cereals. Another compound pyracarbolid is active against smuts and rusts in cereals and coffee rust. These fungicides inhibit succinate dehydrogenase in complex II in fungal respiration.

Benzimidazoles such as benomyl and carbendazim, among the so called modern fungicide groups, introduced in late 1960s (Delp and Klopping, 1968), attracted maximum attention of phytopathologists the world over and were tried against a variety of crop diseases. Compounds of this class had the ability to control a large number of pathogens at relatively low dose rates in field crops, fruits, vegetables, plantation crops etc. These are used as foliar fungicides, seed treatment and as post-harvest applications. Other similar compounds such as thiabendazole, fuberidazole, thiophanate and thiophanate methyl were also developed later. These have a common mode of action and inhibit fungal mitosis by interfering with spindle formation at the level of tubulin biosynthesis. These fungicides are effective against most of the fungal pathogens belonging to Ascomycetes and some of the Basidiomycetes and Deutermomycetes. These are not effective against Oomycetes and dark coloured demataceous fungi. Compounds like benomyl, carbendazim and thiophanate methyl have been widely used against crop pathogens such as *Botrytis, Cercospora, Colletotrichum, Gloeosporium, Fusarium, Erysiphe, Phoma, Phomosis, Pellicularia, Penicillium, Phyllosticta, Rhizoctonia, Sclrotinia, Uromyces, Venturia* etc. in different crops. These have protectant, curative and eradicant action against different pathogens. Benomyl has now been withdrawn for use by the company because of certain undesirable toxicological problems. Thiabendazole is used to control various types of post-harvest infections by fungi like *Penicillium* spp in apple, citrus and other fruits.

Organophosphorus compounds were developed during 1970s in Japan mainly for the control of rice diseases. Iprobenphos and edifenphos, two phosphorothiolate compounds were extensively used in Japan against rice blast (*Pyricularia grisea*). These are also effective against sheath blight and brown leaf spot of rice and act through inhibition of chitin synthesis in target fungi. Triamiphos is active against powdery mildew of apple, pear and grapes. Pyrazophos, another organophosphorus compound, has major uses in the control of powdery mildew of cucurbits, cereals, ornamentals and hops and inhibit conidial germination and appressorium formation.

Dicarboximides or cyclic imides such as iprodione, vinclozoline, procymidone were introduced in early 1970s. Though largely non-systemic in nature with limited penetration except procymidone which has been reported to be systemic in nature, these compounds possess highly selective activity against members of Sclerotiniaceae (species of *Sclerotinia, Monilinia, Botrytis*) and also show good action against pathogens like *Alternaria, Phoma, Penicillium, Septoria, Stemphylium* and *Rhizoctonia* spp. in grapevine, oilseed rape, hops, ornamentals, fruits, legumes, cereals and vegetables (Pommer and Lorenz 1987).

Hydroxypyrimidines developed during early to mid 1970s are highly specific to the members of Erysiphaceae that cause powdery mildews in diverse crops. While ethirimol is more active against powdery mildew in cereals like barely and wheat (*Eryspihe graminis*), dimethirimol provides effective control of powdery mildew in cucurbits (*E. cichoracearum* and *Sphaerotheca fuliginea)*. Bupirimate is used to control powdery mildew in apple (*Podosphaera leucotricha*) and ornamentals (*S. pannosa*). These fungicides act through inhibition of adenosine deaminase in purine salvage pathway of nucleic acid metabolism.

Third generation fungicides

A number of new compounds of different chemical classes were developed during this period that helped in the successful control of several damaging fungal pathogens.

Phenylamides and related compounds

Phenylamides introduced in 1977 onwards became much talked about compounds due to their unique potential activity against destructive plant pathogens belonging to class Oomycetes such as *Phytophthora, Pythium*, downy mildew pathogens etc. Development of metalaxyl by the then Ciba-Geigy Agrochemicals in 1977 came as a landmark discovery and a breakthrough was achieved in the effective control of diseases caused by oomyceteous fungi (Urech et al 1977). This was followed by the introduction of other similar compounds such as benalaxyl, furalaxyl (acylalanines), ofurace, cyprofuram (butyrolactones) and oxadixyl (oxazolidionones). Some other fungicide having different chemistry but similar anti-oomycete activity spectrum were also introduced soon in this period notably prothiocarb and propamocarb (carbamates), hymeazol (isoxazoles), cymoxanil (cyanoacetamide-oimes), fostyl-Al (alkylphosphonates) and dimethomorph (cinnamic acid derivates). Among these, metalaxyl is the most extensively used fungicide on the group with protective and curative activity against majority of Peronosporales. It is taken up readily by plants and shows high acropetal translocation. Metalaxyl M (mefenoxam) is more active of the two enantiomers (Nunninger et al 1996). These compounds act curatively after initial penetration by inhibiting the synthesis of ribosomal RNA (Davidse and Van der Berg-Valthus 1989).

Hymexazole is used as seed dressing or soil drench to control soil borne *Pythium* spp. and *Aphanomyces* spp. in sugarbeet and other crops and its primary mode of action is on DNA synthesis. Prothiocarb (supersede) and propamocarb, apart from controlling *Aphanomyces, Pythium* and *Phytophthora* spp., also check downy mildew fungi like *Peronospora, Pseudoperonospora* and *Bremia* spp. and their activity is related to membrane function. Cymoxanil has important uses similar to metalaxyl but is more effective against hyphal growth than early growth stages. Fosetyl or fosetyl aluminium is a unique compound having strong basipetal (Phloem mobile) systemicity and is considered to act through mobilization of natural defense mechanisms in host plants. It has demonstrated high activity against downy mildew in grapes and hops and *Phytophthora* spp. in citrus and pinapple. Dimethomorph, a cinnamic acid derivative, shows activity against all major Oomycete genera except

Pythium spp. It inhibits cell wall formation and is more of a protectant in nature with little systemicity. Its use against potato late blight is well documented.

Sterol biosthnthesis inhibitors

Sterol or ergosterol biosynthesis inhibitors (SBs) or EBIS) are the largest group of modern fungicides and contain compounds that have different chemical structures but similar mode of action i.e. inhibition of ergosterol synthesis for fungal cell walls. These are constituted of seven sub groups with triazoles as the largest sub-group. Most of the EBIs were developed between 1970s and 1980s and are reported effective against a wide range of plant pathogens belonging to Ascomycetes, Basidiomyetes and Deuteromycetes (Scheinpflug and Kuck, 1987) but are ineffective against Oomycetes. These are used against diverse diseases such as powdery mildews, rust fungi, smuts, bunts and other pathogens like *Venturia, Septoria, Mycophaerella, Neovossia, Penicillium, Fusarium, Alternaria* etc. in several crops and are effective at quite low doses. The SBIs contain the most commercially used groups of fungicides like triazoles (triadimefon, bitertanol, propiconazole, penconazole, tebuconazole, hexaconazole, myclobutanil etc.), imidazoles (imazalil, prochloraz), pyrimidines (fenarimol, nuarimol), piperazines (triforine), pyridines (buthiobate, pyrifenox). They inhibit fungal growth by interfering at C-14 demethylation step of ergosterol synthesis.

The spectrum of disease control of morpholines (tridemorph, dodemorph, fenpropimorph, trimorpharmide), and piperidines (enpropidie), which interfere at 8.7 isomerase step, is limited compared with C-14 demethylation inhibitors, their major use being against the members of Erysiphaceaeae causing powdery mildews in cereals and other crops. Apart from controlling powdery mildews, these are also effective against other diseases like coffee rust, blister blight of tea and Sigatoka disease of banana.

Phenylcarbamates

Phenylcarbamates were mostly developed as herbicides, but a few showed unique feature of being effective against benzimidazole resistant fungal strains. Diethofencarb and MDPC have shown negative cross resistance to strains of *Botrytis cinerea, Cercospora beticola, Fusarium nivale, Penicillium epansum, Venturia nashicola* and *Pseudocercospora herpotrichoides* that had developed resistance to benomyl and carbendazim. The mode of action of phenylcarbamates is quite similar to that of benzimidazoles (Fuzimura et al 1990).

Melanin biosynthesis inhibitors

Among melanin biosnythesis inhibitors (MBIs) tricyclazole was developed by Elanco Products Company in USA in 1976 primarily for use against rice blast (*Pyricularia oryzae*). These are anti-penetrants acting on the pigmented pathogens to prevent them from entering through the host epidemis. The discovery of tricyclazole initiated the development of other such compounds like pyriquilon, chlobenthiazone and phthalide, displaying a novel mode of action in pigmented Ascomycetes and

Deuteromycetes. The inhibition of melanin synthesis by blocking polyketide pathway prevents melanization of appressoria and subsequent formation of infection peg apparatus. These fungicides provide excellent control of *Pyricularia grisea* and other fungi like *Colletotrichum* spp.

New generation fungicides

During the decade of 1991-2000, fungicide research has produced a diverse range of fungicidal products with novel modes of action which had a significant impact on plant disease control. An important aspect has been the identification of novel target sites. Earlier potent fungicide classes such as SBIs and phenylamides etc. have been extensively exploited over the past years and many of them have faced one or other problem of desired field performance. Some of these new compounds and their disease controlling potential are discussed below:

Strobilurins

Of the new compounds, strobilurins (also called methoxyacrylates) are the most important and are effective against a diverse range of plant pathogens and diseases (downy mildews, powdery mildews, *Phythopthora* and *Alternaria* blights, rusts, apple scab, *Rhizoctonia* infections, *Septroia* and *Cercospora* leaf spots, rice sheath light and blast etc. caused by Oomycetes, Ascomycetes, Basidiomycetes and Fungi imperfecti (Jensen 1997, Thind 2004 a, b). These compounds are analogues of strobilurin-A originally obtained from *Strobilurus tenacellus*, a wild mushroom growing in temperate forests. Their synthesis was initiated by BASF and Zeneca (now Syngenta) to prepare analogues of storbilurin-A, possessing desired fungitoxicity, good light stability and systemic properties without phytotoxicity. The first analogue was registered for large scale applications against various diseases in 1996 which was soon followed by kresoxim methyl and trifloxystrobin was developed in Japan in 1998 for use against rice diseases (Koizumi 1998). Their fungicidal activity results from the prevention of electron transfer between cytochrome b and c1 in complex III of the mitochondrial electron transfer chain (Leroux, 1996). These are also effect against fungal strains having developed resistance to other groups of fungicides.

Oxazolidinediones

Famoxadone, a representative of this group, is a broad spectrum fungicide for use in fruits and cereals. It shows remarkable activity against *Phytophthora infestans* and *Plasmopra viticola*. It also controls cereal pathogens like *Puccinia, Septoria* and *Pyrenophora* spp. The compound inhibits the activity of ubiquinol cytochrome c oxido-reductase at complex III. It mainly controls diseases through protectant action with significant translaminar movement. Lately, famoxadone has been introduced in combination with cymoxanil (KX007-42 SC) to enhance disease control and as resistance management step.

Anilinopyrimidines

These are broad spectrum novel fungicides having a potential use in variety of crops. Mepanipyrim and pyrimethanil possess remarkable activity against *Botrytis cinerea* on grapevine and other fruits and *Venturia inaequalis* on apples (Daniels et al 1994). Cyprodanil, another such compounds, has additional activity *against Pseudocercoporella herpotrichoides, Erysiphe gramins, Helminthosporium gramineum, Pyrenophora teres* and *Septoria nodorum* on cereals. These compounds are considered to inhibit methionine biosynthesis (Leroux et al 1996).

Phenylpyroles

These compounds were developed from secondary metabolities produced by *Pseudomonas pyrrocina* having fungicidal properties. Of the two commercial fungicides, fenpiclonil is used as seed dressing while fludioxanil is applied as seed dressing and foliar spray against *Botrytis cinerea* (Gullino et al 2000). These fungicides have a broad disease control spectrum but are inactive against Oomycete pathogens. Though primary target is not yet certain, they appear to affect glucose phosphorylation in target fungi. Another phenylpyrrole pyrrolnitrin is unsuitable for use in practical disease control because of its unstablity in light.

Phenoxyquinolines

These were introduced in 1996 for control of powdery mildews in cereals and other crops. C Quionoxyfen (DE0795) shows marked activity against *Erysiphe gramiins* (cereals) and *Uncinula necator* (grapevine) and is a systemic protectant in action providing long term control (Loghurst et al 1996). It has vapour phase redistribution and inhibits spore germination and appressorium formation by affecting dihydro – orotate- dehydrogenase in pyrimidine biosynthesis pathway. It has also shown good activity against *Botrytis cinerea* in different crops.

Spiroketalamines

Sproxamine is a novel ergosterol biosynthesis inhibitor introduced in 1996 and is particularly effective against powdery mildew of cereals and grapevine (Dutzmann et al 1996). Apart from inhibiting "14 reductase, it has additional activity against "8.7 isomerase, squalene synthase and squalene cyclase and hence shows cross resistance to morpholine and piperdine compounds.

Mefenoxam (Metalaxyl_M), an enanitomer of metalaxyl, was introduced by Syngenta in 1996 and is considered more potent than metalaxyl against Oomycete pathogens (Nunninger et al 1996). Carpopamid (KTU 2616), another MBI compound, was developed by Bayer Crop Science in 1997 for potential use against rice blast (Koizumi 1998) and is a scytalone dehydratase inhibitor.

Benzamides

Two benazamide compounds viz., fluopicolide and zoxamide have been developed recently for use against diseases caused by Oomycete pathogens such as downy mildew and late blight. These show their activity by interfering with microtubule skeleton similar to benzimidazoles though at a different step (Egan et al 1998). These are generally used in combination with mancozeb. Fluopicolide has recently been formulated with propamocarb for better action and modified the cellular localization of a spectrin like protein (Toquin et al 2007).

Valinamides

Two compounds of this group, also known as amino acid amide carbamates, viz., iprovalicarb and benthiavalicarb have been introduced recently by Bayer and Kumiai Chemicals in 2002 and 2003, respectively. These are quite effective against potato late blight and grape downy mildew (Reueni 2003). Combination of iprovalicarb with propineb (Melody Duo) has been introduced to increase its disease control capability and as a measure to check resistant build-up.

Although most of the fungicides introduced during this period were site-specific systemic fungicides belonging to different chemical classes, few fungicides with protectant activity were also developed. Among these, compounds like fenitropan, fluzinam, pencycuron etc. are worth mentioning. Fluainam, a phenylpyridylanine fungicide developed in mid-1990s, is a protectant fungicide used in the control of pathogens like *Botrytis, Colletotrichum, Phytophthora* and *Venturia* in fruits and vegetables and acts by disrupting oxidative phosphorylation in target fungi. Fenitropan, a nitroparaffin compound, is a broad spectrum protectant fungicide applied as seed treatment in cereals, rice and sugarbeet and exhibits some activity as foliar fungicide against apple scab and powdery mildew. With the introduction of pencycuron, a phenylurea-based fungicide, the control of seed and soil-borne *Rhizoctonia* infections would become easy in many crops (Thind et al 2002b).

Imidazolinones

Also known as imidazoles, this group is represented by fenamidone. Introduced by Bayer Crop Science in 2003 for use against Oomycete pathogens, fenamidone has been commercialized in combination with mancozeb (Secure 68 WG) for the control of downy mildew and late blight.

All the new fungicides in combination with mancozeb or similar compounds are reportedly highly effective in checking potato late blight and increasing tuber yield and are comparable in efficacy with Ridomil MZ (Thind et al 2004). These fungicides possess different mode of action and there are less chances of cross resistance to metalaxyl resistant populations of *P. infestans.*

Other compounds developed in the recent years for use against Oomycete pathogens are cyazofamid (Cyanoimidazoles, ISK Biosciences), mandipropamid (Mandelamides, Syngenta), ethaboam (Thiocarbamates, LG Biosciences) (Huggenburger et al 2005). Similarly, strobilurin compounds developed later up to 2006 are pyraclostrobin, picoystrobin, fluoastrobin and dimoystrobin and have the potential to control diverse fungal pathogens.

Few more compounds with different chemistry and activity have also been developed recently. These include tolyfluanid (Phenylsulfamides, Bayer) for *Botrytis* and powdery mildews, metrafenone (Benzophenones, BASF) for cereal powdery mildew, boscalid (Anilides, BASF) for Septoria eye spot and Fusarium scab in wheat, cyflufenamid (Amidoximes, Nippon Soda) for powdery mildews, epoxyconazole (Triazoles, BASF) for sigatoka of banana, ipoconzole (Triazoles, Crompton Corp.) for seed treatment in cereals and prothioconazole (Triazolinthiones, Bayer) for wheat foliar diseases.

Newer active ingredients introduced recently are benthiavalicarb (from Kumiai) and mandipropamid (Syngenta), from thecarboxylic acid amide (CAAA) fungicide group, and fluopicolide (Bayer),metrafenone (BASF), proquinazid (DuPont), and zoxamide (DowAgro) for managing various diseases in different crops. For a more detailed technical description of modern fungicides, including chemistry, the reader may refer to the treatise edited by Kramer and Schirmer (2007). More recently during 2010 and 2011, new fungicides with novel modes of action have been launched by agrochemical companies for use against different diseases. These include fluopyram (pyridinylethylbenzamides) by Bayer against gray mold, powdery mildew, *Sclerotinia* and *Monilinia*; penthiopyrad (carboxamide) by Du Pont against rust, *Rhizoctonia*, gray mold, powdery mildew, apple scab and fluxapyroxad (carboxamide) by Agrow-BASF for seed treatment/protectant for different crops.

Fungicides in Indian agriculture

A number of fungicides belonging to different chemical groups have been registered in India and these are being used against diverse diseases in fruits, vegetables, plantation crops and some field crops. Fungicides such as sulphur, dithiocarbamates, cropper-based, phthalimides, benzimidazole, phenylamides, triazoles etc. are now regularly used in Indian agriculture, with the use of costly specific fungicides in high value crops only. As on 20.01.2012, a total of 65 antifungal products have been registered in India (CIB, 2012) including 4 antifungal/antibacterial antibiotics and four biocontrol agents.

Among various fungicides used in India, consumption of mancozeb is the highest followed by sulphur compounds, copper oxychloride, copper sulphate, carbendazim and thiram as per the demand pattern (Agnihotri, 2000). These six fungicides constitute more than 85 per cent of the total fungicides used with mancozeb alone accounting for 25 per cent of the total. Hexaconazole, propiconazole, metalaxyl-M+ mancozeb, cymoxanil + mancozeb, edifenphos, flusilazole, triadimefon, tricyclazole and azoxystrobin are few other fungicides that account for substantial fungicide market in India. Agrochemicals market in India is worth about Rs.4800 crore, out of which, fungicides account for nearly 12 per cent of the sales. Crop-wise consumption of fungicides in India is maximum on pome fruits, followed by potatoes, rice, tea, coffee, chillies, grapevines, other fruits, vegetables and field crops (Thind 2005).

Apart from contact compounds like sulphur, dithiocarbamates, copper-based, mercurials, phthalimides etc., several of the site-specific fungicides of the groups like benzimidazoles, oxathins, thiophanates, organophosphorus, triazoles and related sterol inhibitors, phenylamides, cyanoacetamide oximes, cinnamic acid derivatives and some other modern fungicides are also being used in India for controlling diseases on a number of crops.

Some of the new generation fungicides have also been registered in India for use against different diseases. These include Strobilurins (azoxystrobin, kresoxim methyl, trifloxistrobin, pyrachlostribin against powdery mildew, downy mildew in grapevine, cucurbits, rice sheath blight), Valinamides (iprovalicarb, benhiovalicarb against oomycete diseases in grapevine, potato, tomato and curcubits), Oxazolidinediones (famoxadone, against potato late blight), Imidazoles (fenamidone against potato late blight and grape downy mildew), Phenyl-ureas (pencycuron, against rice sheath blight, black scurf of potato), Mandelamides (mandipropamid, against late blight of potato, downy mildew of grapevine), Melanin biosynthesis inhibitors (carpropamid, against rice blast), Triazolinthiones (protioconazole against rice blast). A few recently developed compounds such as zoxamide and fluopicolide (bezamides), and cyazofamid (Cyanoimidazoles) are under evaluation against Oomycete diseases.

Some of the old fungicides have either been banned or withdrawn in India by Central Insecticides Board or their use has been made restricted due to certain ill effects on environment and ecology. Mercury compounds ethyl mercury chloride and phenyl mercury acetate are banned because of their longer persistence in the environment and food chain while there is restricted use for methoxy ethyl mercury chloride. Quintozene (PCNB) and pentachlorophenol are also banned due to adverse effects on ecology. Captafol is now allowed to be used as seed treatment only and its use as spray is banned. Likewise, ferbam and nickel chloride have been withdrawn (Anonymous 2006).

Changes in R&D Process

The last half century has seen new scientific approaches in pesticide research, yet last 20 years have recorded numerous discoveries of novel and highly efficient products (Leroux, 1996, 2003, Gullino et al 2000). Revolutions in chemistry in the synthesis of novel molecules, in particular combinatorial and robotic chemistry and a dramatic increase in the knowledge of the genetic and genomic information have led to the designing of novel screening methods to assess more than a hundred thousand novel compounds per year (Corran et al 1998). In addition, biochemistry has developed tools to identify compounds having a novel mode of action very early in the research process so as to carry out early screening. Significant progress has also been made to identify the biochemical targets of these compounds (Beffa 2004). The discovery of new modes of action that result into improved activity of fungicides against commercially important targets, combined with assured environmental and public safely, is a critical step in safeguarding food security. By 1990's there were total 113 active ingredients registered as fungicides worldwide (Knight et al 1997).

Much of these new chemistry arose as a result of the basic manufacturers having moved away from *in vitro* and *in vivo* screenings. In these *in vivo* screens, young potted plants were sprayed or drench-treated with the test compounds and inoculated with a major pathogen soon thereafter. The plants then were placed in a greenhouse or growth chamber under ideal disease conditions until disease symptoms appeared and they could be rated for contact or systemic efficacy as well as for phytotoxicity. This process was usually repeated on a weekly basis with large numbers of test compounds.

Presently, discovery platform for most multi-national crop protection chemical companies consists of a combination of several different approaches including high volume chemical screens, chemistry based and biological target based synthesis programs, and natural products research (Wenhua 2009). Key elements of research and discovery of MNCs include high throughout screening, combinatorial chemistry, advanced formulations and molecular and environmental studies which operate in combination to develop new products and technologies. The focus area of research in fungicide discovery is combinatorial chemistry, high throughput screening, advanced formulation, molecular toxicology and environmental safety.

High-throughput automated screening can assist tremendously in testing a large number of different chemicals to detect potential research leads. Today's miniaturized screens allow scientists to run millions of tests a year, both on the whole organism, and on the isolated a year, both on the whole organism, and on the isolated biochemical target sites. The secret of this massive increase in throughput is the adoption of a small micro-titer plate. These plates contain many individual wells or compartments, which allow many options to be tested at once. A major advantage is that these new screens only require a few micrograms of a test compound. This new technology allows the screening of several hundred thousand compounds per year, with automation being employed throughout the process. For high throughput screening, many chemical compounds must be tested to find starting points, called leads, which help to focus further research into specific areas of chemistry. Locating hundreds of thousands of novel and diverse chemical entities per year to test in the high-throughput screens to find promising chemicals to investigate is a major challenge. However, a variety of approaches are used by MNCs including in-house chemical synthesis, collaborations with third parties, the testing of chemical collections called libraries and robotic synthesis of designed libraries of chemicals.

Although several novel products had been discovered by the high volume screening approaches, the difficulties to obtain reliable source for novel chemicals to feed the biological screening system make this approach unsustainable for a long run. The target based approach is the principle discovery platform for the pharmaceutical industry; however, the success of this approach to pesticide discovery has yet to be achieved. Another approach is the natural product platform. The discoveries of crop protection chemicals from natural products are well documented (examples, pyrethroids, abamectins, spinosyns, etc.), the availability of unique biological samples from new natural sources is limited and unreliable. Finally, the chemical based approach is the most productive pesticide discovery platform today. This platform requires experienced and insightful chemistry teams and the support of rapid biological screening system (Oerke et al 1994, Wenuha 2009).

Significant advances have been made in recent years with regard to the formulation chemistry and delivery mechanisms used to apply new products. For example, new micro-encapsulated formulation and water-soluble bags have been developed. These and other technologies have advantages such as controlling the release of the chemical, hazard reduction, mixture compatibility and improved bio-performance of the product combine with added convenience for the user.

Role of seed treatment

Good seed should be pure and free of weed seeds, seed-borne pathogens, insects or other matter. Seed treatment with chemical, biological or physical agents provides protection to seed and helps in the establishment of healthy crops. It provides good protection to the seed from seed and soil-borne pathogens during the germination period and soil-borne pathogens during the germination period and supports early plant development. A good range of seed treatment fungicides are now available ranging from the traditional chemicals like thiram, captan manocozeb, carbendazim and carboxin to the more effective and safer compounds such as mefenoxam + fludioxanil and azoxystrobin (Syngenta), pyraclostrobin + metconazole (BASF) developed during 2008-10. More recently, two new seed treatment fungicides sedaxane (syngenta) and fluxapyroxad (Agrow-BASF have been introduced in 2011). The global seed treatment market is estimated at over $2.5 billion with USA accounting for more than 50% of the global markets and it will continue to grow as it is considered a cost-effective method to protect the increasingly valuable seed.

As per the estimates of Directorate of Plant Protection, quarantine and Storage, Faridabad, nationwide 70 per cent requirement of seed is met from the farmer's own stock, most of which is sown without any seed treatment. Even if seed is purchased from the private or public sector agencies large percentage of such seed is untreated. The estimates further reveal that on an average, 80 per cent of the seed sown in the country is untreated, as against the 100 per cent seed treatment practice in developed countries. This may be one of the several reasons for our lower productivity in many crops compared to developed nations. There is thus an ample seed treatment practices. The future seed treatment may move towards multiple compound products for complete spectrum of pest and disease control as also abiotic stress tolerance.

Fungicide mixtures

Fungicide mixtures, containing two or more fungicides with different modes of action, have been developed with the twin objectives of broadening the activity spectrum against diverse plant diseases and to check the development of resistance in the target pathogens. Often, these compounds have synergistic effects and enhance the efficacy levels of the individual compounds together. Some of the recently developed and commonly used fungicide mixtures are Ridomil Gold (mefenoxam + mancozeb) against downy mildews, *Phytophthora* and *Pythium* diseases, Prosaro (prothioconazole + tebuconazole) against Fusarium head scab in cereals, Input (porthioconazole + spiroxamine)) against powdery mildew, Fusarium, *Septoria*, rusts in cereals, Twinline (pyraclostrobin + metconazole) against cereal rusts, Amistar Top (azoxystrobin + difenoconazole) against various diseases on diverse crops and Luna Sensation (fluopyram + trifloxystrobin) for control of gray mold, powdery mildew, *Sclerotinia* and *Monilinia* in fruits.

Decision support systems

Reaching an informed and national decision for disease management is a complex task based on epidemiological parameters, Decision support systems (DSS) integrate all relevant information to generate spray recommendations that could help reduce fungicide usage and as a consequence maintain the durability of fungicides effectiveness. An important factor related to disease management decision is the potential of the disease to cause damages. Some diseases are more destructive than other and under favourable conditions may develop rapidly eventually leading to complete destruction of the crop, for example potato late blight (*Phytophthora infestans)* or apple scab (*Venturia inaequalis*). Thus, the decision as to apply a fungicide mainly depends on the value of the crop, the type of diseases and its potential to cause losses and on the cost of control measures (Carisse et al 2010). Many DSS such as Wheat Disease Manager for wheat diseases (UK), More Crop for rusts and other diseases of wheat (Pacific North-West USA), Crop Protection Online for wheat diseases (Denmark), Blight Management and NegFry for potato late blight (UK, USA) RIMPro for apple scab (Netherland) and many more are now available which help the growers reduce fungicide applications upto 50 per cent. A web-based DSS has also been developed for deciding fungicide applications against late blight of potato in Punjab (Thind 2011a).

Fungicide residues

The residue levels of certain fungicides in the soil or edible plant parts may sometimes be well above the acceptable limit and this varies with the dose of the fungicide used and with total number of sprays done. With the use of systemic fungicides for managing fungal diseases in agricultural crops, there are ample changes of their residue build up in harvested produce. Not adhering to the recommended pre-harvest interval may further compound the problem. This not only poses health risks to consumers, but may also affect the export potential of certain commodities to the developed countries. Detection of residues of carbendazim and isoprothiolane above MRLs in basmati rice consignments exported from India to Germany in 2010 and that of tricyclazole in the consignments exported to USA in 2011 has sent warning signals among the agriculturists and traders in India. Modern HPLC technology makes it possible to detect very small amounts (in ppb) of the fungicides in the food products. Farmers need to be educated to follow the recommended guidelines while using fungicides and other pesticides.

Non-target effects

Though fungicides are primarily used to manage plant diseases caused by different kinds of fungal pathogens, these may, sometimes, exert adverse effects on other microbial flora and fauna some of which may be beneficial to the crop plants. Farmers often use over dose of fungicides, regardless of their ecological impact. Many organic fungicide products can provide effective disease control with minimal ecological impact. However, improperly using any fungicide (organic or otherwise) will result in poor control and may harm non-target organisms.

Besides affecting plants, some fungicides can adversely affect native soil microflora. The fungicides, triadimefon and propiconazole, on repeated use, could alter the soil microbial ecology over a longer period of time (Jui et al 2009). Soil microbial populations including various fungi, denitrifying bacteria and aerobic diazotrophs are also greatly affected by the application of mancozeb (Pozo et al

1994). Similarly, fungicides like benomyl, fenpropimorph, fenhexamid, mepronil, furalaxyl, thiram also affect symbiotic relationship between the plant and mycorhizae. Elemental sulphur, benomyl and formulations of difenconazole are reported to be toxic to non-target parasitic and predatory arthropods. A transient negative effect of propiconazole, a widely used fungicide, on the overall abundance of higher collembolan taxa such as springtails has also been reported. Fungicides may also pose potential environmental and human health hazards.

Problem of fungicide resistance

Although fungicides play an important role in crop production, their intensive and faulty use can have adverse impacts on crops, useful antagonistic organisms and environment. However, the most serious problem is the development of fungicide resistance strains leading to failure of disease control. Fungicide resistance is now a widespread problem in global agriculture (Thind 2002, Thind 2008).

Resistance builds up through the survival and spread of initially rare mutants, during exposure to fungicide treatment. This development can be discrete (resulting from a single gene mutation) or gradual (considered to be polygenic). Resistance mechanisms vary, but mainly involve modification of the fungicides within the fungal pathogen. Fungicide resistance is often regarded as a problem for the agrochemical industry rather than growers but in fact, resistance has wider implications for sustainable farming. Fungicide resistance not only threatens the usefulness of individual of fungicides, but also the farm economy because of potential yield losses from poor disease control. Hence, it is in everyone's interest to ensure the efficacy of the best available products in commercial use to be continued. Fungicide resistance problems in the field have been documented for more than 100 diseases (crop – pathogen combinations), and within about half of the known fungicide groups (Brent and Hollomon 2007). Many more cases of resistance are suspected but have not been documented. Therefore, a management strategy should be implemented before resistance becomes a problem. Strategies for managing fungicide resistance are aimed at delaying its development. Specific strategies for resistance management vary for the different fungicide groups, the target pathogen(s), and the crop (Thind 2011b). Monitoring is vital, to determine whether resistance is the cause in cases of lack of disease control, and to check whether resistance management strategies are working. It must start early, to gain valuable base-line data before commercial use begins. Results must be interpreted carefully, to avoid misleading conclusions. Resistance management should integrate cultural practices and optimum fungicide use patterns. The desired results are to minimize selection pressure through a reduction in time of exposure or the size of the population exposed to the at-risk fungicide.

Resistance management

A major consideration for the design of fungicide use strategies is the threat of fungicide resistance. There have been considerable efforts by the industry to conduct research in the areas of mode of action, resistance risk, field monitoring for baseline sensitivity and sensitivity variation in treated fields. Based on the results, strategies are designed that reduce the risk of fungicide resistance building up or worse, the loss of efficacy of whole fungicide classes.

This threat of fungicide resistance and the fact that cross-resistance often exists to related products from different manufacturers has lead to a close collaboration between them in **Fungicide Resistance Action Committee** (FRAC). Results from research in mode of action, resistance risk and field monitoring are pooled and strategies are developed to minimize the risk of resistance build-up. FRAC has produced several monographs on various aspects of fungicide resistance and has grouped the available fungicides according to various criteria that facilitate the understanding of the resistance risk of the different fungicide groups. Such assessments of the resistance risk are made difficult by the unpredictability of cross-resistance. Among different fungicide groups, some like benzimidazoles, dicarboximides, phenylamides and strobilurins carry high level of resistance risk. The FRAC working groups and fora for the at risk fungicides meet from time to time and suggest ways and means for managing risk of resistance build up in important pathogens. It is emphasized to follow a proactive approach at the very beginning of the introduction of a new fungicide compound.

Fungicides as inducers of host resistance

Increasing environmental and regulatory pressures build interest in fungicides or chemicals that act indirectly by interfering with processes involved in fungal penetration into plant or by enhancing host resistance rather than the fungus such as carpropamid (Morton and Staub 2008). Probenazole is a systemic compound that indirectly controls rice blast and some bacterial rice disease by stimulating the accumulation of toxins and enzymes associated with systemic acquired resistance in rice, but are ineffective in other cereals. Acibenzolar-S-methyl (Actigard, Bion) has the widest spectrum of activity among the non-fungitoxic compounds developed to date. It is active against various fungi, bacteria and viruses in several crops and stimulates plant's natural defense system when applied a week before infection (Leadbeater and Staub 2007). Actigard has performed best when incorporated into a program of chemical sprays, as the inherent level of disease control has seldom been sufficient when applied alone. This product has initiated a whole new field of research into utilizing peptides for controlling diseases, and other means of stimulating SAR and Jasmonic acid pathway with chemicals and biological agents in plants. Because these compounds do not put selection pressure directly on the pathogen population, these are unlikely to encounter resistance problem. Such compounds with indirect action against the pathogens are likely to gain prominence in future crop protection.

Concern for environment and public safety

The discovery of new modes of action that provide improved activity of fungicides against commercially important targets, combined with assured environmental and public safety, is a critical step in safeguarding food security. Fungicides in future will be characterized by reduced application rates and lower toxicity to non-target organism (Hewitt 2000). Such products must also contribute to our ability to manage the development of fungicide resistance and thereby extend the utility of established fungicides. An important aspect of current fungicide research has been the identification of suitable and novel target sites. Fungicides that interrupt fungal development before the host penetration are very important. In the last decade already many of new products had shown specific activity before the host penetration e.g. Carpropamid against rice blast, Quinoxyfen against powdery mildew etc. In addition, naturally occurring substances found in fungi, bacteria and higher plants are important sources of molecules with antifungal properties. They may be developed either as products per se or used as starting points for synthesis. Already many natural antibiotics have been successfully developed like blasticidin S, kasugamycin and validamycin against rice blast and sheath blight. Another group of antibiotics, the polyoxins, have been used to protect fungal diseases of fruit trees and vegetables.

Novel modes of action

In order to effectively protect plants from diverse fungal pathogens, we have to develop new anti-fungal compounds with high biological activity, i.e. low dose rate application, specificity and low ecological impact. Moreover, novel modes of action are necessary to overcome resistance to existing products. In addition, public education and awareness is necessary in order to change the negative perception of chemical control that meets the toxicological and environmental standard. An interesting new area of research is the use of antimicrobial peptides (AMP) for improving resistance to pathogens using transgenic plants as bio-factories for fungicides or bactericides.

The future of fungicides

It is expected that fungicides will continue to play a role in disease management programs, especially in intensive production systems. However, to maintain their effectiveness and minimize their effect on human health and on the environment, they should be used in a rational and informed way.

A look at the history of fungicides should give us some idea of what to expect in the future. The major changes in fungicide use have usually been associated with changes in the spectra of pathogens as well as in crop intensities, practices or prices. The migration of tobacco blue mold into Europe or soybean rust into the Americas had a dramatic impact on the fungicide use on these crops. Cereals, especially the foliar applications, represent a relatively recent market when compared to fruits and vegetables. The soybean market was negligible until a few years ago when rust reached the large soybean acreages in North and South America. Such shifts in pathogen spectra could not be foreseen and will continue to occur in the future due to increased global trade with plant material. More generally, a growing world population that wants to be fed better than today will lead to increase areas of intensive cropping and hence, most likely, fungicide use will also increase (Collins 2007).

Future thrust

The trend towards a more judicious use of fungicides in conjunction with disease forecast systems that has been observed in the past can be expected to continue in the future. This will help reduce the risk of adaptation by the target fungi and at the same time will reduce residues in the environment and on the produce. The efforts of breeding for disease resistance will also continue and possibly increase by utilizing the tools of genetic engineering. Both genetic resistance and selective fungicides are prone to adaptation by the pathogen. The balance between genetic and chemical control will, therefore, most likely continue and research in both areas will complement each other to assure the availability of effective combinations of host resistance and fungicides for crops that should produce ever higher yields of ever better quality.

References

Agnihotri NP 2000 Pesticide consumption in agriculture in India –an update, *Pesticide Research Journal* **12(1)** 150-155

Anonymous 2006 List of pesticides/pesticides formulations banned in India. *Pesticide Research Journal* **18(2)** 243

Beffa R 2004 Genomics and biochemistry in the discovery process of modern fungicides. *Pflanzenschutz-Nachrichten Bayer* **57** 46-61

Brent KJ and Hollomon DW 2007 Fungicide Resistance in Crop Pathogens: How Can It Be Managed? 2nd Revised Edition Online, Fungicide resistance Action Committee (FRAC), Crop Life Int'l. Brussels, Belgium, 56pp

Carisse Odile, Trembley David-Mathieu, Jobin Tristan and Walker Anne Sophie 2010 Disease Decision Support Systems: their impact on disease management and durability of fungicide effectiveness, in *Fungicides* (ed) Odile Carisse, Tech Publishers, 177-200pages

CIB 2012 Fungicides registered under section 9(3) of the Insecticides Act, 1968 as on 20/01/2012. http://cibrc.nic.in/

Collins JC 2007 Challenges and opportunities in crop production over the next decade, in *Pesticide Chemistry, Crop Protection, Public Health, Environmental Safety* (eds) H Ohkawa, H Miyagawa and PW Lee, Wiley VCH Verlag, Weinheim, Germany, 3-12 pages

Corran AJ, Renwick A and Dunbar SJ 1998 Approaches to *in vitro* lead generation for fungicide invention. *Pesticide Science* **54(4)** 338-344

Daniels A, Birchmore RJ and Winter EH 1994 Activity of pyrimethanil of *Venturia inaequalis*, 525-532 In *The Proceedings of Brighton Crop Protection Conference – Pests Diseases*, Vol. 2, BCPC, Farnham,

Davidse LG and Van der Berg-Valthius GCM 1989 Biochemical and molecular aspects of the phenylamide fungicide – receptor interaction in plant pathogenic *Phytophthora* spp., in *Signal molecules in plants and plant-microbe interactions*. Springer –Verlarg, Berlin, 261-278 pages

Delp CJ and Klopping HL 1968 Performance attributs of a new fungicide and mite ovicide candidate, *Plant Disease Reporter* **52** 95-99

Dutzmann S, Berg D, Clausen NE, Kramer W, Kuck KH, Ponzen R, Tiemann R and Weissmuller J 1996 KWG 4168: a novel foliar fungicide with a particular activity against powdery mildew, 47-52 *In The Proceedings of Brighton Crop Protection Conference – Pests and Diseases,* Vol. 1 BCPC, Farnham

Egan AR, Mechelotti EL, Young DH, Wilson WL and Mattioda H 1998 R.H. 7281: a novel fungicide for control of downy mildew and late blight, 335-442 *In The Proceedings of Brighton Crop Protection Conference – Pests and Diseases*, vol. 1, BCPC, Farnham

Frazer AC 1963 Balance of pesticides: Benefits and risks, 3-11, In *The Proceedings of the 2nd British Crop Protection Conference*

Fujimura M, Oeda K, Inoue H and Kato T 1990 Mechanism of action of N-phenylcarbamates in Benzimidazole – resistant *Neurospora* strains, in *Managing resistance to agro-chemicals* (eds) MB. Green, H.M Lebaron, WK Moberk, ACS Washington, DC, pages 224-236

Gianessi L and Reigner N 2006 The importance of fungicide in U.S. crop production, *Outlook on Pest Management* **10** 209-213.

Gianessi, L.P. and Reigne,N. (2005). The Value of fungicides in U.S. Crop Production, Crop Life Foundation (http://www. Croplifefoundation.org/upload/)

Gullino ML, Leroux P and Smith CM 2000 Uses and challenges of novel compounds for plant disease control *Crop Protection* **19** 1-11

Hewitt G 2000 New modes of action of fungicides. *Pesticide Outlook* (Feb. issue) 28-32, Invention, *Pesticide Science* **54** 338-344

Huggenburger F, Lamberth C, Iwanzik W and Knauf-Beiter G 2005 Mandipropamid a new fungicide for oomycete pathogens, *BCPC International Congress on Crop Science and Technology, 31 October -2 November, 2005, Glasgow, U.K.*

Jensen H 1997 Amistar a broad spectrum fungicide. *14th Danish Plant Protection Conference Pests and Diseases* **8** 59-71

Jui-Hung Y, Jin-Shu C, Pin-Jui H and Yei-Shung W 2009 Effects of fungicides triadimefon and propiconazole on soil bacterial communities, *Journal of Environmental Science Health* **44** 681-689

Kittleson AR 1952 A new class or organic fungicides, *Science* **115** 84-86

Knight SC, Anthony VM, Brady AM, Greenland AJ, Heaney SP, Murray DC, Powell KA, Schulz MA, Spinks CA, Worthington PA and Youle D 1997 Rationale and perspectives on the development of fungicides, *Annual Review of Phytopathology* **35** 349-372

Koizumi S 1998 New fungicide for use on rice in Japan, 7th International Congress on Plant Pathology, Edinburgh Scotland, 9-17 August, 1998, Abstract No. 5.6.3S

Kramer W and Schirmer U 2007 *Modern Crop Protection Compounds*, Vol. 2. Wiley-VCH-Verlag, Weinheim, Germany

Krishna VV, Byju NG and Tamizheniyan S 2004 Integrated pest management in Indian agriculture: A developing economy perspective, IPM World (ipmworld.umn.edu) 1-8 pages

Leadbeater A and Staub T 2007 Exploitation of induced resistance: A commercial perspective, in *Induced Resistance for Plant Defence* (eds)D Walter, A Newton and G Lyon, Blackwell, Oxford, UK, pages 229-242

Leroux P 2003 Modes of action of agrochemicals against plant pathogenic organisms. C.R. Biol. **326** 9-21

Leroux P, Colas V, Fritz R and Lenon C 1996 Interaction of the anilinopyrimidine fungicide pyrimethanil with amino acids and sulphur containing metabolites in *Botrytis cinerea*, in *Modern Fungicides and Antifungal Compounds* (eds) H Lyr, PE Russel and HD Sisler, Intercept, Andover, 61-67 pages

Leroux, P. (1996). Recent developments in the mode of action of fungicides, *Pesticide Science* **47** 191-197

Longhurst C, Dixon K, Mayr A, Bernhard U, Prince K, Sellars J, Prove P, Richard C, Arnold W, Dreikorn B and Carson C 1996 DE 795- a novel fungicide for the control of powdery mildew, 27-32. *In The Proceedings of the British Crop Protection Conference 1996 – Pests and Diseases*, Vol. 1, BCPC, Farnham

Millardet PMA and Gayon U 1887 Researches nouvelles sur 1' action que les preparation cuivren ses exercent sur le *Peronospora* de la vigne. Journal of Agricultural Prat. **51** 123-139, 151-161

Morton V and Staub T 2008 A Short History of Fungicides, APSnet Features, doi: 10.1094/APSnetFeature-2008-0308 (http://www.apsnet.org/publications/apsnetfeatures /Pages/ Fungicides. aspx).

Nunninger C, Watson G Leadbitter N and Ellgehausen H 1996 CGA329351: Introduction of the enantiometric form of the fungicide metalaxyl, 41-46 *In The Proceedings of the British Crop protection Conference 1996 – Pests and Diseases*, Vol. 1, BCPC, Farnham

Oerke EC, Dehne HW, Schonbeck F and Weber A 1994 *Crop Protection and Crop Production*, Elsevier, Amsterdam

Pommer EH and Lorenz G 1987 Dicarboximide fungicides, in *Modern Selective Fungicides – Proterties, Applications and Mechanisms of Action* (ed) H Lyr, VEB Gustav Fischer, Verlag Jena, 91-106 pages

Pozo C, Rodelas B, Salmeron V, Martinez-Toledo MV, Vela GR and Gonzalez-Lopez J 1994 Effects of fungicides maneb and mancozeb on soil microbial populations, *Toxicological Environmental Chemistry* **43** 123-132

Reuveni M 2003 Activity of new fungicide Benthiavalicarb against *Plasmopara vitcola* and its efficacy in controlling down mildew in grapevines, European Journal of Plant Pathology **190** 243-251

Scheinpflug H and Kuck KH 1987 Sterol biosynthesis inhibiting piperazine, pyrimidines and azole fungicides, in *Modern Selective Fungicides – Properties, Applications and Mechanisms of Action* (ed) H Lyr, VEB Gustav Fischer Verlag, Jena, 173-204 pages

Schlundt H 2002 Risks and benefits of biological and chemical plant protection strategies – food safety aspects, 3021. *In The Proceedings of British Crop Protection Conference 2002*

Singhal S 2000 The pesticides scenario, *India Chemistry* 1-2pages

Thind T S 2012 Fungicide in crop health security-the road ahead, *Indian Phytopathology* **65** 109-115

Thind TS 2004 Significant achievements and current status: Fungicide research, in *One Hundred years of Plant Pathology in India: an Overview* (eds) SS Chahal, RK Khetarpal and TS Thind, Scientific Publishers (India), Jodhpur, pages 267-305

Thind TS 2007 Changing cover of fungicide umbrella in crop protection, *Indian Phytopathology* **60(4)** 421-423

Thind TS 2008 Fungicide resistance: A perpetual challenge in disease control. Indian Journal of Mycology and Plant Pathology **38(3)** 407-417

Thind TS 2011a Development of a web-based decision support system for late blight diseases of potato. *Final Technical Report submitted to DST, New Delhi*, Pages112

Thind TS 2011b Fungicide Resistance in Crop Protection: Risk and Management, CABI, UK

Thind TS, Mohan C and Kaur S 2002a Promising activity of pencyuron, a phenylurea-based fungicide, for effective management of black scurf of potato. *Indian Phytopathology* **55** 39-44

Thind TS, Mohan C and Raj Prem 2002b Competitive fitness of metalaxyl resistant population of *Phytophthora infestans* and its cross resistance to strobilurins and other fungicides, *Indian Journal of Mycology and Plant Pathology* **32** 122-124

Thind TS, Mohan C, Arora JK and Raj Prem 2004a Potential of some recently developed fungicides and their combination products for effective management of late blight of potato, *Indian Phytopathology* **57** 95-98

Thind TS, Mohan C, Raj Prem and Arora Jk 2004b Activity spectrum of strobilurins, a new generation of eco-friendly fungicides, against some fungal pathogens, *Indian Phytopathology* **57** 104-106

Thind, T.S. (2002). *Fungicide Resistance in India*. Indian Crop Protection Association, Mumbai, 50pages

Tisdale WH and Flenner AL 1942 Derivatives of dithiocarbamic acid as pesticides, Industrial and Engineering Chemistry **34** 501-502

Toquin V, Gamer S, Bajra F, Sirven C, Jundel JL, Schmitt F and Beffa R 2007 Modification of the cellular localization of a spectrin-like protein by fluopicokid. A new mode of action for an antioomycete fungicide, *International Reinhardsbrunn Symposium, Friedrichroda, Germany.*

Urech PA, Schwinn F and Staub T 1977 CGA-48988, a novel fungicide for the control of late blight, down mildews and related soil-borne diseases, *In The Proceedings of Brighton Crop Protection Conference – Pests and Diseases*, **42** 623-631

Von Schmeling B and Kulka M 1966 Systemic fungicidal activity of 1, -4, oxathin derivatives, *Science* **152** 659-660

Wenhua, L. (2009). Discovery and development of Crop Protection Chemicals – Strategy and Opportunities. Lecture delivered at '2009 International Conference on Crop Science & Technology' (http://www.echinachem.com/EN/News/ 0910/200910 201153 54620. shtml).

14

Monitored Control Strategies for Apple Scab

V.S. Thakur

Regional Horticultural Research Station, Dr. Y.S. Parmar University of Horticulture and Forestry, Mashobra, Shimla (H.P.)

Apple scab is an important disease of apple trees. In India, the disease was first detected in 1930 on the native cultivar Ambri in Kashmir Valley (Nath 1935). It is most severe in those places which have cool, moist spring and summer. It makes the tree unproductive and weak by mid-season defoliation. Affected fruits lose their market value due to their deformed, scabby and knotted appearance. In very dry or warm climates, it is of minor importance. In India, the disease is no more confined to Kashmir Valley and is now prevalent in all apple growing areas of the country which include Himachal Pradesh, Uttrakhand, Arunachal Pradesh, Sikkim and Conoor and Kodaikanal areas of Tamil Nadu. The pathogen completes its life cycle in two stages, the imperfect (*Spilocea pomi*) stage on living plant parts and perfect or perithecial (*Venturia inaequalis)* on dead fallen leaves. Infections of the leaves and fruit are most common and obvious. Primary (ascospore) infections are usually limited to one or two distinct spots per leaf, whereas secondary (conidia) infections are often much more numerous. Secondary infections occasionally are so numerous that the entire surface of the leaf appears covered with scab, a condition commonly referred to as sheet scab.

In 1973, the disease appeared in epidemic form in the Kashmir valley whereby about 70,000 acres orchard area got infected. This epidemic resulted in a loss of Rs.54 lakh in a season (Anonymous, 1975; Joshi et al 1975). This large scale damage and spread created an alarming situation and the problem promptly engaged the attention of the Indian Council of Agricultural Research and the Directorate of Plant Protection, Quarantine and Storage, Government of India, therein declared it as one of the five main problems of national importance.

In 1977, scab was detected in Himachal Pradesh in few localized pockets (Gupta 1978). Subsequently, the disease spread was so quick that the entire apple growing area of the state was affected resulting in severe epiphytotic in 1983 (Gupta 1989). The affected area increased year after year to 56,881 ha out of the total 75,469 ha in 1996. However, the status of the disease was inconsistent during last 15 years from 1982 to 1996. After 1983 epidemic, the disease intensity during 1984-85 was less due to severe drought or dry weather conditions from February to May end due to which the inoculum did not survive to cause infection but prolonged wet conditions in 1986 -1988 scab reappeared in many apple orchards and resulted in 8 per cent scabbed fruits. The intensity was again low during 1989 and 1990 but during 1991, 1993-1999 the disease affected 5-35 per cent of plant foliage in all apple growing areas. This status of the disease was called 'Endemic', however, in subsequent years probably changes in autumn climate as dry and hot, early leaf fall due to premature leaf fall disease complex and late winter with late or less snowfall conditions affected the perpetuation of scab pathogen. This resulted to affect the life cycle particularly the saprophytic (sexual stage) chain development and primary inoculum could not survive or coincide with the primary scab infection at silver tip and green tip stage. Above all good forecasting and management strategy with curative and eradicant sprays of fungicides and other integrated disease management practices affected the scab disease.

In Himachal Pradesh the disease during past few years particularly from the year 2000 onward was observed but not to the status of economic damage to either on leaves or fruit infections. The disease is endemic in few isolated areas having high humidity and rainfall during the spring and early summer months in general and pathogen remained perpetuating under typical microclimatic conditions particularly in shady and moist conditions (Thakur et al 2005b; Thakur 2013).

Symptoms: Scab may occur on leaves, fruit, petioles, and green twigs. Infections of the leaves and fruit are most common and obvious. Early season infections usually occur on the underside of the blossom cluster leaves because these are the first tissue surfaces to emerge from buds in the spring. Once the cluster leaves have unfolded and terminal leaves begin to develop, infections become evident on the upper surface of the leaves. Individual infections appear as rough, circular, and brown to dark olive-green spots (lesions), which often appear slightly fuzzy or velvety in texture. Lesions along the veins or margins often cause affected areas to distort or crinkle. Primary (ascospore) infections are usually limited to one or two distinct spots per leaf, whereas secondary (conidia) infections are often much more numerous. Secondary infections occasionally are so numerous that the entire surface of the leaf appears covered with scab, a condition commonly referred to as 'sheet' scab.

Lesions on young fruits initially resemble those on leaves but turn dark brown to black and become corky or scab-like with time (Plate 1). Cells near lesions on young fruit may be killed, causing these regions to become deformed or cracked as they fail to grow and expand along with the remainder of the apple. Primary infections are usually limited to one or two distinct spots per fruit, often near the blossom end because it is upturned during the early stages of fruit development. Secondary infections are more frequent and numerous and the spots may be clumped or grown together, particularly if the fruit is directly beneath a concentrated source of secondary spores such as an infected leaf. Secondary infections that occur in late summer or early fall are often numerous and relatively small in size, a symptom referred to as 'pinpoint' scab. Infections that occur just before harvest may be symptom-less at picking yet develop into 'storage' scab lesions after harvest.

Plate 1. Symptoms of apple scab on fruits

Disease Cycle The fungus *Venturia inaequalis* causes apple scab, which is now endemic to all apple- growing countries of the world, is particularly severe in areas having high relative humidity and rainfall. The population variation of *V. inaequalis* isolates from Asia and Europe was studied and it was found that Indian isolate is different from the isolates prevalent in other countries (Xu et al 2008). In India, the Himachal Pradesh and Kashmir isolates are same but they are different from the North East isolate of the fungus.

In winter, the scab fungus over-winters in infected leaves fallen to the ground. During autumn, the fungus begins to form tiny fruiting bodies, which are embedded in the leaves near the surface. Sacs filled with the primary or spring spores of the fungus start to develop within the pseudothecia by late winter or early spring. The ascospores continue to develop and mature as spring progresses. A few spores are usually mature at the time of bud break (green tip), and maturity progresses slowly until about the tight cluster stage of blossom development. After this time, the percentage of mature spores begins to increase rapidly whenever temperatures are favourable for tree growth. Most ascospores have matured by the end of bloom. Mature ascospores are discharged into the air during periods of rain. In daylight, discharge usually begins within 30 minutes after the start of the rain and is largely completed within 3 to 6 hours. When rainfall begins at night, discharge is often delayed until daybreak, although significant night discharge can occur under some conditions. The number of spores discharged during any one rain is determined by both the size of the potential ascospore "crop" for the season (how many leaves were infected the previous year) and the percentage of these spores that have matured since the last discharge. Ascospore discharge usually peaks in the time from pink through bloom, and nearly all ascospores are discharged within 1 or 2 weeks after petal fall.

Ascospores are blown to nearby trees by wind currents, and they germinate in a film of water on the surface of leaves and fruit. If surface wetness continues long enough at prevailing temperatures, growth from the germinated spore penetrates and infects the organs just beneath the outer cuticle. Typical lesions, each bearing tens of thousands of secondary or summer spores, appear about 9 to 17 days later depending on temperature, although long periods of low humidity can delay their development. Conidia are dispersed by splashing rain throughout the rest of the season and are capable of causing new infections. This is due to numerous additional conidia are produced on each new lesion; repeated secondary infections have snowball or epidemic effect on disease development.

Incidence of infection is affected by the age of leaves and fruit; young leaves generally are most susceptible in 1 to 5 days after unfolding and become completely resistant from the time they finish expanding until shortly before leaf drop in the autumn. Fruit are highly susceptible until about 3 to 4 weeks after petal fall, but much longer wetting periods are required for infection to occur after this time. Precise requirements for infection of mature fruit are not known, but limited data indicate that wetting periods must last at least 48 hours for significant infection to occur immediately before harvest.

Epidemiology

This is a weather driven polycyclic disease and is more severe in areas of high rainfall and relative humidity.

(i) Perennation and buildup of primary inoculums

The fungus remains alive through winter in the dead and fallen leaves of apple on the orchard floor and produced pseudothecia mainly in leaves which fell in the September, October and November as the leaves of earlier dates did not survive in the hot autumn days and got decomposed quickly. Cool and moist climatic conditions due to rain or plenty of snow as were available at higher elevations and in shady portions of the orchards were congenial for the development of the pseudothecia in overwintering leaves within 30-35 days after leaf fall when temperature ranged in between 4 - 8°C (Gupta 1979; Gadoury and MacHardy 1982). A gradual rise in temperature favours profuse pseudothecia development. The thermal requirement of pseudothecia for different growth stages are different, 10°C for early development and 15 to 20°C for ascospore maturation. Normal development of pseudothecia was found to be hampered by the prolonged dry weather conditions in the winter and early spring with meager snowfall in late February. After completion of the dormant period, the asci and ascospores start developing within the ascus in the lumen of the pseudothecium (Thakur et al 2008). However, temperature has no significant effect on ascus production after the maturation of first ascospore (Gadouri et al 1984). A pseudothecium contained 100 to 150 asci each with eight ascospores, thereby yielding 800 to 1200 ascospores by the time the apple trees broke dormancy and continued to mature throughout the spring and into early summer. Once mature, they were forcibly ejected from the pseudothecia upon wetting of overwintering leaves on the orchard floor and could then be wind dispersed within the orchard. Maturation and discharge of ascospores usually lasted 4 to 9 weeks (Thakur 2000, Thakur et al 2004 a,b).

(ii) Climate and Infection

Under the prevailing conditions in spring (April and May) the number of hours of wetting necessary for infection (referred to as Mills infection periods) varied (Mills 1944). Wetness due to rain was essential for the discharge of ascospores and for their germination on the cuticle of plant tissue to start primary infection. After such process, the scab lesions became visible within 9 to 17 days. Occurrence of large number of infection periods, 47 and 64 within two and half months in 1982 and 1983, respectively, had led to the buildup of high disease pressure (Thakur 2000; Thakur et al 2005a). Conidia produced on these primary scab lesions served to spread the disease to uninfected tissues throughout the summer and early autumn. Many secondary cycles of the conidial stage occurred during the growing season, and depending on the cultivar at risk, each scab lesions produced thousands (50,000 to 1,00,000) of conidia. High temperature conditions (30°C or above) not accompanied by rain in the later summer months were detrimental to the viability of conidia but inspite of this, even air borne conidia could create problem in affecting disease management particularly when dew was of sufficient length or keeping leaves wet for initiation of new infection in valley areas (Thakur et al 2004a).

(iii) Infection Period

W.D. Mills determined the temperature and leaf wetting requirement for light, moderate and heavy infection by ascospores and conidia in the orchard conditions and developed the so called "Mills Table" (Mills 1944). The Mills table is being adopted for calculating infection periods throughout the world, as such. The interpretations of this Table found that the minimum leaf wetness required for ascospores infection was 9 h and for conidial infection 5.9 h and symptoms expression would take place within 9 days at optimum temperature range of 18.2-23.8°C. However, in Himachal Pradesh conditions, minimum wetting requirement for infection to occur was found to be longer. In an controlled experiment, a wetting of <13.14 h at 14.6°C in 1989 and 10.4 h at 17.6°C in 1990 could not result in the infection of the foliage within 9 days but were worked out 1-3 days later. In our conditions, an infection period of 19.0 h leaf wetting at 24.4° C resulted in the symptoms development on 10th day and an infection period of 17 h leaf wetting at 27°C expressed the symptoms on 9th day. Higher wetting period of 23.0 h at 27-30°C resulted in symptom expression at much shorter time of 7 days. This indicates that higher temperature conditions required wetting periods thereby resulting in symptoms expression within a shorter period of 7-9 days. Hence the Mills Infection period was revised (Thakur 2000; Thakur et al 2005b) to the prevailing weather conditions and infection process in Himachal Pradesh (Table 1).

(iv) Temperature and germination of spores

Conidia are capable of germinating over a range of temperatures from 2 to 30°C with an optimum of 22°C, but the maturity of ascospores is slow at lower temperature but fair between 12-20°C and ceases at 24°C. They germinate well between 10-18°C. Similarly the production and germination of conidia is also temperature dependent. Their production per lesion is maximum at temperature around 30°C and it continues at reduced rate over a wide range. Temperature also affects the infection by the pathogen, its incubation and latent periods.

Table 1. Criteria for primary and secondary scab infection based on Mills Infection Table for Himachal Pradesh

Average daily temperature	Min. wetting hours of leaves for infection to start with		Days required (after infection for symptom appearance)
	Ascospore	Conidia	
17.2-23.8	9	5.9	9
16.6-16.1	9	5.9	10
15.5	9	6.3	11
15.0-14.4	10	6.6	12
13.8-13.3	10	7.0	13
12.7-12.2	11	7.5	14
11.6-11.1	12	7.9	15
10.5	13	8.7	16
9.4-9.8	15	9.8	17
8.2	17	11.3	17
7.7	19	12.6	Infection does not take place

Mashobra Disease Forecast

Based on the information validated on the over wintering stages in the development processes of scab fungus, the saprophytic life and primary disease cycle of the scab pathogen is drawn to detailed out the sequential changes important to estimate the inoculum threshold. The prevailing weather conditions are defined to play deciding role in the life cycle at different time intervals. The scab pathogen has highly synchronized behaviour to the host and the weather conditions. Hence a computer based 'Mashobra Disease forecast' programme is developed to predict the annual epidemic and non-epidemic conditions with an input of integrated management options. The fungicides use at different susceptible stages based on primary and secondary disease cycles are recommended for the cure of existing infection and eradication of the pathogen with minimum sprays in a season. The computer programme analyses inter relationship of scab biological and weather parameters conclude that the scab pathogen did mature the ascospores in 132 days, which predict the reoccurrence of the disease. The forecasting module validated that the leaf fall that occurred during 25 October to 15 November had successful over wintering scab development within 55-68 days. This process is completed between 20 December – 20 January wherein mycelial incubation and colonization is occurred on the orchard floor. After mycelia colonization and incubation in the leaf subcuticular tissues, rain/moisture is required to initiate the fertilization around 4°C. Under these conditions, the pathogen completes the process within 55-68 days i.e. 20 December-20 January to enter the subsequent stages. Successful fertilization takes 35-40 days at 8-10°C for the formation of pseudothecial initials. Pseudothecial development passes through 11 stages, which are usually completed within 45-50 days. However, depending on an average temperature above 8°C, it is predicted that during the pseudothecial development stages the temperature must be in transition from a lower to a higher range. The asci and ascospores maturation takes 132-140 days. The pseudothecia matured more rapidly in leaves, which are alternately wet and dry than in leaves which are continuously wet. Matured asci and ascospores require light rains about 2 mm for the release of ascospores. Under the set of suitable weather and time interval conditions, this process usually starts in the first week of March. The ascospores maturation and release ceases around 25th May. The primary infection cycle begins after the infection/ incubation for 11-15 days and continues for 38-52 days and thereafter many secondary disease cycles repeat.

Concerted efforts have been made to utilize the periodic infection on maturation and discharge of ascospores in relation to tree bud development stages and the occurrence of infection periods based on temperature records and virtual observations of leaf wetness due to rain and/ or dew (Gupta et al 1984).The efforts on the influence of weather and orchard parameters on the scab outbreak and severity estimates are being monitored and processed for interpretation (Khosla 1990, Thakur and Malhotra 1992, Thakur et al 2005b). This forecasting system accounts to determine the potential ascospore dose for discharge following wetting of overwintering pseudothecia by rain and to drive a correlation with the components, air borne ascospore dose, deposited ascospore dose and infective ascospore dose which are essential for the initiation and development of lesions in the spring. This information has to be processed into data base files with weather conditions and host phenological stages to disease severity levels. The sequential development in the overwintering inoculums and weather conditions for the maturation, productivity and release of primary inoculums and its potential for disease development has been simulated and are being broadcasted through All India Radio, Shiml;a to help the orchardists in the initiation of the prudent and need-based spray programme.

Pseudothecial formation and development depend on the weather conditions beginning from the leaf fall stage to late spring dates. Occurrence of leaf fall before October 20 inhibits the pseudothecial formation in the usual dry spell during the following months up to December. Whereas, leaf fall followed by low temperature (below 10ºC) for 60 to 75 days, initiate sexual reproduction from the first rainy day or under optimum moisture condition. The process of sexual reproduction completes within next 52 to 68 days. The estimated period of pseudothecia formation and development has been observed for making predictions on the outbreaks in the following spring season (Thakur and Sharma 1996). For instance, when the general leaf fall occurs between 20th October to 20th November and the temperature ranges between 2-8°C for the following months, the sub-cuticular mycelium get matured in 60-75 days to initiate sexual reproduction up to 20th December to first week of January every year. During this period the conditions are usually wet in apple growing areas. Subsequently the pseudothecial formation takes 35-40 days more after sexual reproduction at temperature around 8°C and this process completes in between 20th January to 20th February every year. Thereafter, pseudothecia require an average temperature of more than 8°C to the formation and emission/ release of ascospores within 45-50 days, i.e. usually between 7th March to 10th April every year in Himachal Pradesh. The summary of the fungus development in relation to prevailing weather for the prediction of apple scab disease has been given (Table 2) to make an account on disease outbreak every year in Himachal Pradesh.

Table 2. Prediction criteria for apple scab disease outbreak in orchard every year in Himachal Pradesh

S.N.	Sequential development of the pathogen	Time and Criteria	Actual sequence and criteria
1.	Time of leaf fall	20 Oct.-20 Nov.	date
2.	Temperature after leaf fall	2-8°C for 60-75 days	Actual temperature and its duration
3.	Fertilization process and its time interval	52-68 days, i.e. between 17 Dec.-11Jan.	Sr. No. 1-3 duration
4.	Temperature after fertilization	8°C	Actual Temperature
5.	Pseudothecia formation after fertilization`	35-40 days, i.e. between 21 Jan.-20 Feb.	Sr. No. 3-5 duration
6.	Temperature after pseudothecia formation	8-12°C	Actual Temperature
7.	Time between pseudothecia formation and its maturation	45-50 days, i.e. between 7 March to 10 April	Sr. No. 5-7 duration
8.	Maturation of ascospores and their release	7 March – 10 April	Predict the primary infection day/date of start in the respective orchards.

(Thakur 2000)

It has been observed that low temperature plays a limiting factor during the process of pseudothecial formation because of snowfall. Just after 20th February when pseudothecial development is generally at the mid stage as temperature around 0°C delays the ascospore emission/release for the following spring season. Since last many years it has been observed that the snowfall occurs irregularly and most of the times after 20th February and temperature remains around 0°C for 10 days or so there was no ascospore formation and when the temperature was around 1-4°C it resulted into delayed pseudothecial maturation by 28 to 35 days and consequently the ascospore emission/release extended up to 15th May and when the temperature was between 4-8°C the process delayed the maturation by 16-22 days and accordingly ascospore emission/release extended to 5th May. This type of weather conditions do occur in Himachal Pradesh and other hilly states, which describes for the 'on' and 'off' scab year. Therefore, it is important to monitor the weather conditions after leaf fall stage to ascospore release every year to estimate the inoculums buildup and disease out break every year (Thakur and Sharma 1996).

Every year in spring and early summer months, similar information as gathered with the help of predictive devices (Rotorod spore sampler, Dewit Leaf Wetness Recorder and Electronic Apple Scab Predictor) and microscopic differential counts have helped in carrying out limited number of sprays without the risk of scab getting established in the orchards. But, prospectively it is yet to be established for how much area, such predictions would be valid due to variations in the infection periods from time to time in different areas due to topography of apple orchards.

Management

Based on the information validated on the overwintering stages in the development processes of scab fungus, the saprophytic life and primary disease cycle of the scab pathogen is drawn to detail out the sequential changes important to estimate the inoculum threshold. The prevailing weather conditions are defined to play deciding role in the life cycle at different time intervals. The scab pathogen has highly synchronized behaviour to the host and the weather conditions. Hence a computer based 'Mashobra Disease forecast' programme predict the annual epidemic and non-epidemic conditions with an input of integrated management options. The fungicides use at different susceptible stages based on primary and secondary disease cycles are recommended for the cure of existing infection and eradication of the pathogen with minimum sprays in a season.

(i) Primary infection

Practical approach of controlling scab pathogen, involves breaking the overwintering cycle by eliminating the ascospores, which is the source of primary infection. The study undertaken on the application of Urea @ 5 per cent at leaf abscission on scab affected leaves indicated that besides accelerating decomposition of fallen leaves, it has encouraged the microbial population, which helps in preventing of pseudothecia formation (Gupta 1985). As many as 59 microorganisms were isolated and identified to parasitize saprophytically on apple scab leaves and/or fungus (Thakur 1993).These microorganisms compete with the scab fungus for nutrients and act antagonistically. The potential of antagonists for biological control of *V. inaequalis* has been established. The effect of *Athelia* sp. and *Chaetomium globosum* supplemented with Urea treatment on pseudothecial density indicated significant reduction in the production of asci and ascospores (Thakur and Sharma 1996).

(ii) Secondary infection

Secondary infection of scab can be substantially contained by timely application of chemical sprays and according to situation different strategies like protective, curative or eradicative action are used to combat the ravages of this disease.

(a) Protective spray

A protective spray programme is being adopted in apple orchards in Himachal Pradesh which includes 6 to 7 sprays of a combination of non-systemic and systemic fungicides during the growing season alongwith one spray of Urea in autumn (Gupta and Thakur 1992). The doses of fungicides are based on experiments both under controlled glasshouse conditions and in the orchard (Thakur 1987). The first spray of protective fungicides must start at the silver tip to partially green tip stage. This spray is important to the low elevation areas because there is early onset of season for the development of plant growth stages and release of primary inoculums. However, 3rd or 4th sprays are required from bud break to petal fall to cover susceptible host tissues continuously. A single application technique (SAT) had been suggested to give continuous protection to susceptible tissue for a longer duration. Only those fungicides are applied under SAT which have good retention and redistribution activity. Earlier, captafol (Difolatan or Foltaf) at massive doses were applied at silver tip stage (Thakur and Gupta, 1990). Since captafol is banned for foliar application and only available for seed treatment in the country, other fungicides like captan, dodine, chlorothalonil, mancozeb have been tested and recommended as alternative. The optimum time of spray of such massive dosage is at the appearance of green bud tissue but not later as such application may result into moderate damage to the cluster leaves and cause fruit russetting. It has been suggested that the protection of primary scab is more important to suppress the disease, therefore, by petal fall or later, if there are lesions of scab on the leaves and fruits, a spray of fungicides like carbendazim or thiophanate methyl/ or sterol inhibiting fungicides viz., hexaconazole, penconazole, difenconazole, flusilazole, myclobutanil may be applied.

(b) Post-infection (after infection and pre-symptom)

In this strategy, the fungicides are applied only after occurrence of an infection period determined either by visual observations or predictive devices. Only systemic fungicides have curative activity against scab fungus. Among the benzimidazole compounds, carbendazim may owe its activity as a curative fungicide to the fact that it is absorbed and translocated into the fruits. In recent years, benzimidazoles have been considered as the most effective curative fungicide in the control of scab in most countries. In Himachal Pradesh, the after infection activity of carbendazim and thiophanate methyl was found to be up to 72 hours after inoculation both under glasshouse and orchard conditions (Thakur 1987). Most of the EBI fungicides have excellent after–infection curative activity against scab. These fungicides, particularly bitertanol, myclobutanil, prochloraz, etaconazole, penconazole, hexaconazole prevent the formation of visible lesions when applied within 3 days after inoculation (Thakur, 1987, Thakur and Gupta 1991, 1992) and only chlorotic specks or spots developed if they were applied later than 3 days after the onset of infection (Thakur and Gupta 1990). At times it may not be possible to carry on sprays even at long lapse of time because of continuous rainy weather. In such cases, certain fungicides can be applied late in the incubation period also but not beyond 2-3 days prior to symptom expression. In such activity the term pre-symptom to describe prevention, or inhibition, of spore production from chlorotic scab lesions that follow a spray of certain fungicides before the occurrence of such lesions is used. The practical significance of this activity is in the reduction of inoculum potential and spread of secondary infection. The developing lesions appear to be chlorotic and devoid of sporulation. An equally high level of pre-symptom activity for EBI fungicides viz., fenarimol, bitertanol, prochloraz, triforine, fenaphnil, hexaconazole and etaconazole has been reported (Thakur and Gupta 1990, 1991, 1992, 1992). The scab lesions that developed on the foliage after spraying fungicides were atypical i.e. either chlorotic or as reddish brown flecks. Irrespective of lesions, conidia were produced, either at margins or on the lesions itself. After the application of myclobutanil, however, no conidia were developed (Thakur and Gupta, 1992). Besides impairing conidia production, germination of conidia from fungicides induced lesions was also reduced in comparison to the germination of conidia from the typical scab lesions (Thakur and Gupta 1991). The post infection spray programme with well timed application of bitertanol and myclobutanil after 72 and 96 h of predicted infection periods is not only more effective but is also more efficient requiring fewer sprays in a season (Thakur 1987; Thakur and Gupta 1990). In the spray schedule recommended in Himachal Pradesh, these fungicides are being recommended at proper time to reduce the inoculums production based on testing of fungicides annually.

(c) Post-symptom (eradication)

Interruption in carrying on sprays at regular intervals in inclement weather or delay in application by few days may result in the establishment of scab fungus on foliage and fruits. If this is accompanied by continuous wet weather, inoculum quickly builds up to increase the disease severity. To avoid this, fungicides having high level of eradicant or antisporulant activity or ability to completely inhibit the conidia production on the established lesion after 5 to 7 days are used. Benzimidazoles (carbendazim) have this post-symptom activity and almost completely inhibit the production and germination of conidia in sporulating lesions. Variable results were reported on the post –symptom activity of EBI fungicides. The EBI fungicides (hexaconazole, penconazole, fenarimol, myclobutanil, beitertanol) have distinct inhibitory effect on the growth of sub-cuticular hyphae and the conidia formation of apple scab fungus thereby preventing development of scab symptoms (Thakur and Gupta 1991; Thakur et al 1994; 1995). The fungicides have eradicant activity for short duration and are required to be repeated by the 5th or 7th day to maintain a continuous level of good anti-sporulant activity (Thakur and Gupta 1990).

Monitored control strategy

In this strategy, instead of initiating protective spray programme at the bud break in the spring and continuing at 14 days interval, possibility has been established for prudent and need based monitored spray programme. The programme concept was developed at Regional Horticultural Research Station, Mashobra include:

First Spray

This concept of SAT in primary scab control was introduced in USA. Use of a protectant fungicide at a massive initial dose for retention and redistribution of the fungicide to achieve protection for extended period of time is the underlying concept of this technique. In Himachal Pradesh, captafol (Difolatant 0.3%) was able to provide excellent control of primary scab under orchard conditions. The bio-efficacy of dodine, mancozeb, dithianon, chlorothalonil and captan was tested by various workers throughout the world and established their efficacy for SAT could give protection to apple scab primary infection upto 36 days as the residues of these fungicides are persistent, stable and redistributive on plant foliage. Based on the fungicides performance and time of application, the SAT has been recommended at silver tip stages as first spray under monitored control programme in Himachal Pradesh (Table 3). The spray schedule prepared for the growers of Himachal Pradesh for the year 2013 have both types of fungicides having protective and curative activity (Table 3).

Table 3. Spray schedule against apple disease taking apple scab as the base disease

S.N.	Tree Stage	Chemical	Quantity/ 100 L water
1.	Silver tip to Green tip	Captan or	400 g
		Dodine or	100 g
		Ziram	300 ml
2.	Pink bud stage	Mancozeb or	300 g
		Combi product	
		(Carbendazim+mancozeb) or	50 + 250 g
		Difenconazole	15 ml
3.	Petal fall stage/ Fruit pea size	Carbendazim or	50 g
		Thiophanate methyl or	50 g
		Hexsaconazole or	50 ml
		Biorational molecule	300 ml
		2.62% copper sulphate or	
		Combi product	
		(Captan+ Hexaconazole) or	100 g
		Flusilazole	25 ml
4.	Fruit Development stage (Walnut size)	Carbendazim +	50 g
		Mancozeb or	250g
		Combi product	
		(Carbendazim+mancozeb) or	250 g
		Myclobutanil or	40 g
		Iprodion +carbendazim or	150 g
		Metiram or	250 g
		Metiram +Pyraclostrobin	100 g
5.	Fruit Development stage	Dodine or	75 g
		Propineb or	300 g
		Zineb or	300 g
		Combi product	
		(Zineb+Hexaconazole) or	300 g
		(Dodine + Hexaconazole) or	60 g
		Krisoxim methyl	50 ml
6.	20-25 days before fruit harvesting	Captan or	300 g
		Mancozeb flowable or	350 ml
		Ziram	300 ml
7.	After Fruit Harvest	Copper hydroxide or	300 g
		Copper oxychloride	300 g
8.	Pre- leaf fall	Urea	5 kg

Second spray

This spray at pink bud stage has been recommended to control primary scab infection as well as powdery mildew infection on the flowering buds. However, according to monitored spray strategy, this spray is considered to be optional and should be given only if: i) SAT spray was not given, and ii) weather is continuously wet after single application spray. The fungicides recommended at this stage in the schedule (Table 3) are both protective and curative and are effective against scab and powdery mildew diseases of apple. High altitude locations (above 6500 ft a.m.s.l.) where primary infection generally initiates after pink bud stage, the single application spray at silver tip to green tip stage is inappropriate. In such areas, use of carbendazim having protective, curative and broad spectrum mode of actions at pink bud stage (as first spray) to control invisible infections caused by scab, canker, powdery mildew and leaf blotch (premature leaf fall disease) is recommended. In addition, to its antifungal properties, it also help to enhance fruit set and can be used safely during blooming.

Third Spray

This spray at petal fall stage is very important to protect the newly emerged fruitlets. This stage of the fruitlets is susceptible for primary scab infection and damages the fruits. Moreover, the earlier spray given either through SAT or at pink bud could not provide protection althrough the long flowering period. The fungicides recommended at this stage in the spray schedule are having both protective and curative actions (Table 3). This spray schedule is meant for heavy disease pressure.

Fruit Development Sprays

At fruit development stages the actual monitored spray strategy implies where curative fungicides sprays are recommended after an infection period instead of routine sprays. Curative fungicides kill the fungus in host tissues after infection has occurred. Such action of most fungicides is limited to a few hours to 5-6 days after infection. Weather monitoring equipments that can record temperature and leaf wetness or electronic automatic apple scab predictor are needed to determine when conditions are favourable for scab infections. Once the scab infection period has been recorded, a fungicide with sufficient post-infection (kickback or back-action) activity (Table 3) should be applied if no other fungicide has been applied in the past 10 days. This way, it has been observed that the climatic conditions that favour the scab infection are very uncertain during the fruit development stages, therefore, it is necessary to observe the primary scab infection in the orchard before adopting monitored control programme for the control of secondary infection.

This method of control is suited to only those orchard areas where i) infection periods can be clearly defined and warning quickly circulated, ii) orchardists are aware to monitor weather conditions and infection correlation iii) curative fungicides are readily available, iv) summer weather is dry followed by high temperature condition, (v) primary infection scab of is not visible, vi) no scab incidence during the past years, and vii) the spraying machinery is effective enough to cover the whole orchard within 24-48 h from the beginning of an infection period. Such an approach is definitely a risky proposition as at times it may not be possible to carry on sprays soon after the start of infection period, particularly when weather conditions continue to be rainy. Thus, the best way is to use routine sprays as an Insurance against infection. Every year state Department of Horticulture in collaboration with Dr. Y.S. Parmar University of Horticulture prepare a spray schedule particularly for the growers of Himachal Pradesh have both types of fungicides having both protective and curative actions.

Concerted efforts have been made to utilize the outcome of the long and dedicated research work on different relevant aspects of apple scab with an aim to develop a system to which apple farmers can use by submitting the basic information of the orchards. The farmers (those who have scab infection during the last three years) were asked to develop data for temperature and rainfall of their orchards from the leaf fall date to the maturation of the over wintering inoculum on the orchard floor. The sequential development of the pathogen is standardized from the research database. The farmers are providing information of their orchards particularly temperature for comparison with the standard criteria devised to predict the primary inoculum or disease for the coming season. This infection criteria of an individual orchard if coincides with the standardized method, the farmers of the sampled area are advised for the outbreak of scab or otherwise also. Similarly, the analysis of an on- orchard conditions for the representative area would describes the 'on' and 'off' scab conditions in coming year. In practice the general forecast of the weather situation and chances of the scab appearance are broadcasted through All India Radio on 10th March every year to all the apple growers to follow the instructions and recommendations.

The computer programme also has option on the control measures that relates to the initial forecast on the inoculum threshold. However, the computer programme with the data for nine epidemic and non-epidemic years resulted that the scab does not cause infection on first three stages i.e. dormant, bud break and silver tip stages. These stages in most of the apple growing areas generally appear between 12-20 March. At these stages the matured inoculum on the orchard floor requires rains to spread out on to the plant parts. However, the rains during the period generally lower the mandatory temperature above 10ºC, which help to initiate the infection. Hence in such conditions, the computer programme would display 'no infection' and 'no spray' till such conditions met out. This programme is in use and for collecting the farmer's data, farmers are provided with low cost weather monitoring device i.e. simple minimum & maximum thermometer and self made three layered newspaper pad (which can hold rain moisture for 1-10 hours after rain) to measure the comparable moisture holding by the apple leaves required for successful infection. This microclimatic input of the orchard of the farmers is in use at specified intervals to predict the scab infection conditions of their orchards. Accordingly, the necessary instructions and recommendations are suggested for the period. The technique has guided the farmers and achieved up to 40 per cent reduction in the fungicides use over the years.

Future thrust

Since scab is not a disease which can be eradicated as with the advent of wet weather, it may reappear in the orchard to epiphytotic nature even after being absent for last many years. No doubt, technologies have been developed to minimise the onslaught of this disease, yet it is not adequate enough to have an assured and healthy crop particularly when condition as favourable for longer period. In view of the abvoe economic ramifications increased on the control of various diseases and pets, an interdisciplnary approach may be followed to develop integrated control programme which should be need based economic and environmentally safe.

References

Anonymous 1975 Hot Spots of Diseases and Pests of Major Field and Horticultural Crops, ICAR, New Delhi

Gadoury D M and MacHardy W E 1982 Effects of temperature on the development of pseudothecia of *Venturia inaequalia, Plant Disease* **66** 468

Gadoury DM, MacHardy W E and Hu C 1984 Effect of temperature during ascus formation and frequency of ascospore discharge on pseudothecial development of *Venturia inaequalia, Plant Disease* **68** 223-225

Gupta GK 1978 Present status of apple scab (*Venturia inaequalia*) in Himachal Pradesh and strategy for its control, *Pesticides* **12** 13-14

Gupta GK 1979 Some observations on apple scab in Himachal Pradesh, *Indian Phytopathology* **32** 172

Gupta GK 1985 *Apple scab* (*Venturia inaequalia*), E Merck (India) Ltd., Bombay

Gupta G K 1989 Apple scab-economics and technology, *Pesticides* **23** 33-39

Gupta GK, Verma KD, and Pal J 1984 Some observations on the prevalence and severity of apple scab in Himachal Pradesh during the year 1978 to 1983, *Indian Journal of Mycology and Plant Pathology* **14** XII-XIV (Abstr.)

Gupta VK and Thakur VS 1992 *Integrated management of apple scab disease*, Information Bulletin, ICAR Publication

Joshi NC, Mallik AG, Kaul ML, and Anand SK 1975 Some observations on the epidemic of scab disease of apple in Jammu and Kashmir during 1973, *Indian Phytopathology* **28** 288-289

Khosla K 1990 To study the influence of bio-meteorological parameters in apple scab development and its management, Ph.D. Thesis, Dr. Y.S. Parmar University of Horticulture and Forestry, Solan

Mills WD 1944 Efficient use of sulphur dusts and sprays during rain to control apple scab, *Cornell University Extension Bulletin* **630** 4

Nath P 1935 Studies in the diseases of apple in Northern India, II, A short note on apple scab due to *Fusicladium dendriticum* Fuckel, *Journal of Indian Botanical Society* **14** 121-124

Thakur VS 1987 Studies on the effect of fungicides on apple scab (*Venturia inaequalia*) (Cke.) Wint. pathogenesis Ph.D. Thesis, Dr. Y.S. Parmar University of Horticulture and Forestry, Solan

Thakur VS 1993 Changes in the microbial population of apple leaf litter associated with the inhibition of the perfect stage of *Venturia inaequalis* in relation to Urea treatment, DST 10th Group Monitoring workshop on DST funded Young Scientists projects, Department of Science and Technology, New Delhi

Thakur VS 2000 Apple scab forecasting and management, in *Horticulture Technology Vision 2000 and Beyond Vol. 2 Protection and Preservation* (eds) VK Sharma and KC Azad, Deep & Deep Publications Pvt Ltd, New Delhi 37-63 pages

Thakur VS 2013 Apple diseases and their management, in *Microbial Resources for Crop Improvement* (eds) B Chakraborty and U Chakraborty, Satish Serial Publishing House, New Delhi, 29-56 pages

Thakur VS and Gupta GK 1990 Note on development of spray schedule for controlling apple scab in Himachal Pradesh, *Indian Journal of Horticulture* **47** 60-63

Thakur VS and Gupta GK 1990 Evaluation of pre-symptom activity of fungicides on symptom expression, conidia production and viability of *Venturia inaequalis, Indian Phytopathology* **43** 520-526

Thakur VS and Gupta GK 1991 Studies on the effect of sterol-inhibitors on apple scab, *Venturia inaequalis* pathogenesis, *Indian Journal of Plant Protection* **19** 185-190

Thakur VS and Gupta GK 1992 Curative and protective action of ergosterol biosynthesis inhibiting fungicides in relation to infection periods in the control of apple scab, *Pestology* **16** 32-37

Thakur VS and Gupta GK 1992 Post-infection fungicidal inhibition of apple scab (*Venturia inaequalis*) sporulation, *Indian Journal of Agricultural Science* **62** 629-636

Thakur VS and Gupta GK 1992 Evaluation of preventive and curative activity of organic fungicides against apple scab (*Venturia inaequalis*), *Pestology* **16** 38-44

Thakur VS, Gupta GK, and Garg ID 1992 Effect of post –infection application of fungicides on apple scab pathogen, *Venturia inaequalis* (Cke.) Wint, *Journal of Phytopathology* **135** 160-166

Thakur VS, Gupta GK, and Malhotra R 1994 Management of symptoms developed after infection of scab (*Venturia inaequalis*) in apple (*Malus domestica*), *Indian Journal of Agricultural Science* **64** 167-172

Thakur VS, Gupta GK, and Malhotra R 1995 Anti-sporulant activity of ergosterol biosynthesis inhibiting fungicides in the management of apple scab, *Venturia inaequalis, Indian Phytopathology* **48** 35-39

Thakur VS and Malhotra R 1992 Impact of meteorological factors on the outbreak of foliar fungal diseases of stone fruits in Himachal Pradesh, Golden Jubilee National Seminar on *Emerging trends in Temperate fruit production in India*, Horticulture Society of India, 4-5 September, 1992.

Thakur VS and Sharma RD 1996 Effect of meteorological parameters on saprophytic stages of apple scab pathogen, *Venturia inaequalis*, Proceedings of National Symposium on Molecular approaches in Plant Disease Management and Annual Conference of Indian Phytopathological Society held on 14-15 November, 1996 at CPRI, Shimla

Thakur VS and Sharma RD 1996 Microbial antagonism to the perfect stage of the apple scab pathogen *Venturia inaequalis*, Proceedings of National Symposium on Molecular approaches in Plant Disease Management and Annual Conference of Indian Phytopathological Society held on 14-15 November, 1996 at CPRI, Shimla

Thakur VS and Sharma RD 1996 Apple scab- forewarning and management, Information Bulletin, Published by the Directorate of Extension Education, Dr YS Parmar University of Horticulture and Forestry, Solan

Thakur VS, Verma S, and Sharma N 2004a Role of meteorological factors in the development of *Venturia inaequalis* pseudothecial protoplasmic content in Himachal Pradesh, *Advances in Plant Science* **21** 269-274

Thakur VS, Verma S, and Sharma N 2004b Epidemic analysis of saprophytic stages of *Venturia inaequalis* to forecast primary inoculum and infection conditions in Himachal Pradesh, *Plant Disease Research* **20 (2)** 131-137

Thakur VS, Sharma N and Verma S 2005a Effect of meteorological factors on the development of lesion density in apple scab epidemic in Himachal Pradesh, *The Horticultural Journal* **18(1)** 4-9

Thakur VS, Sharma N, and Verma S 2005b Present scenario of apple scab in various apple orchards of Himachal Pradesh, *The Horticultural Journal* **18(3)** 134-136

Thakur VS, Verma S and Sharma N 2008 Role of pseudothecial contents in pseudothecial of *Venturia inaequalis* to monitor apple scab epidemic in Himachal Pradesh, *Advances in Plant Science* **21(1)** 265-270

Xu X, Yang J, Thakur VS, Roberts A, and Barbara DJ 2008 Population variation of Apple scab (*Venturia inaequalis*) isolates from Asia and Europe, *Plant Disease* **92** 247-252.

15

Rational Use of fungicides in the management of Vegetable Diseases

T.S. Thind[1] and S.K. Gupta[2]

[1]Department of Plant Pathology, Punjab Agricultural University, Ludhiana-141 004
[2]Department of Plant Pathology, Dr. Y.S. Parmar University of Horticulture & Forestry, Nauni, Solan-173 230 (H.P)

Vegetables provide much needed nutritional security through balanced diet. India is the second largest producer of vegetables after China. Vegetable production in our country is estimated around 50.09 million tones from an area of 4.5 million hectares at an average yield of 11.3 tonnes/ha. India shares about 12 per cent of the world output of vegetables from about 2 per cent of the cropped area in the country. However, the per capita consumption of vegetables in India is only about 140 gm which is far below the minimum dietary requirement of 240 g/day/person. It calls for increase in production of vegetables to meet the dietary demands of ever increasing population. This production is bound to increase with the availability of hybrids in most of the vegetables which are high yielder besides having some resistance to various insect pest and diseases.

India is endowed with varied climatic conditions suitable to the production of various vegetables in one or the other part of the country round the year. During cultivation, these crops are affected by various diseases which necessitate the role of plant protection. Since, hybrids are available in most of the commercial vegetables which are becoming popular among farmers because they are more responsive to fertilizer, quality is good besides being the produce of equal shape and size and more suitable for intensive cropping system. But this intensive cropping system, especially with high level of densities and higher doses of nitrogenous fertilizer has resulted in enhanced incidence of insect pest and diseases. A number of new diseases have appeared in different areas and many diseases which were of minor importance have become a major threat under this system. More over the pathogen (s) associated with these diseases perennate from one season to other in different ways in diverse climatic conditions, their management in a scientific way is a must to get higher yields.

The management strategies are generally categorized as:

(i) Exclusion: prevention of introduction of disease into new locality.

(ii) Eradication: prevention of dissemination of infection and disease.

(iii) Prevention: prevention of disease causing organisms from causing infection.

(iv) Immunology: plant disease control through host plant resistance.

Out of these various disease management strategies, however, protection or eradication through application of chemicals has been dominated resulting in many fold increase in usage of fungicides. The use of chemicals in plant protection for controlling plant pathogens has gained popularity and importance despite the fact that hazards of their use are far more because of site specific systemic fungicides. Some of these hazards include:

1. Side effects against non-target organisms thus exerting negative effect on natural ecosystems.
2. Development of fungicide resistance arising out of repeated and irrational use of systemic fungicides.
3. Environment pollution and problems related to fungicide residue to plant parts.

Without proper knowledge, farmers are indiscriminately using agrochemicals to combat the ravages of insect-pest and diseases, which has led to an enormous increase in the environmental pollution. The proper and effective management of a disease depends on the proper identification and how its causal agent perpetuates from one season to another. In India, major share of fungicide utilization is in vegetable and fruits which are directly consumable commodities. Strangely, about 70 per cent of the amount of chemicals sprayed on field crops does not stick to the plants. The enormous quantities of chemicals, poisonous to microbial life that fall on earth get mixed well with soil. Some problems like non-target effects of chemicals as well as iatrogenic diseases are being experienced. Therefore, the shift to integrate non-chemical strategies is much needed approach to correct the imbalance in our approach and subsequently improve our disease control strategies to restrain the disease epidemics for better plant health necessary for higher production. Keeping in view the above problems, there is an urgent need to rationalize their use in order to get adequate disease control and residue free produce.

Proper understanding of epidemiology is necessary for the prediction of plant diseases which, in turn, can help in advocating need - based application of fungicides for their management. Integration of epidemiology of the disease along with adoption of suitable and relevant cultural practices can be helpful in rationalizing the use of fungicides. Adoption of cultural practices with precision helps in reducing the incidence/ severity of plant diseases in plants. These practices are mainly targeted to take care of the plants suitably formulated with all considerations to the prevailing conditions of the area. They are important non-chemical methods to manage the plant diseases. They are effective individually and an important component of integrated disease management practices when applied in association with other control measures collectively. By adopting cultural practices, the environment of the pathogen can be modified. Disease control through cultural practices is mainly preventive and these practices reduce the inoculum density of soil borne pathogens and making the microclimate unfavourable for the spread of the pathogen within crop canopy thus favouring the better plant growth. However, in polycyclic diseases if limited initial inoculum is present that can easily multiply and cause epidemics if cultural practices are not supported with the application of fungicides. Before advocating rational use of fungicides in the management of a particular disease, one should have the knowledge of environmental conditions favourable for the disease, incidence and severity, in addition to survival and spread of plant pathogens. Study of these factors can indicate about the week links in the disease cycle where it can be managed by minimal use of fungicides.

1. Disease Surveillance

Knowledge of plant disease incidence/ severity and spatial pattern is becoming increasingly important to assess their economic importance and for taking adequate control measures (James 1974). Regular monitoring of plant diseases gives vital information about the likely build-up of an epidemic. Based on this the farmers can be given timely advise for initiating spray programme in order to economize on unnecessary and often wasteful blanket applications of fungicides. The disease progress curve generated over the years can help in building information on the pattern of disease development under particular weather conditions in the area. Buckeye rot (*Phytophthora nicotianae* var. *parasitica*) of tomato is an endemic disease in Solan area of Himachal Pradesh and overwinters in the form of oospore and chalamydospores (Shyam and Gupta 1998). In the presence of high soil moisture and temperature (20-25°C), these start germinating by producing mycelium and sporangia. The sporangia in turn produce biflagellate zoospores which after swimming for some time in film of water, loose flagella and become encysted. These encysted zoospores are splashed by the spattering rains to the immature green fruits and cause infection (Gupta and Upmanyu 2004). In the monsoon season of 2009, very limited rains were there. Since infection takes place only by spattering rains, so the infection is limited. Number of sprays can be reduced either by increasing the interval of sprays or immediately after rains by application of systemic fungicides like metalaxyl, cymoxanil or dimethomorph. This type of studies can form an important basis for the development of the disease forecasting models (McCartney 1998). Such studies are more relevant to polycyclic pathogens, which are influenced by the size and distribution of initial pathogen populations and the environmental factors.

2. Prediction of Disease

(i) Prediction based on initial inoculum

Both monocyclic and polycyclic diseases can be predicted from the knowledge of initial inoculum which is potentially large for the later type. Spore trapping has been used to forecast early blight of potato caused by *Alternaria solani*. Early in the season, more spores are produced from the overwintering mycelium in plant debris and later spores from mycelium in infected tissue contribute to the aerial population.

As the plants mature they become more susceptible to the disease and fungicidal sprays are necessary to suppress disease if the pathogen population is large.

(ii) Prediction based on weather conditions favourable for secondary inoculum

In diseases like late blight of potato and tomato, the initial inoculum is usually low and generally too small to detect and measure directly. Even with low initial inoculum, the initiation and development of late blight epidemic can be forecasted with reasonable accuracy if the moisture and temperature conditions in the field remain within certain ranges favourable to the pathogen. Weather conditions play key role in build-up of late blight epidemic. Normally a severe late blight attack can be expected when the leaf wetness period exceeds 8-10 days for several days and temperature ranges between 10-24°C. Typical blight weather is characterized by cool cloudy days with frequent rains. In the recent years several computerized prediction system for potato late blight epidemic have been developed. BLIGHT CAST, a computerized system was developed by Krause et al (1975) in USA. It is an integral linking and modification of late blight forecasting methods based on two earlier late blight models and takes in to consideration rainfall, relative humidity and temperature. It is widely used in north-eastern USA for timing fungicide application. Based on temperature and humidity, severity values are assigned.

Late blight is expected to appear within 1 to 2 weeks after 18 severity values have occurred and hence fungicide spray is recommended. The scheduling of fungicide application should be timed so that the first protective spray is applied before infection occurs. Similar blight prediction rules have also been developed in other countries like Netherlands and England.

In India also much work has been done on epidemiology of late blight of potato. Development of late blight in the field is generally associated with the prevailing environmental conditions. It appears that the temperature at the time of planting is an important factor in the epidemiology of late blight disease. Generally a temperature range of 10-20° C with high relative humidity and cloudy weather are considered favourable for the rapid development of the disease (Singh and Bhatt 2003).The fungus grows best at 20°C while sporangia formation and germination occur best at 15°C. Wind and rain splashes help in the dispersal of sporangia and heavy rain and dew wash them down into the soil where tubers may be infected. .

In hilly areas, late blight forecasting parameters based on 7 days moving rainfall have been ascertained (Bhattacharya et al 1983) and late blight warning system for Shimla hills is in operation since 1974. The blight favourable periods guide in giving initial application of protective fungicides such as mancozeb or copper based fungicides. Once the first fungicide spray has been applied the decision has to be taken whether to follow a routine spray schedule with fixed intervals or to continue according to prevailing weather conditions by following disease forecasting and decision support systems which can put a fungicide to use when and where needed and avoided when not required. Under high disease pressure conditions, spray programme with protectant fungicides would normally be carried out on 7 day schedule while in the seasons which experience moderate disease pressure spray interval can be extended from 10-21 days. Fungicide mixtures containing phenylamides such as metalaxyl are recommended for application at 7-14 days interval, the shorter being followed under high disease pressure (Thind et al 1989, 1999). Since isolates of *P. infestans* resistant to metalaxyl have appeared both in India and other countries and have reduced the life of this fungicide (Arora 2009). Now the use of metalaxyl based fungicides should be avoided in areas where the resistant pathogen population has been encountered. Cymoxanil in mixture with the contact fungicides can be used to manage the metalaxyl resistant isolates of the pathogen. In hilly areas both tomato and potato crops are grown simultaneously during April to September-October and the crop can be affected any time after July due to availability of sufficient moisture and moderate temperatures. In some areas potato gets infection from earlier infected tomato (Singh 1987). Forecasting for potato in hilly areas can be utilized in tomato also for reducing the number of sprays.

Downy mildew of cucurbits (*Pseudoperonospora cubensis*) is another polycyclic disease which is greatly influenced by prevailing weather conditions. Maximum temperature and leaf wetness duration were identified as key weather variables for development of muskmelon downy mildew. Singh et al (1996) have used humid thermal ratio (HTR) as an important parameter for development of the downy mildew in Punjab. At diurnal temperature combination of 30°C/ 20°C the disease development was maximum. The favourable HTR values for a particular area can be worked out well in advance on the basis of forecast values of temperature and relative humidity through synoptic weather chart analysis. The amount of rainfall and number of rainy days in April- May seems to have a definite impact on downy mildew development due to increased humidity and moderate temperature. However, in Himachal Pradesh, moderate temperature (23°C) coupled with high rainfall and relative humidity (>80%) in the month of July significantly favoured downy mildew development in cucumber (Sharma et al 2003). Similarly, relative humidity (morning) had a positive and significant effect on Phomopsis leaf blight and fruit rot (incidence and intensity) whereas, relative humidity (evening) and temperature (maximum and minimum) had negative but significant effect on disease development (Sharma and Razdan, 2013). Multiple correlation coefficients between leaf blight and fruit rot and weather parameters clearly indicated that >98 per cent variation in disease was due to these weather parameters.

Rainfall/ humidity relationship with disease development may help to schedule timing of fungicide application for effective management of this disease, thereby, reducing the unwarranted use of fungicides. The information on favourable temperature and relative humidity prevalent in the area can guide us in giving first protective spray of fungicides like dithiocarbamates and subsequent application of phenylamide mixtures under high risk situations (Thind et al 1991, Thind and Mohan 2001, Sharma et al 2003).

Powdery mildew diseases, in general, are much influenced by temperature, relative humidity and light which are interrelated. These diseases are favoured by relatively dry soil and atmospheric conditions, moderate temperature and reduced light on plants growing succulently in dry soil. Study on powdery mildew of pea in Himachal Pradesh, India, indicated that mean temperature (19-23°C) appeared to be the only significant factor which is important in the prediction of disease severity in the field (Kumar and Gupta 2006). So when ever favourable temperature is available, the application of effective fungicides like kresoxim methyl, difenoconazole or hexaconazole can be timed (Gupta and Kumar 2008) since these fungicides have both protective and erradicant activity and the spray programme can be planned according to the disease situation. Similarly, powdery mildew of cucurbits occurs when the temperature is moderate (20-30°C). Keeping this in view, the application of effective mildewcide can be timed (Hussain et al 1997).

(iii) Prediction based on initial and secondary inoculum

In pea rust (*Uromyces viciae fabae*), the repeating stage in its infection cycle is predominantly the aeciospores and to a lesser extent the urediniospores in contrast to most other fungi (Chauhan and Singh 1994) and this disease generally start appearing in the field when 20 $\pm$ 2°C temperature and leaves remain wet for longer periods due to dew deposition at pod formation stage. With the availability of aeciospores in air and prevalence of favourable weather the application of fungicides can be timed. Fungicides like mancozeb or chlorothalonil provide good protective activity against this disease (Gupta and Shyam 1998) but EBI fungicides like cyperconazole, flusilazole, penconazole and hexaconazole have longer post infection activity (>72 h) which can be utilized in planning different spray programme in the management of this disease (Gupta and Shyam 2000).

In case of angular leaf spot (*Phaeoisariopsis griseola*) of French bean, diurnal periodicity of conidial release revealed a distinct rhythm at different times of the day with maximum conidial density being observed between midnight and early hours and the conidial release is found to be favoured by temperature (23-24°C), high relative humidity (>90%), prevalence of dew (20.00 h to 08.00 h) and a dark period (Gupta *et al* 2005). High relative humidities (>90%) and temperature (25°C) favoured the disease development under field conditions. These conditions can be utilized in planning the spray programmes by including various fungicides with different activities. Carbendazim and mancozeb exhibited excellent protective activity but poor post infection activity (18 h) whereas EBI's (bitertanol, hexaconazole and flusilazole) demonstrated excellent post infection activity i.e. upto 72 h (Mathew et al 1998). Similarly, Carbendazim and three EBI's showed excellent pre-symptom activity by producing reddish brown spots as compared to typical sporulating spots in mancozeb when these fungicides were applied two days before the expected appearance of symptoms. When these fungicides evaluated under post-infection (curative) spray programme by using weather monitoring devices was found effective and required two less sprays than protective spray programme (Mathew et al 1998).

(iv) Prediction based on meteorological factors

The effect of various meteorological factors like temperature, relative humidity, total rainfall, number of rainy days, leaf wetness, duration of sunshine etc. has been determined in many plant diseases and linear regression equations have been developed to predict the severity of various diseases such as maydis leaf blight and Erwinia stalk rot in maize and Ascochyta blight of chickpea. Based on severity of Ascochyta blight, the crop seasons were categorized as high disease severity years, moderate disease severity years and low disease severity year (Chaube 1987). During the crop season, number of rainy days and total rainfall has significant effect on disease severity, while ambient temperature, relative humidity and sunshine duration had less pronounced effect. Based on partial regression co-efficient a linear equation was derived to predict the severity of disease depending on weather conditions prevailing during crop season. Accordingly, fungicide spray programme can be initiated depending upon the prediction of disease severity. A high soil moisture (80%) and 25°C temperature were the most favourable for root rot (*Rhizoctonia solani*) development in French bean while web blight was best favoured at >85 per cent relative humidity coupled with 25°C temperature (Upmanyu and Gupta 2005). Similarly in tomato, bacterial leaf spot (*Xanthomonas campestris* pv. *vesicatoria*) in tomato is favoured by mean temperature of 28°C, >90 per cent RH and high soil moisture with intermittent rainfall (Shukla and Gupta 2005). The regression equation between disease severity and environmental factors explained 90.84 per cent change in disease severity was due to above mentioned environmental factors. Higher relative humidity (>90%) and temperature (25°C) favoured angular leaf spot development in French bean (Gupta, *et al* 2005). The regression equation between disease severity and environmental factors explained 69.33 per cent change in disease severity was due to above mentioned environmental factors.

(3) Role of automatic weather stations

To relate weather factors to plant diseases weather variables may be averaged over different periods of time. Daily values of air temperature, humidity, rainfall and wind speed are commonly available from the standard meteorological records. More frequently (hourly) records can be obtained from automatic weather stations by interpolating between daily maximum and minimum value. In some circumstances, weather factors averaged over several days or weeks may be appropriate for disease prediction. Sensors for temperature, relative humidity, surface wetness, rainfall and wind speed etc. attached to data logger on an automatic weather station can provide more accurate records which can help in predicting favourable periods for infection and disease development. These are more useful for polycyclic diseases like downy mildews, late blight of tomato and potato, rusts etc. Depending on the weather records farmers can be advised well in advance for initiating fungicide sprays against a particular disease. Careful interpretation of disease weather relationship has a potential to improve the efficacy of disease control fungicides (Butler 1998). Appropriate software is used to convert weather data to management advice to growers. Record of wind velocity is also being used in managing spray drift of fungicides.

(4) Decision Support Systems

Several systems have been devised to help farmers make effective decision to use, or not to use, a fungicide. Many computer based systems are available these days for decision support. A good system comprises crop/ pathogen/ growth simulation models to present a series of management options to the use while others also incorporate diagnostic modules based on photographed examples on fungal diseases. The major uses for a computerized system in disease management are in top fruits and cereals (Hewitt 1998). Some examples of computerized expert systems used in disease management are given in Table 3.

Table 3. Examples of computerized expert systems in disease management

	Target crops						
System	**Apple**	**Maize**	**Melon**	**Peanut**	**Soybean**	**Various**	**Developed in**
Arico						X	Canada
FSR Expert					X		Asia
Maize		X					USA
Maryblyt	X						USA
MSDS			X				USA
Plam				X			USA
Plant/ds					X		USA
PSAOC	X						USA

Source: Hewitt (1998)

In Europe, several systems for use in temperate cereals have been launched. One example is that of EPIPRE (EPIdemiology, PREdiction and PREvention). EPIPRE was designed to advise on fungicide use in winter wheat for the control of rusts and other foliar pathogens (Zadoks 1981). In Holland, EPIPER was able to reduce fungicide inputs and improve disease control overall. About 50 per cent of the quantity of fungicides could be saved without any loss of grains (Forrer 1992). Similar reduction in fungicide use was obtained in Switzerland.

PRO_PLANT (Frahm and Volk 1993) operating in Germany, integrates application frequency, application timing and fungicide rate for all the major cereal pathogens (Obst 1995). Studies at two sites demonstrated the potential of the PRO_PLANT system to deliver a cost benefit -42 per cent above a standard fungicide programme. In USA, MORECROP has been developed for use in wheat disease management (Cu and Line 1994). Many decision support systems are somehow difficult to operate because of the problems associated with disease monitoring as the procedures involved to represent a significant additional work load for the grower.

Infection periods for apple scab can be determined quickly with the Reuter Stokes Apple Scab Predictor. Early prediction of of infection period can help in saving 2-3 fungicide sprays with curative spray programme in comparison to standard protective spray programme (Thakur 1987). Some EBI's are highly effective in warding off the late infections in the infection periods. Similar studies in vegetables are also warranted. A web-based Decision Support System (DSS) based on weather data input from automatic weather stations installed in potato growing areas in Punjab State has been developed by Thind (2010) for taking a decision for need-based application of fungicides.

Future thrust

Epidemiology of a disease can, therefore, play an important role in advocating need based application of fungicides, thereby, avoiding their excessive use. Feedbacks from the field about initiation of the disease can be helpful in timely application of an effective fungicide. The first fungicide application can be timed as soon as the optimum climatic conditions become favourable for the particular disease eliminating the need for wasteful blanket applications. Studies are warranted to use of decision support systems with latest computer-based soft wares which will help the farmers in minimizing the use of fungicides so as to make it more rational and need-based.

References

Bhattacharya SK, Raj S, Singh DS, Khanna RN and Sharma SR 1983 Forecasting late blight of potato in Indian hills in *Potato in developing countries* (eds)B.B. Nagaich et al, Indian Potato Association. CPRI, Shimla 20-23 pages

Butler DR 1998. Daily weather and plant diseases in *Proceedings of 7th International Congress of Plant Pathology,* Edinburg, Scotland, Vol. 1, p. 2.5.1S

Chaube HS 1987 Ascochyta blight of chickpea. Technical Report, Expt. Station, G.B. Pant Univ. of Agric. & Tech., Pantnagar, 110p

Chauhan RS and Singh BM 1994 Effect of different durations of leaf wetness on pea rust development . *Plant Disease Research* **9(2)** 200-201

Cu RM and Line RF 1994 An expert advisory system for wheat disease management. *Plant Disease* **78** 209-215

Farham J and Volk T 1993 PRO-PLANT- a computer-based decision support system for cereal disease control. *EPPO Bulletin* **23** 685-694

Forrer HR 1992 Experience with the cereal disease forecast system EPIPRE in Switzerland and prospects for the use of diagnostics to monitor the disease state in *Proc. Brighton Crop Prot. Conf. Pests and Dis.* 2, BCPC, Fornham, 711-720 pages

Gupta SK and Kumar A 2008 Management of *Erysiphe pisi* through strobilurin and EBI fungicides. *Indian Phytopathology* **61(2)** 184-191

Gupta SK and Shyam KR 1998 Control of powdery mildew rust of pea by fungicide. *Indian Phytopathology* **51** 184-186

Gupta SK and Shyam KR 2000 Post-infection activity of Ergosterol biosynthesis inhibiting fungicides against pea rust. *Journal of Mycology and Plant Pathology* **30** 414-415

Gupta SK, Mathew KA and Shyam KR 2005 Influence of inoculums density and environmental factors on angular leaf spot of French bean. *Plant Disease Research* **20(1)** 25-28

Gupta SK and Upmanyu S 2004 Buck eye fruit rot of tomato and its management. *Annual Review of Plant Pathology* (ed.) R K Khetarpal, A Lal and HN Gour, Scientific Publishers (India), Jodhpur, 293-298 pages

Hewitt HG 1998 *Fungicides in crop protection.* CAB International, Oxm, UK. 221p

Hussain SI, Akram M and Khan W 1997 Epidemiology and management of powdery mildew in *Management of threatening plant diseases of national importance* (eds) V.P. Agnihotri et al, MPH, New Delhi, 263-278 pages

James W 1974 Assessment of plant diseases and losses. *Annual Review of Phytopathology* **12** 27-48

Krause RA, Massie LB and Hyre RA 1975 BLIGHT CAST: A computerized forecast of potato late blight. *Plant Disease Reporter* **59** 96-98

Kumar A and Gupta SK 2006 Influence of abiotic environmental factors on pea powdery mildew. *Journal of Mycology and Plant Pathology* **36(2)** 182-184

Mathew K.A., Gupta, S.K. and Shyam, K.R. 1998. New strategies in fungicidal management of angular leaf spot (*Phaeoisariopsis griseola*) of French bean. *Journal of Mycology and Plant Pathology* **28** 123-133

McCartney, H.A. 1998. The effect of environment on plant pathogens dispersal in *Proceedings of the 7th International Congress Plant Pathology*. Edinburgh, Scotland, Vol. 1, P. 2.5.2S.

Obst, A 1995 Practical experiences with new diagnostic and predictive systems for wheat diseases in *A vital role for fungicides in cereal production* (eds) H.G. Hewitt et al, BIOS Scientific, Oxford, 221-230 pages

Sharma DR, Gupta SK and Shyam KR 2003 Studies on downy mildew of cucumber caused by *Pseudoperonospora cubensis* and its management. *Journal of Mycology and Plant Pathology* **33 (2)** 246-251

Sharma M and Razdan VK 2013 Impact of weather parameters in the development Phomopsis blight and fruit rot of brinjal (*Solanum melongena*). *Indian Journal of Agricultural Sciences* **83(6)** 633-638

Shukla Arti and Gupta SK 2005 Role of epidemiological factors on the development of bacterial spot (*Xanthomonas vesicatoria*) of tomato. *Indian Phytopathology* **58(3)** 319-322

Shyam K R and Gupta SK 1998 Buck eye rot of tomato, Research information Bulletin, No. 1, Directorate of Research and Department of Mycology and Plant Pathology, Dr. Y.S. Parmar University of Horticulture and Forestry, Nauni, Solan, 20 pages.

Singh BP and Bhat N 2003 Emerging trends in the epidemiology of late blight f potato. *Annual Review of Plant Pathology* **2** 43-84

Singh PP, Jhorar OP, Singh R and Sokhi SS 1996 Prodcution model for downy mildew of muskmelon. *Vegetable Science* **23** 180-194

Singh RS 1987 *Diseases of Vegetable Crops*, Oxfors & IBH Publishing Co Pvt Ltd, New Delhi

Thakur VS 1987 Studies on the effect of fungicides on apple scab (*Venturia inaequalis*) (Cke) Wint) pathogenesis, Ph.D. thesis, Dr. Y.S. Parmar University of Horticulture and Forestry, Solan

Thind TS 2010 Development of web-based decision support system for late blight disease of potato. Final Report of DST Project, 138 pages

Thind TS and Mohan C 2001 Disease-weather relationship and relative activity of some new fungicides in different application schedules against muskmelon downy mildew. *Indian Journal of Mycology and Plant Pathology* **31** 174-179

Thind TS, Mohan C and Sekhon PS 1999 Advances in chemical control of potato late blight in *Vistas in Mycology and Plant Pathology* (eds) L.V. Gangawane et al. Common Wealth Publishers, New Delhi, 128-139 pages

Thind TS, Mohan C, Bedi JS, Sokhi SS and Grewal RK 1989 Role of application time of fungicides for the control of late blight of potato. *Plant Disease Research* **4** 113-117

Thind TS, Singh PP, Sokhi SS and Grewal RK 1991 Application timing and choice of fungicides for the control of downy mildew of muskmelon. *Plant Disease Research* **6** 49-53

Upmanyu S and Gupta SK 2005 Influence of environmental factors on the progress of root rot and web blight (*Rhizoctonia solani*) of French bean. *Indian Phytopathology* **58(1)** 79-83

Zadoks JC 1981 A disease and pest management system for winter wheat developed in Netherlands *EPPO Bulletin* **11** 365-365.

16

Integrated Disease Management: Retrospect and Prospect

Mamta Sharma[1], SK Gupta[2] and Suresh Pande[1]

[1]*International Crops Research Institute for the Semi-Arid Tropics, Patancheru, Andhra Pradesh-502 324*

[2]*Department of Plant Pathology, Dr YS Parmar University of Horticulture & Forestry, Nauni, Solan-173230*

Integrated disease management (IDM) came under focus in 1960's when chemicals especially fungicides and insecticides came under the attack from environmentalists due to the overuse of chemicals that created the problems of environmental pollution, chemical residues in food stuffs, land, water and air and the associated health hazard. Though the consumption of chemicals in India is very less as compared to other developing or developed countries but these are widely used in crops like cotton, apple, vegetables, rice, wheat etc. However, fungicides share the 20 per cent consumption as compared to other herbicides (3%). In the view of the challenges from diseases and threats to the environment through the use/overuse of chemicals, the IDM strategy takes the front seat. In general, the crop protection has two basic objectives: i) to have maximum production and optimum income per unit area and ii) supply of food and livelihood for the farmers from harvest to harvest with least health hazards and minimum damage to the environment. This can be achieved by reducing the crop losses due to biotic (diseases, insects, mites and nematodes etc.) and abiotic (heat, cold injury, water stress etc.). Thus, the final goal of crop protection is development of economical and eco-friendly approaches for minimizing the losses without harming the environment, nature, livestock or human life.

1. IDM definition and rationale

An IDM can be defined as the disease management system that in the context of associated environment and population dynamics of microorganisms utilizes all suitable techniques and method in a manner as compatible as possible and maintains the disease below economic level. Thus IDM involves a total systems approach to the suppression of pathogen populations to a level where higher yields can be obtained which provide the farmer with maximum economic return (Jeger 2000). In general, it is the integration of all possible and suitable management techniques for the control of disease where the individual components of disease management such as host plant resistance (HPR), agronomic practices, judicious use of fungicides, bio-pesticides etc. need to be compatible.

Despite good intentions, the distance between scientists and farmers is great. It is greater in developing countries than in developed countries. There is a need to narrow down the distance between the technologies generated in the field of IDM, their refinement, validation, transfer and adoption. It is with this hypothesis we followed "the optimization of disease control measures in an economically and ecologically sound manner, accomplished by coordinated use of multiple tactics to assure stable crop production and to maintain disease damage below the economic injury level whilst minimizing hazards to humans, animals, plants and environment".

2. Methods and factors affecting IDM

Methods of IDM

The three methods of IDM are as follows:

1. **Combined control method:** This is a combination of control methods like adjustment in sowing time + seed treatment + use of resistant variety + chemical spray schedule etc. This type of IDM is widely practiced as a package of practice where the occurrence of disease is certain and sure.
2. **Surveillance based method:** This is an advanced method of IDM approach based on crop health monitoring and surveillance, and to a great extent takes in to account the economic threshold levels or economic damage levels. This is the actual state of practical IDM.
3. **Advanced method:** It involves the high input technology like computer supported forecasting, remote sensing, scouting, multiple pathogen thresholds, information on life cycle of pathogens, epidemiology of diseases, environmental factors and knowledge based decision making.

Factors affecting IDM

The IDM depends upon the four key factors (i) **Knowledge -** the understanding of biology and ecology of pathogens and crop, the interaction with agricultural ecosystems is very important. IDM draws the fundamental knowledge of the plant and pathogen affecting the crop (ii) **Monitoring-** Monitoring on broader scale to predict the disease outbreak and forewarn the farmers to take necessary action. In India monitoring and surveillance for the wheat rust, apple scab are done in a very systematic way (iii) **Economic threshold-** This takes in to account the revenues losses resulted from crop damage due to disease and cost of treatment to prevent the damage. Below economic threshold, the presence of disease can be tolerated. Only when the disease level increases above the threshold does the farmers to take necessary action to control it (iv) **Adaptability-** Proper knowledge of area climate so that management strategies can be adapted to changing circumstances. Research scientists too, must aim to keep at least one step ahead of the pathogens which is also undoubtedly changing and adapting over time.

3. Components of IDM

Main components of IDM used alone or in combination to control diseases are as follows:

- Host plant resistance (HPR)
- Biological control
- Judicious use of chemicals
- Use of biopesticides
- Cultural practices
- Induced systemic resistance
- Plant nutrition
- Use of biotechnological tools

3.1 Host plant resistance (HPR)

Resistance can be defined as the ability of a plant to avoid, tolerate or recover from the damage inflicted by pathogen population that would cause greater injury to other plants of same species grown under similar environmental conditions. Plants resistance to specific pathogen was discovered as early as 1792, when the `Underhill' variety of wheat showed resistance to Hessian fly, *Mayetiola destructor* (Ortman and Peters 1980).The HPR is a result of interaction between the plant and pathogen. The optimum conditions under which the plant species are growing are favourable for the growth and development of pathogen so that plant species is accepted by the pathogen. The HPR significantly contributes to sustainable crop production and environment conservation. It is most durable, economical, and technically feasibly at farmers level. However, development of pathogen resistant variety takes 5-10 years and HPR can put a selection pressure on pathogen to the extent that it may lead to trigger the evolution of new biotype of pathogen or new virulence in case of pathogens. Farmers in India and elsewhere have adopted HPR as the major component of IDM. Major successes have been reported in the management of diseases using HPR in pulses like chickpea, pigeonpea, mungbean and groundnut etc. and in wehat for rust (Pande et al 2007). A combination of available levels of HPR and weather based minimal fungicide application has led to the rehabilitation of chickpea in Botrytis gray mold prone areas in India, Nepal and Bangladesh (Pande et al 2005). Similarly for the management of foliar diseases of groundnut, the HPR and economical use of fungicides has been validated with large number of farmers in the states of Andhra Pradesh, Karnataka and Tamil Nadu in India (Mayee 1998)

3.2. Biological control

Bio-control agents are used as a core component on integrated diseases management system. The science and art of using living organisms as bio-control agents is an important component of environment friendly disease management procedures. These bio-control agents are of enormous value in IDM for sustainable agriculture where they often replace the need of fungicides. The bio- control agents either suppress the pathogen growth either by the antibiotic production, hyperparasitism or by competition. Various bio-control agents used in control of various diseases are *Bacillus subtilis, Pseudomonas fluorescens, Gliocladium* spp., *Trichoderma* spp. *Chaetomium globosum, Pseudomonas cepacia, Bacillus cereus, Agrobacterium radiobacter* etc. There are several examples where biological control has played a major role in controlling the diseases. *Ampelomyces quisqualis* Ces (AQ10) occurs most widely in nature and it was discovered in vineyard of Israel that it can control the powdery mildew of vegetables and fruit crops. The germinating spores of *A. quisqualis* suppressed the development of powdery mildew by hyperparasitism. *Coniothyrium minitans* is a phylloplane fungus and used to control the *Sclerotinia* in oil seed rape and lettuce. It is good invader of leaf surfaces and rapidly covers the foliage of treated plants and prevents the invading of *Sclerotinia. Fusarium oxysporum* Schechtendal strain Fo 47 is used to suppress the growth of *Fusarium oxysporum* and *Fusarium moiliforme. Gliocladium catenulatum* is used to control *Pythium* spp., *Rhizoctonia* spp. *Botrytis* spp. *Didymella* spp. and *Helminthosporium* spp. by parasitizing the pathogen, producing antibiotic and competing for nutrients with pathogens. Similarly *Trichoderma harzianum, T. viride* are used against *Armillaria mellea, Phytophthora* spp. *Chondrostereum purpureum, Pythium* spp., *Botrytis cinerea, Fusarium* spp. *and Rhizoctonia* spp.

Several bacterial species including, *Bacillus subtilis, Agrobacterium radiobacter* strain K-84, *Burkholderia cepacia, Pseudomonas syringae, P. fluorescens* have been found highly effective in controlling soil borne seedling diseases, fruits, vegetables and pulses pathogens (Prashad and Gupta 2012)

Viruses are not only the causal organisms of diseases but also used to control the pests which causes great economic losses to the farmers. Baculoviruses are used to control the population of pests. Insects are attacked by many different type of viruses but when insect injected the Baculoviruses, the protective protein coat of polyhedral occlusion bodies of viruses dissolved in mid- gut and then occlusion bodies enter the haemolymph of insects. These virus particles invade all the cells of insects and multiply there leading to the death of insects. Baculoviruses includes *Spodoptera exigua* NPV against Beet army worm, *S. littoralis* NPV against Egyptian cotton leaf worm, *Anagrapha falcifera* NPV Celery looper against lepidipterous larvae, *Helicoverpa Zea* NPV against cotton bollworm and tobaccobud worm, and *Anticarsia gemmatalis* NPV against Valvet bean caterpillar (*Anticarsia gemmatalis*).

In general it was demonstrated that biological control works, but it was less effective and more variable than chemical protection. The key to overcoming this difficulty lies in the consideration of the growth, survival and behaviour of the bio-protectant in the introduced niche. In nearly every case that we know of, biological control agents must grow well and survive long after their application if they are to be effective. Microbial competition and the levels of appropriate nutrient sources are known factors influencing the rate of growth and survival of bio-protectants. Pressure of strongly competitive microflora can totally eliminate the activity of otherwise effective strains. In addition, biological control agents need frequently to be grown by a convenient method, and then applied in systems similar to those utilized for chemical protectants, largely ignoring the physiological and genetic nature of the organism in question.

3.3. Judicious use of fungicides

Diseases have been ravaging the crops from the time immemorial. The intensive cultivation of horticultural crops like fruits, vegetables, flowers, medicinal plants and mushroom and the cereals as well as pulses has posed serious problems of diseases resulting in heavy crop losses. Chemicals have been used successfully to combat the ravages of these diseases for many years. The over use of these chemicals resulted in water pollution, residues on fruit and vegetable crops, effect on non- target organisms and development of resistance in pathogens against the chemicals thus have drawn the attention toward the rational use of fungicides by including monitored control strategies and cultural practices. Different fungicides have different physical modes of action like protective (broad spectrum fungicides), post infection activity (EBI), pre- symptom and post symptom (benzimidazoles and EBI). Curative activity of fungicides has been utilized only in those diseases in which epidemiological parameters are well studied. The IDM of Botrytis gray mold in chickpea involving tolerant genotype, soil application of diammonium phosphate, wider row spacing, seed treatment with carbendazim + thiram and need based foliar application of carbendazim has proved more effective than any of the individual disease management components in large scale, on-farm studies conducted in India, Nepal and Bangladesh (Pande et al 2006).

3.4. Biopesticides

Pesticides of plant origin are derived from plant parts and genes used to transform crops to express resistance to insects, fungal and viral attack. The plant parts and their extracts with their antifungal properties play an important role in plant disease management (Copping and Menn 2000). Plant with pest killing properties has been recorded as early as Indian Reg Veda during second millennium B.C. (Hoddy 1991). Garlic (*Allium sativum*) has long history of reputed value and actual use for its medicinal, antimicrobial and pesticidal properties (Timonin and Thexton 1951, Amonkar and Benerji 1971 and Fliermans 1973). The growth of *Rhizoctonia solani* can be reduced with ethanolic extracts of *Eucalyptus* sp. (Hardtl 1986), *Chenopodium ambrosioides* leaves (Dubey et al 1983), *Lippia alba*, *Aegle marmelos* and *Cestrum diurnnm* (Kishore et al 1983 and Renu 1983), *Vitex negundo* L and lemon grass (Michael et al 1985; Handique and Singh 1990), *Ocimum sanctum* (Tiwari and Nayak 1991), *Allium sativum, A. cepa, A. indica, Calotropis procera, Datura stramonium, Polyatlthia longifolia, Tagetes erecta, Vinca rosea* and *Withania somifera* (Shivpuri et al 1997).The seed extract of *Piper nigrum* was found to be effective against *R. bataticola* (Alice 1984). The root bark extract of *Plumbago* zeylanica L. was effective against *R. bataticola, Curvularia lunata, Helminthosporium sativum, Alternaria tenius, H. spiciform* and *Fusarium moniliofome* (Bambode and Shukla 1974). Leaf juices of *Iberis amara, Adhatoda vasica* and *Eupatorium cannabinnum* and *Taxus* sp. caused complete inhibition of *R. solani* (Kumar and Tripathi 1991; Wagner and Flores 1994). The sclerotia production of *R. solani* was completely inhibited with cold water extracts of *Calophyllum inophyllum* cake, *Prosopis juliflora* and *Thevetia peruviana* (Ezhilan et al 1994). *O. sanctum* leaf extracts was inhibitory to the spore germination and growth of *Botryodiplodia theobromae* and *Rhizopus arrhizus* and this is due to the presence of polyamine ornithine inhibitor(s) in the extract which block the ornithine decarboxylase pathway in these fungi (Patil et al 1992). *Allium sativum* and *Curcuma longa* caused maximum inhibition of *Xanthomonas campestris* pv. *vesicatoria* under *in vitro* conditions (Prashad and Gupta 2012).

Different leaf extracts were found effective in reducing the incidence of various viral diseases. Leaf extract of *Mirabilis jalapa* was found to reduce the incidence of pepper veinal mottle virus in bell pepper (Kapoor and Thakur 2012). Antiviral activity of plant extracts could be due to de novo synthesis of antiviral substances, production of virus inhibiting agents and production of mobile inducing signal that bind to host plant surface which produces virus inhibiting agents (Verma et al 1998).

3.5. Induced systemic resistance

Plants actively respond to a variety of environmental stimuli, including gravity, light, temperature, physical stress, water and nutrient availability. Plants also respond to a variety of chemical stimuli produced by soil- and plant-associated microbes. Such stimuli can either induce or condition plant host defenses through biochemical changes that enhance resistance against subsequent infection by a variety of pathogens. Induction of host defenses can be local and/or systemic in nature, depending on the type, source, and amount of stimuli. The systemic acquired resistance (SAR) is mediated by salicylic acid (SA), a compound which is frequently produced following pathogen infection and typically leads to the expression of pathogenesis-related (PR) proteins. These PR proteins include a variety of enzymes some of which may act directly to lyse the invading cells, reinforce cell wall boundaries to resist infections, or induce localized cell death. The induced systemic resistance (ISR) is mediated by jasmonic acid (JA) and/or ethylene, which are produced following applications of some nonpathogenic rhizobacteria. Interestingly, the SA- and JA- dependent defense pathways can be mutually antagonistic, and some bacterial pathogens take advantage of this to overcome the SAR. For example, pathogenic strains of *Pseudomonas syringae* produce coronatine, which is similar to JA, to overcome the SA-mediated pathway (He et al 2004). Because the various host-resistance pathways can be activated to varying degrees by different microbes and insect feeding, it is plausible that multiple stimuli are constantly being received and processed by the plant. Thus, the magnitude and duration of host defense induction will likely vary over time. Only if induction can be controlled, i.e. by overwhelming or synergistically interacting with endogenous signals, host resistance will be increased.

A number of strains of root-colonizing microbes have been identified as potential elicitors of plant host defenses. Some biocontrol strains of *Pseudomonas* sp. and *Trichoderma* sp. are known to strongly induce plant host defenses (Harman et al 2004). In several instances, inoculations with plant-growth-promoting rhizobacteria (PGPR) were effective in controlling multiple diseases caused by different pathogens, including anthracnose (*Colletotrichum orbiculare*), angular leaf spot (*Pseudomonas syringae* pv. *lachrymans* and bacterial wilt (*Erwinia tracheiphila*). A number of chemical elicitors of SAR and ISR may be produced by the PGPR strains upon inoculation, including salicylic acid, siderophore, lipopolysaccharides, and 2,3-

butanediol, and other volatile substances (Van Loon et al 1998, Ongena et al 2004, Ryu et al 2004).

3.6. Plant nutrition

The nutrition of crop plants has direct effect on the diseases and is an important component of integrated disease management (IDM). Both deficient and over nourished plants by imbalanced fertilization invite the high incidence of diseases as well as loss in yield and quality of produce and products. The essential elements are those which compensate the toxic effects of other elements. The essential elements are required for growth, reproduction and other metabolism of plants. The essential elements cannot be substituted.

The amount, proportion, time and mode of application of fertilizers affect the metabolism of plants and thus occurrence and severity of diseases. Fertilization with both P and K significantly reduces the leaf rust damage and powdery mildew infection in wheat (Singh 2008). The deficiency of macronutrients may also affect the incidence of many diseases. Potassium (K) plays an important role in survival of crop plants under environmental stress conditions. Potassium also affects the reaction of plants to pest or diseases by having direct effect on the pathogen number, development, multiplication, survival, vigour and length of life cycle. It has direct effect on severity of host: by altering internal metabolism of the plant affecting food supply for the pathogen and by modifying the microclimate through changes in habitat, density of growth. Secondly, it has effect on the establishment of the pathogen and spreading within the plant – through effects on plant structure (e.g. cell wall and cuticle thickness, tissue firmness) – and through the functioning of stomata and also by affecting the plant's ability to recover from pest or diseases attack through the repair of damage or, by general improvement of growth.

Judicious use of potash fertilizers helps in minimizing the susceptibility of plants to stem rust (*Puccinia graminis tritici*). Integrated treatment of maneb and zinc (Zn) gave the maximum control of *Septoria tritici* and *P. recondita* and increased the yield by 32% in Brazil (Luz 1976). However, the severity of powdery mildew and septoria leaf spot increased by application of N and was more in early application of N. The nitrogen status of wheat crop affects the level of stripe rust (*P. striiformis*). This appears to be independent of the interaction of increased biomass production creating humidity within the canopy that is more conducive to the development of stripe rust. The incidence of stripe rust was minimum in nitrogen deficient crops The increase in doses of nitrogen (N) in the form of ammonium nitrate increased the incidence and severity of stripe rust on wheat in Netherland (Daniel and Parlevliet 1995). The severity of brown rust increases with the increase in N application (180 Kg/ha) whereas K application reduces the rust incidence (80 Kg/ha) (Sinha 1998).

The incidence of powdery mildew (*Erysiphe graminis*) was high in wheat with increased application of N fertilizers (Keblikas 1997, Odorfer and Pommer 1997). The phosphorus (P) deficiency (Fernandez et al 1998) and application of N (Sharma et al 2006) in wheat decreased the leaf spot caused by *Bipolaris sorokiniana*. The incidence of Rhizoctonia root rot in wheat can be reduced with application of zinc (Thongbai et al 1993). Similarly, the incidence of flag smut (*Urocystis agropyri*) was reduced with increased contents of phenols and sugars with application of K fertilizers (Prashar et al 1995).

3.7. Use of genetically engineered plants

Breeding for resistance to some economically important diseases in different crops has not been very effective. Application of genetic engineering holds a promise to alleviate some of these constraints to crop productivity in horticultural crops as well as in pulses and cereals. Genetic engineering of plants makes it feasible to transfer genes from totally unrelated organisms thus breaking species barriers otherwise not possible by conventional genetic enhancement. Genes from plants, microbes and animals can be combined and introduced in to the living cells of other organisms. The organisms that have genes from other species inserted in to their genome are called transgenic. The production of pathogen resistant transgenic plant has been achieved by this method; certain genes are inserted in to plant genome that confers resistance to pathogens such as viruses, fungi and insects. These transgenic plants reduced the pesticides use and thereby provide environmental benefits while reducing farmers cost. Genetically modified plants are generally used to control the viral diseases. Transgenic plants have been developed from genes derived from pathogen and after insertion in to the host plant (Table 1) and the transformed plant showed the resistance against that specific pathogen from that gene was derived.

Table 1. Virus resistant transgenic plants generated in various crops for pathogen derived resistance

Crop	Transgene	Mode of action	Virus	Reference
Potato	Replicase gene	Competition for enzyme	PVY	Audy et al (1994)
Pea	Replicase gene	Competition for enzyme	Psb MV	Jones et al (1998)
Rice	Replicase gene	Competition for enzyme	RYMV	Pinto et al (1999)
Tobacco	Movement Protein	Interference with transport	TMV	Cooper et al (1995)
Potato	Movement Protein	Interference with transport	PLRV, PVX, PVY	Tacke et al (1996)
Tobacco	Transport Protein	Interference with transport	TEV	Cronin et al (1995)
Potato	Viral Protease	Poly protein processing	PVY	Vardi et al (1993)
Potato	Anti Sense RNA	Block viral RNA and prevent translation	PLRV	Kawchuck et al (1991)
Tomato	Ribozyme	Cleave viral RNA	CEVd	Atkin et al (1995)
Tobacco	Satellite RNA	Compete for capsids	CMV	Harrison et al (1987)
Potato	Anti Viral Potein ribonuclease	Degrade dsRNA (Viriod)	PSTV	Sano et al (1997)
Tobacco	Pokeweed Antiviral Protein	Inhibit rRNA of 60 S subunit	TMV, PVX	Wang et al (1998)

4. Status of IDM in India

The research and development to combat plant diseases in different crops has not changed much and the main emphasis was on host resistance and chemical control where ever applicable, and quite often these components of disease management were practiced in isolation to each other. Recently a shift in scientific thinking and practice in the management of diseases has been seen and greater emphasis is on identifying, evaluating, and integrating location specific components of IDM. In general IDM has followed the principles of IPM (Jeger 2000). The location specific IDM in different crops is primarily based on host plant resistance (HPR) or genetic resistance; additionally other components of disease management were identified and integrated depending upon the need, scope and affordability of the end users (Fig 1 and Table 2). In several countries and environments components of IDM were either used alone (for example HPR) or in combination with other components (such as fungicide seed treatment) to combat diseases. Several management practices have been developed and tested in farmers' fields in India. Some examples are given in Table 2.

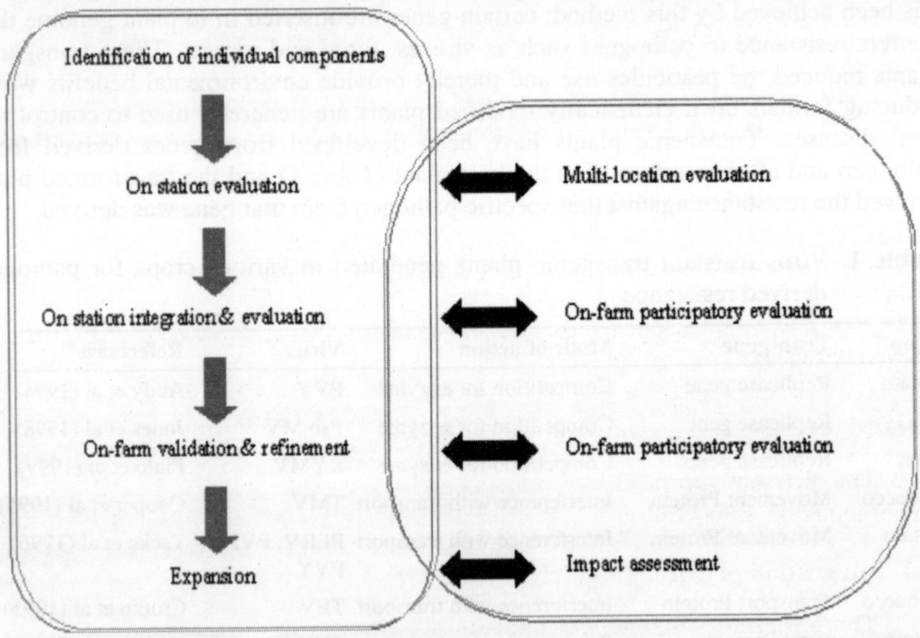

Fig. 1. Thematic structure for the development of IDM technology (*Source:* Pande et al 2007)

Table 2. Progress in IDM of foliar diseases in legume crops.

Crop	Disease	Components of IDM				Reference
		HPR	AP	CC	BC	
Chickpea	Ascochyta blight	**	*	*	*	Vishwa Dhar et al 2004; Pande et al 2005, Bretag et al 2008
	Botrytis gray mold	**	*	**	*	Davidson et al 2004; Vishwa Dhar et al 2004; Pande et al 2006
Pigeonpea	Sterility mosaic disease	***	*	*	X	Vishwa Dhar et al 2004; Kumar et al 2006

Crop	Disease	HPR	AP	CC	BC	Reference
Lentil	Stemphylium blight	**	*	**	X	Huq and Khan 2007 Taylor et al 2007
	Ascochyta blight	***	*	**	X	Baaya and Erskine 1998
	Rust	***	*	**	X	Taylor et al 2007
Field pea	Powdery mildew	***	*	***	*	Bretag et al 1995, Grunwald et al 2007
	Downey mildew	***	*	***	X	Grunwald et al 2007
	Rust	***	*	**	X	Grunwald et al 2007
Faba bean	Ascochyta blight	***	*	**	X	Jellis et al 1998
	Chocolate leaf spot	***	*	**	X	Jellis et al 1998
Mungbean and black gram	Yellow vein mosaic	***	X	***	X	Singh et al 2002
	Cercospora leaf spot	*	***	*	**	Singh et al 2001
	Powdery mildew	**	*	**	X	Vishwa Dhar et al 2004
Cowpea	Cercospora leaf spot	***	*	**	X	Saxena et al 1998, Allen et al 1998
	Cowpea golden mosaic	***	**	*	X	Saxena et al 1998, Allen et al 1998
	Cowpea aphid-borne mosaic	***	**	*	X	Saxena et al 1998, Allen et al 1998

HPR=Host plant resistance, AP=Agronomic practices, CC= Chemical control, BC= Biological control;
*, **, ***=low, medium and high potential, respectively; X = Not in common use
(*Source:* Pande et al 2007)

5. IDM method: Integrated approach

The different approaches recommended for the control of diseases when applied in combination are as follows:

5.1. Integration of different cultural practices

The different cultural practices like crop rotation, mulching, tillage, soil amendments, soil solarization, soil sterilization, change in date of sowing, plant spacing etc. when applied alone are able to control diseases up to some extent but when these cultural practices are combined with each other, they not only control the disease but also increase the yield of crop. The use of intercropping in field vegetable production in tropical and temperate regions is emphasized as a potential means of pest and disease reduction (Alegbejo and Uvahi 1986, Theunissen 1997). The use of intercropping in field vegetable production in tropical and temperate regions is a potential means of disease reduction (Alegbejo and Uvahi 1986, Theunissen 1997). The intercropping of maize and sorghum with peppers served as barriers against the aphid, vector of pepper veinal mottle virus and reduced virus spread. Significant decrease in tomato wilt occurrence was observed in treatment of soil with urea, fly ash and active bleaching powder singly or in combination (Kanjilal et al 2000).

Robak and Naurocka (2005) reported the efficacy of natural compounds to control diseases in cucumber, lettuce, cabbage, cauliflower, broccoli, onion and leek. The Alternaria blight (*Alternaria brassicae* and *Alternaria brassicicola*) can be controlled by combination of three cultural practices i.e. spacing, date of planting and nutrient management. Alternaria blight in cabbage can be controlled by planting cabbage on 1st November at 40 x 40 cm spacing in the field treated with S-Zn-Mg-Mo-B at 30-5-1-1-1 kg/ha and one spray with Rovral (0.2%) and increased the seed yield by 207.8% over the normal management practices (Hossain and Mian 2005).

Bacterial wilt is one of the most devastating diseases of solanaceous vegetable crops affecting more than 450 plant species. The bacterium *Ralstonia solanacearum* exists in different races and biovars. Reduction of initial inoculum is of primary importance in the management of this disease. The disease can be controlled by following different practices such as growing certified disease-free planting material, resistant cultivars, application of soil amendments and soil solarization (Sood 2005). Soil solarization for 40 days along with the addition of cabbage, cauliflower, broccoli and sarson leaf residues controlled the gladiolus wilt (*Fusarium oxysporum* f.sp. *gladioli*) by 74.6% whereas soil solarization (40 days) alone reduced the gladiolus wilt by 67.3% compared to the unsolarized control (Raj and Upmanyu 2005).

5.2 Integration of cultural and chemical control

The integration of chemicals and cultural practices has resulted in reduced anthracnose caused by *Colletotrichum orbiculare* in watermelons (Peregrine et al 1978, Peregrine and Ahmad 1983), *Pseudomonas syringae* pv. *lachrymans* and *Pseudoperonospora cubensis* (Alexandru, 1989), African cassava mosaic Gemini virus (Fauquet and Fargette 1990), Verticillium wilt (*Verticillium dahliae*) of cool vegetables, (Vallad et al 2004), club root (Cheah and Fallon 2005), damping-off, Sclerotinia rot, downy mildew, and black rot (Kapoor 2005). Covering the tomato seedling nursery with nylon net for 25-30 days plus 4 sprays of monocrotophos (as Nuvacron) at 10-days intervals after transplanting, delayed the spread of TLCV for 3-5 weeks and increased tomato yields (Saikia and Muniyappa 1989). Viral diseases spread more slowly in the early and mid-growing season in *Capsicum* plots mulched with silver striped black vinyl.

The use of polythene shelters, intercropping, or high tomato planting densities each combined with Dithane M-45 (mancozeb) gave consistently higher numbers of fruits and yield than any of the cultural practices, with or without sanitation (Tumwine 1999). In field, vectors spread viral diseases and aphids are main vectros. The combination of barrier crops (wheat, barley and oat) and insecticides like malathion, monocrotophos, lambda-cyhalothrin, dimethoate and oxydemeton-methyl and the organic compound, Virex-H controlled the Radish mosaic virus (Sharma et al 2005). Similarly in cucurbits change in sowing date, depth and use of barrier crops i.e. sunflower, gobhi sarson (*Brassica campestris* var. *sarson*) and raya (*Brassica juncea*) in combination with sprays of malathion and Thiodan (endosulfan) reduced the cucurbit mosaic incidence caused by watermelon mosaic virus and potato virus Y. The sowing of crop in the second fortnight of March, upto 9.00 cm depth, sunflowers barrier crop and spray of Thiodan reduced the disease incidence to (37.24%) and severity (40.34%) as compared to control (Kaur and Kang 2005). Early sowings along with sprays of different fungicides like Ridomil MZ, Indofil M-45, metalaxyl etc. reduced the disease incidence of different diseases Alternaria blight (*Alternaria brassicae*), white rust (*Albugo candida*) and stem rot, (*Sclerotinia sclerotiorum*) in mustard to minimum as compared to those beds where sowing was done in last week of November (Yadav 2004). Integration of soil solarization (6-8 weeks) with cereal straw compost and fungicides such as benomyl, chlorothalonil, maneb, triadimefon and sulfur controlled early blight (*Alternaria solani*), leaf mould (*Fulvia fulva*), powdery mildew (*Leveillula taurica* and *Erysiphe cichoracearum*); some pests such as bollworm (*Heliothis armigera*), tomato russet mite (*Aculops lycopersici*) and root-knot nematode (*Meloidogyne javanica*) and blossom end rot in tomato cultivars Ferguson, Kalimba, Ovacik and Target (Bicici et al 2000). Mango malformation is most important and destructive disease in cultivation of this crop. Culutral practices including the pruning of malformed branches and spray of Phosphamidon 85 EC (0.05%) has the lowest incidence of floral malformation to 10 - 15% (Shamshad et al 2004). The IDM has positive effect on the yield and quality of sugarcane. The sett treatment with 0.1% Bavistin (carbendazim) for 15 minutes, rouging of infected canes, application of N:P:K (250:100:60 kg/ha), 80 cm row-to-row spacing and one spray of insecticide controlled the grassy shoot disease (disease caused by a Phytoplasma) (Mishra et al 2003). Sowing of DDK 1029F, a moderately resistant wheat cultivar, during the last week of November as normal sown crop along with two foliar sprays of Tilt 20 EC (0.1%) were found highly effective in the management of Alternaria leaf blight and increased grain yield (Samanta et al 2012).

5.3 Integration of chemical and biological control

Bio-control agents such as *Pseudomonas fluorescens, Trichoderma viride, T. harzianum, Bacillus subtilis, P. putida, P. cepacia, Talaromyces flavus* and *Agrobacterium radiobacter* strain K 84 etc. can be used with integration of chemicals for the effective control of diseases. The foliar spray of *P. fluorescens* strain RPB14 at the seedling (10^7 cfu/ml) and flowering (10^6 cfu/ml) stages followed by single spray of carbendazim (0.05%) at pod development controlled the fuscous blight (*Xanthomonas axonopodis* pv. *phaseoli* var. *fuscans*) and floury leaf spot (*Ramularia phaseoli*) in rajmash bean (*Phaseolus vulgaris*) cv. Baspa and increased the seed yield by 81.3% than control (Mondal 2004).The integration of Roko (0.1% thiophanate-methyl) with *Trichoderma harzianum* (Th) reduced the leaf blight (*Phyllosticta spinaciae*) by 92.8% and leaf spot (*Fusarium roseum*) by 93.9% over the control in spinach (Kadam and Raut 2006).

5.4 Integration of HPR, cultural, biological and chemical control

The integration of resistant cultivars, cultural practices, fertilizers, soil fumigation, solarization, pesticides and biocontrol agents are used for the control of club root (*Plasmodiophora brassicae*) of vegetables (Cheah and Falloon 2005). The IDM package to control Potato late blight include the use of resistant varieties, tuber treatment with boric acid, rouging, dehaulming, spray of Metasystox and mancozeb and reduced the disease severity of late blight (6.0- 7.0%), mosaic (4.5- 5.5%), leaf roll (4.0- 5.0%) and common scab (4.5- 6.0%). This treatment also increases the tuber yield by 23.94% as compared to untreated one (Basu and Srikanta 2003). Hou and Hang (2004) gave IDM package for the control of pests and weeds in the production of organic vegetables. Seed soaking in warm water, crop rotation and vegetable rotation, using organic fertilizer, adopt proper cultivation procedures, using separation methods such as planting a shelter belt or setting up a net for protection against insects, applications of certain expel methods such as using a mulching film, using appropriate trap and kill methods for different pests, biocontrol, rationalize the use of chemical control and develop the proper field management procedures. Pande et al (2005) reported the IDM package involving use of resistant cultivar, need based application of fungicides and cultural practices for the management of Botrytis gray mold and Fusarium wilt in chickpea.

Adjustments of temperature, RH, pH and electrical conductivity of the nutrient solution, periodical washing of the substrate, spraying chemicals, pruning and elimination of diseased plants were effective in managing the diseases caused by *Erwinia carotovora, P. solanacearum, F. oxysporum, Sclerotinia sclerotiorum, Phoma* spp., *Alternaria* spp., *Phomopsis* spp. *Botrytis* spp. and *Oidium* spp. in tomato hybrid grown in hydroponic culture (Trotin 1998). The eradication of primary infection sources by stubble removal and burning, and interception of the initial infection route by mulching with manure or compost was most effective means of controlling asparagus stem blight caused by *Phomopsis* spp. (Sakai et al 1992). Viana et al (2000) reported that more bean plants (*Phaseolus vulgaris*) survived damping off infection (*S. sclerotiorum*) with liquid swine manure, *Bacillus subtilis* suspension (10^8 cfu /ml) and iprodione (0.1% a.i.). Management of the disease by integration of various methods like soil solarization, soil amendment with mustard cake, seed treatment and foliar spray with biopesticides and biological control agents (eg. *Bacillus subtilis*, *Trichoderma* spp.) alone or in combination along with standard treatment like Bavistin were found effective in reducing the incidence of root rot and severity of web blight (Sharma and Gupta 2003). The organic amendments of cattle manure compost (CMC) and grape marc compost (GMC), and their integration with *Trichoderma harzianum* and mycorrhizas increased the yield, root weight and reduced the incidence of damping off caused by *Rhizoctonia solani* and *Pythium aphanidermatum* in tomatoes grown under greenhouse conditions (Tsror et al 2003). The integration of organic amendments with fungicides and *Trichoderma virens* and *T. harzianum* reduced the root rot and web blight (*R. solani*) in French bean (Upmanyu et al 2002). The pre emergence root rot, post emergence root rot and web blight disease severity was minimum in integrated treatment of Mustrad cake amendment + Bavistin (ST) + Bavistin foliar spray (Fig. 2) as compared to control.

Crop rotation is used to reduce the inoculum level of many pathogens. Soil treatment with garlic oils, flooding and dipping transplants of onion in fungicide solutions have been utilized to control white rot. Seed treatments with fungicide to control seed borne Botrytis neck rot, smut and damping-off diseases. Seed furrow treatments with fungicides control Fusarium, Pythium damping off and onion smut. Fungicide sprays control the Botrytis and Stemphylium leaf blights, Botrytis flower blight, Alternaria purple blotch and downy mildew. Copper and antibiotic sprays may reduce the incidence of bacterial leaf blights and bulb decays. Undercutting, wind rowing, and/or heating in storage to dry the inside tissues of onion bulb necks prevents Botrytis neck rot (Poll 1996). Diseases are outlined of vegetables grown in greenhouses in Russia, including root rot (caused by *Fusarium, Pythium,*

Aphanomyces cladogamus, Rhizoctonia solani, Sclerotinia sclerotiorum, Ascochyta cucumis (Didymella bryoniae) and *Erwinia carotovora and wilt (caused* by *Clavibacter michiganensis, Pseudomonas corrugata, Fusarium and Botrytis cinerea*). Observations conducted in greenhouses without annual disinfections showed that the most common *Fusarium* spp. was *F. oxysporum*, followed by *F. moniliforme*, *F. nivale* and *F. solani*. Control measures are outlined, including the use of seed treatments, antibiotics, beneficial soil microorganisms, green manures, soil steaming, fumigation, pre-sowing seed heating, greenhouse disinfection and crop rotations (Rudakov and Rudakov 2000, Munshi and Sokhi 2000).

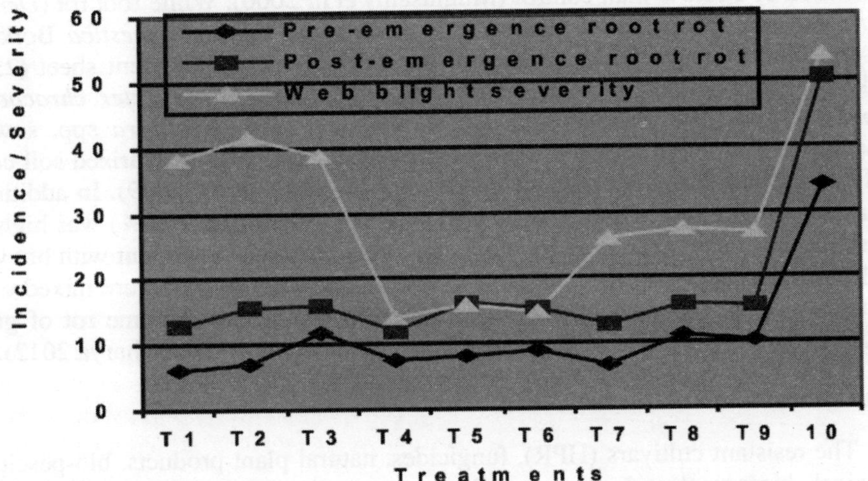

Fig. 2. Integrated effect of soil amendments with fungicides, bioagents on root rot incidence and web blight severity
[T1- M. cake amendment + Bavistin (ST); T2- *Th. virens* (ST) + C. cake amendment + Vitavax (ST); T3- *Th. virens* (ST) + Raxil (ST); T4- M. cake amendment + Bavistin (ST) + Bavistin (F S); T5-*Th. virens* (ST) + C. cake amendment + Vitavaz (ST) + Bavistin (FS); T6- *Th. virens* (ST) + Raxil (ST) + Bavistin (FS); T7-M. cake amendment + Bavistin (ST) + *Th. harzianum* I2(FS); T8-*Th. virens* (ST) + C. cake amendment+ Vitavax (ST) + *Th. harzianum* I2 foliar spray; T9-*Th.. virens* (ST) + Raxil (ST) + *Th. harzianum* I2 (FS); T10- Control] (Upmanyu *et al* 2002)

The integration of long term soil solarization with a biological control agent (*Trichoderma virens*), chemical fungicide (pentachloronitrobenzene [PCNB]), organic amendment (chicken litter) or physical method (black agri plastic mulch) reduced the incidence of southern blight (*Sclerotium rolfsii*) and southern root-knot (*Meloidogyne incognita*) diseases. The soil solarization treatment, and soils amended with chicken litter, stimulated a shift in the soil microbial population dynamics of rhizobacteria of *Bacillus* spp. and fluorescent pseudomonads increased significantly in the rhizosphere, rhizoplane, and interior root tissues of tomatoes and sweet potatoes, in solarized soil compared to nonsolarized soil (Stevens et al 2003).

Gladiolus is very important ornamental crop but it suffered from Fusarium yellows caused by *Fusarium oxysporum* f.sp. *gladioli*. The integrated application of *neem* cake amendment, carbendazim foliar spray and *Trichoderma harzianum* provided the highest control of Fusarium yellows, increased the corm yield and improved the plant health (Sharma et al 2005). The intercropping of pigeon pea with sorghum, application of farmyard manure (FYM), green manure, zinc sulphate, seed dressing with carbendezim, thiram and *Trichoderma* reduced the incidence of *Fusarium udum* (Bharathi et al 2006).

Basal stem rot (BSR) disease of coconut caused by *Ganoderma lucidum* is a major limiting factor for coconut production in India. The pathogen is soil borne and can be controlled by integration of frequent irrigation through basin method of irrigation, neem cake amendment, application of *Trichoderma viride, Pseudomonas fluorescens* and root feeding of fungicides controlled the vertical spread of pathogen with a reduction of 121.6% than control (Muthusamy *et al* 2006). White root rot (*Dematophora necatrix* (Mart.)) is a serious disease of apple (*Malus domestica* Borkh) in nurseries and orchard. Soil solarization with transparent polyethylent sheet (25 μm thick) for 40 days and inoculation of apple seeds with *Azotobacter chrococcum, Glomus fesiculatum, G. macrocarpum, G. mosseae* and *Gigaspora* spp. sown in solarized soil, *A. chrococcum* coated seeds and their planting in solarized soil completely controlled the white root rot incidence (Raj and Sharma 2009). In addition to disease control, shoot length (9.6-10.6%) and root length (9.2-16.0%) was higher in solarized plots as compared to non-solarized plots. Rhizome treatment with hot water at 51°C for 10 min + soil application of *Trichoderma harzianum* culture mixed with 1 kg neem cake at the time of planting resulted in the lowest rhizome rot of ginger incidence and increased rhizome yield (Bandyopadhyay and Bhattacharya 2012).

Future thrust

The resistant cultivars (HPR), fungicides, natural plant products, bio-pescitides, botanical, biological and agronomical practices will remain the potentially viable options for IDM. Considerable progress has been made in identifying disease resistance sources and developing IDM packages for some of the diseases. Further, IDM strategies need to be refined for emerging pathogens taking in to account effect of climate, variability in pathogen dynamics that is contributing to emergence and resurgence of new strains/pathogens, necessitating the appropriate refinement for disease control. In addition, future module will incorporate the latest advancements being made in the identification of biocontrol agents and biotechnological tools such as marker-assisted selection, genetic engineering and wide hybridization to develop crop cultivars with resistance to diseases and will play a key role in future disease management programs. Some progress has been made in transferring resistance genes in to high yielding, improved and agronomically superior genetic backgrounds in some of the crops. Disease modelling, decision support systems, and remote sensing could contribute to up-scaling and dissemination of IDM technologies. Thus, our future activities need to be focused on the on-farm testing and validation of IDM components, development of IDM packages for different crops and cropping systems, and expansion of IDM technologies by increasing the awareness of farmers.

References

Alegbejo MD and Uvahi H 1991 Effect of intercropping pepper with tall crops on the incidence of pepper veinal mottle virus disease on pepper, *Nigerian Journal of Entomology* **7 (1-2)** 82-87.

Alexandru D 1989 Control of bacteriosis and fungi in greenhouse cucumber in the second cultivation cycle, *Productia-Vegetala Horticultura* **38** (12) 8-10

Alice D 1984 Studies on antifungal properties of some plant extracts, M. Sc. thesis, Tamil Nadu Agricultural University, Coimbatore

Allen DJ, Thottappilly G, Emechebe AM, and Singh BB 1998 Diseases of cowpea, in *The Pathology of Food and Pasture Legumes* (eds) DJ Allen and JM Lenne, CAB International, Wallingford, Oxon, UK, 267-308 pages

Amonkar SV and Benerji A 1971 Isolation and characterization of larvicidal principle of garlic, *Science* **174** 1334-1344

Atkins D, Young M, and Gerlach WZ 1995 The expression of antisense and ribozyme genes targeting citrus exocortis viriod in transgenic plants, *Journal of General Virology* **16** 1781-1790

Audy P, Palukaitis P, Slack SA, and Zaitlin M 1994 Replicase mediated resistance to potato virus Y in transgenic tobacco plants, *Molecular Plant Microbe Interaction* **7** 15-22

Bambode RS and Shukla VP 1974 Studies on fungitoxic properties of *Plumbago zeylanica* L. extract, *Punjabrao Krishi Vidyapeth Research Journal* **2** 110-112

Bandyopadhyay S and Bhattacharya PM 2012 Management of Rhizome rot of Ginger using physical, chemical and biological methods, *Journal of Mycology and Plant Pathology* **42** 314-316

Basu A and Srikanta D 2003 Integrated management of potato (*Solanum tuberosum*) diseases in Hooghly area of West Bengal, *Indian Journal of Agricultural Science* **73**(12) 649-651

Bayaa B and Erskine W 1998 Diseases of lentil, in *The Pathology of Food and Pasture Legumes* (eds) DJ Allen and JM Lenne, Wallingford, Oxon, UK: CAB International 423-471 pages

Bharathi V, Rao KC, and Reddy MV 2006 Management of pigeonpea wilt incited by *Fusarium udum*, *Indian Journal of Plant Protection* **34** (1) 108-112

Bicici M, Toker S, Canihos Y, and Erkilic A 2000 Integrated disease management in tomato crops grown in high tunnels, *Bulletin-OILB/SROP* **23**(1) 15-19

Bretag TW, MacLeod WJ, Kimber RBE, Moore KJ, Knights EJC, and Davidson JA 2008 Management of ascochyta blight in chickpeas in Australia, *Australasian Plant Pathology* **37** 486-497

Bretag TW, Keane PJ, and Price TV 1995 Effect of ascochyta blight on the grain yield of field pea (*Pisum sativum* L), *Australian Journal of Experimental Agriculture* **35** 531-536

Cheah LH and Falloon RE 2005 Integrated disease management for club root of vegetable brassicas, *Vegetables: growing-environment-and-mineral-nutrition* 126-137.

Cooper B, Lapidot M, Heike JA, Dodds JA, and Beachy RN 1995 A defective movement protein of TMV in transgenic plants confer resistance to multiple viruses whereas functional analysis increases susceptibility, *Virology* **206** 307-313

Coping LG and Menn JJ 2000 Review Bio- pesticides: a review of their action application and efficacy, *Pest Management Science* **56** 651-676

Cronin S, Verchot J, Haldeman Cachill RS, and Carrington JG 1995 Long distance movement factor: A transport function of the potyvirus helper component proteinase, *Plant Cell* **7** 549-559.

Daniel DL and Parlevliet JE 1995 Effect on nitrogen fertilizers on disease severity and infection type of yellow rust on wheat genotypes varying in quantitative resistance, *Journal of Phytopathology* **143** 679-681

Davidson JA, Pande S, Bretag TW, Lindbeck KD, and Kishore GK 2004 Biology and management of *Botrytis* spp., in legume crops in *Botrytis: Biology, Pathology and*

Control. (eds) Y Elad., B Williamson, P Tudzynski and N Delen, Netherlands: Kluwer Academic Publishers, 395-318 pages

Dubey SK, Kishore N, Srivastava OP, Diskshit A, and Singh SK 1983 Fungitoxicity of some higher plants against *Rhizoctonia solani, Plant and Soil* **72** 91-94

Ezhilan JC, Chandrasekar V, and Kurucheve V 1994 Effect of six selected plant products and oil cakes on the sclerotial production and germination of *Rhizoctonia solani, Indian Phytopathology* **47**(2) 183-185

Fauquet C and Fargette D 1990 African cassava mosaic virus: etiology epidemiology, and control, *Plant Disease* **74(6)** 404-411

Fernandez MR, Zentner RP, McConkey BG, and Campbell CA 1998 Effect of crop rotation and fertilizer management on leaf spotting disease of spring wheat in southwestern Saskatchewan, *Canadian Journal of Plant Science* **78** 489-496

Fliermans CB 1973 Inhibition of Histoplasm capsulatum by garlic, Mycopathologia et Mycologia Applicata **50** 227-231

Grunwald NJ, Chen W, and Larsen RC 2007 Pea diseases and their management, in *Diseases of Fruits and Vegetables*, Vol. II., Kluwer Academic Publisher, Netherlands, 301-331 pages

Handique AK and Singh HP 1990 Antifungal action of lemon grass oil on some soil borne plant pathogens, *Indian Perfumer* **34**(3) 232-234

Hardtl AK and Konde BK 1986 Microbial antagonism to fungal soil borne plant pathogens of chickpea, *.Journal of Maharashtra Agricultural University* **11**(2) 236-238

Harman GE, Howell CR, Vitarbo A, Chet I, and Lorito M 2004 *Trichoderma* species - opportunistic, avirulent plant symbionts, *Nature Reviews Microbiology* **2** 43-56

Harrison BD, Mayo MA, and Balcombe DC 1987 Virus resistance in transgenic plants that expression cucumber mosaic virus satellite RNA, *Nature* **328** 799-801

He P, Chintamanani S, Chen Z, Zhu L, Kunkel BN, Alfano JR, Tang X, and Zhou JM 2004 Activation of a COI1-dependent pathway in *Arabidopsis* by *Pseudomonas syringae* type III effectors and coronatine, *Plant Journal* **37** 589-602

Hoddy E 1991 Nature's Bitter Boon, the neem tree a substitute for pesticide, *Develop Cooperation* **5** 29-30

Hossian MS and Mian IH 2005 Integrated approach for the management of Alternatia blight of cabbage seed crop, *Bangladesh Journal of Plant Pathology* **21**(1/2) 19-23

Hou Xi Lin and Hang Zhi Wu 2004 The prevention and control of plant disease and elimination of pests and seeds in non-pollution vegetable production, *China Vegetables* **1** 58-62

Huang X, Suyama K, Buchanan JA, Zhu AJ, and Scott MP 2005 A Drosophila model of the Niemann-Pick type C lysosome storage disease: dnpc1a is required for molting and sterol homeostasis, *Development* **132(22)** 5115-5124

Huq IM and Khan AZM 2007 Effect of sowing dates on the incidence of Stemphylium blight of lentil during 1998-2001, *Bangladesh Journal of Scientific and Industrial Research* **42(3)** 341-346

Jeger MJ 2000 Bottlenecks in IPM, *Crop Protection* **19** 787-792

Jellis GJ, Bond DA, and Boulton RE 1998 Diseases of faba bean, in *The Pathology of Food and pasture Legumes*. (eds) DJ Allen and JM Lenne, Wallingford, Oxon, UK: CAB International 371-410 pages

Jones AL, Johansen IE, Bean SJ, Bach I, and Maule AJ 1998 Specificity of resistance to pea seed borne mosaic potyvirus in transgenic peas expressing the viral replicase (NIb) gene, *Journal General Virology* **79** 3129-3137

Kadam PB and Raut SP 2006 Integrated disease management of spinach, *Annals of Plant Protection Sciences* **14**(1) 255-257

Kanjilal S, Samaddar KR, and Samajpati N 2000 Field disease potential of tomato cultivation in West Bengal, *Journal of Mycopathological Research* **38**(2) 121-123

Kapoor KS 2005 Integrated disease management strategies in temperate vegetables, 153-166. In *Challenging problems in horticultural and forest pathology*, Solan, India

Kapoor P and Thakur PD 2012 Efficacy of plant extracts for the management of pepper veinal mottle virus in bell pepper, *Plant Disease Research* **27** 242-243

Kaur R and Kang SS 2005 Integrated management of cucurbit mosaic, 187-194. In *Challenging problems in horticultural and forest pathology, Solan, India*

Kawchuck LM, Martin RR, and Pherson JMC 1991 Sense and antisense RNA mediated resistance to potao leaf roll virus in Russet Burbank potato plants, *Molecular Plant Microbe Interaction* **4** 247-253

Keblikas D 1997 Incidence of powdery mildew of different spring wheat varieties at various fertilization with nitrogen fertilizers, 76-79. In *Integrated Plant Protection: achievements and Problems. Proceeding of Scientific Conference devoted to the 70th anniversary of plant protection science,* Lithuania

Kishore N, Dubey NK, Srivastava OP, and Singh SK 1983 Volatile fungitoxicity in some higher plants as evaluated against *Rhizoctonia solani* and some other fungi, *Indian Phytopathology* **36** 724-726

Kumar A and Tripathi SC 1991 Evaluation of the leaf juice of some higher plants for their toxicity against soil borne pathogens, *Plant and Soil* **132** (2) 297-301

Kumar PL, Jones AT, and Waliyar F 2006 Virus diseases of pigeonpea, with particular reference to sterility mosaic and yellow mosaic diseases, in *Characterization, Diagnosis and Management of Plant Viruses*: Vol 3. *Vegetable and Pulse Crops* (eds) G P Rao, P L Kumar and R J H Penna, Texas, USA: Studium Press LLC, 235-258 pages

Luz NK 1976 Testing of fungicides to control Septoria diseases and leaf rust of wheat, *Agrononia Sulriograndense* **12**73-83.

Mayee CD 1998 Diseases of groundnut and their management, in *Diseases of Field Crops and their management*. (ed) TS Thind, National Agricultural Technology Information Centre, Ludhiana, India, 115-134 pages

Michael G, Soleem A, Wallace C M and Hylin WJ 1985 Plant species reportedly possessing pest control properties. An EWC/UH Database Resource System Institute, Honolulu, University of Hawaii.

Mishra MK, Mishra PK, Panda SK, and Mishra PJ 2003 Effect of integrated disease management practices on the yield and quality of sugarcane, *Indian Sugar* **53** (5) 337-338

Mondal KK 2004 Integrated management strategies for fuscous blight and floury leaf spot of rajmash, *Indian Phytopathology 57*(2) 135-139

Munshi GD and Sokhi SS 2000 Alternatives to fungicides in the management of diseases of vegetables, *Plant Disease Research* **15**(2) 162-167

Muthusamy K, Krishnan R, Ramanujam B, Subramanian M, Ramasamy S, and Rethinasamy V 2006 Rapid detection of Ganoderma disease of coconut and assessment of inhibition effect of various control measures by immunoassay and PCR, *Plant Protection Science* **42**(2) 49-57

Odorfer A and Pommer G 1997 Correation between growing intensity and infection with wheat diseases. 4 th Scientific Meeting on Ecological Agriculture, Rheinische Friedrich-Wilgelms- Universitat Bonn, *Scientific Institut Fur Organischen Landbau* **4** 302-307

Ongena M, Duby F, Rossignol F, Fouconnier ML, Dommes J, and Thonart P 2004 Stimulation of the lipoxygenase pathway is associated with systemic resistance induced in bean by a nonpathogenic *Pseudomonas* strain, *Molecular Plant Microbe Interaction* **17** 1009-1018

Ortman EE and Peters DC 1980 Introduction, in *Breeding plants Resistant to insects* (eds) M S Maxwell and PR Jennings, New York: John Wiley and sons Inc

Pande S, Sharma M, and Rao JN 2007 Etiology, biology, and management of diseases of food legumes, in Food Legumes for Nutritional Security and Sustainable Agriculture (ed) MC Kharkwal, Indian Society of Genetics and Plant Breeding, New Delhi, India, 1-15 pages

Pande S, Sharma M, and Rao JN 2005 Etiology, biology, and management of diseases of food legumes, in Food Legumes for Nutrional Security and Sustainable Agriculture (ed) M C Kharkwal, Proceedings of the Fourth Imternational Food Legumes Research Conference (IFLRC-IV), New Delhi, India

Pande S, Galloway J, Gaur PM, Siddique KHM, Tripathi HS, Taylor P, MacLeod MWJ, Basandrai AK, Bakr A, Joshi S, Kishore GK, Isenegger DA, Narayana RJ, and Sharma M 2006 Botrytis grey mould of chickpea: A review of biology, epidemiology and disease management, *Australian Journal of Agricultural Research* **57** 1137-1150

Pande S, Stevenson J, Narayana RJ, Neupane RK, Grzywacz D, and Kishore GK 2005 Reviving chickpea production in Nepal through integrated crop management, with emphasis on Botrytis gray mold, *Plant Disease* **89** 1252-1262

Pande S, Siddique KHM, Kishore GK, Baya B, Gaur PM, Gowda CLL, Bretag T, and Crouch JH 2005 Ascochyta blight of chickpea: biology, pathogenicity, and disease management, *Australian Journal of Agricultural Research* **56** 317-332

Peregrine WTH, Ahmad K, and Momin M 1978 Chemical and cultural control of anthracnose disease of watermelon, 82-86. *In Proceedings Plant Protection Conference, Kuala-Lumpur*

Peregrine WTH and Ahmad KB 1983 Chemical and cultural control of anthracnose (*Colletotrichum lagenarium*) in watermelon, *Tropical Pest Management* **29**(1) 42-46

Pinto YM, Kok RA, and Baulcombe DC 1999 Resistance to rice yellow mosaic virus (RYMV) in cultivated African rice varieties containing RYMV transgenes, *Nature Biotechnology* **17** 702-707

Poll JTK 1996 Effect of soil cover on the yield and quality of vegetables, Publicatie Proefstation voor de Akkerbouw en de Groenteteelt in deVollegrond, Lelystad **81** 103-109

Prashad D and Gupta S K 2012 *In vitro* efficacy of biocontrol agents and plant origin pesticides against *Xanthomonas campestris* pv. *vesicatoria*, the incitant of bacterial spot of bell pepper, *Plant Disease Research* **27** 234-236

Prashar RD, Sindhan GS, and Hooda I 1995 Effect of fertilizers in relation to phenol and carbohydrate metabolism ans susceptibility of wheat plants to flag smut disease, *Plant Disease Research* **10** 43-46

Raj H and Sharma SD 2009 Integration of soil solarization and chemical sterilization with beneficial microorganisms for the control of white root rot and growth of nursery apple, *Scientia Horticulturae* **119** 126-113

Raj H and Upmanyu S 2005 Integration of soil solarization with cruciferous leaf residues for the control of Fusarium wilt pathogen of gladiolus, 215-220. *In Challenging problems in horticultural and forest pathology, Solan, India*

Renu 1983 Fungitoxicity of leaf extracts of some higher plants against *Rhizoctonia solani* Kuhn, *National Academy Science Letters* **6** 245-246

Robak J and Nawrocka B 2005 Actually available tools for disease and pests management in organic farming, Nowosci *Warzywnicze* **41** 129-139

Rudakov OL and Rudakov VO 2000 Protection of vegetable crops in greenhouses from root rots and wilts, *Zashchita I Karantin Rastenii* **10** 27-29

Ryu CM, Farag MA, Hu CH, Reddy MS, Kloepper JW, and Pare PW 2004 Bacterial volatiles induce systemic resistance in *Arabidopsis*, *Plant Physiology* **34** 1017-1026

Saikia AK and Muniyappa V 1989 Epidemiology and control of tomato leaf curl virus in southern India, *Tropical Agriculture* **66**(4) 350-354

Sakai Y, Itho T and Tanaka A 1992 Cultural control of stem blight of asparagus (*Asparagus officinalis* L.), *Bullettin Hiroshima Prefectural Agricultural Research Centre* **55** 109-119

Samanta S, Devi NS, Mukherjee S, and Basu A 2012 Evaluation of host resistance, cultural and chemical methods to manage Alternaria leaf blight of wheat in the plains of West Bengal, *Journal of Mycology and Plant Pathology* **42** 189-193

Sano I, Nagayama A, Ogawa J and Okada Y 1997 Transgenic Potato Expressing a double stranded RNA specific ribonuclease is resistant to potato spindle tuber viroid, *Nature Biotechnology* **15** 1290-1294

Saxena M, Saxena DR, Bhale MS, and Khare MN 1998 Diseases of cowpea and their management, in *Diseases of Field Crops and their Management* (ed) T S Thind, National Agricultural Technology Information Centre, Ludhiana, India, 239-252 pages

Shamshad A, Kumar R, and Singh RP 2004 Integrated management of floral malformation in mango, *Indian Journal of Plant Protection* **32**(1) 118-122

Sharma M and Gupta SK 2003 Eco- friendly methods for the management of root rot and web blight (*Rhizoctonia solanii*) of French bean, *Journal of Mycology and Plant Pathology* **33**(3) 345-361

Sharma P, Duveiller E, and Sharma RC 2006 Effect of mineral nutrients on spot blitch severity in wheat and associated increases in grain yield, *Field Crops Research* **95** 426-430

Sharma SN, Chandel S, and Tomar M 2005 Integrated management of Fusarium yellows of gladiolus caused by *Fusarium oxysporum* f. sp. *gladioli* Snyder & Hans. under polyhouse conditions, 221-229. *In Challenging problems in horticultural and forest pathology*, Solan, India

Sharma U, Chowfla SC, Thakur PD, Handa A, and Gupta B 2005 Management of radish mosaic disease in Himachal Pradesh, 195-198. *In Challenging problems in horticultural and forest pathology*, Solan, India

Shivpuri A, Sharma OP, and Jhamaria SL 1997 Fungitoxic properties of plant extracts against pathogenic fungi, *Journal of Mycology and Plant Pathology* **27**(1) 29-31

Singh DP 2008 Plant nutrition in relation to management of diseases in wheat, 62-65. *In Integrated pest management in wheat based cropping system*, winter school

Singh RA, De RK, Gurha SN, and Ghosh A 2001 Management of Cercospora leaf spot of mungbean, *In National Symposium on pulses for Sustainable Agriculture and National Security*, New Delhi, India

Singh RA, De RK, Gurha SN, and Ghosh A 2002 Yellow mosaic virus of mungbean and urdbean, 395-408, in *IPM System in Agriculture, Key pathogens and Diseases*, (eds) RK Upadhyay, DK Arora and OP Dubey, New Delhi: Aditya Books Pvt. Ltd

Sinha AP 1998 Effect of potash and nitrogen fertilizers on brown rust of wheat, *Indian Phytopathology* **51** 100-101

Sood AK 2005 Bacterial wilt of solanaceous vegetable crops and its management, 167-182. *In Integrated plant disease management challenging problems in horticultural and forest pathology,* Solan, India

Stevens C, Khan VA, Rodriguez KR, Ploper LD, Backman PA, Collins DJ, Brown JE, Wilson MA, and Igwegbe FCK 2003 Integration of soil solarization with chemical, biological and cultural control for the management of soilborne diseases of vegetables, *Plant and Soil* **253**(2) 493-506

Tacke E, Salamani F, and Rohde W 1996 Genetic engineering of potato for broad spectrum protection against virus infection, *Nature Biotechnology* **14** 1597-1601

Taylor P, Lindbeck K, Chen W and Ford R 2007 Lentil diseases, in *Lentils: An Ancient Crop for Modern Times* (ed) SS Yadav, CAB International 291-313 pages

Theunissen J 1997 Intercropping in field vegetables as an approach to sustainable horticulture, *Outlook on Agriculture* **26** (2) 95-99

Thongbai P, Graham RD, Neate SM, and Webb MJ 1993 Interaction between Zn nutritional status of cereals and Rhizoctonia root rot severity II. Effect of Zn on disease severity of wheat under controlled conditions, *Plant and Soil* **153** 207-214

Timonin MI and Thexton RH 1951 The rhizosphere effect of onion and garlic on soil microflora, *Soil Science Society of America Proceedings*, **15** 186-189

Tiwari SN and Nayak M 1991 Activity of four plant leaf extracts against three fungal pathogens of rice, *Tropical Agriculture* **68** 373-375

Trottin CY and Capy A 2000 Integrated control in sweet pepper crops under cover, *Infos Paris* **158** 32-35

Trottin CY 1998 Botrytis in tomato crops: situation and control methods, *Infos Paris* **146** 36-39

Tsror L, Lebiush MS, and Shapira N 2003 Control of soilborne pathogens in organic tomato, 71-74. *In Colloque international tomate sous abri, protection integree agriculture biologique,* Avignon, France

Tumwine J 1999 Towards the development of integrated cultural control of tomato late blight (Phytophthora infestans) in Uganda, in Towards the development of integrated cultural control of tomato late blight Phytophthora infestans in Uganda, 152 pages

Upmanyu S, Gupta SK, and Shyam KR 2002 Innovative approaches for the management of root rot and web blight (*Rhizoctonia solani*) of French bean, *Journal Mycology and Plant Patholology* **32**(3) 317-331

Vallad GE, Qin QM, and Subarao KV 2004 Verticillium wilt of cool season vegetable crops: their distribution, impact and management, *Recent Research Developments in Plant Pathology* **3** 189-210

Van Loon LC, Bakker PAHM, and Pieterse CMJ 1998 Systemic resistance induced by rhizosphere bacteria, *Annual Review of Phytopathology* **36** 453-483

Vardi E, Sela I, Edelbaum O, Livneh O, and Stram Y 1993 Plants transformed with a cistron of potato virus Y protease (Nia) are resistant to virus infection, *Proceedings of National Academy of Science, USA* **90** 513-7517

Verma HN, Baranwal VK, and Srivastava S 1998 Antiviral substances of plant origin, in *Plant Virus Disease Control* (eds) A Hadidi, RK Khetarpal and H Koganezawa, St. Paul, Minnesota: APS Press, 154-162 pages

Viana FMP, Kobori RF, Bettiol W, and Athayde SC 2000 Control of damping-off in bean plant caused by *Sclerotinia sclerotiorum* by the incorporation of organic matter in the substrate, *Summa Phytopathologica* **26**(1) 94-97

Vishwa D, Singh RA, and Gurha SN 2004 Integrated disease management in pulse crops, in *Pulses in New perspective* (eds) Ali Masood, B B Singh, Shiv Kumar and D Vishwa,. Indian Society of Pulses research and development, IIPR, Kanpur, 324-344 pages

Wagner LJ and Flores HE 1994 Effect of taxol and related compounds on growth of plant pathogenic fungi, *Phytopathology* **84** (10) 1173-1178

Wang P, Zoubenko O, and Tumer NE 1998 Reduced toxicity and broad spectrum resistance to viral and fungal infection in transgenic plants expression pokeweed antiviral protein II, *Plant Molecular Biology* **38** 957-964

Yadav MS 2004 Integrated management of mustard diseases in Punjab, *Plant Disease Research* **19**(2) 176-178.

17

Integrated Managment of Root Rot and Wilt Complex in Scented Geranium

D. Prasad and K.P. Singh*

*Department of Plant Pathology, College of Forestry & Hill Agriculture, *College of Agriculture, G B Pant University of Agriculture & Technology, Pantnagar, UdamSingh Nagar, Uttarakhand, India.*

The Himalayan region is a reservoir of a large number of medicinal and aromatic plants. This is largely because of the diverse ecological and climatic conditions existing in the area. Uttarakhand is the major drug and aromatic oil producing state of India. Garhwal Himalayas, in Uttarakhand is a treasure house of medicinal and aromatic plants, and it's temperate and sub temperate climate is best suited for cultivation of scented geranium (*Pelargonium graveolens* L. Herit). In Uttarakhand, scented geranium is the most important temperate aromatic crop and about 150 hectare of land is under its cultivation. More than 10,000 small and marginal farmers are actively engaged in its cultivation and depend for their survivability on this aromatic crop. Essential oil extracted from geranium is becoming more and more popular in the treatment of dysentry, hemorrhoids, inflammation, heavy menstrual flows, and even cancer. The French medicinal community is currently treating diabetes, diarrhoea, gall bladder problems, gastric ulcers, jaundice, liver problems, sterility and urinary stones with this oil. The leaves are used as a form of herbal tea to de-stress, fight anxiety, ease tension, improve blood circulation and to cure tonsillitis (Peterson et al 2005, Anonymous 2006). There are several fungi, nematodes, bacteria, viruses and other physiological disturbances inflicting the production and oil contents partly or wholly to the crops (Table 1). But the most serious disease of the scented geranium is root rot and wilt complex which has been observed as the main constraint in the extensive cultivation. The disease has already plagued geranium cultivation in Tamil Nadu, Karnataka and Uttar Pradesh and is on the increase in Uttarakhand, which happens to be the third largest geranium growing state of the country. Yield losses during previous years were up to 36 per cent or even more. Geranium wilt complex, a dreaded disease of scented geranium, is threatening the cultivation of the crop in both hills and plains. Depending on the stage of growth and the prevailing environmental conditions, the symptoms of wilt complex may vary somewhat on the different crops and even on the same host. On the scented geranium plants, however, *Rhizoctonia solani* and

Fusarium solani complex appear in the form of drooping leaves, particularly on the lower ones, and the plant collapse within a week. When uprooted, roots of such plants show severe rotting, indicating that wilt results from root rot. Stems and roots turn black showing browning or blackening of the xylem vessels when split open and the plant ultimately die (Bhaskaruni Rao et al 1999). *F.* oxysporum and *R. solani* Kuhn have been reported to be associated with this disease in several states of India (Narayana et al 1986, Kalra and Parameswaran 1988) and leads to the significant losses every season (Kalra et al 1992, 1998). Geranium wilt caused by *F. oxysporum* var. *redolens* (Wollenweb.) Gordon, a dreaded disease of scented geranium, is threatening the cultivation of the crop in plains as well as hills. In Garhwal Himalayas, the incidence of Geranium root rot and wilt ranges from 24.92 to 37.33 per cent (Prasad et al 2008, Prasad and Singh, 2010, 2013). Wilt disease resulted in plant mortality of 2.6 to 72.2 per cent in one year old plantation and 0-18 per cent in three months old plantation (Bhaskaruni Rao et al 1999).

Table 1. Most common diseases of scented geranium

Diseases	Causal Organisms
	Fungal diseases
Root rot	*Rhizoctonia solani and Fusarium solani; F. oxysporum*
Stem rot	*Rhizoctonia solani and Fusarium solani*
Leaf spot	*Alternaria solani and Botrytis sp.*
Anthecnose	*Colletotricum acutatum*
Tip rot disease	*Gleosporium* spp.
Bacterial diseases	
Bacterial leaf spot	*Xanthomonas and Pseudomonas*
Bacterial leaf spot	*Ralastonia sp.*
	Plant parasitic nematodes
Root knot disease	*Meloidogyne hapla* and *M. incognita*
	Virus diseases
	Many viruses occur singly and in various combinations in plant tissue
	Nutritional imbalance
Edema	Physiological disorder is favoured by cool, cloudy weather
Leaf Burn	Iron excess
	Disease caused by insects
Termite attack	Termite

Prevalence and severity

Extensive surveys were carried out on pre-designated circular routes covering nearly 6,000 hectares of four principle geranium growing districts in the Garhwal valley, viz., Tehri, Uttarakashi, Dehradun and Chamoli during the growing periods from summer to winter of 2005 to 2007. Root rot severity was recorded on the basis of the percentage of root tissue affected on a 0-4 scale, where 0 = no disease 1=<10%, 2 = 10-25%, 3 = 25-50% and 4 = >50% root tissue affected.

Based on three year data of the scented geranium growing area of Garhwal hills, mortality of the plants has been observed up to 72 per cent due to wilt complex. The experimental as well as commercial fields of geranium at Ranichauri, on farmers and NGO demonstration plots were found to be severely affected by stem and root rot wilt (RRW) during summer and winter season. In hilly area (650- 2400 m asl), the geranium terminal cutting were planted in nursery bed during the month of February to March and September to October where they showed maximum stem rot incidence (60.55%) and plant mortality (17.21%) in winter season crops. Well rooted cuttings transplanted to the field during March - April and October - November crops showed maximum root rot and wilt complex incidence (38.33%) and plant mortality (12.22%) in summer season crops (Fig. 1). Three years pooled analysis revealed that the plant mortality was initially more severe in the summer planted crop at Ranichauri experimental block. It has been observed that the plant root system is highly susceptible to *R. solani* at the initial stages of growth (Shephard and Wood 1963, Bateman and Lumsden 1965) and at higher temperature (Khan 1950, Bateman and Dimock 1959, Mildenhall and Williams 1973, Benson and Baker 1974, Hollins et al 1983, Shehata et al 1984, Elango 1986, Kalra et al 1992). This was clearly reflected in the summer planted crop of 2006 and 2007 during which all the plantation of geranium performed badly due to high mortality rates. Multiple regression equation between root rot mortality and month variables exhibited similar relationships (R^2 = 0.965 and 0.950) in summer and winter crops. Similar results have been obtained in the NGO demonstration plot at Khari also.

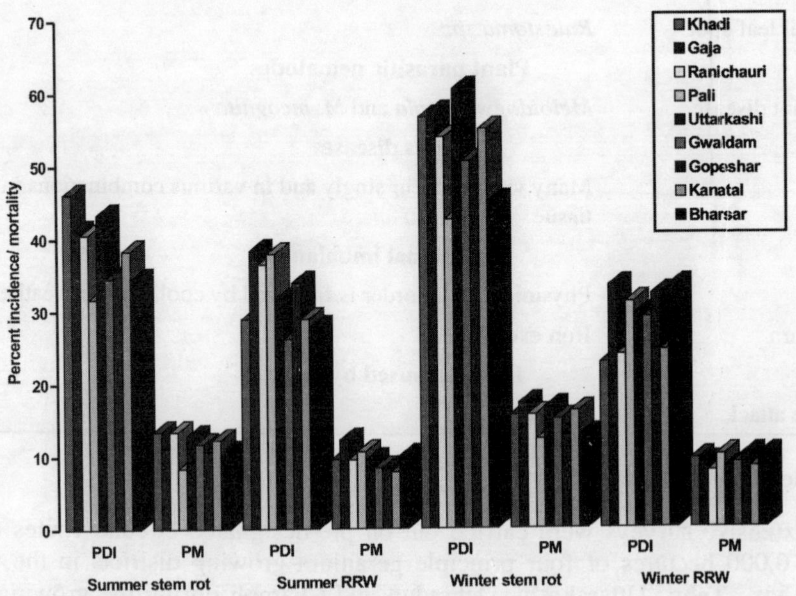

Fig. 1. Prevalence and severity of wilt complex disease of Geranium in Garhwal Himalayas

Our observations agree with that of Kalra et al (1992) who reported that mean mortality data over the three clones are more severe in the summer crop. It is evident (Fig. 2) that the average stem rot mortality was comparatively low (16-20%) in summer crop but increased 18 to 26 percent in winter crop by the 49 days after transplanting at experimental block, Ranichauri. The percentages of root and stem rot mortality in plantation of geranium were initially low but had increased to 37.35 per cent (summer crop) and 26 per cent (winter crop) at Ranichauri. The constant value appeared positive (y = 0.285x + 8.571, y = 0.336x + 3.238) reflected by high disease pressure and the relationship remained varied (97.77 and 92.46%) from season to season. Multiple regression equation between stem rot mortality and days after transplanting exhibited strong relationship among the different compo-nents during both the year. The high mortality percentage at Ranichauri may be due to low temperature and high humidity prevailing in the mid hilly regions (1600 to 1700 m asl.) of district Tehri (Prasad and Singh, 2010, 2013). Therefore, it should not be recommended for cultivation in areas where root rot and wilt is a major problem.

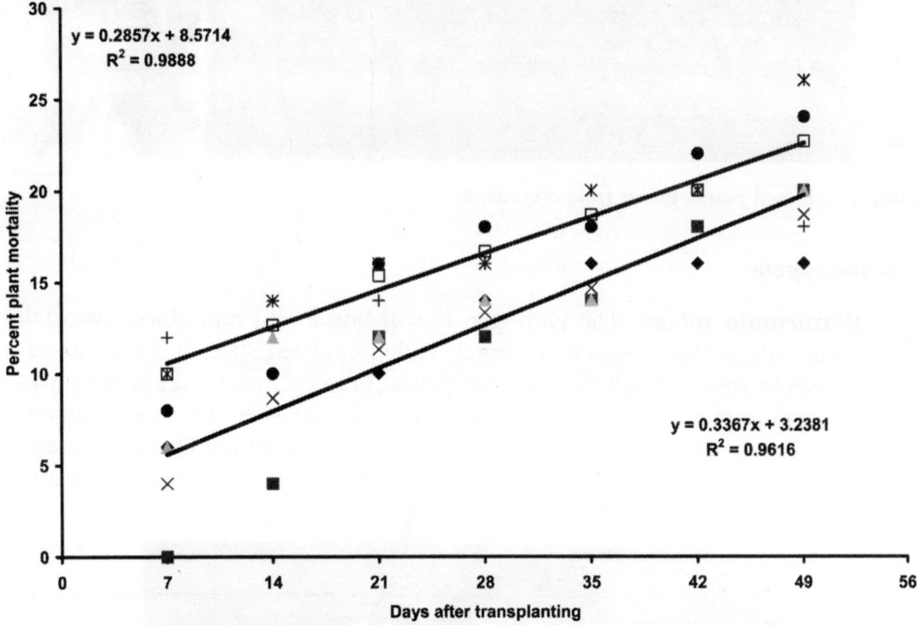

Fig. 2. Percentage plant mortality due to stem rot in winter and summer nursery crop

Symptoms: The symptoms of wilt complex may vary in different crops and even on the same host. In initial stages of infection, infected plants show yellowing of the leaves. Later on such plants show sudden drooping, withering and drying of leaves followed by complete drying of the plant or some of its branches. Due to stem rot at the base of the plant, whole plant wilts and die. Lesions developed on the stem at or below the soil line, and their edges often had pink or red discolorations. The main roots and base of the stem tissues are blackened either uniformly, or especially in the early stages. If the stem is examined after breaking it open, the black streaks may be traced up to a height of several inches on the stem. The tap roots of young plants at first show a slightly reddish discoloration, this later becomes darker red to brown and larger, more or less covering the tap root and the stem below the soil line without a definite margin, or appearing as streaks extending up to the soil line. Longitudinal fissures appear along the main root, while the small lateral roots are killed. Plant growth is generally retarded, and in dry weather the leaves may turn yellow and even fall off. Sometimes, infected plants develop secondary roots and a large number of rootlets just below the soil line (Plate 1).

Plate 1. Infected plants under field conditions

The pathogens

Rhizoctonia solani: The pathogen is soil borne and reproduces asexually and exists primarily as thin, vegetative strands called hyphae (Plate 2). *Rhizoctonia solani* Kuhn (Teleomorph: *Thanatophorus cucumeris* (Frank) Donk) is a cosmopolitan plant parasite fungus that can attack a broad range of host (Ou, 1985) and therefore, is regarded as an unspecilized fungus. *R. solani* forms part of the Deuteromycetes class with perfect stage *Thanatephorous* which belongs to Basidiomycetes class and order Ceratobasidiales.

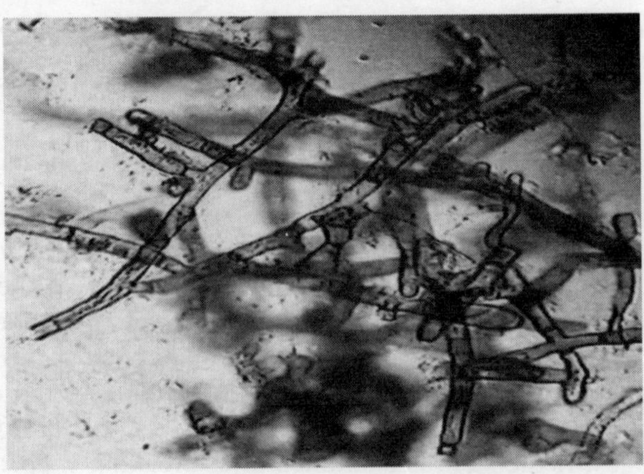

Plate 2. Hyphae of Rhizoctonia solani

Morphological and cultural characters: The mycelium of *R. solani* is septate and the colour of hyphae varies from hyaline when young to brown on maturity. The fungus produces a fast growing stout mycelium with septate hyphae in artificial culture media. Young colonies may look white but older mycelium is invariably have some shade of brown (no other colour, a distinctive feature). The diameter of hyphae ranges between 5 and 14 μm. The septa are typically of dolipore type. The lateral branches form the main hyphae are invariably constricted at the point of origin and a septum occurs in the branch near the junction with the main axis. In young advancing hyphae, the branching almost invariably occurs near the distal septum but it may occur at any point in older hyphae. The dimension of vegetative hyphae were found to vary with the age and also on different media (Pelter 1916). Dimensions ranging between 3-17 μm have been recorded by various workers (Duggar 1915, Matz 1921, Valdez 1955, Flentje and Saksena 1957, Saksena 1960, Hansen 1963). In addition to ordinary vegetative hyphae, *R. solani* produces barrel shaped pyriform cells called as moniloid cells. The formation of moniloid cells has been illustrated by Duggar (1915) and Townsend and Willelts (1954). However, Saksena and Vaartaja (1961) referred them as chlamydospores.

Fusarium solani: The soil borne fungus *Fusarium solani* (Mart.) Sacc. (Teleomorph: *Nectria haematococca* Berk. et Br.) is one of the most frequently occurring causal agent of root rot on many host plants. *F. solani* belongs to subdivision Deuteromycotina and form-class Deuteromycetes (Alexopoulus and Mims 1979). It is a very variable specie having both saprophytic and parasitic forms. Now a days, *F. solani* is defined as a complex taxons having a wide morphological and biologically active spectrum, containing the different genetical mating groups. Three types of asexual spores are produced by the fungus microconidia (Plate 3), macroconidia, and chlamydospores (Agrios 1988). Diseases caused by *Fusarium spp.* and its several formae speciales on agricultural crops include rot of roots, root crowns, vascular wilts and post-harvest decay. During their development on host tissues or in culture, fungi can produce a wide range of toxic compounds with varied biochemical structures and modes of action. These include polypeptides, glycoproteins, amino acid derivatives, polyketides, terpenoids, sterols and quinones (Kono et al 1981, Stoessl 1981).

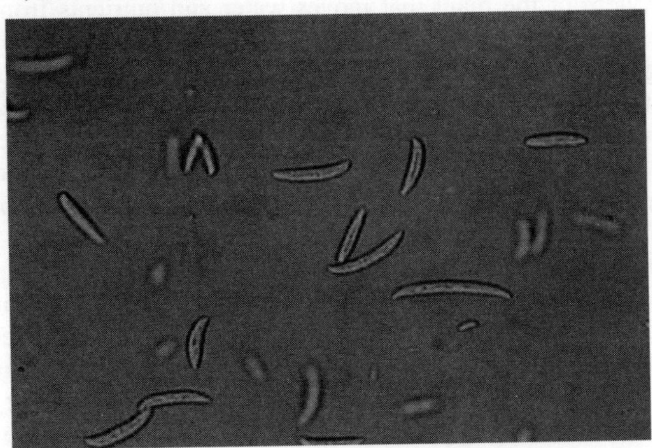

Plate 3. Micro and macro conidia of *Fusarium solani*

Morphological and cultural characters: Hyphae of *Fusarium solani* is septate and hyaline. Conidiophores are simple (non-branched) or branched monophialides (phialides with a single opening). Macroconidia are moderately curved, stout, thick-walled, usually 3-5 septate, measure 4-6 x up to 65 μm long, and are borne on short conidiophores that soon form sporodochia. Microconidia are borne from long monophialides, are one to three-celled, 2-5 x 8-16 μm long, and occur in false heads only (in clusters of conidia at the tip of the phialide). Macroscopic morphology may vary significantly on different media, Colonies are woolly to cottony with cream to white aerial mycelium and a cream reverse. The macroconidia of *F. solani* tend to be cylindrical in the central area, the walls appearing to be parallel and are comparatively heavy and strong looking. Macroconidia are not often strongly curved; some individuals being nearly straight. They are borne on comparatively long phialides, produced in sporodochia and sometimes are so numerous in cultures that they merge to form a large mat on the surface. These spores have blunt rather than sharply pointed ends, although foot cells are usually quite evident. Very often the macroconidia contain an insoluble blue, green or yellowish pigment, which appears to be firmly attached to the inside of the conidial wall. Cultures appear as blue, green or creamy, depending on spore colour. Microconidia appear either numerously or sparsely in culture, and the same is true of chlamydospores. Microconidia oval in shape, but some are quite wide, almost approaching a sphaerical form.

Disease cycle: Both fungi survive in soil or on infected root stock. *R. solani* survives in the form of sclerotia while *F. solani* overwinters as chlamydospores in soil. Viability of both fungi in soil is greatly reduced by exposure to hot summer temperatures. The fungus mainly survives as spore and /or mycelia in diseased plant debris left in the field. The viability of fungus in soil is influenced by environmental conditions. Wilt symptoms usually appear when the plants are about 5-6 weeks old. The wilting and senescence occur as a result of water stress induced by the combined effect of mycelial growth in the xylem, fungal toxins and vascular occlusion by the first in an attempt to prevent systemic spread of fungus.

***Rhizoctonia solani*:** The pathogen is soil borne surviving in soil mainly as sclerotia formed by the hyphal and conidial cells. Infection is always through injured secondary or tertiary roots. After penetration of pathogen in host tissue it travels through the xylem of the plant that moves water and nutrients from the root to the crown of the plant. One of the chief methods of distribution of the pathogen is the use of cuttings raised in infested soil. When cuttings with broken roots are planted in infested soil the fungus grows quickly through the wounds and slowly progresses up to the stem through xylem vessels. The optimum temperature for infection and lesion development on geranium root is 20ºC. The damage to the host is less at 24ºC. Soil moisture is regarded as secondary to soil temperature in the development of root rot and stem rot. Wilt disease develops more in sandy soil than clay. The viability of sclerotia in soil is influenced by environmental conditions. In undisturbed land, sclerotia are present only up to 2 cm depth. Population of sclerotia in ploughed land is higher at 6-12 cm depth than in the upper layers. Survival of sclerotia is better in upland (dried) fields than in flooded fields. After infection, mycelium moves up the plant by surface hyphae and develops new infections/structures over the entire plant. The pathogen becomes aggressive when suitable temperature and moisture conditions are available. Exposure to sunlight inhibits infection, mycelia development, and sclerotia formation. Optimum temperature for growth of the fungus and scletotia formation is about 23-35°C and humidity 96-97 per cent. If the temperature of the soil remains at 37°C for more than few days the fungus is killed.

Fusarium solani: The pathogen is soil borne, initially surviving in diseased crop debris and then as free chlamydospores. The fungus has been found up to depth of 1 meter or more below the soil surface but infection is most severe in first 40 cm. The conidia and chlamydospores germinate in the rhizosphere of 1-3 weeks old plants by germ tubes. These form hyphal strands on the root surface and grow along it forming spores. The fungus enters the roots directly, grows through the cortex to the stele and invades the xylem where it starts sporulating. It becomes systemic and proliferates in the xylem vessels. The conidia are transported upwards in the transpiration stream. The most abundant and commonly produced spores under all conditions are micro-conidia, which are one - or two -celled. Macro-conidia are three- or more celled, with pointed and curved ends; they are frequently seen on the surface of infested plants killed by the fungus (Gonsalves and Ferreira 1993). Micro and macro-conidia are produced for short-term survival and dispersal of the fungus. Chlamydospores are thick walled spores, filled with lipid-like material and are able to survive in soil under adverse climatic conditions in the absence of host plant (Agrios 1997, 1988). Fusaric acid, a toxin responsible for wilt, is known to be produced by the fungus in the plant tissues as well as in the rhizosphere and in soil. The disease is favored by high soil temperature and low to moderate soil moisture. Optimum temperature for development of geranium wilt is 20- 27°C, the maximum being 31°C. However, temperatures of 25°C to 30°C favour germination, germ tube elongation, growth and sporulation of the fungus (Upadhyay and Rai 1984). Although different reports suggests soil moisture levels of 20-40 per cent moisture holding capacity or 80-90 per cent saturation favourable for the disease and a warm and wet weather favors activity of the pathogen, there is no significant correlation between geranium wilt incidence and soil moisture.

Disease management

Root rot/ wilt complex generally found widely distributed in moderate to high intensity in Garhwal Himalayas. The disease was reported as a serious threat to the crop due to monoculture and inappropriate management practices followed by the farmers.

Organic amendment: Soil amendments are one of the viable and eco- friendly propositions available in modern agriculture which can considerably minimize the disease incidence if applied at least 30-40 days before transplanting and allowed to decompose. Use of soil solarization with mustard and neem cake (1 kg/ plot) amendments is popular and commercially viable approaches for management of root rot and wilt complex (Prasad, 2010) (Fig. 3).

Organic amendments, such as animal's manures and compost, were commonly used in agricultural production for their fertility value prior to the availability of chemicals fertilizers. In India, soil solarization with integration of organic wastes has been studied for controlling chickpea, tomato, pigeonpea, castor and gladiolus wilt (Arya and Mathew 1993, Dange 2003, Anand and Gautam 2006, Usha Rani and Satheesh 2007). These amendments also provide other benefits such as improved plant health due to reduction of pathogens (Lazarovits et al 2000). This scientific technique is gaining importance as another alternative to chemical fungicides in recent times. Plant debris, green manures, farm yard manure (FYM), compost, oil cakes and fertilizers are known to improve crop productivity by improving nutrient status and soil tilt besides, increasing the microbial activity in the rhizosphere to suppress certain soil-borne diseases (Padmodaya 2003). A reduction in infection by *Rhizoctonia* spp. was also reported by amending the soil with saw dust, wheat straw and rice hulls (Satischandra et al 1979). Population of *Fusarium oxysporum* reduced by 59-100 per cent at varying depth of 0-25 cm and *Verticillium* wilt reduced 25-95 per cent in egg plant and tomato with increased plant growth and yield (Katan et al 1976, 1990).

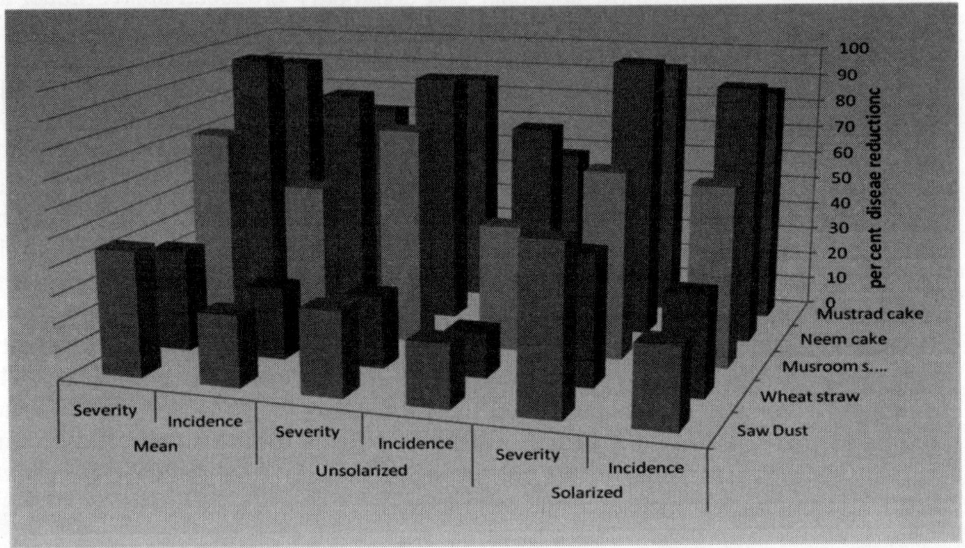

Fig. 3. Disease management through organic amendments in solarized and unsolarized soil

Plant extract: In modern agriculture, use of plant extracts is non–chemical and eco-friendly approach which can considerably minimize the disease incidence. Many plants and plant products showed pest control compounds and properties (Fig 4) (Dwivedi and Shukla 2000, Behura et al 2000, Bowers and Locke 2000, Liu et al 2001, Sungeun et al 2001, Ramanathan et al 2004). Such compounds being biodegradable and selective in their toxicity are considered valuable for controlling some plant diseases (Singh and Dwivedi 1987, Bhatanager et al 2004).

The effect of an extract on the target pathogen population over time also is an issue when determining whether repeated root treatment is necessary throughout the growing season. Botanical like neem formulated products have been shown to be effective in reducing soil population of *Fusarium*; *Rhizoctonia* and other pathogens and increased productivity (Jayarajan et al 1986, Bhattacharya and Pramanik 1998, Tomar and Chandel 2006). The least pathogen population was found in *haldi*, (*Curcuma longa*) and Nimbecidine root treated plots with reduced plant mortality (97.89%) and increased survivability (97.81%) (Prasad 2010) (Fig. 4). Extracts of kalabasa (*E. adenophorum*) are also found most effective natural fungicide against *R. solani* and *F. solani* (Prasad 2010). The inhibitory effect of the plant extracts on the radial growth might be attributed to the presence of some antifungal ingredients in them. The antifungal activity of *C. longa* extract and neem formulated nimbecidine are potential, safe and economical alternative to prevent geranium root rot and wilt complex.

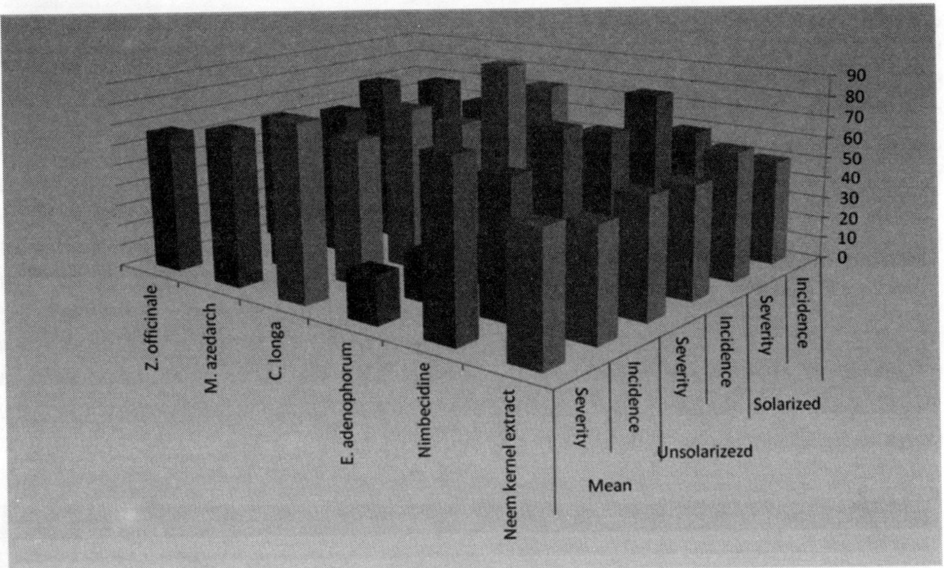

Fig. 4. Disease management through use of botanicals

Biocontrol rol : This is one of the viable and ecofriendly propositions available in modern agriculture which can considerably minimize the disease incidence (Cook, 1985). It is a natural phenomenon and nature's own way to check diseases. In the recent years, several biocontrol agents such as *Trichodrma, Pseudomonas, Bacillus, Aspergillus* etc. are available in the market and research centres. All the fungal and bacterial biocontrol agents produced considerable amount of volatile antibiotic compounds which inhibited the mycelial growth of most of the pathogens (Singh et al 1993, Bahadur et al 2007). *F. solani* was inhibited (50.00 and 55.60%) independently the *P. fluorescens* and *Bacillus* sp. tested, while *R. solani* was inhibited (57.77 and 59.26%) by non-volatile metabolite (Prasad 2010) (Fig. 5). Allippi and Monaco (1994) observed hyphal vacuolization and deformation in *R. solani*, *F. solani* and *Pythium ultimum* as the result of treatment with a *Bacillus subtilis* strain which secreted a volatile metabolite with fungicidal properties. It also secretes several antifungal metabolites such as subtiline, bacitracin, bacillin and bacillomycin, which belong to the iturine family (Allippi and Monaco, 1994). The inhibition of pathogen by the volatile antibiotic compounds of antagonists is more advantageous as they may diffuse through air filled pores in soil and actual contact between pathogen and antagonists may not be necessary for inhibition of pathogen. Diffusible (non-volatile) metabolite activity of the antagonists was more potent against *R. solani* and *F. solani*. Species of *Trichoderma*, *Pseudomonas* and *Bacillus* are known to produce volatile and non- volatile antibiotics that are responsible for the antagonism of *R. solani* and *F. solani* (Allippi and Monaco, 1994, Dubey 1995, Silva et al 2001, Montealegre et al 2003, Mishra and Sinha 2004). *T. harzianum* and

T. viride and *T. virens* have been reported to be most effective antagonists against *R. solani, F. solani* and *F. oxysporum* (Rini and Sulochana, 2007, Maliesh et al 2008). The efficacy of *T. harzianum* (*Th*) and *P. fluorescens* (*Pf*) for the management of root rot and wilt complex of geranium was evaluated and the treatment with *Th* + *Pf* (10 g/l) showed least incidence and severity and gave 70.33 and 83.51 per cent disease control in solarized plots (Prasad 2010). The potential of *Trichoderma* species as biocontrol agents of plant diseases was first recognized in the early 1930s (Weindling 1932), and in subsequent years, control of many diseases has been reported (Howell 2003). This has culminated in the commercial production of several *Trichoderma* species for the protection and growth enhancement of a number of crops in the United States (Gardener and Fravel 2002), in India, Israel, New Zealand, and Sweden (Howell 2003, Kim et al 2008). Maintaining an adequate population level of a biocontrol agent at the target site is essential for successful disease control (Weller 1988). The use of biocontrol agents (BCAs) offers good control of many soil borne pathogens like *Fusarium* and *Rhizoctonia* spp. (Mukhopadhaya and Mukherjee 1996, Sharma 2003, Sharma and Gupta 2003, Gaur et al 2005, Bautista 2007, Singh and Chaurasia 2008). Antagonistic bacteria especially fluorescent *Pseudomonas* and *Bacillus* sp. have also been used against *Rhizoctonia solani* and *Fusarium oxysporum* (Kim et al 2003, Rini and Sulochana 2007, Abeysinghe 2007, 2009) on different crops.

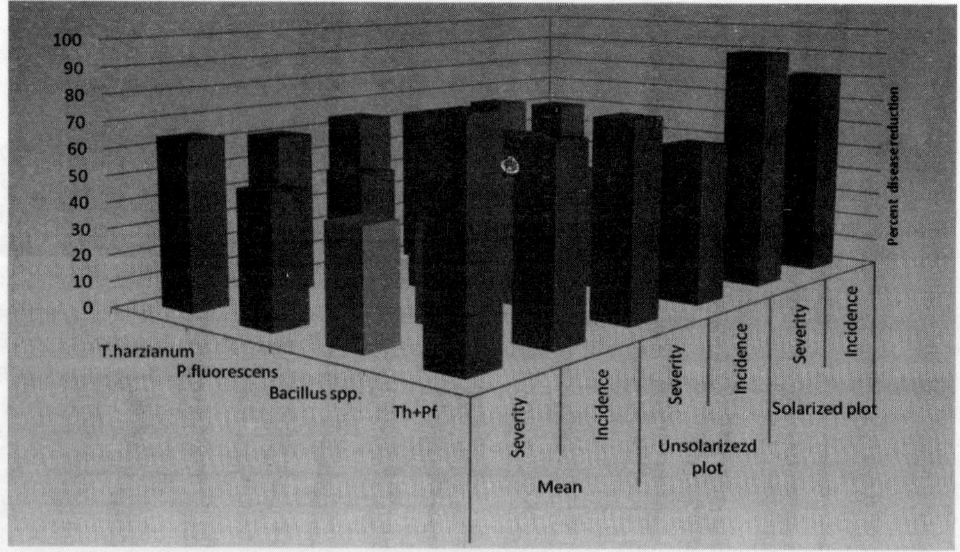

Fig. 5. Disease management through use of antagonists

Chemical control: Management of this soil borne disease is difficult through application of fungicides alone as the pathogens have wide host range. For *R. solani*, many chemicals like benomyl, captan, thiram, thiophanate methyl, carbendazim, carboxin, terrachlor, pencycuron, iprodion + thiram have been reported effective both as seed treatment and foliar sprays (Mathew and Gupta 1996, Prasad et al. 2011) (Fig. 6). Bavistin, Benlate Topsin- M and Tecto applied as seed treatment, seed soaking and soil drenching were reported effective in the management of *F. solani*.

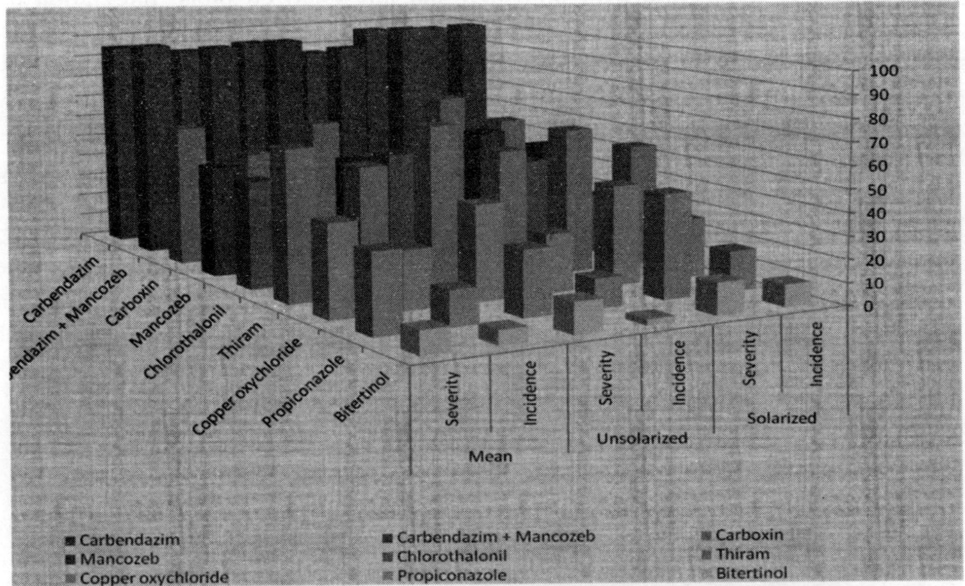

Fig. 6. Disease management through use of chemicals

Integrated management: Among the proposed alternatives, integration of soil solarization technique with other management options offers a rich opportunity to manage soil borne pathogens. Combining solarization with other treatment (Antagonist + amendment + fungicides) could increase the performance of biocontrol agents or other treatments (Yucel et al 2000, Mukherjee and Tripathi 2000, Mukherjee et al 2001, Misra 2006, Nawar 2007) and could be effective against wilt complex (*F. solani*, *R. solani*, *Sclerotium rolfsii*, *F. oxysporum*) of different crops. In the absence of single effective methods, integration of different methods is highly desirable for the management of this disease. Based on the *in vitro* and *in vivo* studies on efficacy of fungicides, bioagents, plant extracts and soil amendments against the pathogens, the best treatments were then integrated with soil solarization to find out their combined effect on the incidence of root rot/ wilt complex.

Integration of soil solarization with amendments, fungicides, plant extracts and formulation of *Th-Pf* proved effective in reducing the incidence of root rot/ wilt complex of geranium. Soil solarization with prior amendment of neem cake and cutting dip treatment in Sixer was the most effective and provided 100 per cent control of the wilt complex, followed by 88.25 per cent control in soil solarization with prior combination of mustard cake and cutting dip in Sixer (Table 2). Soil solarization in combination with mustard cake/neem cake amended plots and cutting dip in plant extract (*Curcuma longa*) showed 82.34 per cent control of the wilt complex in comparison to un- amended and un- solarized plots (Fig. 7). Cuttings treatment in Sixer followed by plantation in amended soil and prior solarized soil was the effective against wilt complex of geranium. However, treatment of soil solarization alone was found more effective (35.30%) than soil amendment with farmyard manure (5.87%) alone. Among bioagents, geranium rooted cuttings dip in *Th-Pf* formulation along with soil amendment of neem cake and mustard cake treatments followed by plantation in solarized plots provided 64.69 and 58.82 per cent respectively, protection over untreated and un- solarized plots, Integration of soil solarization with other methods of disease management have been reported more effective. Root dip of strawberry runners and gladiolus corm dip in SAAF/ Quintal with prior amendment of cabbage leaves followed by plantation in solarized plots resulted in complete control of root rot / wilt complex (Bhardwaj and Gautam 2004, Anand and Gautam 2006).

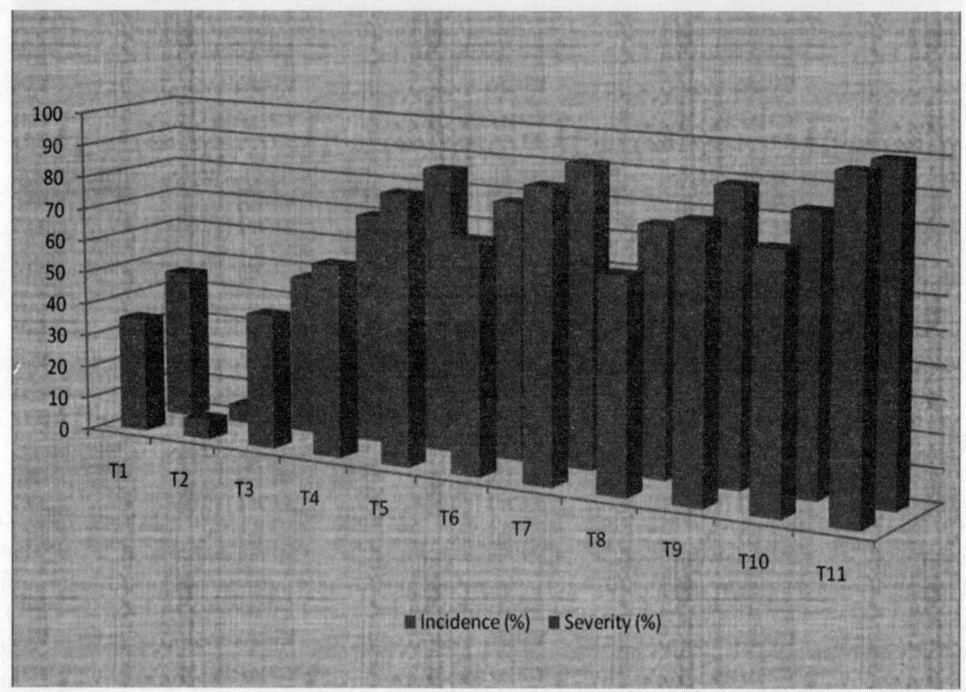

Fig. 7. Effect of different treatments on rose scented geranium in field condition

The area under disease progress curve and apparent infection rate (r/ unit/day) values, reflected the real disease stress. Area under disease progress curve for 12 treatments (Table 2) indicated that the lowest AUDPC and infection rate (00.00) was observed in geranium cutting treatment in Sixer (carbendazim + mancozeb) with prior amendment of neem cake followed by plantation in solarized plots. Area under disease progress curve (AUDPC) was less than 100 in treatments which includes soil solarization with prior amendment of neem cake/ mustard cake and rooted cutting dip in fungicide/ biopesticide/ antagonist/ plant extract. Soil amendment of farmyard manure, soil mulch with single polytheene and soil solarization with amendment of FYM showed the maximum AUDPC value 214.7 to 347.5. soil solarization with prior amendment of mustard cake and cutting dip in Sixer, soil solarization with amendment of mustard cake/ neem cake and rooted cutting dip in *Curcuma longa* showed a moderate level AUDPC value 039.0 to 054.6. Infection rate was higher in FYM amended plots (0.20). The results thus, suggested that geranium rooted cuttings dip treatment in Sixer (0.2%) with prior amendment of neem cake (2 kg/plot) followed by plantation in solarized plot (30 days) can successfully control of root rot/ wilt complex of geranium in Garhwal Himalayas.

Table 2. Integrated management of wilt complex of geranium

S.N.	Treatments	Incidence (%)	Disease control (%)	Severty (%)	Disease control (%)	AUDPC	Infection rate (r)/ unit/day
1	T1 (Mulch with single polythene (25 μm))	22.91	35.30	10.41	45.95	222.5	0.013
2	T2 (FYM)	33.33	05.87	18.22	05.39	347.5	0.020
3	T3 (Solarization + FYM)	20.83	41.17	09.89	48.65	214.7	0.012
4	T4 (Solarization + Mustard cake + *Th-pf*)	14.58	58.82	05.72	70.30	097.5	0.011
5	T5 (Solarization + Mustard cake + *Curcuma longa*)	06.25	82.34	02.60	86.50	054.6	0.011
6	T6 (Solarization + Mustard cake + Nimbecidine)	10.41	70.60	04.16	78.40	081.9	0.013
7	T7 (Solarization + Mustard cake + Sixer)	04.16	88.25	01.56	91.90	039.0	0.009
8	T8 (Solarization + Neem cake + *Th-pf*)	12.50	64.69	04.68	75.70	089.7	0.010
9	T9 (Solarization + Neem cake + *Curcuma longa*)	06.25	82.34	02.08	89.20	054.6	0.008
10	T10 (Solarization + Neem cake + Nimbecidine)	08.33	76.47	03.12	83.80	085.8	0.009
11	T11 (Solarization + Neem cake + Sixer)	00.00	100	00.00	100	000.0	0.000
12	T12 (Control)	35.41	-	19.26	-	394.4	0.025
SEm ±		2.45		0.93		21.74	0.35
CD ($P = 0.0$ 5)		7.19**		2.75**		63.78**	0.10*

Future thrust

Geranium is an important crop for mid-himalayan regins of the country. During cultivation, it is infected with wilt complex which is difficult to manage. If some degree of resistance found in any cultivar may provide the boost in integrated disease management.

References

Abeysinghe S 2007 Biological control of *Fusarium solani* f.sp. *phaseoli* the causal agent of root rot of bean using *Bacillus subtilis* CA 32 and *Trichoderma harzianum* RU01, *Ruhuna Journal of Science* **2** 82-88

Abeysinghe S 2009 Effect of combined use of *Bacillus subtilis* CA 32 and *Trichoderma harzianum* RUO1 on *Solanum melongena* and *Capsicum annuum*, *Journal of Plant Pathology* **8** 9-16

Agrios GN 1988 *Plant Pathology,* 3rd. ed. Academic Press, Inc, New York

Agrios GN 1997 *Plant Pathology*, San Diego, CA, USA: Academic Press

Alexopoulus CJ and Mims CW 1979 *Introductory Mycology*, 3rd ed. Wiley Eastern Limited

Allippi A and Monaco C 1994 Antagonismo in vitro de especis de Bacillus contra Sclerotium rolfsii, Fusarium solani. Revista de la Facultad de Agronomia, La Plata **70** 91-95

Anand S and Gautam H R 2006 Use of solarisation, biocontrol agents, fungicide corm dip and soil amendments for management of *Fusarium* wilt pathogen of gladiolus, *Journal of Mycology and Plant Pathology* **36** 201-204

Anonymous 2006 Geranium oil, www.frontiercrop.com.auracacia

Arya A and Mathew DS 1993 Studies on rhizosphere microflora of pigeonpea: qualitative and quantitative incidence of microorganisms after solarization, *Indian Phytopathology* **46** 151-154

Bahadur A, Singh UP, Sarma B K, Singh D P, Singh KP and Singh A 2007 Foliar application of plant growth-promoting rhizobacteria increases antifungal compound in pea (*Pisum sativum*) against *Erysiphe pisi, Mycoobiology* **35** 129 -134

Bateman DF and Dimock AW 1959 The influence of temperature on root rots of poinsettia caused by *Thielaviopsis basicola*, *Rhizoctonia solani* and *Pythium ultimum, Phytopathology* **49** 641–647

Bateman DF and Lumsden RD 1965 Relation of calcium content and nature of pectic substances in bean hypocotyls of different ages to susceptibility to an isolate of *Rhizoctonia solani*, *Phytopathology* **55** 734–738

Baustista G, Mendoza H and Uribe D 2007 Biocontrol of *Rhizoctonia solani* in native potato (*Solanum phureja*) plants using native *Pseudomonas flourescens*, *Acta- boil Colombia* **12** 19-32

Behura C, Ray P, Rath CC, Mishra RK, Ramachandra DS and Charyulu J K 2000 Antifungal activity of essential oils of Curcuma longa against five rich pathogens *in vitro*, *Journal of Essential Oil* **3** 79-84

Benson DM and Baker R 1974 Epidemology of *Rhizoctonia solani* pre-emergence damping off of radish, *Phytopathology* **64** 957–962

Bhardwaj H and Gautam HR 2004 Mulching with transparent polyethylene and rot dip in fungicides for the management of collar and root rot of strawberry, *Indian Phytopathology* **57** 48-52

Bhaskaruni R, Rajeswara, Bhattacharya AK, Singh Harikesh B and Mallavarapu GR 1999 The impact of wilt disease on oil yield and quality of two cultivars of rose- scented Geranium (*Pelargonium graveolens*), *Journal of Essential Oil Research* **11** 769–775

Bhatnagar K, Sharma BS and Cheema HS 2004 Efficacy of plant extract against *Fusarium oxysporum* f.sp. *cumini* wilt in cumin, *Journal of Mycology and Plant Pathology* **34** 360-361

Bhattacharya J and Pramanik M 1998 Effect of different antagonist rhizobacteria and neem products on clubrootof crucifers, *Indian Phytopathology* **51** 87-90

Bowers JH and Locke JC 2000 Effect of botanical extracts on the population density of *Fusarium oxysporum* in soil and control of *Fusarium* wilt in the greenhouse, *Plant Disease* **84** 300-305

Cook RJ 1985 Biological control of plant pathogen: theory of application, *Phytopathology* **75** 25

Dange SRS 2003 Wilt of castor- an overview, *Journal of Mycology and Plant Pathology* **33** 333-339

Dubey SC 1995 Evaluation of fungal antagonists against *Thanatephorus cucumaris* causing banded blight of rice, *In 1st international symposium on Rhizoctonia ISR'95* at Leeuwenhorst Congress Centrum Noordwijkerhout, Netherlands

Duggar BM 1915 *Rhizoctonia crocorum* (Pers.) D.C. and *Rhizoctonia solani* Kuhn (*Corticium vagum* B. and C.) with notes on other species, *Annals of Missouri Botanical Garden* **2** 403-458

Dwivedi BP and Shukla DN 2000 Effect of leaf extract of some medicinal plants on spore germination of some *Fusarium* species, *Karnataka Journal of Agricultural. Science* **13** 153-154

Elango F 1986 The role of high soil temperatures in the damping off of true potato seedling in the lowland tropics, *Tropical Agriculture* **63** 66–67

Flentje NT and Saksena HK 1957 Studies on *Pellicularia filamentosa* (Pat.) Rogers.II. Occurrence and distribution of pathogenic strains, *Transactions of British Mycological Society* **40** 95-108

Gardener MBB and Fravel DR 2002 Biological control of plant pathogens: Research, commercialization, and application in the USA, *Plant Health Progress*, Online publication in doi: 10.1094/php-2002-0510-01-RV

Gaur RB, Sharma RN and Singh V 2005 Manipulations in the mycoparasite application techniques against *Rhizoctonia* root rot of cotton, *Indian Phytopathology* **58** 402 - 409

Gonsalves A.K. and Ferreira SA 1993 *Fusarium oxysporum,* http://www.extento.hawaii.edu/kbase/crop/Type/f_oxys.htm.

Hansen LR 1963 Skarp oyeplekk pa korn parasapet av *Rhizoctonia solani* Kuhn. *Meld. Narges Landbrittojsk* **42** 1-12

Hollins TW, Jellis GJ and Scott PR 1983 Infection of potato and wheat by isolates of *Rhizoctonia solani* and *Rhizoctonia cerealis, Plant Pathology* **32** 303–310

Howell CR 2003 Mechanism employed by *Trichoderma* species in the biological control of plant diseases: the history and evolution of current concept, *Plant Disease* **87** 4 - 10

Jayarajan R, Doraiswamy S, Bhaskaran R, and Jayarajan S 1986 Proceedings III Neem Conference, Nairobi, 635-644 pages

Kalra A and Parmeswaran TN 1988 Root rot and wilt of *Pelargonium graveolens, Indian Journal of Plant Pathology* **6** 82-83

Kalra A, Parameswaran T N and Ravindra N S 1992 Influence of planting date on plant losses and yield responses of geranium (*Pelargonium graveolens*) to root rot and wilt, *Journal of Agricultural Science* **118** 309-314

Kalra A, Parmeswaran TN and Ravindra NS 1998 Stem rot of cutting of scented geranium caused by *Rhizoctonia solani, Journal of Mycology and Plant Pathology* **28** 338 – 339

Katan J, Green Berger A, Alon H and Grinsten A 1990 Solar heating by polythene mulching for control of diseases caused by soil borne pathogens, *Phytopathology* **76** 683-688

Katan J, Greenberger A, Alon H and Grinstein A 1976 Solar heating by polyethylene mulching for the control of diseases caused by soil borne pathogens, *Phytopathology* **66** 683-688

Khan AM 1950 Temperature and development of *Rhizoctonia solani* on legumes(ab), *Phytopathology* **40** 14

Kim HS, Park J, Choi SW, Choi KH, Lee GP, Ban SI, Lee C H and Kim CS 2003 Isolation and characterization of *Bacillus* strains for biological control, *Journal of Microbiology* **41** 196 - 201

Kim YC, Jung H, Kim KY and Park S K 2008 An effective biocontrol bio fumigation against Phytophthora blight of pepper using growth mixtures of combined chitinolytic bacteria under different field conditions. *European Journal of Plant Patholology* **120** 373 - 382

Kono Y, Knoche HW and Daly JM 1981 Structure: Fungal host specific, in *Toxin in Plant Disease* (ed) RD Durbin, Academic Press. 221-257 pages

Lazarovits G, Conn KL and Tenuta M 2000 Control of *Verticillium dahliae* with soil amendments: efficacy and mode of action. *In: Advances in Verticillium research and disease management, In* EC Tjamos, RC Rowe, JB Heale and DR Fravel (eds), *The Proceedings of. VIIth International Verticillium Symp*osium, Oct. 1997, Cape Sounion, Athens. American phytopathological Society Press, St. Paul, Minn., 274-291 pages

Liu CH, Mishra AK, He B and Tan RX 2001 Composition and antifungal activity of essential oils from *Artemisia princeps* and *Cinnamomum camphora. International Pest Control* **43** 72-74

Maliesh SB, Narendrappa T and Ramanujam B 2008 Evaluation of microbial antagonists and organic amendments on root rot of sage (*Salvia officinalis*) by *Fusarium solani* and *Rhizoctonia solani, Karnataka* Journal of Agricultural. Sciences **21** 301 - 302

Mathew KA and Gupta SK 1996 Studies on web blight of French bean (*Phaseolus vulgaris*) caused by *Rhizoctonia solani* Kuhn and its management, *Indian Journal of Mycology and Plant Pathology* **26** 171-177

Matz J 1921 Una enfermendad de la haleichuela. Porto Rico Insular Sta. Circular **57** 8.

Mildenhall JP and Williams PH 1973 Effect of soil temperature and host maturity on infection of carrot by *Rhioctonia solani, Phytopathology* **63** 276–280

Mishra DS and Sinha AP 2004 Compatative antagonistic potential of some biocontrol agents against *Rhizoctonia solani* Kuhn, the cause of sheath blight of rice, Indian Journal of Plant Pathology **22** 91 - 96

Misra AK 2006 Wilt of Guava- a disease of national importance, *Indian Phytopathology* **59** 269-280

Montealegre JR, Reyes R, Perez L M, Herrera R, Silva P and Besoain X 2003 Selection of bioantagonistic bacteria to be used in biological control of *Rhizoctonia solani* in tomato, *Electronic Journal of Biotechonology* **6** 106 – 127

Mukherjee S and Tripathi HS 2000 Biological and chemical control of wilt complex of French bean, *Journal of Mycology and Plant Pathology* **30** 380-385

Mukherjee S, Tripathi HS and Rathi YPS 2001 Integrated Management of wilt complex in French bean (*Phaseolus vulgaris* L *Journal of Mycology and Plant Pathology 31* 213-215

Mukhopadhyay AN and Mukherjee PK 1996 Fungi as fungicides, *International Journal of Tropical Plant Diseases* **14** 1 - 17

Narayana MR, Prakasa Rao EVE, Rajeswara Rao BR and Sastry KP 1986 Geranium cultivation in India: Potentials and prospects, *Pafai Journal* **8** 25-30

Nawar L S 2007 Pathological and rhizospherical studies on root rot disease of squash in Saudi Arabia and its control, *African Journal of Biotechnology* **6** 219- 226

Ou SH 1985 *Rice Diseases.* 2nd ed. Kew, Commonwealth Mycological Institute

Padmodaya B 2003 Effect of soil amendments at different concentrations and incubation periods of decomposition against *Fusarium* wilt of Tomato. *Journal of Mycology and Plant Pathology* **33** 317-319

Pelter GL 1916 Parasite Rhizoctonia in America, Illinoise Agricultural Experimental Staation Bulletin **189** 281-390

Peterson A, Goto M, Machmuah SB, Roy BC, Sasaki M and Hirose T 2005 Extraction of essential oil from geranium (*Pelargoium graveolens*) with supercritical carbon dioxide, *Journal of Chemical Technology and Biotechnology* **81**167-172

Prasad D, Singh KP and Bisht S 2011 Fungicidal management of wilt complex of scented geranium, *Archieves of Phytopathology and Plant Protection* **44** 1519–1533

Prasad D, Singh KP and Kumar J 2008 Root rot and wilt complex of rose-scented geranium (*Pelargonium graveolens*) in Uttarakhand, *Indian Phytopathology* **61** 365-366

Prasad D and Singh KP 2010 Root rot and wilt complex of geranium: An Overview, in *Microbial Diversity and Plant Disease Management* (eds) KP Singh and DK Shahi, VDM Verlag Dr. Muller, Germany, 123- 151 pages

Prasad D 2010 Integrating practices to manage scented geranium (*Pelargonium graveolens* L. Herit) wilt complex in Garhwal Himalayas, Ph.D. thesis, HNBGU Uttarakhand

Prasad D and Singh KP 2013 Root rot and Wilt Complex of Scented Geranium and its Management, in *Plant Disease Management for food security* (eds) CR Prajapati, KP Singh and AK Gupta,. Lambert Academic Publishing, 215- 244 pages

Ramanathan A, Marimuthu T and Raguchander T 2004 Effect of plant extracts on growth in *Pythium aphanidermatum. Journal of Mycology and Plant Pathology* **34** 315-317

Rao BR, Rajeswara, Bhattacharya, Singh AK, Harikesh B and Mallavarapu GR 1999 The impact of wilt disease on oil yield and quality of two cultivars of rose- scented geranium (*Pelargonium graveolens*), *Journal of Essential Oil Research* **11** 769–775

Rini CR and Sulochana KK 2007 Usefulness of Trichoderma and Pseudomonas against Rhizoctonia solani and Fusarium oxysporum infecting tomato, Journal of Tropical Agriculture **45** 21-28

Saksena HK 1960 The damping-off organism, *Pellicularia praticola* (Pat) Flentje in India, *Indian Phytopathology* **13** 165-167

Saksena HK and Vaartaja O 1961 Taxonomy, morphology and pathogenicity of *Rhizoctonia spp*. from forest nurseries, Canadian Journal of Botany **38** 931- 945

Satischandra KM, Hiremath RV and Hegde RK 1979 Effect of organic amendments and fungicides on the saprophytic activity of *Rhizoctonia bataticola* causing root rot of beans, Indian phytopathology **32** 543-546

Sharma M 2003 Management of root rot and web blight (*Rhizoctonia solani* Kuhn) of French Bean through eco- friendly methods, M.Sc. thesis, YSPUHF, Solan.

Sharma M and Gupta SK 2003 Eco-friendly methods for the management of root rot and web blight (*Rhizoctonia solani*) of French bean, *Journal of Mycology and Plant Pathology* **33** 345-361

Shehata MA, Davis DW and Anderson NA 1984 Resistance to *Rhizoctonia solani* stem rot in peas as influenced by temperature, watering method and period of disease development, *Plant Disease* **68** 22–24

Shephard MC and Wood RKS 1963 The effect of environment and nutrition of pathogen and host in the damping off of seedlings by *Rhizoctonia solani*, *Annals of Applied Biology* **51** 389–402

Silva P, Montealegre J, Besoain X and Perez LM 2001 Efecto de bacterias biocontroladoras sobre germinacion y desarrollo de tomate y sobre expression de preteinas de defensa, *In Congreso Sociedad Chilena de Fitopatolagia*. (XI, 4th-5th December, 2001, Santa cruz, Chile). Abstracts, 2001, 29 page

Singh RK and Dwivedi RS 1987 Effect of oils on *Sclerotium rolfsai* causing root rot of barley, *Indian Phytopathology* **40** 531-533

Singh T and Chaurasia DK 2008 Biological control of tomato wilt, *Journal of Eco-friendly Agriculture* **3** 183-184

Singh UP, Mishra GD and Singh KP 1993 Control of powdery mildew (*Erysiphe pisi*) of pea (*Pisum sativum*) by the application of culture filtrate of *Trichoderma harzianum*, *Phytophylactica* **25** 269 - 270

Stoessl A 1981 Structure and biogenetic relations: Fungal non host specific, in *Toxin in Plant Disease* (ed) RD Durbin, Academic Press. London, 110-197 pages

Sungeun L, Byeoungsoo P, Mookey K, Wonsik C, Heungtae K, Kwangyun C, Sangguel L and Hoiseon L 2001 Fungicidal activity of pipernonaline, a piperidine alkaloid derived from long pepper, *Piper longum* L., against phytopathogenic fungi, *Crop Protection* **20** 523-528

Tomar M and Chandel S 2006 Use of phytoextracts in the management of gladiolus wilt, *Journal of Mycology and Plant Pathology* **36** 142-144

Townsend BB and Willelts HJ 1954 The development of sclerotia by certain fungi. *Transactions of British Mycological Society* **37** 213-221

Upadhyay RS and Rai B 1984 Ecology of *Fusarium* species causing root diseases, in *Progress in Microbial ecology*(eds) KG Mukerji, VP Agnihotri and RP Singh Print House, Lucknow (India), 37-72 pages

UshaRani S and Satheesh G.2007 Management of damping- off of tomato caused by *Pythium aphanidermatum* (Edson) Fitz. through biocontrol agents, organic amendments and soil solarization, *Plant Disease Research* **22** 27-29

Valdez RB 1955 Sheath spot of rice, *Philippine Agriculture* 317-336

Weindling R 1932 *Trichoderma lignorum* as a parasite of other soil fungi, *Phytopathology* **22** 837 - 845

Weller DM 1988 Biological control of soil borne plant pathogen in the rhizosphere with bacteria, *Annual Review of Phytopathology* **26** 379 - 407

Yucel S, Pala H, Cali S, Erkilic A, Albajes R and Sekeroglu E 2000 Combination of *Trichoderma* spp. and soil solarization to control root rot diseases of cucumber in greenhouse conditions, *Bulletin OILB-SPOR* **23** 77-81.

18

Intergated Management of Pea Powdery Mildew

Om Gupta, Usha Bhale, Madhuri Mishra and S.P. Tiwari

Department of Plant Pathology, JNKVV Jabalpur 482 004 (M.P.)
E-mail: omgupta_jnkvv@rediffmail.com, omgupta_jnkvv@yahoo.com

Pea (*Pisum sativum* L.) is one of the important legume crops of leguminosae (Febaceae) family. Wild and primitive forms are found in the East and some parts of Central Asia, the Mediterranean and Ethiopia. The Migration of humans and their other activities spread or disperse the peas throughout the world. Record from history says the peas were grown for dry seed as they are rich source of protein and carbohydrate. The word 'pisos' was used by Greeks and 'Pisum' by Romans. It became 'Peason' then 'pease' and finally 'peas' by the English (Hedrick et al 1928). Peas fall into two main categories; those that are harvested at the green , immature stage and those harvested at the green are further subdivided into 'Vining peas' which are harvested and shelled mechanically for processing as canned and frozen and some are picked by hand and marketed as fresh peas. Peas used for human consumptions and harvested at full maturity are called dried peas.

Peas suffer from a wide range of diseases, are specific while other may also attack other crops in rotation. The important diseases of peas are of bacterial, fungal, viral and nematode origin which can significantly decrease both yield and quality. The present chapter deals with the integrated management of pea powdery mildew. This is one of the important diseases, which occur almost every year in severe epidemic form wherever this crop is grown throughout the world. In favourable weather conditions, it causes considerable yield reductions.

Powdery mildew is the most widely distributed disease of pea throughout the world. Its occurrence has been reported from Afganistan, Alaska, Agrentia, Australia, Canada, China, France, Germany, Hungry, Italy, Iraq, Japan, Libya, Malaya, Mauritius, Pakistan, Philippines, Portugal, Russia, Spain, Siberia, South Africa, Sudan, Switzerland, Tanzania, Tucuman, Turkey, U.S.A., Zambia and many other countries (Singh and Singh 1988). In India, the disease was first reported by Butler (1918) from Dehradun (U.P.) its report of common occurrence all over the pea growing areas of the country have been received only after 1935 (Uppal et al 1935).

Losses

The disease usually appears late in the season, reaching its maximum intensity during pod formation. However, the damage due to this disease varies, depending on many factors including the varietal susceptibility, sowing time, stage of the crop at the time of infection and prevailing temperature and humidity from middle of February onwards. However, the reduction in yield is mainly due to reduced photosynthetic activity in infected plants. Uppal et al (1935) observed heavy reduction in the pod formation due to severe infection of powdery mildew in Bombay and even one picking was not possible as compared to 6-7 pickings in normal healthy crop. Munjal et al (1963) reported that in a hundred per cent infected crop the losses were 21-31 per cent in terms of pod number and 26-47 per cent in terms of pod weight. Gritton and Ebert (1975) observed about 50 per cent reduction in yield due to attack of this pathogen. Epidemic development of powdery mildew is very fast and it progresses as compound interest. Since the disease attains severity at the end of season, seed crops and dry edible types that mature later than the vegetable types are more likely to become infected and sustain a heavy loss in productivity. In a crop badly affected by powdery mildew, the reduction in number of pods/plant is estimated to be 28.6 per cent (Rathi and Tripathi, 1994).

Host range

Powdery mildews exhibit some degree of host specialization. There are conflicting reports regarding the range of parasitism of *Erysiphe pisi* from peas. According to some workers, it has a narrow host range and confined to *Pisum sativum, P. arvens, Lathyrus* spp. and *Lens esculenta* only (Yarwood 1957, Singh and Singh 1983a, Jhooty and Munshi 1990). On the other hand, different workers have listed a large number of leguminous species as host of *E. pisi*. These include *Cajanus cajan, Cicer arietinum, Clitoria biflora, Crotolaria junicia, Lathyrus aphaca, L. odoratus, L. sativus, L. eseculentus, Macroptilium atropurpureum, Medicago polymorpha, Pisum arvense, P. sativum, Trigonella foenumigraecum, Vicia sativa, Vigna mungo, V. umbellata, V. sinensis, Sesbania aculeata, S. aegyptica, Phaseolus mungo, P. vulgaris* (Singh and Singh 1983a, Bhardwaj and Singh 1984).

Symptoms

Symptoms first appear as minute discoloured patches with thin fine meshwork of white mycelium arising at many places on the upper surface of lower leaves. These white patches soon join together to form larger whitish floury areas. As the disease progress, the upper leaves of the plants also express similar symptoms. In severe cases of disease development both upper and lower surfaces of the leaf are infected. The conidia and condidiophores are produced in such profusion that the foliage looks as if dusted with flour and the whole crop in the field appeared white from a distance (Plate 1). When the conidial production is nearly over, the leaf surface becomes patchy yellow brownish and the foliage is then killed.

With the progress of the disease, the symptoms are also noticed on tendrils, petioles and stems. In severe cases entire tendrils and petioles are covered by white floury growth. The basal portion of the stem near the ground level is first infected by the pathogen and shows white floury patches. These patches on petioles and stem later turn light brown to dark brown and necrotic.

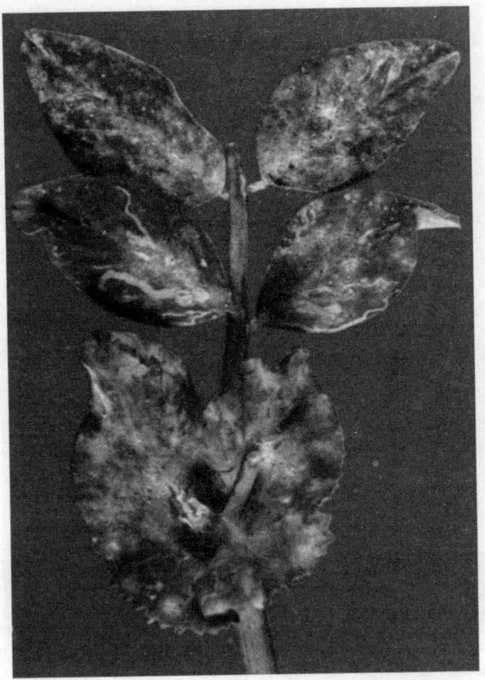

Plate 1. Symptoms of pea powdery mildew (Courtesy T.S. Thind)

Pods are infected in all the developmental stages, but immature pods are more prone to rapid infection. The affected pods show white floury patches consisting of white powdery mass. After infection the pods remain small in size, shrivelled and pod number and pod weight is also reduced. These floury patches consist of mycelium and spores of the fungus. These patches then turn light brown and finally dark brown in colour and become necrotic. Such infected pods are either unmarketable or fetch minimum price in the open market.

Etiology

The causal agent of this disease is *Erysiphe pisi* DC.ex Saint-Amans (sometimes also referred as *Erysiphe polygoni* DC.). The fungus is ectoparasitic, spreading on the surface of the host and sending haustoria into the epidermal cells to draw out nourishments. The haustoria lie outside the cytoplasm of the host and special structures are formed between the haustoria and the cytoplasm. These interfacial structures distinguish haustoria from hyphae and therefore, their structures and physiology are important factors in parasitism.

The conidiophores arise vertically from the superficial hyphae on the host surface (Plate 2). Each conidiophore bears conidia either singly or in chains. Of these, only the last one in the chain matures as the spores are formed in regular order from the tip downwards. The mature conidia fall-off quickly and are disseminated by wind. The conidiophores are septate and their cells are not much different from conidia in shape. The conidia are ovate, barrel-shaped, hyaline, and unicellular and measure 25-35 x 13-16 μm.

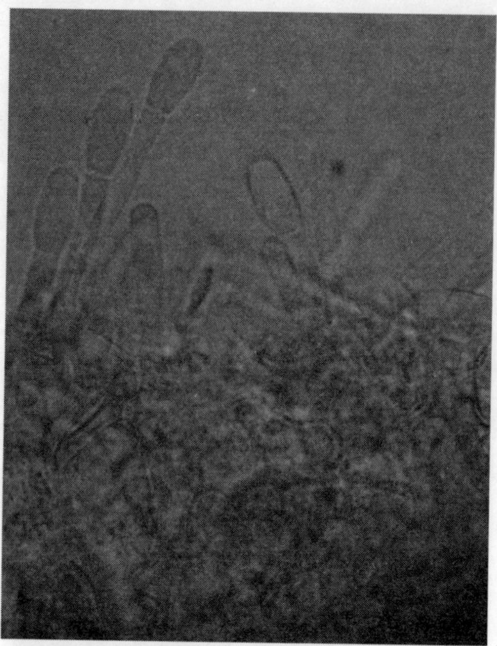

Plate 2. Conidiophore and conidia of *Erysiphe pisi*

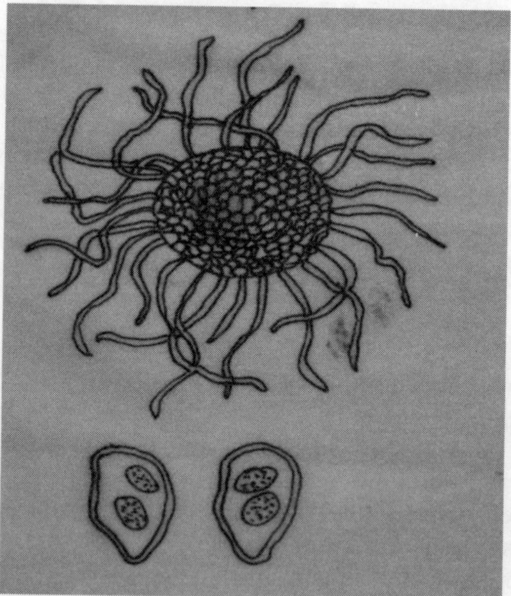

Plate 3. Cleistotheciumand ascuss of *Erysiphe pisi*

Later in the season, cleistothecia (sexual stage) appear as dark coloured, round, minute bodies scattered in the mycelial web and measuring 85-180 μm in diameter (Plate 3). The peridium is composed of distinct polygonal cells and is provided with a number of myceloid appendages. These outgrowths may be few or many, free or interwoven and resemble the ordinary hypae. Each cleistothecium contains usually 2-8 asci. These asci originate from a single point in the fruit, appearing in a fanlike manner. They are ovate, nearly sessile and 46-72 x 30-45 µm in size. The asci contain 3-8 ascospores, which are elliptical, hyaline, and unicellular, and measure 10-25 x 9-14µm. Cleistothecia production appears to be related with host nutrition and environmental conditions.

Variability

Some work on the variability and specificity for host species has been reported in *E. pisi*, however, there is no published information on the occurence of races. Studies over a decade have failed to establish existence of physiological races in pea growing areas of Uttar Pradesh (Singh 1985). However, some variability in 10 isolates from Haryana has also been indicated (Rathi and Tripathi 1992).

Disease cycle

The fungus is an obligate parasite and perennates either through cleistothecia formed mostly on the infected fallen debris and sometimes on standing crop in Bangalore area of Karnataka (Venkatakrishna 1951), Lahaul, Spiti and Kinnaur districts of Himachal Pradesh (Kapoor and Chaudhary 1995) or in seed as dormant mycelium (Uppal et al 1935) or in the form of mycelium and conidia on the host (Jhooty and Munshi 1990) and on annual and perennial hosts. However, the concept of seed borne nature of the disease could never be confirmed (Yarwood 1978). The primary infection of the lower leaves is caused by wind blown ascospores (whereever cleistothecia are present) (Griffee 1969) or by conidia formed on the crop sown earlier in the neighbouring areas (Jhooty and Munshi 1990). In the process of infection, the germination of conidia occurred after 2.5 h of incubation while appressoria were formed after 3h on susceptible pea varieties and the germination was evident from both the poles of a conidium by maximum of six germ tubes (Singh et al 2000). Colonization of the entire leaf may occur in 150 hours under conducive weather conditions. Conidia of *E. pisi* not only germinate without liquid water but can do so even in completely dry environment. The conidia formed abundantly on primary infections are blown by wind and air currents and cause secondary infections. In areas where favourable conditions for cleistothecia formation are prevalent, the fungus reproduces sexually and form ascospores.

Epidemiology

The powdery mildew of pea becomes a serious problem under warm and dry conditions during the day and cool nights. Irrigation at flowering to pod formation stages can prolong the vegetative phase, thereby increasing period for susceptibility. Crop raised in low and wet areas with high soil moisture is more vulnerable to the disease as these conditions tend to maintain the plants in vegetative phase for relatively longer periods. The disease is more prevalent in dry weather and moderate temperatures. A temperature range of 10-30°C is favourable for conidia germination with optimum being at 20ºC. Highly significant and positive correlations were observed between temperature and disease severity while high relative humidity was not a pre-requisite for development of pea powdery mildew (Banyal and Tyagi 1997). A fairly dry soil and heavy application of nitrogenous fertilizers increase the disease incidence (Singh 1987). Maximum disease occurred in plants supplied with high N and Fe combined with low doses of P, K, Zn and Cu while disease intensity was significantly reduced by double doses of P, K and Zn combined with low N and Fe (Sindhan and Parashar 1990).

MANAGEMENT

Pea powdery mildew can be managed by using cultural practices, fungicides, plant extracts and oils and host resistance.

Cultural practices

Collection and destruction of plant debris and avoidance of close planting can be helpful in reducing disease incidence (Singh 1987). Change in date of sowing has also been recommended. Early sown crop (late October and early November) escapes the onslaught of the disease, whereas in the late planting (late November and December) the disease intensity was very high resulting in heavy reduction in yield (Schroeder and Previndeti 1965, Shukla and Amin 1991).

Biological control

Srivastava and Bisht (1983) sprayed the infected plants with 8 phyllosphere fungi and maximum suppression of the disease was obtained with *Cladosporium cladiosporiodes* followed by *Curvularia lunata* (*Cochliobolus lunatus*). The Coccinellid, *Psylloboranana* was observed feeding on the mildew *Erysiphe polygoni* in Cuba and can be used as biocontrol agent (Cruz et al 1990). Sprays of plants with *Trichoderma hamatum* was most effective followed by *T. viride* and *Pseudomonas fluorescence* as the disease severity ranged in between 13.84 to 27.68 per cent (Jagtap and Khalikar 2012).

Plant extracts and oils

Control of this disease by plant extracts and oils is superior to chemicals because they are environment firendly and safe from residue point of view. Singh and Singh (1983b) and Singh (1985) screened various plant extracts and oils against conidial germination of *E. pisi* and reported that the conidial germination was significantly reduced in all the extracts and oils at 100 ppm or more whereas, neem and garlic oils proved to be highly effective. Volatile oils and extracts of plant parts also have strong effect on germination of conidia or conidial density (Singh et al 1984). Methanolic extract of cashewnut (*Anacardium occidentale*) shell extract (300 ppm) was found highly effective in reducing the conidial germination under *in vitro* conditions and reduced diseases severity to 39 per cent under field conditions as compared to 96.53 per cent in control (Bahadur et al 2008).

Host Resistance

Many workers have investigated varietal resistance of pea to powdery mildew but very few varieties have been reported to be resistant (Singh et al 1988). Pal et al (1980) screened 292 pea accessions but found only five entries such as P431, P438, Glorire de Quimper, Sel 18, and Sel 30 resistant against powdery mildew. Cultivars P-185, P-388, P-6583, P 6587, P-6588 and T-10 (Singh 1985), Pant P-5 (Singh et al 1988), P-185, P6583, NDPI, JP6, JP886 and HUP-9 (Gupta 1990) P6588, Rachna, P6587, P383, P388, DPS-3, DPS-6, HUP-2, DMR9, P-185 and P6583 (Mathur et al 1991), Groach Laser (Mahmood et al 1991), UVS-1, DPH-13, DPH-19, DPH-46, DPH51, DPH54, DPH62, DPH64, DPH65, DPH71, GC477, HFP4, JP4, JP71, JP83, JP169, JP514-A, JP829, P6587, Pant P-5, PLP555, PLP582, PLP590, PM5, PMR3, PMR4, PMR334, PLP555, PLP582, PLP590, PM5, PMR3, PMR4, PMR334, PRS4, PRS18-6, PRMK-SM-1, Sel8-1, Sel8-12-1, Sel35, Sugar Giant Tall, VG1, VG2, VG9, VG7906 (Jindal et al 1995), Pant P-9, Rachana, DMR 11 (Khare and Lakpale 1997) have been reported resistant to this disease.

Some of the cultivars are reported to have a longer latent period and reduced number and area of mildew specks compared with other cultivars. These parameters are responsible for the lower rate of disease spread in these varieties. Such slow mildewing varieties reported from different parts of the country are IPS-207, IPS-77, IPS-293, IPS-198 and IPS-214 (Krishna and Mishra 1988) and P185, JP4, JP753, NDP1 and HUP2 (Gupta 1990a). Krishna et al (1989) also reported less losses in pod length, number of seeds/pod, 100 seed weight and total seed yield in slow mildewing varieties than the fast mildewing.

Resistance in powdery mildew is controlled by a single gene pair 'er' (Singh et al 1983). Singh and Singh (1991) also suggested that one recessive gene controls resistance in T_{10} but three pairs of recessive genes have been involved in var. PRS15 and P388 (Ram 1992). Vaid and Tyagi (1997) studied genetics of powdery mildew resistance using five isolates of *E. pisi* and found that resistance is controlled by a single recessive gene in all the pea cvs. HPPC-63, HPPC-95, DPP-26, DPP-54, Mexique-4, SVP-950, Wisconsin-7104 and P-6588 included in the study.

Munshi et al (1987) reported that resistant cultivars contained greater quantities of orthodihydroxy phenols and polyphenols while cuticle thickness, germination of conidia and formation of appresoria were the same in resistant and susceptible ones.

Chemical control

Since last century wettable sulphur in general is recommended for the control of this disease (Spencer 1978, Yarwood 1978). Among the large number of fungicides tested by various workers, high level of control of the disease was achieved with the use of Calixin (tridemorph), Karathane (dinocap) and Sulfex (wettable sulphur) as foliar spray (Bhardwaj and Sharma 1984). However, Sulfex is the most economical fungicide till date.

Foliar sprays of EBI fungicides like Vigil (diclobutrazol), Bayton (triadimenol), Bayleton (tridemefon), Tilt (propiconazole), Rubigan (fenarimol), Punch (flusilazole), Contaf (hexaconazole), Topas (penconazole), Score (difenconazole) and Saprol (triforine) have also shown very good control of this disease at much lower doses than sulphur and also possess good antisporulant activity (Rana et al 1991, Maheshwari et al 1992, Upadhyay and Gupta 1994, Gupta and Shyam 1996, Kapoor and Thakur 1997). Since, these are systemic and site-specific fungicides, continuous use of these may result in the development of resistant strains of the pathogen. Rotation of fungicides and their use in mixture with non-systemics will be a good strategy to prolong the life of these fungicides. Score and Topas when sprayed in combination either with Benlate or Dithane M-45 showed synergistic effect and managed the powdery mildew effectively and increased yields (Ahmad et al 2007).

Gupta and Kumar (2008) also evaluated different fungicides in protective and eradicative spray programmes and reported that carbendazim caused complete conidial germination inhibition followed by wettable sulphur and kresoxim methyl while hexaconazole was least effective under *in vitro* conditions whereas difenconazole (10 ug/ml) caused >50 per cent inhibition of gertube elongation besides affecting the morphology of germtubes which appeared distorted. In protective spray programme, three sprays of kresoxim methyl (0.1%) at fortnightly intervals were found highly effective followed by difenconazole, penconazole, diniconazole and hexaconazole while wettable sulphur was least effective and increased the green pod yield.

Some insecticides viz. triazophos, quinalphos, phoxim, phosphamidon, dimethoate and fenitrothion have also been found effective in controlling this disease (Singh 1986, Shyam et al 1995) and would appear to be economic alternatives. Combined application of Karathane + endosufan, Bavistin + monocrotophos and Sulfex + monocrotophos were not only biologically compatible but also provided effective control of powdery mildew and pod borer (Shukla and Lal 1988). Shyam et al (1995) also reported the efficacy of combination of Sulfex + demeton-o-methyl in keeping the crop free from powdery mildew and leaf miner.

Integrated Disease Management

Though the disease can be kept under check with elaborate chemical spray programme but they are costly and cause pollution hazards. If the chemical sprays are combined with botanicals and bioagents, the disease severity can be reduced with minimum number of sprays. Surwase et al (2009) evaluated the efficacy of five fungicides, four botanicals and four bioagents against the disease and found that hexaconazole (0.05%), penconazole (0.05%), botanical NSKE (5%) and bioagent *Trichoderma harzianum*, (0.5%) sprays were highly effective and economical for controlling this disease and provided better seed yield.

Future thrust

Being an important disease of pea crop, efforts should be made to combine all available methods to keep the revages of this disease in check. Though the disease can be kept in check by chemical sprays but they have harmful residues. Emphasis should be given on host resistance, botanicals and bioagents alone and their combination to check this disease.

Referencres

Ahmad S, Mohammad I and Irfan UD 2007 Synergistic effect of fungicides application on powdery mildew control and yield in pea, *Sarhad Journal of Agriculture* **23(3)** 733-736

Bahadur A, Singh UP, Singh DP, Sarma BK, Singh KP, Singh A and Aust HJ 2008 Control of *Erysiphe pisi* causing powdery mildew of pea (*Pisum sativum*) by cashewnut (*Anacardium occidentale*) shell extract, *Mycobiologia* **36(1)** 60-65

Banyal DK and Tyagi PD 1997 Role of climatic factors in the development pf powdery mildew of pea, *Journal of Mycology and Plant Pathology* **27(1)** 64-66

Bharadwaj SS and Sharma SL1984 Field evaluation of mildewcides against powdery mildew (*Erysiphe polygoni*) of peas (*Pisum sativum*) in Himachal Pradesh, *Indian Journal of Mycology and Plant Pathol*ology **14** 103-104

Bhardwaj CL and Singh BM 1984 Host range of *Oidium* state of *Erysiphe pisi* on some leguminous hosts of Kangra Valley of Himachal Pradesh, *Indian Phytopathology* **37** 732-733

Butler EJ 1918 *Fungi and Diseases in Plants,* Calcutta: Thacker Spink and Company

Cruz B, Chianglok ML, and Castanedaruiz RF 1990 *Psyllobora nana* (Coleoptera Coccinellidae), biological control agent, *Ciencias de la Agricultura* **40** 168

Griffee PJ 1969 Cleistocarps of Erysiphe polygoni on Pisum sativum in Zambia, Transactions of the British Mycological Society **53** 491-493

Gritton ET and Ebert RD 1975 Interaction of planting date and powdery mildew on pea plant performance, *Journal of American Society of Horticultural Sciences* **100** 137-142

Gupta RP 1990 Evaluation of pea germplasms for their reaction to powdery mildew and rust, *Indian Journal of Pulse Research* **3(2)** 186-188

Gupta RP 1990a Evaluation of slow mildewing resistant varieties of peas in the field, *Indian Journal of Pulses Research* **3(1)** 52-55

Gupta SK and Kumar A 2008 Management of Erysiphe pisi through strobilurin and EBI fungicides, *Indian Phytopathology* **81(2)** 184-191

Gupta SK and Shyam KR 1996 Fungicidal control of powdery mildew and rust diseases of pea, *Indian Journal of Mycology and Plant Pathology* **26(1)** 126-127 (Abstract)

Hedrick UP, Hall FH, Hawthorn LR, and Berger A 1928 Vegetables of New York Volume I: Peas of New York Report of the New York State Agricultural Experiment station, Albany, New York, USA: JB Lyon Company

Jagtap GP and Khalikar PV 2012 Integrated management of pea powdery mildew caused by *Erysiphe polygoni* DC, *Scientific Journal of Agriculture* **1(2)** 33-38

Jhooty JS and Munshi GD 1990 Mode of perpetuation of powdery mildew of pea in North-Western India, *Indian Phytopathology* **43(4)**: 455-456

Jindal KK, Gupta SK, Shyam KR, and Rastogi KB 1995 Evaluation of pea germplasm against bacterial blight and powdery mildew, *Plant Disease Research* **10(1)** 35-38

Kapoor AS and Chaudhary HK 1995 Mode of perpetuation of *Erysiphe pisi* in dry temperate zone of Himachal Pradesh, *Indian Phytopathology* **48** 77-78

Kapoor AS and Thakur BR 1997 Fungicidal spray schedule for the economical management of pea powdery mildew, *Indian Phytopathology* **50(1)** 115-118

Khare N and Lakpale N 1997 Sources of resistance to powdery mildew in field pea, *Journal of Mycology and Plant Pathology* **27(2)** 219-220

Krishna A and Mishra SP 1988 Slow mildewing in pea, *Indian Phytopathology* **41(2)** 204-207

Krishna A, Mishra SP, Mandloi KC, and Khosla HK 1989 Assessment of yield losses in slow and fast mildewing pea cultivars, *Indian Journal of Mycology and Plant Pathology* **18(2)** 177-179

Maheshwari SK, Gupta PC, and Suhag LS 1992 Management of powdery mildew of pea through fungicides, *Legume Research* **15(1)** 51-52

Mahmood T, Ahmad I, Haq IU, and Aslam M 1991 Varietal resistance of pea to powdery mildew, *Indian Phytopathology* **44(1)** 52-54

Mathur K, Shekhawat KS, and Bhata LK 1991Performance of pea varieties against powdery mildew, *Indian Phytopathology* **44(1)** 137

Munjal RL, Chenulu VV, and Hora TS 1963 Assessment of losses due to powdery mildew (*Erysiphe polygoni* DC) on pea, *Indian Phytopathology* **16** 268-270

Munshi GD, Jhooty JS, and Bajaj KL 1987 Basis of resistance in peas to powdery mildew caused by *Erysiphe polygoni, Indian Journal of Plant Pathology* **17(3)** 280-283

Pal AB, Brahmappa, Rawal RD, and Ullasa BA 1980 Field resistance of pea germplasm to powdery mildew (*Erysiphe polygoni*) and rust (*Uromyces fabae*), *Plant Disease* **64** 1085-1086

Ram RS 1992 Inheritance of powdery mildew resistance in peas *(Pisum sativum* L), *Experimental Genetics* **8(1-2)** 37-40

Rana DPS, Bhardwaj PK, Rao MVB, and Chatterjee D 1991 Field evaluation of fenarimol and triadimefon for control of powdery mildew (*Erysiphe polygoni* DC) of pea, *Indian Journal of Plant Protection* **19(1)** 31-35

Rathi AS and Tripathi NN 1994 Variability in *Erysiphe polygoni* DC on peas, 71 (Abstract). *In 44th Annual Meeting of Indian Phytopathology Society*, Bhavanagar University, Bhavanagar, Gujarat

Schroeder WT and Previndeti R 1965 Breakdown of the er resistance to powdery mildew in *Pisum sativum, Phytopathology* **55** 1075

Shukla P and Amin KS 1991 Integrated disease management of powdery mildew of field peas (*Pisum sativum* L.) caused by *Erysiphe polygoni* DC, *Legume Research* **14(2)** 59-63

Shukla P and Lal SS 1988 Effect of combined application of fungicides and insecticides on the powdery mildew and pod borer of field pea, *Pesticides* **22(3)** 5-7

Shyam KR, Gupta SK, and Verma AK 1995 Management of powdery mildew and leaf miner in pea through fungicide and insecticide applications, in *Integrated disease management and plant health* (eds) VK Gupta and RC Sharma, Jodhpur: Scientific Publishers (India), 119-122 pages

Sindhan GS and Prashar RD 1990 Effect of macro and micro nutrients on the development of powdery mildew of pea, *Indian Journal of Mycology and Plant Pathology* **19(2)** 219-221

Singh D and Singh D 1991 Development of white rot of pea (*Sclerotinia sclerotiorum*) in relation to meteorological factors, *Indian Journal of Mycology and Plant Pathology* **21(2)** 192-195

Singh HB 1985 *In vitro*, glasshouse and field evaluation of some plant extracts and oil for the control of powdery mildew of pea, 13. *In Proceedings of Seminar on Chemistry, biology, cultivation and utilization of medicinal and aromatic plants*

Singh HB and Singh UP 1988 Powdery mildew of pea (Pisum sativum L), International Journal of Tropical Plant Diseases **6** 1-18

Singh RB, Jain AK, Jat RR, and Mathur SS 1988 Varietal screening of pea to powdery mildew, *Plant Pathology Newsletter* **6(1-2)** 20

Singh RB, Singh MN, Singh UP, and Singh RM 1983 Inhertance of resistance to powdery mildew in pea and its use in breeding, *Indian Journal of Agricultural Science* **48** 86-88

Singh RN 1986 Field evaluation of certain organophosphorus compounds for the control of powdery mildew of peas, *National Academy of Scienence Letters India* **9(11)** 329-330

Singh RS 1987 *Diseases of vegetable crops,* Oxford and New Delhi: IBH Publishing

Singh UP and Singh HB 1983a Differential responses of *Erysiphe polygoni* conidia to living and non-living substrata, *Phytopathology Z* **107** 140-145

Singh UP and Singh HB 1983b Comparative efficacy of some commercial fungicides, plant extracts and oil for the control of powdery mildew (*Erysiphe polygoni* DC) of pea (*Pisum sativum* L.), *Australian Plant Pathology* **12** 22-24

Singh UP, Prithiviraj B, Mishra GD, and Singh HB 2000 Modes of germination of *Erysiphe pisi* conidia on pea (*Pisum sativum*) leaves and glass slides, *Indian Journal of Plant Pathology* **18 (1 and 2)** 57-59

Singh UP, Singh HB, and Chauhan VB 1984 Effect of some fungicides, plant extracts and oil on inoculum density of different nodal leaves of pea infected by *Erysiphe polygoni, Zeitschrift fur Pflanzenkrankheiten und Pflanzenschutz (Journal of plant diseases and protection)* **91** 20-26

Spencer DM 1978 *The powdery mildews*, London: Academic Press

Srivastava SL and Bisht S 1983 Studies on biological control of *Erysiphe polygoni* DC on pea, *Geobios* **10** 117-119

Surwase AG, Badgire DR and Suryawanshi AP 2009 Management of pea powdery mildew by fungicides, botanicals and bio-agents, *Annals of Plant Protection Science* **17**(2) 384-388

Upadhyay AL and Gupta RP 1994 Fungicidal evaluation against powdery mildew and rust of pea (*Pisum sativum* L), *Annals of Agricultural Research* **15(1)** 114-116

Uppal BN, Patel MN and Kamat MN 1935 Pea powdery mildew in Bombay, *Bulletin of Bombay Department of Agriculture* **177** 12 pages

Vaid A and Tyagi PD 1997 Genetics of powdery mildew resistance, *Euphytica* **96** 203-206

Venkatakrishnaiah NS 1951 The perithecial stage of *Erisiphe polygoni* DC on pea (*Pisum sativum* Linn.), *Current Science* **20(5)** 137-138

Yarwood CE 1957 Powdery mildews, *Botanical Reviws* **23** 235-293

Yarwood CE 1978 History and taxonomy of powdery mildews, in *The powdery mildews* (ed) DM Spencer, London: Academic Press, 1-37 pages.

19

Integrated Management of Fungal Diseases of Pome and Stone Fruit Nurseries

Satish K. Sharma[1] and Jitender Kumar[2]

[1]Regional Horticultural Research Station, Sharbo, Dr Y S Parmar University of Horticulture and Forestry, Nauni, Solan,
[2]NBPGR, Regional Station, Phagli, Shimla-4
E-mail: satishsharma_27@yahoo.com

The pome and stone fruits are mainly grown in the North-Western Indian States of Jammu and Kashmir (J&K), Himachal Pradesh (H.P.) and in Uttarakhand. The North-Eastern Hill region, comprising of the States of Arunachal Pradesh, Nagaland, Meghalaya, Manipur and Sikkim also grow some of the temperate fruits on a limited scale. Pome (apple and pear) and stone fruit (peach, plum, apricot and cherry) orchards are generally raised through the plants developed in the nurseries. Due to introduction and adaptation of low chilling cultivars of crops like apple, peach, plum and pear, they are also now being grown commercially in certain low heights of Himalayan region of India for home fruit or landscape purposes. However, several diseases of these fruit crops in nursery seriously reduce plant quality, get transmitted to planting sites and subsequently tree health is affected in the orchards.

These diseases are one of the limiting factors in production of temperate fruits nurseries successfully. Diseases attack at various seedling growth stages and reduce both quality and quantity of graftable/transplantable plants. With the cultivation of temperate fruits over a wide range of elevation including lower warmer areas the demand for nursery plants is increasing every year and biotic stresses have multiplied tremendously and are major obstacle in making available healthy plants to the new developing areas of the states. Major diseases are caused by three classes or types of micro-organisms: fungi, bacteria, and viruses. Fungi are plants; however, unlike our familiar green plants, they cannot make their own food from water, carbon dioxide, and the sun's energy. The fungi obtain their food from organic matter of dead plants or from living plants. When the latter occurs, the symptoms of disease are produced on the host plant. Many fungi are mold-like organisms existing in the form of microscopic threads. Most are transmitted from plant to plant by tiny seed-like bodies called spores, which are carried by air currents, splashing water, or tools.

Fungal diseases viz. white root rot (*Dematophora necatrix*), seedling blight (*Sclerotium rolfsii*), powdery mildew (*Podosphaera* spp., *Sphaerotheca* sp), leaf spots (*Phyllosticta* sp., *Cercospora* sp., *Alternaria* sp.), frosty mildew (*Cercosporella* sp.), leaf rust (*Tranzschelia* spp.), shot hole (*Stigmina* sp.) etc. are the major constraints in production of disease free healthy plants. Further, the soil borne diseases are more damaging and difficult to manage as compared to foliar diseases. Every year large number of seedlings/grafted plants are destroyed resulting in huge losses to the nursery grower and upsetting the planting targets. The incipient infection of apparently healthy looking plants/roots help in the dissemination of pathogens to the planting sites in different geographical locations of temperate and sub-temperate regions. Most of these diseases are managed by soil or foliar application of chemicals viz. formaldehyde, carbendazim, thiophanate methyl, thiabendazole, mancozeb, aureofungin, copper sulphate, metalaxyl etc. However, due to high cost of chemicals, low levels of disease control, and growing environmental awareness, attention is shifting to develop an integrated approach consisting both chemical and non-chemical approaches. These practices either alone or in combination reduce the survival of propagules of soil borne pathogens and consequently reduce disease incidence and increase shoot length, root length etc. In this compilation efforts have been made to gather the available information on different aspects of a particular disease and its best possible management.

Why Nursery Management

1. Tender in nature and require more care and attention
2. Upsetting planting targets specially in fruit crops
3. Dissemination of pathogens to planting sites
4. Easy and cheap management
5. Post planting dyeing of plants
6. Quality deterioration and bad reputation
7. Poor growth of seedlings
8. Source of introduction of disease into existing plantations

During nursery raising of pome and stone fruit crops following two types of fungal diseases attack seedlings and reduce number of healthy seedlings substantially.

(a) Soil borne diseases

(i) Seedling blight
(ii) White root rot
(iii) Collar rot

(b) Foliar diseases

(i) Powdery mildews
(ii) Leaf spots
(iii) Frosty mildews

(iv) Rusts

(v) Stigmina blight/Shot hole

(a) Soil borne diseases

(i) Seedling Blight

Host: Apple

The disease also known as southern blight, incited by *Sclerotium rolfsii* Sacc. is a destructive disease of many horticultural crops in warmer regions of Himachal Pradesh. Southern blight was first reported in apple nursery in 1952 (Anonymous 1952) and the incidence of the disease has been observed upto 40 per cent (Bhardwaj and Agarwala 1986).

Apple trees or root stocks are usually grown for 2 to 3 years in the nurseries to attain the desired saleable size. *Sclerotium rolfsii* can infect and kill seedlings up to 3 years age. Sandy loam soils are more suitable for disease development as compared to clay loam soils. Soil temperature of 30-32°C, pH of 6.0 and moisture content of 38 per cent favoure seedling blight (Bhardwaj and Agarwala 1986). Fungal growth is confined primarily to root tissues which may be killed rapidly. But the tissue slightly above the soil surface may also be invaded. During monsoon months, white to brown sclerotia form on the mycelia mat (Plate 1) and the entire plant may wilt and dye.

Management

Application of fungicides to the base of the seedlings is necessary every year till the seedlings are in the field. Brassicol (120 g/bed), and Aureofungin (8.5 g/bed) were best for the control of disease. Out of 15 rootstocks screened 3 were found resistant and 2 moderately resistant. Soil drenching with thiram (0.35%) or Brassicol (0.3%) or Aureofungin (400 ppm) was found most effective in controlling the disease (Bhardwaj 1987). Sprays of Benlate (0.2%), Calixin (0.3%) prevent the development of sclerotia (Sindhan et al 1979).

The addition of a food base in formulations of biological control agents has been reported to be essential to achieve control of several soil borne pathogens. The use of biological control agents has a great potential as they colonize around the crown area of seedling and protect against infection caused by *S. rolfsii*. The manner in which these agents are formulated can have a great impact on their efficacy. Formulations of *T. harzianum* with a food base (oat seed) were very effective in controlling southern blight, while those using either conidia or chlamydospores were not as effective.

Furthermore, incidence of this disease can dramatically increase if dead seedlings are not removed during the winter season and the infected residues are left in the field. On the basis of field observations disease incidence may be reduced by proper weed control, avoiding infested fields, and rouging diseased seedlings including those exhibiting latent infections in the fall and spring.

(ii) White Root Rot

Hosts: Apple, stone fruits, strawberry, kiwifruit, etc.

The *Dematophora necatrix* Hartig is a soil inhabiting non-sporulating fungus attacking all the temperate fruits viz. apple, apricot, cherry, kiwifruit, pear, plum, peach, pecan nut, persimmon, strawberry and walnut. The perfect stage of the fungus, *Rosellinia necatrix* (Hart.) Berl. is known to occur only in few geographical regions of the world otherwise the serious losses are only due to imperfect stage. The pathogen is widely distributed in temperate and sub-temperate regions of the world attacking more than 160 plant species belonging to 45 families. The disease is of major importance in temperate nurseries. The fungus was first noticed in India in Uttaranchal hills in 1929 (Bose and Sindhan 1976) and later on in Himachal Pradesh (Agarwala 1961).

Plate-1. Seedling blight of apple

The onslaught of the fungus can be easily diagnosed by observing under and above ground parts of the seedlings in the nursery. The earliest above ground manifestation is bronzing of the leaves (Plate 2a) due to infection of lateral roots which turn dark brown and are covered with a greenish gray or white mycelial mat having a flocculent web of whitish strands or ribbons during monsoon season (Plate-2b). As the disease progresses, all the roots are attacked and fibrous root system almost disappears. During rainy season white mycelial mat can be seen in the soil adhering to the roots. White flocculent web disappears after the death of the host, leaving the root surface dotted over with small rounded black sclerotia. The most distinguishing feature of *D. necatrix* is the pear shaped swellings at each septum of the fungal hyphae. The fungus produces scattered black, microsclerotia. The fungus survives either in the form of mycelium or sclerotia in the old infected roots or in soil for several years. Primary infection takes place from diffused mycelium present in root debris or soil or by contact of new plant roots with old dead roots. The fine roots are attacked first. The disease is favoured by high soil moisture and temperature (20-25°C) in acidic soils at pH 6.1 to 6.5. The pathogen spreads very fast along drip irrigation lines. In ill drained soil, the disease develops very fast and mycelium spreads from infected to healthy roots, through soil particles.

Plate 2a. Root rot infected seedling

Plate 2b. Underground web of mycelium on root rot infected seedling

Management

Select nursery site in well drained sites, preferably with light clay loam soils. Do not over fertilize with nitrogen. Prune out diseased and week roots at the time of uprooting nursery plants. Destroy the infected root system/plant debris. Fallowing, deep ploughing and hot water treatment of transplants (45°C for one hour), improving ill drained soils, applying lime in acidic soils help in controlling the disease. The disease can also be controlled by soil solarization particularly in the early stages of infection and it has long term effect in controlling the pathogen in naturally infested soils. The pathogen is sensitive to temperatures higher than 40°C. Soil amendment with neem cake and deodar needles has also been reported to reduce the incidence of the disease. Green amendments such as soybean, black gram, mung bean along with VAM i.e. *Glomus heterospoum* and *Glomus* sp. help in reducing the disease and in improving plant health. Application of fungicides, such as carbendazim (0.1%) or combination of aureofungin (0.02%) + copper sulphate (0.02%) improve the disease control. The treatment is done 3-4 times every year during the rainy season for suffering nursery fields. The antagonists such as *Trichoderma viride*, *T. harzianum* and *Enterobacter aerogenes* have been found to protect the plants from *D. necatrix* and improve plant health and prevent the appearance of the root rot symptoms.

(iii) Collar Rot

Hosts: Apple and pear

Collar rot continues to be a major cause of seedling death in higher hill regions where the summer temperature fluctuates around 20°C and is often observed in low-lying areas where the soils are heavy and poorly drained. Agarwala (1966) recorded the occurrence of collar rot from Himachal Pradesh.

Collar rot symptoms mostly appear near the soil line portion of stem or at graft union if it is near to soil line, in the nursery plants. The bark of the diseased plants is discoloured and turns into a moist rot in which necrotic tissues eventually turn reddish brown to dark brown and develop an alcoholic odour (Plate 3). Wood beneath the necrotic bark is stained dark brown and such staining may extend beyond the edge of the lesion. Such infected seedlings die promptly.

Plate 3. Collar rot infected seedling showing reddish brown discoulouration

Seven Pythiaceous fungi viz., *Phytophthora cactorum, P. citricola, Pythium ultimum, P. debaryanum, P. vexans, P. polytylum* and *Pythium* sp. have been isolated from the soil collected from different apple orchards in Himachal Pradesh near the base of collar rot affected trees. All these fungi are pathogenic on young seedlings. The association of *P. polytylum* was reported for the first time in the world and *P. ultimum, P. debaryanum* and *Phytophthora citricola* from India (Rana and Gupta 1984). *Phytophthora cactorum* (Lebert & Cohn) is the main causal agent. Other pythiaceous fungi namely *Phytophthora syringae, P. megasperma, P. cryptogea, P. cinnamomi, P. drechsleri, P. cambivora, P. citricola, P. pseudosyringae* sp. nov, *Pythium ultimum* and *P. vexans* have also been found to be associated with similar symptoms. The fungus causing crown rot lives in the soil. It can survive long periods of unfavorable conditions in thick-walled spores. It has a wide range of hosts and presumably may be present in most soils suitable for nursery growing. Saturated soil moisture coupled with high soil temperature of 20-25°C is favourable for sporangial production whereas oospores are produced at low moisture level at same temperatures.

Management

A number of preventive measures have been recommended to reduce the primary infection of this disease for its effective control such as elevation of susceptible scions out of the soil splash zone by high grafting at 30cm above soil level, selection of nursery site at properly drained soils, removal of soil from the tree base have been advised to reduce infection where the scion is at or below soil level.

Sanitation practices such as removal of infected plant refuse and dead plants prevent the accumulation of *P. cactorum*. Avoidance of injuries to stem, removal of weeds and long grass from nursery fields are other practices, which are helpful in reducing inoculum. Pre-plant dipping of seedlings in fungicides may help to reduce inoculum density on infested transplants. Drenching of effective fungicides such as metalaxyl and fosetyl-Al into the soil near the crown region of the seedlings help to reduce crown rot. *Enterobacter aerogenes*, a potential bacterial antagonist has been isolated from orchard soil which inhibit *P. cactorum* in culture and is unaffected by the fungicides such as metalaxyl, fosetyl Al and mancozeb. This bacterium reduced root rot of apple seedlings due to *P. cactorum* and the number of viable oospores of the pathogen in sterilized field soil.

(b) Foliar diseases:

(i) Powdery Mildew

Hosts: Wild and cultivated apple, crab apple and pear

Powdery mildew occurs in all major apple growing nurseries, but is especially serious in semi-arid regions. Losses from the disease vary depending upon the inherrent susceptibility of the cultivar, environmental conditions, and management practices. The disease can cause extensive foliar infections in dry years, following mild winters, or on seedlings and grafted plants of highly-susceptible cultivars. Mildew is usually severe on emerging saplings from the seed and young leaves in the nursery. Later on young terminal shoots remain susceptible through out the growing season. Expanding young plant tissues are the most susceptible to pathogen attack.

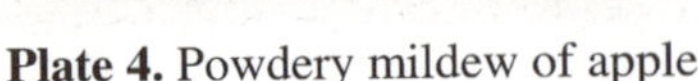

Plate 4. Powdery mildew of apple

Plate 5. Alternaria leaf spot of apple

On the freshly infected leaves, the mildewed area appears on the lower surface as indistinct white patch. Later it is more readily visible and may result in curling of leaves. Eventually, these patches may turn pale purple. The disease symptoms are observed immediately after bud-burst when all the freshly produced tissues are completely mildewed and resulting leaves appear as white rosette (Plate 4). Infected lateral buds may produce completely mildewed side shoots in the same manner. From the mildewed surfaces numerous spores are released which infect young unexpanded leaves, young buds, lateral buds on new shoots and growing points. When the terminal buds begin to grow in early spring, the fungus advances with the new succulent growth. Infected terminal shoots appear stunted and young shoots may be killed outright in the spring, or may survive the season and die in late fall or winter.

The disease is caused by fungus *Podosphaera leucotricha* (Ell. and Ev.) E.S. Salmon. The fungus consists of weft of white hyphal threads and draws nutrition through the haustoria from the infected plant parts. It produces chains of conidia at the host surface giving powdery appearance to the leaf surface. The fungus overwinters as mycelium inside infected buds. As these buds open in the spring, all of their parts become covered with a powdery coating of spores. The spores are easily windblown and serve to infect new leaves and shoots.

Management

The major objectives for the management of powdery mildew in nursery should be to avoid infection of young seedlings, avoid built up of the inoculum, reduce the number of spores produced on newly infected tissues and to prevent secondary infections of new shoots, buds, and leaves during the growing season. Spray coverage is extremely important because infections can occur in the absence of rain and susceptible tissue not covered with the fungicide may become infected. Wettable sulphur fungicides such as Cosan, Solbur, and thiovit were equally effective in controlling the disease (Agarwala and Thirumalachar 1967), however, the introduction of new generation systemic fungicides the management of the disease has become easy however regular sprays are of utmost importance in nursery management. Benomyl, binapacryl, benlate, dinocap and carbendazim fungicides have been reported to provide good control of the disease and also control some other diseases and insects on apple (Sharma and Gupta, 1980; Gupta and Sharma 1981; Gupta et al 1977). Later on Gupta and Gupta (1987) observed Saprol and Calixin having antisporulent activity, however, later was phytotoxic to the foliage. Tridemorph reduced the spore production up to 15 days and effectively controlled powdery mildew (Gupta and Gupta 1991). Some EBI fungicides such as tetraconazole, triadimenol, hexaconazole and bitertanol sprayed at 15 days interval have been reported to control powdery mildew (Mendoza et al 1991). The demethylation inhibiting (DMI) fungicides e.g. flusilazole and myclobutanil effectively control powdery mildew under heavy inoculum pressure (Sholberg and Haag 1994). These fungicides have both inhibitory as well as antisporulent activities against *P. leucotricha* (Climanowski and Bielenine 1996). Winter and summer pruning of silvered twigs up to 1-2 inches before spray is recommended. All silvered shoots should be removed by cutting few millimeters behind the limit of the diseased area. During this practice shaking of twigs should be avoided to prevent the spread of free matured conidia from leaf surface to healthy areas. Coccinellid *Halyzia hauseri* feeds on the mycelium of *Podosphaera leucotricha. Ampelomyces* a common mycoparasite of powdery mildews over-winters as resting hyphae in the dried powdery mildew mycelia covering the shoots and in the parasitized ascomata of *P. leucotricha* on the bark and the scales of the buds. Soon after the bud burst both the fungi start their life cycle but *Ampelomyces* slowly follows the spread of its myco-host on infected apple leaves (Szentivanyi and Kiss 2003).

(ii) Leaf Spots

Hosts: Apple, almond, apricot, cherry, peach, pear, walnut and strawberry

Determining the causes of leaf spots is often difficult because the spotting caused by different pathogens or injuries looks very much alike. In some cases, leaf spots develop as a result of interactions among multiple factors, any one of which would not cause leaf spotting on its own. Causes of leaf spots can sometimes be deduced from disease patterns within the nursery, the time of symptom appearance, and characteristics of the individual leaf spots. Apple scab, rust and powdery mildew cause spots on leaves although they are not usually classified as "leaf-spotting fungi." Alternaria leaf spot looks very similar to frog-eye leaf spot caused by *Botryosphaeria obtusa* exemplifies the more stereotypical leaf spot disease.

Leaf spots of Apple

The leaf spots occurring in nursery plants and orchards are almost common depending upon the prevailing climatic conditions. All types of leaf spots occurring in pome and stone fruits have been discussed in the following text. Normally spots can be grouped in to two types. The less frequent one occurs on the leaf between the veins as a small, light-gray spot with a dark dot in its center. The more common leaf spots occur on the veins, midribs, and petioles as long, narrow, slightly sunken, light-colored lesions. These contain several dark dots, the fruiting structures (pycnidia) of the fungus. When petiole infections are numerous leaf drop may result. New shoot infections first appear similar to the petiole infections except they are longer and are seen more easily. They occur at the junction of the petiole with the shoot (node) or between the nodes. Once the lesion becomes established, it may continue to enlarge for 3 or 4 years, becoming noticeably larger than the diameter of the normal limb. Chaudhury and Nath (1935) reported the occurrence of leaf spot caused by *Oothecium indicum* which produced purplish to rusty spots on apple leaves. Mundkur and Ahmad (1946) reported the occurrence of *Linospora ochracea* and *Canavalis mali* on apple leaves. Chona et al (1956) recorded the occurrence of *Discosia artocreas* on dead and fallen leaves and *Mycosphaerella pyrina* on living leaves of *Pyrus communis* and *Helicotrichum obscurum* on the dead leaves of *Pyrus malus*. Chandra (1964) reported the occurrence of *Sphaeropsis malorum* on the apple leaves for the first time. Gupta and Agarwala (1968) observed the symptoms developed by *Alternaria mali* in Kotgarh area of Himachal Pradesh (Plate 5). Rao (1965) reported the occurrence of *Phyllosticta solitaria* on leaves of *Malus sylvestris*.

Sarbhoy et al (1978) recorded the occurrence of *Pestalotiopsis japonica* on apple leaves from Manali in Himachal Pradesh. Sud and Agarwala (1972) reported the occurrence of *Coniothyrium pyrinum* on *Prunus amygdalus* and *Pyrus pashia* which are new host records of the world. *Pestalotia hartigii* was recorded on living leaves of *Malus pumila.* Various other fungi recorded on different pome and stone fruits are *Sphaeropsis malorum* on *M. pumila; Cladosporium* sp. on *Prunus persica*; *Phoma limitata* on *P. armeniaca; Alternaria alternata* on *P. armeniaca* and *P. avium* and *Phyllosticta prunicola* on *Prunus avium* and *P. salicina* and *Phyllosticta mali* on *Malus pumila* (Sud and Agarwala 1972). Alternaria leaf spot (*A. alternata*) has recently appeared in severe form causing 40 per cent defoliation in Kotkhai area. The disease was characterized by appearance of 1-3 mm dia. reddish brown necrotic spots on young apple leaves resulting in bigger blotches. Dark brown to black elongated lesions appeared on leaf petioles and mid ribs causing yellowing and defoliation of infected leaves (Sharma et al 2005).

Leaf spots of Almond

Alternaria pruni and *A. tenuis* cause leaf spots on almond (Saini, et al 1989, Singh and Singh 1987). *Gloeosporium amygdalinum* was found causing anthracnose (Srivastava et al 1984). *Sphaerotheca pannosa* was reported causing powdery mildew of almond and peach in India (Cooke 1876), however, first ever record of fungi from the Himalayan region is that of *Polystigma ochraceum* (Wahlenb.) Sacc. on *Prunus padus.*

Leaf spot of Apricot

Phyllosticta cerasicola was recorded on apricot from Himachal Pradesh (Gupta and Sohi 1967). Later on *Phoma glomerata, Sphaeropsis* sp., *Botryosphaeria dothidea* and *Microdiplodia pruni* was also recorded from Kullu Valley by Gupta and Verma (1986).

Leaf spot of cherry

Cercospora rubrotincta has been recorded on sweet cherry leaves. It produced small circular, reddish brown spots with dark red margin in early June (Plate 6). These enlarge to 1/8 to 1/4 inch in diameter, becoming red-brown to brown. By then, the spots show brown on the undersides of the leaves and during wet periods tiny, whitish, felt-like patches appear in their centers. Affected leaves develop pronounced chlorosis and subsequently the plants gave a striking yellow appearance (Kaul 1962; 1967). Verma and Gupta (1979) recorded association of *C. circumscissa* and all the cherry cultivars were susceptible to it and Captan (0.2%) was effective for its control.

Plate 6 Cercospora leaf spot on cherry

Plate 7. Leaf spot of pear

Leaf spot of peach

The infected leaves (*Alternaria alternata*) show brown to dark brown spots starting from the outer portion of the leaves in an irregular manner. The spots coaleceed and sometimes produced concentric rings resulting in blighted appearance and defoliation (Madan et al 1979). Target leaf spot caused by *Phyllosticta persicola* was reported for the first time from Arunachal Pradesh in India in 1984 (Sharma and Paul 1986).

Leaf spot of pear

Chona et al (1956) reported the occurrence of *Fusicladium dendriticum* var. *orbiculatum* on living leaves of *Pyrus pashia* in Mandi and Kullu areas of Himachal Pradesh and *Fusicladium diospyri* and *Pestalotia breviseta* were also reported (Munjal et al 1959). Berkeley (1856) reported the occurrence of *Dasyscypha clandestine*, *Vibrissea stilboidea* and *Vibrissea turbinella* on leaves of *Pyrus* spp.

The *Phyllosticta pyricola* survives as pycnidia in the infected plant debris. The disease appeared in third week of July (Plate 7) and whole plant is infected within 25 days. The disease appears early by 9 days on the leaves upto 1 feet height in comparison to the leaves borne at 10-11 feet height (Sharma et al 1986).

Leaf spots of strawberry

Various pathogens such as *Hainesia lythri* (Bose 1953), *Pestalotia jeolikotensis* sp. nov Bose; *Pillidium concavum, Dendrophoma obsurans, Marssonina fragariae* and *Mycosphaerella fragariae* (Plate 8) (Bose 1970), *Ramularia narakandensis* sp. nov (Deighton 1973), *Rhizoctonia bataticola* (Plate 9), *Hainesia lythrion* (Phatak and Payak 1965), *Pestalotiopsis disseminate* (Singh et al 1975) causing various leaf spots in strawberry in Kumaon hills of Uttrakhand and Himachal Pradesh have been reported. *Colletotrichum fragariae* have been found associated with tan to light brown coloured water soaked lesions on strawberry fruits (Singh 1974).

Plate 8. Mycospherella leaf spot of strawberry

Plate 9. Rhizoctonia spot of strawberry

Management

Pre-infection application of Dithane Z-78 (0.25%) was effective for the control of leaf spots caused by *Pestalotia hartigii* or *Phyllosticta mali* (Bhardwaj and Agarwala 1985, 1986). Frog eye leaf spot is favoured by high rainfall and relative humidity (>75%) at 20-22°C (Singh and Agarwala 1987) which can be kept in check by the application of Blitox (Kondal 1977), whereas leaf spots caused by *Conithyrium pyrinum* and *Mycosphaerella* sp. can be best controlled by application of ferbam (0.25%) and Aureofungin (0.004%), respectively (Kondal and Agarwala 1973).

Application of N, P fertilization + Bavistin (0.1%) before disease appearance and subsequently at 20 days interval controlled leaf spot caused by *Phyllosticta pyrina* (Roy et al 1985). Alternaria blight of apple (*A. mali*) can be successfully controlled by applying ziram (0.1%) or ferbam, zineb or captan (0.15%) or Bordeaux mixture (1%) (Sud and Agarwala 1971). Protective application of dodine, mancozeb, zineb, iprodione and mancozeb + hexaconazole exhibited effective control of Alternaria leaf spots (*A. alternata*). All the commercially grown cultivars are highly susceptible; however, pollinizers are tolerant (Sharma et al 2005).

All the pear varieties are susceptible to leaf spot (*Phyllosticta pyricola),* however, Kieffer was least susceptible (Gautam and Gupta 1989). Four sprays of mancozeb (0.25%) or carbendazim (0.05%) control this disease under field conditions (Gupta and Gautam 1988). Two sprays of benomyl at fortnightly interval have been found effective for management of disease (Sharma et al 1986).

In strawberry, sprays of Bavistin, Benlate, Difolatan Ziram and Fernacol have been found effective for reducing disease incidence in strawberry (Sindhan and Roy 1981).

(iii) Rust

Hosts: Apricot, plum, peach, almond

Rust occurs on almost all the stone fruits. It is particularly harmful in warmer areas. It is caused by *Tranzschelia discolor.* Rust pustules start appearing in June at temperature 17.3-32.4°C. Symptoms appear on both surfaces of leaves as pale yellowish spots, which later change into bright yellow islands prominent on upper surface only (Plate 10). On lower surface, cinnamon brown dusty pustules appear. Severe infection leads to defoliation. Twig infection is uncommon on peach, but on plum, the bark gets infected during the early spring when buds are swelling.

Plate 10. Rust of stone fruits

Plate 11. Shot hole/ Stigmina blight of stone fruits

The occurrence of rust in almond has been recorded in 1985 and the pathogen was identified as *Tranzschelia discolor* f. sp. *dulcis* (Sharma et al 1987). The occurrence of Himalayan rust caused by *Gymnosporangium cunninghamianum* and its life history on *Pyrus pashia* was reported by Barclay (1890). This pathogen was also reported on *Cotoneaster bacillaris* from Dalhausie, Himachal Pradesh by Cummins (1943). The occurrence of rust on plum was first recorded in India by Waraitch and Khatri in 1976. Bhardwaj and Shyam (1986) recorded severe out break of rust of Japanese plum (*Prunus salciana*) from Himachal Pradesh and pathogen has been identified as *Tranzschelia pruni –spinosae*. Rust development is positively correlated with humidity and rainfall and negatively with temperature (Ram et al 2003). The maximum temperatures (21.6-28°C) during July-August trigger the disease severity wherein rain and relative humidity had positive and insignificant role.

Management

Following measures have been suggested for the control of rust. Spray Bordeaux mixture (1%) during February and March. Pre-symptom application of Anvil and Impact (0.05%) and Dithane M-45 (0.25%) at 21 days interval provide best control of rust (Bhardwaj 1991). The disease can be controlled by spraying SAN-619 (Cyproconazole) (Ram et al 2000). Application of Delan (0.05%) at bud swell and after petal fall proved promising for controlling rust, leaf spots (*Phyllosticta prunicola* and *Cercospora* spp.) and powdery mildew mildew (*Sphaerotheca pannosa*) on peach, plum, apricot and almond (Thakur et al 1991). Sprays of triadimefon (0.2%), carbendazim (0.1%) or bitertanol (0.2%) at 15 days interval effectively control rust (Sharma and Gupta 1993). The disease is effectively controlled with a single prophylactic spray of mancozeb (0.3%), when temperature is 20-23°C at all locations in H.P. (Thakur and Xu 2004).

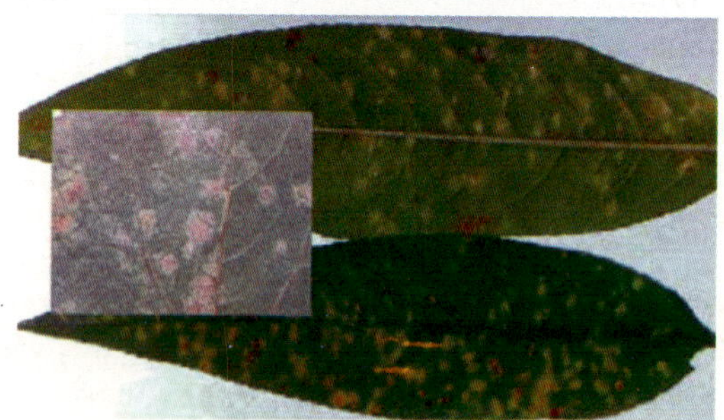

Plate 12. Apricot/peach frosty mildew

(iv) Shot Hole/ Stigmina Blight

Hosts: Almond, apricot, plum, peach, cherry

Shot hole also called as stigmina blight, is caused by *Wilsonomyces carpophilus (Clasterosporium, Coryneum* and *Stigmina).* The disease was recorded first time in India on cherry trees in Kashmir Valley during March-May 1987 and is very serious in Kinnaur district of Himachal Pradesh where the maximum disease incidence has been recorded in Nichar. The fungus attacks the twigs, leaves, and unopened buds.

On leaves, dark brown scattered lesions enlarge rapidly and loss of the dead areas results in the formation of shot holes (Plate 11). On twigs, small, purplish raised spots appear, which expand into necrotic cankers. High temperature and prolonged rain favour the spread of the disease (Gupta et al 1979).

The disease can be effectively controlled by spraying Difolatan (0.2%) followed by Foltaf, Blitox, Dithane M-45 and Dithane Z-78 (Sharma and Gupta 1988). Spray application of Captaf (0.2%) at leaf fall in autumn (November) and at bud swell stage in spring gave 96 per cent control of the disease (Gupta et al 1973; 1979).

(v) Frosty Mildew

Hosts: Peach, apricot

The disease is very serious in stone fruit nurseries and results in premature defoliation. The disease is caused by *Cercosporella persica* and *Gloeosporium laeticolor,* respectively (Sharma and Paul 1986). Sohi et al (1964) recorded 7 new fungal species from Himachal Pradesh, out of which frosty mildew (*Cercosporella persicae*) was recorded for the first time. They also recorded the presence of *Cercospora persicae*. Frosty mildew of plum is caused by *Cercosporella persica*. As a result of infection, fungus produces pale green areas on the upper leaf surface and creamy white fungal growth on lower surface (Plate-12).

Future thrusts

Most of the present knowledge about the management of nursery diseases is mainly through chemicals. The integration of various cultural practices like soil solarization particularly for soil borne diseases is needed to be followed along with biocontrol agents and biopesticides. However the foliage diseases can be effectively managed by the judicious use of chemicals in combination with plant origin pesticides.

References

Agarwala RK 1961 Problem of root rot of apple in Himachal Pradesh and prospectus of its control with antibiotics, *Himachal Horticulture* **2** 171-178

Agarwala RK 1966 Annual progress report of the scheme for intensification of research on root rot and collar rot disease of apple in Himachal Pradesh, 1965-66

Agarwala RK and Thirumalachar MJ 1967 Effect of Aureofungin on powdery mildew of apple, *Proceedings of 1st International Symposium on Plant Pathology*, New Delhi

Anonymous 1952 Fungoid diseases of crops and their remedies, *Mysore Agriculture Calendar* 1952-53, 41-46 pages

Barclay A 1890 On the life history of a Himalayan Gymnosporangium (*G. cunninghamianum* n. sp.), Scientific Memoirs Medical Officers, Army of India **5** 71-78

Berkeley MJ 1856 Decades of fungi, Decas, 1-62, 1-620, In Hooker London, *Journal of Botany* 3-8

Bhardwaj LN 1991 Chemical control of almond rust in dry temperate zone of Himachal Pradesh, *Plant Disease Research* **6(2)** 89-90

Bhardwaj LN and Agarwala RK 1986 Seedling blight of apple and its control, in *Advances in Research on Temperate Fruits* (eds) TR Chadha, VP Bhutani, and JL Kaul, UHF, Solan, 397-401 pages

Bhardwaj LN 1987 Management of seedling blight disease of *Malus domestica* in Himachal Pradesh, *Journal of Tree Sciences* **6** 38-40

Bhardwaj LN and Agarwala RK 1985 Evaluation of fungicides against leaf spot of apple caused by *Pestalotia hartigii* Tub., *Himachal Journal of Agricultural Research* **11(2)** 44-46

Bhardwaj LN and Agarwala RK 1986 Chemical control of leaf spot of apple caused by *Phyllosticta mali* Prill. and Del, *Journal of the Institute of Agriculture and Animal Science* **7** 115-118

Bhardwaj LN and Shyam KR 1986 Occurrence of rust on Japanese plum, *Himachal Journal of Agricultural Research* **12** 62

Bose SK and Sindhan GS 1976 Evaluation of fungicides against powdery mildew (*Podosphaera leucotricha* Sacc.) of apple, *Progressive Horticulture* **8** 15-18

Bose SK 1953 Diseases of strawberry in Kumaon and their control, *Proceedings of 40th Indian Science Congress*, Part iii (Abstract)

Bose SK 1970 Diseases of valley fruits in Kumaon (iii) Leaf spot diseases of strawberry, *Progressive Horticulture* **2(2)** 33-35

Chandra S 1964 Three new fungi to India, *Current Science* **33** 440-441

Chaudhury H and Nath P 1935 Studies on the diseases of apple in Northern India, I, A new leaf spot disease of apples caused by *Oothecium indicum* n. sp., *Journal of Indian Botanical Society* **XIV** 101-107

Chona BL Munjal RL, and Kapoor JN 1956 Notes on miscellaneous Indian fungi-V, *Indian Phytopathology* **10** 148-156

Climanowski IJ and Bielenine A 1996 Antisporulent activity of new apple powdery mildew fungicides, *Journal of Fruit Ornamental Plant Research* **4** 21-25

Cooke MC 1876 Some Indian Fungi, *Grev* **5** 14-17

Cummins GB 1943 Uredinales from North Western Himalayas, *Mycologia* **35** 440-458

Deighton FC 1973 Ramularia narkandensis sp. Nov. on Fragaria in India, Transactions of British Mycological Society **60** 162-164

Gautam HR and Gupta VK 1989 Screening of pear cultivars against leaf spot of pear *(Phyllosticta piricola)*, *Indian Phytopathollogy* **42** 285 (Abstract)

Gupta GK and Agarwala RK 1968 Alternaria blight of apple, *FAO Plant Protection Bulletin* **16** 32

Gupta GK and Gupta SK 1987 Post symptom antisporulent activity of chemicals against apple powdery mildew, *Plant Disease Research* **2** 131-132

Gupta GK and Sohi HS 1967 New records of fungi, *Research Bulletin (N.S.) Panjab University* **18** 405-408

Gupta GK and Verma KD 1986 Fungi associated with die-back of almond (*Prunus amygdalus*) and apricot (*Prunus armeniaca*), *Indian Phytopathology* **39** 315

Gupta GK, Agarwala RK, and Dutt K 1973 Apricot leaf curl caused by *Taphrina deformans* in Kullu Valley, India, *Plant Disease Reporter* **57** 361-362

Gupta GK, Agarwala RK and Dutt K 1979 Stigmina blight of peach and its control in Himachal Pradesh, *Horticultural Advances* **9** 49-51

Gupta SK and Gupta GK 1991 Eradicative activity of fungicides against apple powdery mildew, *Indian Journal of Mycology and Plant Pathology* **21** 70-72

Gupta VK and Gautam HR 1988 Control of *Phyllosticta* leaf spot of pear, in *Tree protection* (eds) VK Gupta and NK Sharma, Solan: Indian Society of Tree Scientists 42 (Abstract)

Gupta VK and Sharma SK 1981 Studies on the effects of spraying fungicides on apple powdery mildew, *Anzeiger Fur Schadlingskde Pflanzenchutz* **54** 8-10

Gupta VK, Verma KD, and Munjal RL 1977 Influence of fungicides on the development of powdery mildew in susceptible cultivars of apple, *Indian Phytopathology* **30** 245-250

Kaul TN 1967 Diseases of stone fruits in Kashmir, *Horticulturist,* Srinagar **2** 52-58

Kaul TN 1962 Occurrence of *Cercospora rubrotincta* ELL. &Ev. on cherry in India, *Current Science* **31** 29-30

Kondal MR 1977 Efficacy of fungicides for the control of frog eye leaf spot of apple, *Indian Journal of Mycology and Plant Pathology* **6** 93

Kondal MR and Agarwala RK 1973 Relative efficacy of fungicides for the control of leaf spot diseases of apple, *Indian Journal of Mycology and Plant Pathology* **3** 122

Madan RL, Suhag LS and Gupta PC 1979 A new leaf spot disease of peach (*Prunus persica*), *Indian Journal of Mycology and Plant Pathology* **9** 108

Mendoza ZC, Pounce GF, and Lepiz LR 1991 Chemical control of apple powdery mildew (*Podosphaera leucotricha*) and apple scab (*Venturia inaequalis*) with inhibitors of ergosterol biosynthesis in Zacatlan, Puebla, *Revista Mexicana de Fitopatologia* **9** 144-147

Mundkur BB and Ahmad S 1946 Revisions and additions to Indian fungi-II, *Mycological Paper, Imperial Mycological Institute* **18** 11

Munjal RL, Chona BL, and Kapoor JN 1959 Notes on some miscellaneous Indian fungi VI, *Indian Phytopathology* **12** 176-181

Phatak HC and Payak MM 1965 *Hainesia lythri* on strawberry-a new record for India, *Indian Phytopatholgy* **18** 237-239

Ram V, Sharma U and Bhardwaj LN 2003 Effect of weather parameters on the development of apricot leaf rust (*Tranzschelia discolor* f.sp. *armeniacae), Plant Disease Research* **18(2)** 180-182

Ram V, Sharma U and Bhardwaj LN 2000 Management of apricot leaf rust, *Plant Disease Research* **15(1)** 113-115

Rana KS and Gupta VK 1984 Occurrence of phythiaceous fungi in collar rot affected apple soils of Himachal Pradesh, *Indian Phytopathology* **37** 39-42

Rao VG 1965 The genus *Phyllosticta* in India, *Sydowia* **19** 117-120

Roy AJ, Seth JN, and Shah A 1985 Comparative effect of fertilizers and different fungicidal treatments on the incidence of leaf spot diseases of apple var. Red Delicious, *Progressive Horticulture* **17(2)** 241-244

Saini SS, Kumari Sudesh, and Atri NS 1989 Two new leaf spot diseases from India caused by *Alternaria, Current Science* **58** 514-515

Sarbhoy AK, Agarwal DK and Varshney JL 1978 Ten new host records of *Pestalotiopsis* from India, *Indian Phytopathology* **31** 535-536

Sharma IM and Gupta VK 1988 Incidence of *Stigmina* blight of apricot caused by *Stigmina carpophila* (Lev.) Ellis in dry temperate zone of Himachal Pradesh and its chemical control, in *Tree Protection* (eds) VK Gupta and NK Sharma, Solan: Indian Society of Tree Scientists 444(Abstract)

Sharma IM and Gupta VK 1993 Field evaluation of some fungicides against almond rust in Himachal Pradesh, *Indian Journal of Mycology and Plant Pathology* **23(2)** 203-205

Sharma JN, Gupta D, Bhardwaj LN and Kumar R 2005 Occurrence of *Alternaria* leaf spot (*Alternaria alternata*) on apple and its management, in *Challenging Problems in Horticultural and Forest Pathology* RC Sharma and JN Sharma, New Delhi: Indus Publication Company 25-31pages

Sharma RC and Paul YS 1986 Target leaf spot of peach - A new disease from India, *Plant Disease Research* **1** 90-91

Sharma RC, Gupta VK, and Agarwala RK 1986 Epidemiology and control of Phyllosticta leaf spot of pear, *National Symposium in social forestry for rural Development,* Delhi University, Delhi, 12 (Abstract)

Sharma RC, Jindal KK and Kaul JL 1987 Reaction of apricot cultivars to *Tranzschella discolor* rust in India, *IUFRO Workshop on Agro forestry for Rural Needs,* New Delhi, 32 (Abstract)

Sharma SK and Gupta VK 1980 Efficacy of fungicides in relation to time of application against powdery mildew of apple, *Indian Journal of Mycology and Plant Pathology* **10** 173-174

Sharma SK, Kishore DK and Pramanick KK 2005 *Dematophora necatrix*-a serious pathogen of temperate fruits, in *Challenging Problems in Horticultural and Forest Pathology* (eds) RC Sharma and JN Sharma, New Delhi: Indus Publication Company, 53-71 pages

Sholberg PL and Haag P 1994 Control of apple powdery mildew in British Columbia by demethylation inhibiting fungicides, *Canadian Plant Disease Survey* **74** 5-11

Sindhan GS and Roy AJ 1981 Screening of different varieties and selections of strawberry against leaf diseases and their control *Indian Phytopatholgoy* **34** 304-306

Sindhan GS, Ghildyal PC, and Shah A 1979 A note on the evaluation of fungicides against *Sclerotium rolfsii* Sacc. *in vitro, Progressive Horticulture* **11(2)** 45-47

Singh D and Agarwala RK 1987 Incidence of apple frog eye leaf spot (*Sphaeropsis malorum*) and its correlation with weather conditions, *Indian Journal of Mycology and Plant Pathology* **17** 169-172

Singh G and Singh RN 1987 Alternariosis of almond leaves, *Indian Journal of Mycology and Plant Pathology* **17** 82

Singh SJ 1974 A ripe fruit rot of strawberry caused by *Colletotrichum fragariae*, *Indian Phytopatholgoy* **27** 433- 434

Singh SJ, Prakash O, and Tewari RP 1975 A leaf spot disease of strawberry from India, *Indian Phytopathology* **28** 303-304

Sohi HS, Jain SS, Sharma SL, and Verma BR 1964 New records of plant diseases from Himachal Pradesh, *Indian Phytopathology* **17** 42-45

Srivastava KK, Singh RK, Shah A, and Roy AJ 1984 Anthracnose disease of the almond (*Amygdalus communis* Linn.), *Progressive Horticulture* **16(1-2)** 140

Sud VK and Agarwala RK 1971 Laboratory and field evaluation of fungicides for the control of Alternaria blight of apple, *Indian Phytopathology* **24** 201-204

Sud VK and Agarwala RK 1972 New records of pome and stone fruit tree diseases from Himachal Pradesh-II, *Indian Phytopathology* **25** 580-582

Szentivanyi O and Kiss L 2003 Over-wintering of *Ampelomyces* mycoparasites on apple trees and other plants infected with powdery mildews, *Plant Pathology* **52** 737-746

Thakur VS and Xu XM 2004 Epidemiology and management of almond rust (*Tranzschelia discolor* Dulcis) in Himachal Pradesh, *Advances in Plant Sciences* **17(1)** 97-107

Thakur VS, Gupta VK, and Garg RC 1991 Management of fungal diseases of stone fruits with fungicides in Himachal Pradesh, *Indian Journal of Agricultural Sciences* **61** 677-681

Verma KD and Gupta GK 1979 Studies on leaf spot of cherry (*Prunus avium*) caused by *Cercospora circumscissa* Sacc. and its control, *Progressive Horticulture* **10(7)** 57-62

Waraitch KS and Khatri HL 1976 The occurrence of pink disease and rust of plum in India, *Indian Journal of Mycology and Plant Pathology* **7** 203-206.

20

Plant Viruses: Causes of Plant Diseases and their Management

Awadhesh Kumar Pal

Department of Plant Breeding and Genetics, Bihar Agricultural University, Sabour, Bhagalpur-813210, Bihar

Agriculture has been the base for human civilization all over the world since long back. Directly or indirectly, plants make up all the food on which human and other animals depend. Plants whether wild or cultivated, also get sick. Sick plants grow and yields poorly, exhibit various types of symptoms and often, parts or whole plants die. The causative agents include pathogenic microorganisms such as viruses, bacteria, fungi, protozoa, nematodes and unfavourable environmental conditions such as drought, salinity, chilling and high temperature and lack of nutrients. Viruses represent significant threats to modern agriculture and their biological control remains a challenge in the twenty-first century (Ritzenthaler 2005).

Although viruses behave like microorganism in that they cause disease, can function genetically and are able to reproduce, they also behave like a chemical molecule, as they are inactive outside the living system. They consist of nucleic acid and a protein, which forms a protective coat around the nucleic acid. The nucleic material is either RNA or DNA, both can be single stranded or double stranded. Therefore, a virus is a nucleoprotein that has the ability to cause disease. It multiplies in a living host and is very small to be seen under light microscope. The shape of the viruses can vary from rod shape to polyhedral. All viruses are parasitic and infect almost all form of living organisms, ranging from bacteria to higher plants and animals. Viruses cannot reproduce themselves by producing any reproductive structures. Instead, they infect and induce host cell to produce more virus particles. They cause disease utilizing the host machinery during multiplication, make space for itself into the host cell and disrupt all the cellular processes. This upset the host cellular metabolic processes showing various symptoms in the infected host cells (Caranta et al 2011).

Plant viruses, viroids, and phytoplasmas are known to cause considerable losses in crop yield, quality of plants, and plant products (Kirkpatrick, 1991; Waterworth and Hadidi, 1998; Randles, 2003). There are hundreds of plant viruses which are classified into 6 orders and 22 families with additional 72 families which are not assigned to any order yet (ICTV webpage, 2011) that infect economically important horticultural, vegetable, field and ornamental crops. They pose a particular risk because they are difficult to detect and identify. Disease caused by plant viruses were illustrated in the form of mosaic type symptoms more than three hundred years ago.

Brief history of plant viruses

Mayer (1886) reproduced the "tobacco mosaic" disease by injecting juice from infected plants to the healthy one. Mayer could not identify the causal organism. However, Ivanowski (1892) gave the first concrete evidence for the existence of a non-bacterial infectious agent, showing that infected sap remained infectious even after filtering through finest Chamberland filter candles, which bacteria cannot. Therefore, he concluded it to be a toxin released by bacteria. It was Beijerinick (1898), who finally concluded that tobacco mosaic was caused not by microorganism but by a living fluid, which he called a virus. The roman meaning of virus is 'poison'. Stanley (1935) added ammonium sulfate to tobacco juice extracted from infected tobacco leaves and obtained crystalline protein as sediment in the flask. This protein when rubbed on healthy tobacco plant caused the mosaic disease. Therefore, he concluded virus to be autocatalytic protein that could multiply within living cells. Bowden et al (1936) demonstrated that the crystalline preparations of the virus actually consisted of not only protein but also a small amount of ribonucleic acid (RNA). The first tobacco mosaic virus particles were seen with the electron microscope by Kausche et al (1939). Gierrer and Schramm (1956) showed that the protein could be removed from the virus and that the RNA carried all the genetic information, which enabled it to cause infection and reproduce more virus particles.

Nomenclature and classification of viruses

The highest level of virus classification recognizes six major groups, based on the nature of the genome (Van Regenmortel and Fauquet, 2000):

- ***Double-stranded DNA (dsDNA):*** It includes those viruses with the largest known genomes (up to about 400,000 base pairs) and there is only one genome component, which may be linear or circular. Well-known viruses in this group include the herpes and pox viruses.
- ***Single-stranded DNA (ssDNA):*** There are two families of plant viruses with single stranded DNA and both have small circular genome components, often with two or more segments.
- ***Reverse-transcribing viruses***: Their replication includes the synthesis of DNA from RNA by the enzyme reverse transcriptase; many integrate into their host genomes. These have dsDNA or ssRNA genomes. The group includes the retroviruses, of which Human immunodeficiency virus (HIV), the cause of AIDS, is a member. There is a single family of plant viruses in this group and this is characterized by a single component of circular dsDNA, the replication of which is via an RNA intermediate.
- ***Double-stranded RNA (dsRNA):*** These groups include some plant viruses and many of the mycoviruses.

- ***Negative sense single-stranded RNA (ssRNA-):*** Some or all of the genes are translated into protein from an RNA strand complementary to that of the genome. There are some plant viruses in this group along with the viruses that cause measles, influenza and rabies.
- ***Positive sense single-stranded RNA (ssRNA+):*** The majority of plant viruses are included in this group. It also includes the SARS coronavirus and many other viruses that cause respiratory diseases (including the "common cold"), and the causal agents of polio and foot-and-mouth disease.

Within each of these groups, many different characteristics are used to classify the viruses into families, genera and species. Typically, a combination of characters is used and some of the most important are:

Particle morphology

Genome properties

Biological properties

Serological properties

1. *Particle morphology*: Amongst plant viruses, the most frequently encountered shapes are:

(i) *Isometric*: Apparently spherical and about 18nm or more in diameter (depending on the species). e.g., *Tobacco necrosis virus*, genus *Necrovirus* with particles 26 nm in diameter.

(ii) *Rod-shaped*: About 20-25 nm in diameter and 100-300 nm long. These appear rigid and often have a clear central canal (depending on the staining method used). Some viruses have two or more different lengths of particles and these contain different genome components. e.g., *Tobacco mosaic virus*, genus *Tobamovirus* with particles 300 nm long.

(iii) *Bacilliform*: Short round-ended rods. These come in various forms up to about 30 nm wide and 300 nm long. e.g., *Cocoa swollen shoot virus*, genus *Badnavirus* with particles 28 x 130 nm.

(iv) *Filamentous*: Usually about 12 nm in diameter and more flexuous than the rod-shaped particles. They can be up to 1000 nm long, or even longer in some instances. Some viruses have two or more different lengths of particle and these contain different genome components. e.g., *Potato virus Y*, genus *Potyvirus* with particles 740 nm long.

(v) *Geminate*: Twinned isometric particles about 30 x 18 nm. These particles are diagnostic for viruses in the family *Geminiviridae* which are widespread in many crops especially in tropical regions. e.g., *Maize streak virus*, genus *Mastrevirus*.

2. *Genome properties*: This includes the number of genome components and the translation strategy. Where genome sequences have been determined, the relatedness of different sequences is often an important factor in discriminating between species (Reed and Rice2000). Important features of genome include:

- *Nature of the genome*: circular (as in all known plant DNA viruses) or linear.

- *Number of genome components*: This varies from a single component (e.g. in the genera *Potyvirus* and *Tobamovirus*) to eleven (in some members of the genus *Nanovirus*). Individual components vary in size from about 1 kb (*Nanovirus*) to about 20 kb (*Closterovirus*).
- *Number of genes*: These vary considerably. Most plant viruses have at least 3 genes; 1 (or more) concerned with replication of the nucleic acid, 1 (or more) concerned with cell-to-cell movement of the virus and 1 (or more) encoding a structural protein that is assembled into the virus particle (usually called the "coat" or "capsid" protein). There may also be some additional genes that have a regulatory function or which may be required for transmission between plants (association with a vector).
- *Translation strategy*: Different strategies are employed to translate the genes from the genome components either directly or via mRNA intermediates and (in some cases) to permit different amounts of protein to be produced from the different genes. Following examples here serve to illustrate some of the variety involved in translation:
    1. Genus *Potyvirus*: There is one ssRNA component that encodes single polyprotein of about 350 kDa. This is cleaved by 3 different proteases into 10 different mature proteins. All of these proteases are encoded by the virus itself. The two proteins at the C-terminus of the polyprotein are respectively an RNA-dependent RNA polymerase (NIb, involved in replication of the virus) and the (single) coat protein (CP).
    2. Genus *Furovirus*: These have two ssRNA components. The 5'-proximal gene on RNA is translated directly from the genomic RNA: on RNA1 (the larger RNA component) which encodes a replication protein and on RNA 2 which encodes for the coat protein. The stop codons of both of these genes are "leaky" and in a small percentage of cases, translation continues to produce a larger ("readthrough") protein.
    3. Genus *Fijivirus*: There are 10 components of dsRNA. Most of the components encode a single protein and at least three of these are structural proteins assembled into the complex virion.
- *Genome relatedness*: The degree of nucleotide identity (or amino acid identity in the protein sequence) between sequences is often used to examine the relationship between different viruses or isolates. For example, in the genus *Carlavirus*, different species have less than 73% nucleotide identity (or 80% amino acid identity) in their coat proteins.

3. *Biological properties*: This may include the type of host and the mode of transmission. In some families, the type of host is a useful feature for classification (Murphy et al 1995). For example, in the family *Reoviridae*, there are 3 genera with plant-infecting members (*Fijivirus*, *Oryzavirus* and *Phytoreovirus*), 1 genus of mycoviruses (*Mycoreovirus*), 1 genus containing viruses of fish and cephalopods (*Aquareovirus*), two genera that are restricted to insects (*Cypovirus* and *Entomoreovirus*) and 5 genera of vertebrate viruses that sometimes also infect insects.

The mode of transmission is an useful characteristic of some groups of plant viruses. For example in the family *Potyviridae*, members of the largest genus (*Potyvirus*) are transmitted by aphids, while viruses in the genera *Rymovirus* and *Tritimovirus* are transmitted by mites of the genus *Abacarus* or *Aceria* respectively, those in the genus *Ipomovirus* are transmitted by whiteflies and those in the genus *Bymovirus* by plasmodiphorids (root-infecting parasites once considered to be fungi but probably more closely related to protists).

4. *Serological properties*: It includes the relatedness (or otherwise) of the virion protein(s). Many viruses are good antigens i.e they elicit strong antibody production when purified preparations are injected into a mammal. This property has been widely exploited to produce specific antibodies that can be used for virus detection and for examining relationships between viruses. Agar diffusion plates used previously have been largely superseded by ELISA (enzyme-linked immune sorbent assay) procedures. Although serological properties are still important, their significance in taxonomy has now declined to some extent due to the availability of nucleotide sequence data (Cheng Mayer and Levy 1988).

Transmission of plant viruses

Plant cells have a robust cell wall as opposed to animals; therefore viruses cannot penetrate them unaided. Most plant viruses are therefore transmitted by a vector organism that feeds on the plant or are introduced through wounds (Blanc and Ducker, 2011). A small number of viruses can be transmitted through pollen to the seed (e.g. *Barley stripe mosaic virus*, genus *Hordei virus*) while many that cause systemic infections accumulate in vegetatively propagated crops. The major vectors of plant viruses are:

1. *Insects*: This forms the largest and most significant vector group and particularly includes:

Aphids: Transmit viruses from many different genera, including *Potyvirus*, *Cucumovirus* and *Luteovirus*. Example: green peach aphid (*Myzus persicae*) is the vector of many plant viruses, including *Potato virus Y*.

Whiteflies: Transmit viruses from several genera but particularly those in the genus *Begomovirus*. Example: *Bemisia tabaci* is the vector of many viruses including *Tomato yellow leaf curl virus* and *Lettuce infectious yellows virus*.

Hoppers: Transmit viruses from several genera, including those in the families *Rhabdoviridaeand Reoviridae*. *Example: Micrutalis malleifera* is the treehopper vector *of Tomato pseudo-curly top virus.*

Thrips: Transmit viruses in the genus *Tospovirus*. Example: *Frankinella occidentalis*, the western flower thrips is a major vector of *Tomato spotted wilt virus*.

Beetles: Transmit viruses fro*m several* genera, including *Comovirus and Sobemovirus.*

2 *Nematodes:* These are root-feeding parasites, some of which transmit viruses in the genera*Nepovirus* and *Tobravirus*. Example: *Paratrichodorus pachydermus* is a vector of *Tobacco rattle virus.*

3. ***Plasmodiophorids:*** These are root-infecting obligate parasites traditionally regarded as fungi but now known to be more closely related to protists. They transmit viruses in the genera *Benyvirus*, *Bymovirus*, *Furovirus*, *Pecluvirus* and *Pomovirus*. Example: *Polymyxa graminis* is the vector of several cereal viruses including *Barley yellow mosaic virus*, growing within a barley root cell.

4. ***Mites:*** These transmit viruses in the genera *Rymovirus* and *Tritimovirus*. Example: *Aceria tosichella* is the vector of *Wheat streak mosaic virus*.

The virus-vector relationships are of several types:

- **Non-persistent**: This relationship occurs with the feeding apparatus of the insect, where the virus can be rapidly adsorbed from one plant cell and then released into a different one. The feeding insect looses the virus rapidly when feeding on a non-infected plant. Such a relationship is termed "non-persistent". The best-studied examples are of *potyvirus* transmission by aphids.
- **Persistent:** In this relationship, the virus is taken up into the vector, circulates within the vector body and is released through the salivary glands. The vector needs to feed on an infected plant for much longer and there is an interval before it can transmit. Once it becomes viruliferous, the vector will remain so for many days and such a relationship is therefore termed as "persistent" or "circulative". The best studied examples are of luteovirus transmission by aphids. In some examples of this type, the virus multiplies within the vector and this is termed as "propagative".

Symptoms, detection and identification of plants viruses

Viruses cause many important plant diseases and are responsible for huge losses in crop production and quality. Infected plants may show variety of symptoms depending on the disease but often there is leaf yellowing (either of the whole leaf or in a pattern of stripes or blotches), leaf distortion (e.g. curling) and/or other growth distortions (e.g. stunting of the whole plant, abnormalities in flower or fruit formation).

Sometimes the virus is restricted to certain parts of the plant (e.g. the vascular system; discrete spots on the leaf) whereas in others it spreads throughout the plant causing a systemic infection (Lindbo and Dougherty, 1992). Infection does not always result in visible symptoms (as witnessed by names such as *Carnation latent virus* and *Lily symptomless virus*, both members of the genus *Carlavirus*). Occasionally, virus infection can result in symptoms of ornamental value, such as 'breaking' of tulips or variegation of *Abutilon*. Some examples of the symptoms caused by different viruses includes Yellow mosaic on lettuce caused by *Lettuce mosaic virus*, Yellow vein-banding on grapevine caused by*Grapevine fanleaf virus*, Fruit distortion on eggplant fruit caused by *Tomato bushy stunt virus*, Bark scaling caused by *Citrus psorosis virus* etc (Lindbo and Dougherty, 1992).

Virus Testing/Detecting methods

I. *Electron microscopy*

Electron microscopy has been used for many years as a 'multi target' assay; however, although very useful for detecting and discriminating rod shaped particles, often the presence of spherical viruses is very difficult to confirm.

II. *Matrix Assisted Laser Desorption Time of Flight mass spectrometry (MALDI TOF)*

Detection of viral coat protein using methods such as Matrix Assisted Laser Dissorption Time of Flight mass spectrometry (MALDI TOF) have been reported for a number of viruses and can be described as a generic.

However, both EM and MALDI TOF mass spectrometry are not suitable for viruses that do not have a protein target, for example viroids or NM isolates of Tobacco rattle virus (Harrison and Robinson, 1978), and both are useful only if the viruses are in a very high titre.

III. *Enzyme-Linked Immunosorbent Assay (ELISA)*

1. *Scope/Purpose*

This protocol describes the testing procedure of Enzyme-Linked Immunosorbent Assay (ELISA) for the detection of plant viruses using microtitre plates. This test employs antisera (virus antibody); universal conjugate (enzyme (alkaline phosphatase) labeled goat anti-rabbit Ig G); and the substrate (p-nitrophenyl phosphate-PNPP) for the detection of viruses (Voller et al 1976; Clark and Adams 1977; Dijkstra and de Jager 1998).

Two types of testing protocol viz., (a) direct antigen coated enzyme-linked immunobsorbent assay (DAC-ELISA) (Gillaspie Jr et al 1995); and (b) double antibody sand-witch enzyme-linked immunosorbent assay (DAS-ELISA). DAC-ELISA usually employed for screening large number of viruses but in case of plant viruses, where specific strain of virus to be detected, the DAS-ELISA will be employed (Edwards and Cooper 1985).

Also at least two negative controls (healthy plant sap) besides buffer controls will be used in order to have more confident background values. It is also advised to cross-adsorb antisera with healthy plant sap to prevent false positives.

2. *Requirements of Test*

2.1 Equipments

ELISA Reader with Printer, Balance, Magnetic Stirrer, Microcentrifuge, Deep Freezers (-20°C, -80°C), Incubator, pH Meter, Refrigerator, Micropippettes (10µl, 50 µl, 100 µl), Multichannel Adjustable micropippette (4-well type), Tissue Grinder/Pestle and Mortar, Polystyrene Microtips

2.2. Supplies

- Microplates (polysterene, 96 wells)/strips
- Antisera (polyclonal/monoclonal)
- Antigen (virus affected plant material)
- Enzyme (alkaline phosphatase) labeled goat anti-rabbit Ig G
- Goat Anti-rabbit Enzyme Conjugate (universal conjugate)
- Substrate (p-nitrophenyl phosphate-PNPP)

2.3. Buffers/Reagents

- Stock buffer (phosphate buffer- saline, pH 7.4): Dissolve NaCl 8.0 g, $Na_2HPO_4.2H_2O$ 1.44 g or $Na_2HPO_4.12H_2O$ 2.90 g or $Na2HPO_4$ 1.50 g, KH_2PO_4 0.20 g, KCl 0.20 g in distilled water to make 1 litre.
- Wash buffer (PBS-Tween, PBS-T): Add 0.5 ml Tween – 20 to 1 litre PBS.
- Coating buffer (Carbonate buffer, pH 9.6): Dissolve Na_2CO_3 1.59 g and $NaHCO_3$ 2.93 g in 1 litre of distilled water.
- Enzyme conjugate diluent/buffer (PBS-T polyvinyl-pyrrolidone and ovalbumin, PBS-TPO): Add 20.0 g polyvinyl-pyrrolidone (PVP, MW 40,000) and 2.0 g egg ovalbumin to 1 litre PBS-T.
- Antibody diluent/buffer: Same as PBS-TPO
- Substrate buffer (diethanolamine buffer, pH 9.8): Mix 97 ml of Diethanolamine and 800 ml of distilled water.Adjust the pH to 9.8 with 1N HCl, and make up the volume to1 litre with distilled water.
- Blocking solution: Add 5.0 g bovine serum albumin (BSA/spray dried milk (SDM) to 1 litre PBS-T.
- Fixing solution: Dissolve 120.0 g of NaOH in 1 litre of distilled water

(All buffers contain 0.02 % sodium azide as a preservative)

3. *Test Protocol: DA C-ELISA*

3.1. Weigh 500 mg test / healthy tissue and grind using autocalved mortar and pestle/tissue grinder in coating buffer containing 2% (w /v) PVP at a ratio of 1:1 (w / v) at room temperature and low speed centrifuge (ca. 12,000 g) for 2 min.

3.2. Dispense 200 µl extract from test as well as healthy samples to each well of the microtitre plate. Cover the plate and incubate at 37°C for 1 h.

3.3. Decant and wash the plate by flooding the wells with PBS-T for about 3 min. Repeat wash and soak operations thrice, and shake out residual liquid draining on a paper towel.

3.4. Dispense 200 µl blocking solution (Bovine Serum Albumin, 1%, w /v) to each well. Incubate at 37°C for 1 h to block polystyrene well reactive surfaces.

3.5. Wash the plate three times as above.

3.6. Dispense 200 μl of antiserum (primary antibody) diluted in PBS-TPO to each well. Incubate at 37°C for 2 h.

3.7. Wash the plate three times as above.

3.8. Dispense 200 μl anti-rabbit immunoglobulin alkaline phosphatase (universal conjugate, Sigma, USA) diluted in PBS-TPO to each well. Incubate at 37°C for 2 h.

3.9. Wash the plate three times as above.

3.10. Dispense 200 μl freshly prepared substrate (p-nitrophenyl phosphate-PNPP, Sigma, USA) (0.5-1.0 mg/ml) solution in substrate buffer to each well. Incubate at 37°C for 2 h.

3.11. Measure the intensity of color in each well at 405 nm by using ELISA reader. Compare the absorbance values of the test samples with healthy control. Take a print out for record (Consider samples showing absorbance (OD_{405}) values more than two times of healthy control as positive).

4. *Testing Protocol: DAS-ELISA*

4.1. Dispense 200 μl specified antibody of desired working dilution in coating buffer to each well and incubate for 1 h at 37°C

4.2. Decant and wash the plate by flooding the wells with PBS-T for about 3 min. Repeat wash and soak operations thrice, and shake out residual liquid draining on a paper towel.

4.3. Weigh 500 mg sample (test/healthy tissue) and grind using autoclaved mortar and pestle in PBS-TPO (1:1 w / v) at room temperature and centrifuge (ca. 12,000 g) for 2 min.

4.4. Dispense 200 μl extract from test as well as healthy samples to each well and incubate at 37°C for 2 h.

4.5. Wash the plate three times as above.

4.6. Dispense 200 μl enzyme labeled (alkaline phosphatase) specific antibody diluted in PBS-TPO to an appropriate dilution and incubate at 37°C for 2 h.

4.7. Wash the plate three times as above.

4.8. Dispense 200 μl freshly prepared substrate (p-nitrophenyl phosphate-PNPP, Sigma, USA) (0.5-1.0 mg/ml) solution in substrate buffer to each well and incubate at 37°C for 2 h.

4.9. Measure the intensity of color in each well at 405 nm by using ELISA reader. Compare the absorbance values of the test samples with healthy control (Consider samples showing A_{405} nm values more than two times of healthy control as positive).

Helpful Hints

- Draw test plan in checkerboard format and include minimum of two replications for each treatment.

- Include negative (healthy) control to validate the assay.
- Determine the optimum dilutions of various reagents to be used.
- Do not wash or rinse microtitre plates prior to use. Once used plates should not be reused.
- Use separate beakers for preparing antigen, antiserum, conjugate and substrate dilutions.
- Dilute antisera according to their titre value and commercial conjugates as per the manufacturer's directions.
- Antisera used should be free from cross-reacting antibodies against host proteins to avoid false positive reactions. Otherwise, cross-absorb the antiserum with sap extracted from healthy plant by incubating required quantity of antiserum in about 1:50 dilution of healthy plant sap extracted in PBS. Incubate at 37°C for 1 h and centrifuge at 10,000 rpm for 5 min. The supernatant is then directly used.
- Always use freshly prepared substrate buffer.
- PNPP is photo-degradable. Hence cover with aluminium foil to avoid direct contact with light. Use as quickly as possible, once PNPP solution is prepared.

IV. Dot Immuno-binding Assay (DIBA)

1. Scope/Purpose

This protocol describes the testing procedure of Dot Immuno-binding Assay (DIBA) for the detection of plant viruses by using nitrocellulose or nylon membranes (Clark and Bar-Joseph 1984; Powell 1987). This test employs antisera, enzyme labeled Ig G and substrate for the detection of viruses.

2. Requirements of Test

2.1 Equipments

- Micropippettes (10 μl, 50 μl, 100 μl)
- Multichannel Adjustable micropippette (4-well type)
- Tissue Grinder/Pestle and Mortar
- Polystyrene microtips

2.2. Supplies

- Nitrocellulose/Nylon membranes
- Antisera
- Antigen (virus affected plant material)
- Enzyme (alkaline phosphatase) labeled anti-Ig G
- Substrate (p-nitrophenyl phosphate-PNPP)

2.3. Buffers/reagents

- *Stock buffer (tris buffer saline, TBS,* pH 7.5: Dissolve 4.84 g Tris (0.02 M) and 58.48 g NaCl (0.5 M) in 1 litre of distilled water and adjust the pH to 7.5 with H Cl and make final volume up to 2 litres.
- *Antigen extraction buffer (TBS + 50mM DIECA)*: Add 11.25 g diethyl dithiocarbamate (DIECA) to 1 litre TBS
- *Blocking solution (TBS + SDM)*: Add 5.0 g spray dried milk (SDM) to 100 ml TBS
- *Antibody and enzyme conjugate diluen/buffer*: Same as TBS-SDM
- *Substrate buffer (pH 9.5)*: Adjust the pH 9.5 with 1 N HCl and make up to 1 litre
- *Substrate solution:* Solution A: 0.1 M Tris (12.11 g in 1 litre), 0.1 M NaCl (5.85 g in 1 litre), 5 mM $MgCl_2 6H_2O$ (1.01 g in 1 litre)

Solution B: Store solutions A and B refrigerated in amber bottles. Add NBT to 0.33 mg/ml and BCIP to 0.17 mg/ml to the substrate buffer just before use

- *Fixing solution (pH 7.5)*: 10 m M Tris (1.21 g in 1 litre), 1 m M EDTA (0.29 g in 1 litre). Adjust the pH to 7.5 with 1 N HCl and make up to 1 litre.

  (All buffers contain 0.02% sodium azide as preservative)

  Bromo chloro indolyl phosphate (BCIP) 50 mg

  Dimethyl formamide (DMFA) 1 ml

  Nitroblue tetrazolium (NBT) 75 mg

  Dimethyl formamide (DMFA) 1 ml

3. Testing Protocol: DIBA

- Weigh test and healthy tissue (100 mg) and grind using autoclaved mortar and pestle in antigen extraction buffer at a ratio of 1:10 (w /v) at room temperature and express through cheese cloth.
- Pipette out 0.8 ml expressed sap into 1.5 ml eppendorf tube. Add 0.4 ml $CHCl_3$, vortex and centrifuge (ca. 12,000 g) for 2 min.
- Pipette out 200 μl clarified sap (upper aqueous layer) into 800 μl antigen extraction buffer and vortex.
- Put on the gloves, cut desired size piece of NCM and draw a lattice of squares of 1 x 1cm each with a soft lead pencil. Always use forceps for handling the membrane.
- Wet the NCM by floating it in TBS and then air dry. Spot the test/control samples (5-10 μl) by hand.
- Allow NCM to air dry and immerse in blocking solution with gentle oscillation for 1 h at room temperature or overnight at 4°C.
- Rinse once in TBS for ca. 10 min.

- Incubate for 1 h at room temperature or overnight at 4°C in crude antiserum (primary antibody) diluted in TBS-SDM.
- Rinse thrice in TBS for ca. 10 min. each.
- Incubate for 1 h at room temperature or overnight at 4°C in secondary antibody (anti-rabbit Ig G alkaline phosphatase) diluted in TBS-SDM.3.11. Rinse thrice in TBS for ca. 10 min. each.
- Incubate in substrate solution at room temperature in the dark and watch for color development (good purple color development take 5-10 min.).
- Rinse thrice in the fixing solution for ca.10 min. each and then air dry NCM between Whatman filter paper sheets.
- Assess the result by visual observation by comparing the intensity of purple color or by using densitometer.
- Photograph NCM (dry or wet) and store protected from light.

Helpful Hints

- Nylon or PVDF membranes in place of nitrocellulose may also be used.
- Wear gloves while handling membranes.
- NCM, being fragile, should be handled with blunt tipped forceps.
- Always use buffer and negative (healthy) controls.
- The antibody and conjugate solution can be re-used for 3-5 times.
- If the antigen concentration is good, colour development starts immediately after addition of substrate solution.
- Antisera used should be free from cross–reacting antibodies against host proteins to avoid false positive reactions. Otherwise, cross-absorb the antiserum with sap extracted from healthy plant by incubating required quantity of antiserum in about 1:50 dilution of healthy plant sap extracted in PBS. Incubate at 37°C for 1 h and centrifuge at 10,000 rpm for 5 min. The supernatant is then directly used.

  i. The membrane once developed may be photographed or preserved between folds of filter paper for several years.

  ii. Extra care should be taken while spotting the samples. See that sample diameter is less as much as possible to ensure concentration of antigen at the same area. Spotting can be done manually or also through the use of dot - blot apparatus. If done manually, do not use more than 2.5 µl for dotting each time. If planning to dot 10 µl of sample, take 2.5 µl of sample and slowly release it on the centre of the grid area without touching the membrane. Allow it air dry. Repeat this process for another three times.

- If one does not have antiserum or other facilities for processing the membrane, the dotted membrane can be sent anywhere in the world where such facility exits. The membrane can also be sent back after processing.

V. Reverse Transcription-Polymerase Chain Reaction (RT-PCR)/PCR

1. Scope/Purpose

The protocol of Reverse Transcription-Polymerase Chain Reaction (RT-PCR) for the detection of plant viruses (RNA Viruses) involves primer-mediated *in vitro* reaction involving amplification of target nucleic acid sequences. Since majority of the plant viruses are RNA viruses, one step reverse transcription–polymerase chain reaction (RT-PCR) is used to amplify a segment of RNA that lies between two regions of known sequences. A one step RT-PCR will have four phases in the process viz., (i) cDNA synthesis using reverse transcription at 42 °C; (ii) denaturation at a high temperature (90-95 °C); (iii) annealing of target specific primers, and (iv) primer extension by a thermostable DNA polymerase. Two-step RT-PCR, which is also known as quantitative real-time PCR, is used for detection of low abundance transcripts. DNA Plant viruses are directly detected by PCR techniques (Sambrook et al 1989).

2. Requirements of Test

2.1 Equipments

- Thermal Cycler
- Horizontal Gel Electrophoresis Unit with Power Pack
- Gel Documentation Unit with Computer and Printer
- Refrigerated Micro-centrifuge
- Deep Freezers (-20°C-80°C)
- Micropipettes (1 μ, 10 μl, 50 μl, 100 μl, 1000 μl, 5000 μl)
- Multichannel Adjustable micropippette (4-well type)
- Tissue Grinder/Pestle and Mortar
- Polystyrene microtips

2.2. Supplies

- PCR Tubes (500 μl/200 μl)
- Virus affected plant tissue/healthy tissue
- Specific primers
- RNeasy Kit
- Taq DNA Polymerase

2.3. Chemicals and Solutions

- Ethidium bromide (10 mg/ml): Dissolve 1 g Ethidium bromide in 100 ml H_2O and transfer to a dark bottle and store at 4°C.
- 0.5 M EDTA (pH 8.0): Dissolve 186.1 g EDTA in 800 ml H_2O. Adjust the pH to 8.0 with 10 N NaOH. Make the volume to 1 litre. Dispense into aliquots and sterilize.

- Running buffer (50 X TAE (Tris Acetate EDTA, pH 8.0): Tris base 242.0 g, Glacial acetic acid 57.1 ml. EDTA (0.5 M, pH 8.0) 100 ml. Adjust pH 8.0 with 1 N NaOH. Make up volume to 1 litre. Dilute 50 X TAE buffer to 1X before use.

3. Testing Protocol- RT-PCR

3.1. Isolation of total nucleic acid

- Weigh test/healthy sample (~100 mg) and cut into strips. Extract total nucleic acid using RNeasy kit according to the manufacturer's instructions. Inactivate the RNA at 70°C for 2 min and then snap chill on ice for 10 min and use this RNA as template for reverse transcription – polymerase chain reaction (RT-PCR).

3.2. cDNA amplification

- Assemble RT-PCR reaction components on wet ice and prepare amplification mix by dispensing into ~ 200 µl microfuge tube in the order listed below (see box).
- Mix RT-PCR reaction assembly by inversion and place the tubes in a thermal cycler and proceed with thermal cycling profile chosen for reaction (see box).
- Analyze the amplified product by electrophoresing 10 µl from the total reaction on 1% agarose gel in Tris-acetate EDTA (TAE) containing ethidium bromide.

3.3. Analysis of amplicons

- Melt 0.5 g agarose in 50 ml 1 X TAE running buffer and add ethidium bromide (0.5 mg/ml) to the agarose after cooling to around 50°C.

Composition of PCR mix

Reagents	Volume required (µl) per reaction	
	Test sample	Negative control
10 X PCR buffer	10.0	10.0
5 X Q solution	20.0	20.0
10 mM Dithiothreitol	10.0	10.0
100mM dNTPs	2.0	2.0
Forward primer, 100 pM (100 pM/µl)	1.0	1.0
Reverse primer, 100 pM (100 pM/µl)	1.0	1.0
RNase inhibitor 2 units (4units /µl)	0.5	0.5
Omniscript reverse transcriptase 2.5 units (5units/µl)	0.5	0.5
Taq DNA polymerase 2.5 units (5 units/µl)	0.5	0.5
Template	44.5	44.5
RNase free water	10.0	10.0
Total	100	100

- Pour the melted agarose into the casting tray for polymerization and fill the buffer tank with running buffer (1 X TAE) and remove the comb.
- Load 10 µl RT-PCR products with 2 µl 6X loading dye and run the gel at 60 volts for 2 h along with marker.
- Examine the gel under ultraviolet transilluminator and photograph.

Temperature profile for RT-PCR

Steps	Temperature	Time	Cycle
Reverse transcription	42 °C	45 min	1
Initial denaturation	94 °C	5 min	1
Denaturation	94 °C	30 s	
Annealing	45-50 °C	1 min	30
Extension	72 °C	1 min	
Final extension	72 °C	10 min	1

Helpful Hints

- Always perform a healthy control reaction.
- Use fresh gloves for RNA isolation and each reaction set-up.
- Wear gloves while handling agarose gels containing ethidium bromide.
- Ensure that the RNA used is inactivated at 70 °C for 5 min and then snap chill on ice.
- Thaw and vortex all the reagents before use (except DNA polymerase and reverse transcriptase).
- Optimize the annealing temperature for the primer pair before setting the reaction.
- Keep all the reagents at -20°C after use for long storage.
- Ensure 50X TAE buffer is diluted to 1X buffer before running the gel.

VI. Nucleic Acid Spot Hybridization (NASH)

1. Scope/Purpose

This protocol describes the testing procedure of Nucleic Acid Spot Hybridization (NASH) for detecting very small amount of virus in the plant material. The detection is based on mobilization of target (virus) NA on to a solid matrix followed by hybridization with NA probes under appropriate conditions. Nitrocellulose or charged nylon membranes are the commonly used solid matrix for hybridization (Baulcombe and Fernandez-Northcote 1988; Singh and Singh 1996). Both radioactive probes (^{32}P DNA probes) as well as non-radioactive probes (chemiluminscent, digoxigenenin-labelled cRNA probes) are available for nucleic acid spot hybridization.

2. Requirements

2.1. Equipments

- Hybridization oven
- Hybridization flask
- Lucite screen, a separate room for handling radioactivity, X-ray cassette, X-ray film developing facility, radioactivity, monitor vacuum oven, shaking water bath, bag sealer, etc.

2.2. Supplies

- Nitrocellulose or nylon membranes
- Pestle and mortar/tissue grinder
- Virus infected and healthy plant samples

2.3. Buffers and Reagents

- *Standard saline citrate (SSC) buffer 20X*, pH 7.0: Dissolve 175.3 g NaCl (3.0 M) and 88.2 g Na Citrate (0.30 M) in 800 ml H_2O. Adjust the pH to 7.0 and make up the volume to 1 litre.

Dispense into aliquots and sterilize before storing at room temperature.

- *Tris HCl- EDTA (TE) buffer* (pH 8.0): Tris 10 mM: Dissolve 121.1 g in 800 ml H_2O. Adjust the pH to 8.0 with concentrated HCl. Make the volume to 1 litre. Dispense into aliquots and sterilize.

EDTA 0.5 M: Dissolve 186.1 g in 800 ml H_2O. Adjust the pH to 8.0 with 10 N NaOH. Make the volume to 1 litre. Dispense into aliquots and sterilize. Make appropriate dilutions to prepare required amount of 1X TE.

- *Oligolabelling buffer (5X)*: Tris Cl (pH 8.0; 250 mM), $MgCl_2$ (25 mM), β-mercaptoethanol (5 mM), ATP, GTP, TTP (2 mM each), HEPES (1 M, adjusted to pH 6.6 with 4 N NaOH) and Oligo nucleotides (1 mg/ml)
- *SSPE 20X* (pH 7.4): NaCl (175.3 g), $NaH_2PO_4.H_2O$ (27.6 g), EDTA (7.4 g). Dissolve in 800 ml H_2O. Adjust the pH to 7.4 with NaOH and make up the volume to 1 litre. Dispense into aliquots and sterilize by autoclaving.
- *10 % SDS:* Dissolve 100 g analytical grade SDS in 900 ml H_2O and heat at 68°C to dissolve. Adjust the pH to 7.2 by adding few drops of concentrated HCl. Adjust the volume to 1 litre. Dispense into aliquots.
- *0.5 N NaOH:* Dissolve 20 g NaOH in 800 ml H_2O and make up the volume to 1 litre (no need to sterilize).
- *3 M sodium acetate*, pH 5.0: Dissolve 408.1 g of tri-hydrate sodium acetate in 800 ml H_2O. Adjust the pH to 5.0 with glacial acetic acid. Adjust the volume to 1 litre and autoclave.
- *Bovine serum albumin* (1 %): Dissolve 10 mg of BSA in 1 ml H_2O. Filter sterilize and aliquot them. Store at -20 °C.
- *Stop buffer:* NaCl 20 mM, Tris HCl 20 mM, EDTA 2 mM, SDS 0.25 %, dCTP 1 mM

- *Denhardt's reagent 50X:* Ficoll (type 400; 5 g), PVP (5 g), BSA (5 g), H_2O (500 ml)
- *Salmon sperm DNA (1%):* Salmon Sperm DNA is dissolved in water at a concentration of 10 mg/ml. The solution is stirred for 2-4 h at room temperature. The concentration of NaCl is adjusted to 0.1 M and the solution is extracted with phenol once and then with phenol chloroform once. DNA is sheared by passing 12 times through a 17 guage hypodermic needle. DNA is precipitated by ice-cold ethanol and then recovered by centrifugation, redissolved at a concentration of 10 mg/ml. Boil the solution for 10 min., aliquot and freeze at -20°C.

3. Testing Protocol-NASH

3.1. Dot blotting

- Put on disposable gloves and cut a piece of membrane (size depends on the number of samples to be tested).
- Draw a lattice of 1 cm square with a soft lead pencil and cut one edge to indicate orientation.
- Wet the membrane in 20 X SSPE or 20 X SSC for 30 min. and place it on a sheet of whatman-3mm filter paper.
- Grind test/healthy samples (~200 mg) in TE buffer (1:1 w/v). Also include a positive and negative control. Press through muslin cloth.
- Take 200μl of this extract and denature DNA by adding 50 μl 0.5 N NaOH for 10 min.
- Add 25 μl 3 M sodium acetate (pH 5.0) and make dilutions of 1:10 and 1:100 in TE.
- Pipette carefully 2.5 μl of each of the samples in each square of the membrane.
- Let the spots dry at room temperature and bake the membrane in vacuum oven for 2 h at 80°C.

3.2. Probe production (Random primer method)

- Mix the following in the given order in eppendorf tube. Protect yourself from radiation while preparing probe.

H_2O	33 μl
Oligolabelling buffer	10 μl
BSA	2 μl
Template	1 μl (20 ng DNA / μl)
32P-dCTP	3 μl (10 μCi)
Klenow enzyme	1 μl

- Incubate the mix behind Lucite green at room temperature overnight. Add 2 μl of 1 mM EDTA to stop the reaction.

3.3. Pre-hybridization

- Pre-hybridize the membrane by treating in 100 ml pre-hybridization solution in a sealable plastic bag by incubating with constant agitation for 15 min. at 60°C.
- Pre- hybridization solution (10 ml):

H_2O	6.3 ml
50 X Denhardt's solution	0.8 ml
Denatured Salmon sperm DNA (10 mg/ml)	0.1 ml
20 X SSC	2.5 ml
10 % SDS	0.3 ml

Heat in boiling water bath for 90 s and chill

3.4. Hybridization

- Add probe at the rate of $1X10^6$ cpm/ml of solution after denaturing by boiling for 5 min in a water bath. Re-seal plastic bag and incubate with constant agitation for overnight at 65°C.

3.5. Washing Procedure

- Remove the membrane from pre hybridization solution and discard the solution into radioactive wastes.
- Transfer the membrane to a plastic bag filled with washing solution. Wash twice with 20 X SSC containing 0.1 % SDS at room temperature with constant agitation for 15-20 min.
- After two washes, place the membrane on a sheet of filter paper and allow it to dry at room temperature.
- After drying, place the membrane between 2 layers of thin polythene sheet in a cassette with intensifying screen.
- Place X-ray film on the top in dark room, close cassette and place at -70 °C overnight. Develop auto radiogram.

Helpful Hints

- Use gloves at all times and frequently wash or change them.
- Cross- contamination during DNA isolation should be avoided.
- Always use clean, blunt forceps and avoid contacting dirty surfaces as binding of probes to contaminants on the membrane may lead to high background.
- Blot can be re-used after stripping the probe. Nitrocellulose blots have been re-used 6-7 times with limited loss in sensitivity.

VII. Microarray Technique

DNA microarrays were introduced for the first time by Schena et al (1995) and since then this technology has attracted great interest among biologists (Afshari

2002). Recent reports on initiating research on the detection of plant viruses and other plant pathogens by DNA microarrays have been limited, generally unspecific and in the form of short abstracts. Research works have been initiated toward detection of plant viruses (Perez-Ortin 2002; Sip 2002; Bystricka et al 2003). It has the ability to display simultaneously the expression of thousands of genes at a time, thus it is a powerful tool for genetic analysis. Microarray technology is leading the way in offering researchers the ability to examine simultaneously the expression levels of hundreds or thousands of genes in a single experiment. Microarrays or 'gene chips' are being used that have thousands of genes arrayed on the surface with current protocols allowing reliable detection of target down to approximately ten copies per cell (Lockhart and Winzeler 2000). So far, the technology has been applied mainly in research fields for expression level profiling, SNP mapping and studying host pathogen interactions. Microarray technology could allow the detection of a large number of different viruses in a single generic assay. Up to 30 000 DNA probes can be arrayed onto a single glass microscope slide, which forms the microarray (Eisen and Brown 1999). The DNA probes arrayed would be gene sequences from each of the viruses that need to be detected in a single assay. The microarray can then be exposed to fluorescently labeled cDNA from the sample to be tested, and finally scanned using a microarray scanner to reveal if any of the targets were present in the sample. With the explosion of information arising from sequencing of plant viruses and viroids, predicted the simultaneous detection and identification of many plant viruses, viroids, and other plant pathogens by DNA microarrays.

Base-pairing of complementary sequences by hybridization is the underlining principle of DNA microarray. This specific binding of DNA allows a target DNA or RNA to hybridize to a specific complementary DNA probe on the array. Each probe is made of thousands of cDNAs or oligonucleotides, each specific for a gene, DNA sequence, or RNA sequence of interest. An array is an orderly arrangement of samples. It provides a medium for matching known and unknown nucleic acid samples based on base-pairing rules (A-T and G-C for DNA; A-U and G-C for RNA) and automating the process of identifying the unknown. Each array is generated by depositing a few nanoliters of DNA probe on a solid support. The printing is performed by a robot, which allows identical spotting serially. In general, arrays are described as macroarrays or microarrays, the difference being the size of the sample spots. Macroarrays contain sample spot sizes of about 300 μm or larger. The sample spot sizes in microarray are typically less than 200 μm in diameters and these arrays usually contain thousands of spots (Cheung et al 1999).

In many cases, a glass surface is utilized for low-density arrays. Nucleic acids bind poorly to plain glass, which is generally treated before use. Poly-L-lysine, which binds molecules by ionic interactions, is often used for this purpose. The glass substrate may also be treated with saline, which binds covalently to the probe DNA, thus preventing removal of the probe DNA during hybridization and washing steps. Glass slides that are used to detect radioactively labeled nucleic acids may be coated with a nylon membrane to which the DNA probes are covalently cross-linked using ultraviolet light. The DNA applied to the surface of the array may consist of plasmids of 500-5,000 bases, complementary DNA of several hundred bases in length, products of the PCR of 100-500 base pair, or synthetic oligonucleotides that may be modified by addition of an amino or thiol group on their 5' end (Hadidi et al 2004).

Microarrays with a density of the order 1,000-10,000 spots per cm^2 are generally known by the term "DNA chip". The substrate used for immobilization of high-density DNA arrays is usually chemically modified glass or silicon.

The steps in any microarray experiment include:

- ***DNA probes with known nucleotide sequences:*** Probes may include short or long ligonucleotides, cDNAs, chromosomes, whole organism, or others with known identity. The probes may be created by using standard technologies, such as restriction enzymes, cloning and PCR.
- ***Application of DNA to the array:*** A wide variety of techniques exist for the immobilization of DNA on the surface of the array, including contact-dip deposition printing, micro contact printing, microfluidics networks and electrocapture. Piezoelectric printing and micro wet printing may be used both for immobilization and for *in situ* synthesis. The photolithographic technique is used for *in situ* synthesis alone. Each array is generated by depositing a few nanoliters of purified DNA on a solid support. The printing is performed by a robot which allows identical spotting serially.
- ***Target preparation:*** Poly (A+) RNA, purified or as total RNA, is reverse-transcribed to the first single-stranded cDNA. Alternatively, reverse transcription is performed with chemically reactive nucleotide analog (amino allyl-dUTP) and the cDNA is "post labeled" with the reactive forms of Cy3- or Cy5-NHS esters, which bind to the modified nucleotides. These cDNAs generate even and bright signal levels in microarray hybridization. When the support is nylon membrane, the cDNA is labeled during reverse transcription with radiolabeled nucleotides, such as ^{32}P-dCTP, and hybridization is detected by autoradiography or chemiluminescence. Alternatively, the cDNA can be labeled by incorporating biotin-16-dUTP or digoxigenin-11-dUTP; following hybridization, b-galactosidase-conjugated streptavidin or alkaline-phosphatase-conjugated anti-digoxigenin antibody is employed and the hybrids are detected by colorimetry.
- ***Hybridization***: The probe DNA is denatured by heating, and then the labeled target solution is applied. In the case the array is used to analyze differential gene expression, control cDNA may be labeled with Cy3-dCTP, and experimental cDNA with Cy5-dCTP. In this case, it is important to make sure that equal amounts of cDNA and of dye are used. The slide or chip is moved to a moist 65 °C hybridization environment for several hours. Finally, the slide or chip is washed to remove any unbound nucleotides in preparation for the detection stage.
- ***Image detection***: Detection is done using either a scanning or a direct imaging system. Scanning uses a laser excitation source (for the detection of fluorescently labeled DNA) that works on each spot one at a time and a detector that measures the resulting light. Direct imaging uses an excitation source that illuminates the entire array at once, and a camera to capture the

resulting image. It is used when radioactive hybridization or colorimetry is chosen. Scanning has higher resolution and sensitivity than direct imaging. Graphic computer files (TIFF, etc.) are created based on the detection results.

- ***Analysis of results***: When using dyes, the resulting image files are analyzed with special software to determine the intensity of light at each of the wavelengths measured. When hybridizations are done with a mixture of cDNAs, labeled for example with Cy3 and Cy5, the slide is scanned to detect Cy3 and Cy5 fluorescence and the ratio Cy3/Cy5 is determined. This requires sophisticated bioinformatics methods. These measurements are then combined by the software to create a graphical representation of the relative expression levels of the experimental and controlled samples. The output can usually be seen in different formats. This include a two dimensional grid that shows red, green, and yellow spots of varying shades that correspond to the spots in the array. Alternatively, the output can be a scatter plot that better shows the difference in expression levels between different genes. Complex data sets need to be analyzed using sophisticated software such as GeneSpring.

Some important disease caused by plant viruses

(a) Rod shaped viruses:

Tobamoviruses: Tobacco Mosaic virus

Tobraviruses: Tobacco Rattle virus

Furoviruses: Wheat Mosaic virus

Hordeiviruses: Barley Stripe Mosaic virus

(b) Filamentous viruses:

Carlavirus: Pea Strick virus and Popler Mosaic virus

Capillovirus and trichovirus: Apple stems grooving vius, Apple chloratic leaf spot virus.

Potyviruses: Bean Common Mosaic and Bean Yellow Mosaic Lettuce Mosaic, Papaya Ringspot, Plum Pox, Potato Virus Y, Sugarcane mosaic, Tobacco Etch, Turnip Mosaic, Watermelon Mosaic, Zucchni Yellow mosaic.

Rhymoviruses and Bymoviruses: Rye Grass Mosaic virus, Oat Necrotic Mottle Virus, Oat Mosaic virus and Rice Necrotic Mosaic virus.

Closteroviruses: Beet Yellows, Citrus Tristeza, Lettuce Infectious Yellows.

(c) Isometric single stranded RNA viruses:

Waikaviruses: Rice Tungro, Maize Chlorotic Dwarf

Luteoviruses: Beet Western Yellows, Potato Leafroll, Barley yellow dwarf.

(d) Comoviridae

Comovirus: Cowpea Mosaic virus, Radish Mosaic virus

Nepoviruses: Tomato ringspot virus, tobacco Rings pot virus, Cherry Leafroll virus, Grapevine Fanleaf and Raspberry Yellow Dwarf, Raspberry Ringspot, Raspberry Leaf Curl

(e) Bromoviridae

Cucumoviruses: Cucumber Mosaic

Ilarviruses: Prunus Necrotic Ring Spot

(f) Isometric double stranded RNA viruses:

Reoviridae: Rice Dwarf virus, Rice Ragged dwarf virus.

(g) Negative RNA (-ssRNA) viruses

Rhabdoviruses: Lettuce Nectrotic Yellows virus, Potato Yellow Dwarf virus, Rice Transitory Yellowing virus and Wheat Striate Mosaic virus.

Tospovirus: Tomato spotted Wilt virus, Watermelon Silver Mottle virus

Tenuivirus: Rice Boja Blanca virus, Rice Grassy stunt virus, Rice Stripe virus.

(h) Double stranded DNA viruses

Caulimoviruses: Cauliflower Mosaic virus, Dahlia Mosaic virus and Carnation Etched Ring virus

Badnaviruses: Rice Tungro Bacilliform virus, Banana Streak virus, Cacao Swollen shoot virus

(i) Single stranded DNA viruses

Geminiviruses: Tobacco Yellow Dwarf virus, Maize Streak virus, Wheat Dwarf Virus, Beet Culrly Top virus, African Cassava Mosaic virus, Bean Golden Mosaic virus, Squash Leaf Curl virus, Tobacco Leaf Curl virus, Tomato Golden Mosaic virus, Tomato Mottle virus, Tomato Yellow Leaf Curl virus.

(j) Isometric single-stranded DNA viruses

Banana Bunchy top, Subterranean Clover Stunt virus, Coconut Foliar Decay virus and Faba Bean Necrotic Yellows virus

(k) Viroids

Potato Spindle Tuber, Citrus Exocortis, Coconut Cadang Cadang

National Certification System for Tissue Culture Raised Plants (NCS TCP)

(NCS TCP) is the unique quality management system of Department of Biotechnology (DBT), Government of India for tissue culture industry, which facilitate certification of the tissue culture raised plants upto laboratory level.

Presently 52 tissue culture production facilities have been recognized based on infrastructure, technical competency and package & practice of production of tissue culture plants. For any case of controversy of any sample tested by accredited test laboratory two referral laboratories each for virus indexing and genetic fidelity has been appointed by DBT. For virus detection, related controversy Indian Agricultural Research Institute at New Delhi would be referred for the same. A list of viruses for various crops/plants has been listed by DBT, the presence of which has to be looked into the plant for being virus free. The list of viruses and the methods of their detection has been listed in the table below.

Nine Test Laboratories have been accredited by DBT to test for virus free plants production and lab status and certify various tissue culture raised plant species of recognized tissue culture companies. List of Accredited Test Laboratories is as follows (source: DBT Webpage 2012a):

S. No	Entity	Address	Accredited for
1	Indian Institute of Spices Research	P.B. 1701, Marikunnu Post, Calicut Kerala-673012	Virus indexing and genetic fidelity testing of spices
2	National Research Centre for Banana	Thogamalai road, Thyanapur post, Tiruchirapalli Tamil Nadu-620120	Virus indexing and genetic fidelity testing of banana
3	Indian Institute of Horticultural Research	Hessaraghatta Lake P O, Bangalore Karnatka-560089	Virus indexing of ornamentals and genetic fidelity testing of anthurium and gerbera
4	Central Potato Research Institute	Shimla (HP)-171001	Virus indexing and genetic fidelity testing of potato
5	Institute of Himalayan Bioresource Technology, Palampur	P.O. Box 6, Palampur (H.P)-176061	Virus indexing and genetic fidelity testing of TC raised plants
6	University of Agriculture Science, Banagalore	Dept. of Plant Pathology Agriculture College, UAS,GKVK Bangalore Karnataka	Virus indexing and genetic fidelity testing of TC raised plants
7	Sugarcane Breeding Institute	Coimbatore Tamil Nadu-641007	Genetic fidelity and virus indexing of sugarcane
8	Agharkar Research Institute, Pune	G.G. Agharkar Road, Pune-411004, Maharashtra	Genetic fidelity testing of tissue culture raised banana and sugarcane
	IARI Regional Station, Pune	Agriculture College Estate, Shivajinagar, Pune Maharashtra-411005	Virus indexing of tissue culture raised banana and sugarcane
9	Central Research Institute for Jute and Allied Fibres	Barrackpore, Kolkata-700120, West Bengal	Virus indexing and genetic fidelity testing of banana, potato, orchids, bamboo and other crops as per requirement

Table 1: List of plant species and viruses covered under national certification system for tissue culture plants and their test protocols. (Source: DBT webpage 2012b)

S.No.	Plant Species: Common Name	Plant Species: Scientific Name	[@]Viruses/Viroids/phytoplasmas/ bacteria covered under NCS-TCP	Test protocol*: ELISA	Test protocol*: DIBA	Test protocol*: RT-PCR/ PCR	Test protocol*: NASH
			Fruit Plant Species				
1	Apple*	*Malus domestica*	Apple chlorotic leaf spot virus (ACLSV); Trichovirus; Flexiviridae.	+		+	
			Apple mosaic virus (ApMV); Ilarvirus; Bromoviridae.	+		+	
			Apple stem grooving virus (ASGV); Capillovirus; Flexiviridae.	+	+	+	
			Apple stem pitting virus (ASPV); Foveavirus; Flexiviridae.	+		+[IC]	
2	Banana	*Musa* spp.	Banana bract mosaic virus (BBrMV); Potyvirus; Potyviridae.	+		+	
			Banana bunchy top virus (BBTV); Babuvirus; Nanoviridae.	+		+	+
			Banana streak virus-Mysore strain (BSMyV); Badnavirus; Caulimoviridae.	+		+	+
			Cucumber mosaic virus (CMV); Cucumovirus; Bromoviridae.	+		+	
3	Citrus*	*Citrus* spp.	Citrus greening (Liberobacter asiaticum)	+		+	
			[1]Citrus exocortis viroid (CEVd); Pospiviroid; Pospiviroidae.			+	
			Citrus tristeza virus (CTV); Closterovirus; Closteroviridae.	+	+	+	
			Citrus mosaic virus (CMBV); Badnavirus; Caulimoviridae.			+	+
			Indian citrus ringspot virus (ICRSV); Mandarivirus; Flexiviridae.	+		+	
4	Straw-berry	*Fragaria* spp.	[$]Raspberry ringspot virus (RpRSV); Nepovirus; Comoviridae.	+		+	+
			[$]Strawberry crinkle virus (SCV); Cytorhabdovirus; Rhabdoviridae.	+			
			[$]Strawberry latent ringspot virus (SLRSV); Sadwavirus; unassigned.	+		+	
			[$]Strawberry mild yellow edge virus (SMYEV); Potexvirus; Flexiviridae.	+			
			Strawberry mottle virus (SMoV); Sadwavirus; unassigned.	+			
			Strawberry pseudo mild yellow edge virus (SPMYEV); Carlavirus; Flexiviridae.	+		+	

			[$]Strawberry vein banding virus (SVBV); Caulimovirus; Caulimoviridae.	+		+	
			Tomato black ring virus (ToBRV); Nepovirus; Comoviridae.	+		+	+
			Spices				
6	Black Pepper*	*Piper nigrum*	Cucumber mosaic virus (CMV); Cucumovirus; Bromoviridae.	+		+	
			Piper yellow mottle virus (PYMOV); Badnavirus; Caulimoviridae.	+		+	
			Piper little leaf (phytoplasmas).	+		+	
7	Carda-mom, large	*Amomum subulatum*	Cardamom mosaic virus (CdMV); Macluravirus; Potyviridae.	+		+	
8	Carda-mom, small	*Eletteria cardamo-mum*	Cardamom mosaic virus (CdMV); Macluravirus; Potyviridae.	+		+	
9	Vanilla*	*Vanilla* spp.	Vanilla mosaic virus (VanMV); Potyvirus; Potyviridae.	+		+	
			Vanilla necrosis virus (Watermelon mosaic virus (WMV)); Potyvirus; Potyviridae.	+		+	
			Cymbidium mosaic virus (CYMMV); Potexvirus; Flexiviridae.	+	+	+	+
			Odontoglossum ring spot virus (ORSV); Tobamovirus; unassigned.	+	+	+	+
			Commercial crops				
10	Potato*	*Solanum tuberosum*	Potato leaf roll virus (PLRV); Polerovirus; Luteoviridae.	+[DAS]	+	+	+
			Potato virus M (PVM); Carlavirus; Flexiviridae.	+[DAS]	+	+	
			Potato virus S (PVS); Carlavirus; Flexiviridae.	+	+	+	+
			Potato virus X (PVX); Potexvirus; Flexiviridae.	+[DAS]		+	+
			Potato virus Y (PVY); Potyvirus; Potyviridae.	+	+	+[IC]	+
			Potato apical leaf curl virus (Tomato leaf curl virus (TlCV)); Begomovirus; Geminiviridae	+		+	
11	Sugarcane*	*Saccharum officinarum*	Sugarcane mosaic virus (SCMV); Potyvirus; Potyviridae.	+	+	+	
			Sugarcane yellow leaf virus (ScYLV); Polerovirus; Luteoviridae			+	
			Sugarcane bacilliform Mor virus (SCBMV); Badnavirus; Caulimoviridae.			+	
12	Alocassia	*Alocassia* spp.	Dasheen mosaic virus (DsMV); Potyvirus; Potyviridae.	+	+	+	

13	Alstro-meria	*Alstomeria* spp.	Cucumber mosaic virus (CMV); Cucumovirus; Bromoviridae.	+		+	
			Alstroemeria mosaic virus (AlMV); Potyvirus; Potyviridae.	+		+	
			Alstroemeria streak virus (Alstroemeria mosaic virus); Potyvirus; Potyviridae.	+		+	
			[$]Alstroemeria carlavirus (Lily symptomless virus); Carlavirus; Flexiviridae.	+	+	+	
			[$]Tomato spotted wilt (TSWV); Tospovirus; Bunyaviridae.	+		+	+
14	Anthurium	*Anthurium* spp.	Dasheen mosaic virus (DsMV); Potyvirus; Potyviridae.	+	+	+	
15	Asparagus	*Asparagus* spp.	Asparagus virus-1 (AV-1); Potyvirus; Potyviridae.	+		+	
			Asparagus virus-2 (AV-2); Ilarvirus; Bromoviridae	+		+	
			Asparagus virus-3 (AV-3); Potexvirus; Flexiviridae	+		+	
16	Caladium	*Caladium* spp.	Dasheen mosaic virus (DsMV); Potyvirus; Potyviridae.	+	+	+	
			Cucumber mosaic virus (CMV); Cucumovirus; Bromoviridae.	+		+	
17	Calla Lily	*Zantedeschia aethiopica*	Dasheen mosaic virus (DsMV); Potyvirus; Potyviridae.	+	+	+	
			[$]Tomato spotted wilt (TSWV); Tospovirus; Bunyaviridae	+		+	+
18	Carnation	*Dianthus caryophyllus*	Carnation etched ring virus (CERV); Caulimovirus; Caulimoviridae.	+		+	
			[$]Carnation Italian ringspot virus (CIRV); Tombusvirus; Tombusviridae.	+		+	
			Carnation latent virus (CLV); Carlavirus; Flexiviridae.	+		+	
			Carnation mottle virus (CarMV); Carmovirus; Tombusviridae.	+		+	
			Carnation necrotic fleck virus (CNFV); Closterovirus; Closteroviridae.	+		+	
			Carnation ringspot virus (CRSV); Dianthovirus; Tombusviridae.	+		+	
			Carnation vein mottle virus (CVMV); Potyvirus; Potyviridae.	+		+	
19	Chrysa-nthemum	*Dendranthema morifolium*	Chrysanthemum virus B (CVB); Carlavirus; Flexiviridae.	+[DAS]		+	
			Chrysanthemum aspermy virus (Tomato aspermy virus); Cucumovirus; Bromoviridae.	+		+	
			Cucumber mosaic virus (CMV); Cucumovirus; Bromoviridae.	+		+	

			[$]Chrysanthemum stem necrosis virus (CSNV); Tospovirus; Bunyaviridae.	+		+	+
			[2]Chrysanthemum stunt viroid (CSVd); Pospiviroid; Pospiviroidae.			+	+
20	Colacasia	*Colocasia* spp.	Dasheen mosaic virus (DsMV); Potyvirus; Potyviridae.	+	+	+	
21	Cordyline	*Cordyline* spp.	Impatiens necrotic spot virus (INSV); Tospovirus; Bunyaviridae	+		+	
			[$]Tomato spotted wilt (TSWV); Tospovirus; Bunyaviridae.	+		+	+
			[$]Tomato spotted wilt (TSWV); Tospovirus; Bunyaviridae.	+		+	+
			[$]Pelargonium flower break virus (PFBV);Carmovirus;Tombusviridae.	+		+	
			[$]Pelargonium vein clearing virus (PVCV); Eggplant mottled dwarf virus; Nucleorhabdovirus; Rhabdoviridae.	+		+	
22	Geranium	*Geranium* spp.	[$]Pelargonium leaf curl virus (PLCV); Tombusvirus; Tombusviridae	+		+	
			[$]Arabis mosaic virus (ArMV); Nepovirus; Cómoviridae.	+		+[IC]	
			[$]Tomato ring spot virus (ToRSV); Nepovirus; Comoviridae.	+		+	
			Tomato black ring virus (ToBRV); Nepovirus; Comoviridae.	+		+	+
			Tobacco rattle virus (TRV); Tobravirus; un- assigned.	+		+	
23	Gerbera	*Gerbera jamesonii*	Gerbera mosaic virus (Tobacco mosaic virus (TMV)); Tobamovirus; unassigned.	+	+	+	+
			Gerbera symptomless virus (GeSLV); unassigned; Rhabdoviridae.	+		+	
			Gerbera phyllody (Phytoplasma).	+		+	
			Cucumber mosaic virus (CMV); Cucumovirus; Bromoviridae.	+		+	
24	Ginger	*Zinziber* spp.					
			Ginger chlorotic fleckvirus (GCFV); Sobemovirus; unassigned	+		+	
25	Lily (Asiatic/ Oriental lily)	*Lilium longifolium*	Cucumber mosaic virus (CMV); Cucumovirus; Bromoviridae.	+		+	
			Lily symptomless virus (LSV); Carlavirus; Flexiviridae.	+	+	+	
			Lily mottle virus (LMoV); Potyvirus; Potyviridae.	+		+	
			[$]Lily virus X (LVX); Potexvirus; Flexiviridae.	+		+	
			[$]Strawberry latent ring spot virus (SLRSV); Sadwavirus; unassigned.	+			

			[\$]Tomato spotted wilt (TSWV); Tospovirus; Bunyaviridae.	+		+	+
			Tulip breaking virus (TuBV); Potyvirus; Potyviridae.	+		+	
			Impatiens necrotic spot virus (INSV); Tospovirus; Bunyaviridae.	+		+	
			Cucumber mosaic virus (CMV); Cucumovirus; Bromoviridae.	+		+	
26	Limonium	*Limonium* spp.	[\$]Clover yellow vein virus (ClYVV); Potyvirus; Potyviridae	+DAS		+	
			Tobacco rattle virus (TRV); Tobravirus; un assigned.	+		+	
			Cymbidium mosaic virus (CYMMV); Potexvirus; Flexiviridae.	+	+	+	+
		Cymbidium spp;	Cymbidium ringspot virus (CymRSV); Tombusvirus; Tombusviridae.	+		+	
27	Orchids	*Dendrobium* spp; *Cattelyia* spp;	Impatiens necrotic spot virus (INSV); Tospovirus; Bunyaviridae.	+		+	
		Phalenopsis spp. Etc.	Odontoglossum ring spot virus (ORSV); Tobamovirus; unassigned.	+	+	+	+
			[\$]Tomato spotted wilt (TSWV); Tospovirus; Bunyaviridae	+		+	+
27	Philode-ndron	*Philodendron* spp.	Dasheen mosaic virus (DsMV); Potyvirus; Potyviridae	+	+	+	
28	Spathy-phyllum	*Spathyphyllum* spp.	Dasheen mosaic virus (DsMV); Potyvirus; Potyviridae.	+	+	+	
29	Syngonium	*Syngonium* spp.	Impatiens necrotic spot virus (INSV); Tospovirus; Bunyaviridae	+		+	
			[\$]Tomato spotted wilt (TSWV); Tospovirus; Bunyaviridae.	+		+	+
30	Weeping fig	*Ficus benjamina*	Ficus chlorosis virus (Fig leaf chlorosis virus) (FigLCV); Potyvirus; Potyviridae.	+		+	
			Cucumber mosaic virus (CMV); Cucumovirus; Bromoviridae	+		+	
			Forest plant species				
31	Bamboo	*Bambusa vulgaris*	Bamboo mosaic virus (BaMV); Potexvirus; Flexiviridae.	+		+	

[\$] not known to occur in India and or/ restricted distribution in India;

[1]Citrus exocortis also can be detected by **sPAGE** (sequential polyacrylamide gel electrophoresis);

[2]Chrysanthemum stunt viroid, Potato spindle tuber viroid also can be detected by **R-PAGE** return gel PAGE); +: Detection of virus by specified test; **+DAS**: Detection of virus by double antibody sandwich-ELISA; +[IC]: Detection of virus by Immunocapture-RT-PCR); **+dsRNA**: Detection by dsRNA; ***ELISA:** Enzyme-linked Immunosorbent Assay; **DIBA:** Dot-immunobinding Assay; **RT-PCR**: Reverse transcription polymerase chain reaction; **NASH**: Nucleic Acid Spot Hybridization; @Viruses/viroids nomenclature is as per the International Taxonomy Committee for Viruses.

* Covered under certification standards established by the Department of Biotechnology in accordance with provisions of Seeds Act, 1966.

Control of plant viruses

Plant viruses cannot be directly controlled by chemical application. The major means of control (depending on the disease) include:

- **Growing resistant crop varieties**: In some crops and for some viruses there are highly effective sources of resistance that plant breeders have been using for many years. However, no such "natural" resistance has been identified for many others. Transgenic resistance has shown considerable promise for many plant-virus combinations following the discovery that the incorporation of part of the virus genome into the host plant may confer a substantial degree of resistance (Fraser 1990). For example, the use of this approach in Hawaii to control *Papaya ringspot virus* has been credited with saving the local papaya industry. However, this technology is controversial, particularly in Europe, and the extent to which it will be used commercially is currently uncertain.
- **Use of virus-free planting material**: In vegetatively propagated crops (e.g. potatoes, many fruit crops) and where viruses are transmitted through seed major efforts are made through breeding, certification schemes etc., to ensure that the planting material is virus-free.
- **Exclusion:** The prevention of disease establishment in areas where it does not yet occur. This is a major objective of plant quarantine procedures throughout the world as well as more local schemes.
- **Meristem tip culture**: Only the meristematic dome and 1 pair of subtending leaves should be excised. If larger pieces are taken, it is likely that the virus will be transmitted. The size of a meristem plus the subtending leaves ranges from 0.1-0.5 mm. The apical dome itself measures from 0.1-0.25 mm depending on the species. The meristem tip must be small enough to eradicate viruses and other pathogens, yet large enough to develop into a shoot. Although roots may form on the shoot directly in the same medium, often the shoot has to be transferred to another medium in order for roots to develop. Meristem tip culture is used successfully to remove viruses, bacteria, and fungi from plants. In a majority of cases heat therapy is combined with meristem tip culture in order to produce the greatest number of plants that are "virus free" (Faccioli and Marani 1998).
- **Thermotherapy/Heat treatment:** Heat treatment is used in those plants in which viruses cannot be eradicated just by meristem tip culture alone. In plant tissue cultures viruses can also be eliminated with incidence of higher temperatures (heat treatment.). In such case explants are exposed to the incidence of higher temperatures, which are not lethal for plant cells, but they are lethal for viruses. Mostly used temperature range is 50-52°C with exposition about 10-30 minutes. In case this method is applied on whole plants, lower temperatures have to be used (32-40 °C) with exposition about 4-30 days (depends on plant species and virus type). Heat therapy combined with meristem tip culture is able to eradicate viruses, bacteria, and fungi but does not remove viroids. Unlike viruses, viroids are

RNA without a protein coat – thus they are known as 'naked' RNA and are very difficult to eradicate.

- **Chemotherapy**: Virus-free plants can be also obtained when antivirus matters are added into nutrient solution (Ribavirin or 2-thiouracil). Ribavirin was shown to be the most active chemical in case of potato viruses (Khuri et al 2000). Mostly combination of thermo-therapy and meristem culture is used for virus-free plants production (e.g., cassava, bananas, citruses, strawberry, Irish potato, apples, chrysanthemas, garlic) (Cieślińska 2007). Usually the infected plant must be destroyed.
- **Shoot tip grafting (STG)/Micrografting:** In a number of species including those of citrus, attempts at meristem culture remained unsuccessful. As an alternative, shoot tips of 0.14-0.18 mm in length isolated aseptically from a diseased plant were grafted on to young etiolated root stock seedlings grown *in vitro*. STG had been used to produce virus free plants in other crops including peach, and apple.
- **Electrotherapy:** Useful in Grapevines. Green cane pieces, three cm long and comprising a bud, were collected from the GVA infected grapevine and washed with tap water for 20 minutes. Their leaves were cut off and then exposed to electric currents of 0, 10, 20 and 30 milliamps (mA) for 10 or 15 minutes using a horizontal electrophoresis apparatus in TBE buffer (90 mM Tris-borate, 2 mM EDTA, pH 8) followed by immediate sterilization in 2.5% (w/v) of sodium hypochlorite for five minutes and were then washed again with sterilized water for 10 minutes (Bayati et al 2011). The explants are inoculated on MS basal media for culturing. After two months, generated plantlets were tested by RT-PCR for the presence of virus.
- **Cryotherapy:** Shoot tips (1 mm in size) were excised from green shoots and suspended in a solution of ¾ Murashige and Skoog (1962) medium supplemented with 3% (w/v) Na alginate, 2 M glycerol, 0.4 M sucrose, and 2 μM benzylaminopurine (BA). The medium and the shoot tips are transferred with a sterile pipette into a 0.1 M $CaCl_2$ solution containing 2 M glycerol and 0.4 M sucrose at room temperature. Shoot-tips are kept for 30 minutes to form beads (about 4 mm in diameter). The beads are then precultured on ¾ MS medium with increasing sucrose concentrations of 0.25, 0.5, 0.75 and 1.0 M for 4 days, one day at each concentration. The beads are then placed on sterilized filter paper in Petri dishes and dehydrated by air-drying in laminar flow at room temperature for 12 h. Following dehydration, beads are transferred into a 5 ml cryotube and immersed directly in liquid nitrogen for one hour. Then they are rapidly thawed in a water bath at 40 ºC for 3 min. Thawed beads are post-cultured on a Petri dish containing solidified ¾ MS medium containing 2 μM benzylaminopurine (BA). Petri dishes are maintained in the dark at 24°C for 2 days and, then, are maintained at a temperature of 24±2ºC under a 16 h photoperiod with a light intensity of 45 μE s^1 m^2 provided by cool-white fluorescent tubes. After 6 weeks, shoots longer than 3 mm are transferred

to solid ¾ MS medium for regeneration of plants (Wang and Valkonen 2009).

- **RNA interference or RNA Silencing:** The related processes of post-trancriptional control of gene expression found in plants, animals and fungi is called RNA silencing. It mediates sequence-specific degradation of target transcripts, recruiting RNA molecules of 21–23 nucleotides as specificity determinants. RNA silencing in higher plants serves as an adaptive, antiviral defense system, which is transmitted systemically in response to localized virus challenge. Plant viruses have elaborated a variety of counter-defensive measures to overcome the host silencing response. One of these strategies is to produce proteins that target the cell autonomous or signalling steps of RNA silencing (Voinet 2001; Tenllado et al 2004).

Plant quarantine

When plant pathogens are introduced into an area in which they did not previously exist, they are likely to cause much more catastrophic epidemics than do the existing pathogens. Some of the worst plant disease epidemics that have occurred throughout the world, for example the downy mildew of grapes in Europe and the bacterial canker of citrus, the chestnut operation of the quarantine regulations. It is extremely difficult to predict accurately whether an exotic organism will become established, and once established, become economically important. Factors, which affect entry and establishment of an organism, are (1) hitchhiking potential compared with natural dispersal; (2) ecological range of the pest as compared with ecological range of its host; (3) weather; (4) ease of colonization including reproductive potential, and (5) agricultural practices including pest management (Ruesink et al 1995). Plant quarantine may, therefore, be defined as 'Rules and regulations promulgated by governments to regulate the introduction of plants, planting materials, plant products, soil, living organisms, etc. with a view to prevent inadvertent introduction of exotic pests, weeds and pathogens harmful to the agriculture or the environment of a country/region, and if introduced, to prevent their establishment and further spread'. Plant quarantine is thus designed as a safeguard against harmful pests/pathogens exotic to a country or a region. Plant quarantine measures aim at providing protection to the agriculture of a country or region against the likely ravages of alien pests/pathogens should they get introduced and established. These measures are of particular importance and relevance to countries like India whose economy is largely based on agriculture (Schrader and Unger 2003). Quarantine not only helps to ward off the threats of exotic pests, but also aim to eliminate and prevent further spread of pests/pathogens (both indigenous and introduced) with restricted distribution within the country (domestic quarantine). Plant quarantine services are charged with the responsibility of preventing entry of hazardous pests, pathogens and weeds, but to deny entry to the valuable genetic resources would be against national interest. These activities are meant to help agricultural development and they are complementary to each other. Too much conservatism on the part of plant quarantine officials and too liberal an attitude on the part of plant introduction officials/breeders would be harmful. Plant quarantine officials must strive to provide adequate safeguards to allow smooth flow of germplasm resources in a healthy state. They should also try to ensure that the germplasm, when received in quarantine station, is processed promptly and that the delays in release, if any, are purely due to biological considerations alone. At the same time, circumvention of plant quarantine must be avoided at all cost even if it means delay in release or rejection of certain materials based on biological consideration. The two should work in unison as members of a single team. Together they should decide the type, quantity and source of the material, and also the required quarantine safeguards. Success or failure of plant quarantine measures would depend, to a great extent, on the ability of plant quarantine officials to detect pests and pathogens that may be associated with the introduced planting material. For quarantine purposes, techniques should be sensitive enough to detect even trace infections (Cronk and Fuller 1995). This is particularly important in case of pests/pathogens with very high multiplication rate like certain pycnidial fungi, downy mildews, bacteria and also viruses when the insect vectors are efficient.

In India, Directorate of Plant Protection, Quarantine and Storage (DPPQS) of Ministry of Agriculture is the nodal agency in India for implementing plant quarantine (Khetarpal and Gupta 2007). DPPQS deals with the commercial import of consignments of grains, plants and plant products for consumption through its network of 35 Plant Quarantine Stations spread across the country including seaports, airports and land frontiers as well as commercial imports of seeds/ plants for sowing or planting through five major stations at Amritsar, Chennai, Kolkata, Mumbai and New Delhi. National Bureau of Plant Genetic Resources (NBPGR) undertakes the quarantine processing of all germplasm including transgenic planting material under exchange for research purposes. The material known/ suspected to carry viruses is grown in isolation and the seedlings showing symptoms are tested for viruses by a combination of biological/ physical/ serological/ molecular tests. As per the recent amendments made under the PQ order, the Advanced Centre for Plant Virology at IARI, New Delhi, Indian Institute of Horticultural Research, Bangalore and Institute of Himalayan Bioresource Technology, Palampur have been identified for ensuring virus-free status in the imported *in vitro* material. Under the DIP Act, there is a provision of Domestic Quarantine to restrict the inter-state movement of nine invasive pests viz flute scale, sanjose scale, coffee berry borer, codling moth, banana bunchy top and mosaic viruses, potato cyst nematode, potato wart and apple scab (Sharma and Thakur 2007). Keeping in view the biosafety concerns associated with the growing of imported transgenic material, a containment Facility of Level -4 has been established at NBPGR, New Delhi for their quarantine processing.

Future thrust

Although these has been tremendous advancement in the virus diagnostics, prevention and control, emergence of new viruses has evinced the requirement of most sophisticated and cost effective techniques. The present day technique are still limited to the laboratories only for which the samples are required to be brought to the laboratories. Future focus should improve the development of cheap, easily operational and less time consuming methods to be used in the field. Generally, the plants are screwed for limited number of viruses but plants may be infected by many viruses. Detection methods for multiple viruses are need to be developed which should be cost effective and easy to operate. A part from these, literature is lacking about the microstructure, metabolism and interaction mechanism with best plant which are pre-requisiste to make every diagnostic method easy.

References

Afshari CA 2002 Perspective microarray technology, seeing more than spots, *Endocrinology* **143** 1983-1989

Baulcombe DC and Fernandez-Northcote EN 1988 Detection of strains of potato virus X and of a broad spectrum of potato virus Y isolates by nucleic acid spot hybridization (NASH), *Plant disease* **72** 307-309

Bawden FC, Pirie NW, Bernal JD, and Fankuchen I 1936 Liquid crystalline substances from virus-infected plants, *Nature* **138** 1051-1055

Bawden F 2006 Plant viruses and virus diseases, New Delhi: Biotech Publishers

Bawden FC 2002 Plant viruses and virus diseases, New Delhi: Daya Books

Bayati S, Shams-Bakhsh M, and Moieni A 2011 Elimination of Grapevine virus A (GVA) by cryotherapy and electrotherapy, *Journal of Agriculture, Science and Technology* **13** 443-450

Beijerinck MJ 1898 Concerning a contagium vivum fluidum as cause of the spot disease of tobacco leaves, *Verhandelingen der Koninkyke akademie Wettenschapppen te Amsterdam* **65** 3-21, Translation published in English as Phytopathological Classics Number 7 (1942) St. Paul, Minnesota: American Phytopathological Society Press

Berg LA and Bustamante M 1974 Heat treatment and meristem culture for the production of virus-free bananas, *Phytopathology* **64** 320-322

Blanc and Ducker 2011 Functions of virus and host factor during vector mediated transmission in *Recent Advances In Plant Virology* (eds) C Caranta, MA Aranda, M Tepfer and JJ Lopez-Moya, Norkfolk, UK: Caister Academic Press 103-120 pages

Bystricka D, Lenz O, Mraz I, Dedic P, and Sip M 2003 DNA microarrays: parallel detection of potato viruses, *Acta Virologica* **47** 41-44

Caranta C, Aranda MA, Tepfer M, and Lopez-Moya JJ 2011 *Recent advances in plant virology*, Norkfolk, UK: Caister Academic Press

Cheng-Mayer C and Levy JA 1988 Distinct biological and serological properties of human immunodeficiency viruses from the brain, *Annals of Neurology* **23(S1)** S58-S61

Cheung VG, Morley M, Aguilar F, Massimi A, Kucherlapati R, and Childs G 1999 Making and reading microarrays, *Nature Genetics* **21** 15-19

Cieślińska M 2007 Application of thermo-and chemotherapy *in vitro* for eliminating some viruses, *Journal of Fruit and Ornamental Plant Research* **15** 117-124

Clark MF and Adams AN 1977 Characteristics of the microplate method of enzyme-linked immunosorbent assay for the detection of plant viruses, *Journal of General Virology* **34(3)** 475-483

Clark MF and Bar-Joseph M 1984 Enzyme immunosorbent assays in plant virology, *Methods in virology* **7**

Cronk QC and Fuller JL 1995 Plant invaders: the threat to natural ecosystems (Vol. 1), Springer

DBT Webpage 2012a Site visited on Sept, 2012, http://dbtncstcp.nic.in/ html/content/ ATLs.html

DBT Webpage 2012b Site visited on Sept, 2012, http://dbtncstcp.nic.in/List%20of%20viruses%20covered%20under%20NCS-TCP.pdf

Dijkstra J and CP de Jager 1998 *Practical Plant Virology: protocols and exercises*, New York: Springer

Edwards ML and Cooper JI 1985 Plant virus detection using a new form of indirect ELISA, *Journal of Virological Methods* **11(4)** 309-319

Eisen MB and Brown PO 1999 DNA arrays for analysis of gene expression, *Methods in Enzymology* 179-204

Faccioli G and Marani F 1998 Virus elimination by meristem tip culture and tip micrografting, *Plant Virus Disease Control* 346-380

Fraser RSS 1990 The genetics of resistance to plant viruses, *Annual Review of Phytopathology* **28(1)** 179-200

Gierer A and Schramm G 1956 Infectivity of ribonucleic acid from Tobacco Mosaic Virus, *Nature* **177** 702-703

Gillaspie Jr AG, Hopkins MS, Pinnow DL, and Hampton RO 1995 Seedborne viruses in preintroduction cowpea seed lots and establishment of virus-free accessions, *Plant Disease,* **79(4)** 388-391

Hadidi A, Flores R, Randles JW, and Semancick JS 2003 *Viroids*, Collingwood, Victoria, Australia: CSIRO Publishing

Harrison BD and Robinson DJ 1978 The Tobraviruses, *Advance Virus Research* **23** 25-77

ICTV webpage 2011 Site visited on Sept, 2011, International Committee on Taxonomy of Viruses, http://www.ictvonline.org/virusTaxonomy.asp?version=2011

Iwanowski D 1892 Concerning the mosaic disease of the tobacco plant, *St. Petersberg Academic Imperial Science Bulletin* **35** 67-70, Translation published in English as Phytopathological Classics Number 7 (1942) St. Paul, Minnesota: American Phytopathological Society Press

Kausche GA, Pfankuch E, and Ruska H 1939 Die Sichtbarmachung von pflanzlichem Virus im Übermikroskop, *Naturwissenschaften* **27 (18)** 292-299

Khetarpal RK and Gupta K 2007 Plant biosecurity in India-status and strategy, *Asian Biotechnology and Development Review* **9(2)** 83-107

Khuri FR, Nemunaitis J, Ganly I, Arseneau J, Tannock IF, Romel L, and Kirn DH 2000 A controlled trial of intratumoral ONYX-015, a selectively-replicating adenovirus, in combination with cisplatin and 5-fluorouracil in patients with recurrent head and neck cancer, *Nature Medicine* **6(8)** 879-885

Lindbo JA and Dougherty WG 1992 Untranslatable transcripts of the tobacco etch virus coat protein gene sequence can interfere with tobacco etch virus replication in transgenic plants and protoplasts, *Virology* **189(2)** 725-733

Lockhart DJ and Winzeler EA 2000 Genomics, gene expression and DNA arrays, *Nature* **405(6788)** 827-836

Mayer A 1886 Über die Mosaikkrankheit des Tabaks, *Die Landwirtschaftliche Versuchs-stationen* (in German) **32** 451-467, Translated into English in *Phytopathological classics* 1942 (ed) J Johnson, St. Paul, Minnesota: American Phytopathological Society No. **7** 11-24

Murphy FA, Fauquet CM, Bishop DH, Ghabrial SA, Jarvis AW, Martelli GP, and Summers MD 1995 Virus taxonomy: classification and nomenclature of viruses, *Sixth report of the International Committee on Taxonomy of Viruses*, Springer-Verlag

Perez-Ortin JE 2002 Microarray methods for the detection of pathogenic bacterias and viruses in plants, in *Abstracts Agricultural Biomarkers for Array Technology, Management Committee Meeting,* Wadenswil 13 page

Powell CA 1987 Detection of three plant viruses by dot-immunobinding assay, *Phytopathology* **7**

Randles JW 2003 Economic impact of viroid diseases, in *Viroids* (eds) A Hadidi, R Flores, JW Randles, and JS Semancik, Collingwood, Victoria, Australia: CSIRO Publishing, 3-11 pages

Reed KE and Rice CM 2000 Overview of hepatitis C virus genome structure, polyprotein processing, and protein properties, in *The Hepatitis C Viruses,* Heidelberg: Springer Berlin, 55-84 pages

Ritzenthaler C 2005 Resistance to plant viruses: old issue, news answers? *Current Opinion in Biotechnology* **16(2)** 118-122

Ruesink JL, Parker IM, Groom MJ, and Kareiva PM 1995 Reducing the risks of nonindigenous species introductions, *BioScience* **45(7)** 465-477

Sambrook J, Fritsch EF, and Maniatis T 1989 *Molecular cloning Volume 2,* New York: Cold spring harbor laboratory press, 14-19

Schena M, Shalon D, Davis RW, and Brown PO 1995 Quantitative monitoring of gene expression patterns with a complementary DNA microarray, *Science* **270** 467-470

Schrader G and Unger JG 2003 Plant quarantine as a measure against invasive alien species: the framework of the International Plant Protection Convention and the plant health regulations in the European Union, *Biological Invasions* **5(4)** 357-364

Sharma S and Thakur M 2007 Role of plant quarantine in the management of pest organisms-A Review, *Agricultural Reviews-Agricultural Research Communications Centre* **28(4)** 235

Singh M and Singh RP 1996 Factors affecting detection of PVY in dormant tubers by reverse transcription polymerase chain reaction and nucleic acid spot hybridization, *Journal of Virological Methods* **60(1)** 47-57

Sip M 2002 Parallel detection of potato pathogens: possibilities and problems, in *Abstracts Agricultural Biomarkers for Array Technology, Management Committee Meeting,* Wadenswil, 24 page

Stanley WM 1935 Isolation of a crystalline protein possessing the properties of tobacco mosaic virus, *Science* **81** 644-645.

Tenllado F, Llave C, and Díaz-Ruíz JR 2004 RNA interference as a new biotechnological tool for the control of virus diseases in plants, *Virus Research* **102(1)** 85-96

Van Regenmortel MH and Fauquet CM 2000 Virus taxonomy: classification and nomenclature of viruses, *Seventh Report of the International Committee on Taxonomy of Viruses*

Voller A, Bartlett A, Bidwell DE, Clark MF, and Adams AN 1976 The detection of viruses by enzyme-linked immunosorbent assay (ELISA), *Journal of General Virology* **33(1)** 165-167

Wang Q and Valkonen J 2009 Cryotherapy of shoot tips: novel pathogen eradication method, *Trends In Plant Science* **14(3)** 119-122

Waterworth HE and Hadidi A 1998 Economic losses due to plant viruses, in *Plant Virus Disease Control* (eds) A Hadidi, RK Khetarpal and H Koganezawa, St. Paul, Minnesota, USA: The American Phytopathological Society Press, 1-13 pages.

21

Protected Cultivation: Nematode Problems and their Management

R. K. Walia and Kum Kum Walia

Department of Nematology, CCS Haryana Agricultural University, Hisar 125 004

Protected cultivation (PC) may be defined as the "***modification of the natural environment to achieve optimum plant growth***" (Singh 2005). Modifications can be made to both the aerial and root environments to increase crop yields, extend the growing season and permit plant growth during periods of the year not commonly used to grow crops in open field. PC may also indicate comprehensive systems of controlled environmental agriculture in which all the aspects of natural environment are modified for maximum plant growth and economic return.

Advantages of PC: Excellent for cultivation of high value vegetable crops, ornamentals, medicinal plants, horticultural nurseries etc. Vegetable crops can be grown under unfavorable conditions; certain vegetable crops can be grown round the year. Management of insect pests/diseases is easier. PC provides excellent opportunity to produce vegetables for export, and production of genetically engineered and micro-propagated vegetable varieties. Further, greenhouses also offer opportunities to grow high-quality products in large quantities on a small surface area. For example, in the Netherlands only 0.5% of the area in use for agriculture is covered with glasshouses. On this small area of 10,000 ha, about 20% of the total value of agricultural products is realized (van Lenteren 2000).

Production Systems and Media for PC: Crops can be grown directly in natural soil under green-house cover (Geoponics), or cultivation may be done in containers having soil-fewer medium comprising various blends of peat, vermiculite, sawdust etc. and supplemented with nutrient solution (Hydroponics). Growing of plants in a container in which roots are suspended and sprayed with a nutrient mist is called Aeroponics. Geoponics is the most widely practiced method in India.

Protected Structures: These are of various kinds, e.g., net houses (shade nets and insect-proof nets), plastic low tunnels/row covers (laid in open fields to cover rows of plants with transparent plastic film stretched over steel hoops), plastic mulch (covering the surface around the plant), trench cultivation (for growing vegetables in extreme winter, even at -15°C), floating plastic covers (plastic sheets to cover open field to protect crops from frost/snow), and green-houses (framed structures of variable shapes and sizes).

The underlying principle of green-houses is to create "green-house effect" whereby the major fraction of sunlight is absorbed by plants; plants emit long wave thermal radiations raising green-house temperature; during summer temperature is brought down by cooling devices. Based on climate control, green-houses can be classified into three types - High-tech or Climate controlled green-house; Semi-climate controlled green-house; and Naturally ventilated or Low cost green-house. High-tech green-houses have controlled temperature, relative humidity, CO_2, photoperiod, soil temperature, plant nutrients etc.

The total world area covered by green-house is about 300,000 ha, 50,000 ha of which are covered with glass, and 250,000 ha with plastic (Albajes et al 1999, Parrella et al 1999). Vegetables are produced on 195,000 ha and ornamentals on 105,000 ha (van Lenteren 2000). According to another source, China leads the comity of nations with 380,000 ha under green-house cultivation, followed by Japan (51042 ha), based on 1999-2000 data. In India green-house cultivation is practiced only in about 1000 ha, but it is picking up fast (Table 1).

Table 1. Area (ha) under green-house cultivation of horticultural crops in various countries

Country	1985-86	1990-91	1999-2000
China	13,130	200,000	380,000
Japan	27,300	45,033	51,042
South Korea	1,300	1,600	2,200
Spain	12,150	35,000	51,000
UK	2,600	2,650	2,500
USA	4,000	5,000	9,250
Morocco	660	5,000	10,000
Israel	400	1,500	5,200
USSR	6,900	4,850	3,250
India	50	100	1,000

Source: International Committee for Plastics in Agriculture, 2000

Pest Scenario in Green-houses-Certain Limitations: Growing vegetables and ornamentals in protected situation is very expensive and pest damage is not tolerated. Pest control is complicated by the virtually year-round culture of crops and by continuous heating during cold periods. These conditions provide excellent opportunities for the survival and development of a pest or disease once it has invaded the greenhouse. Some organisms that normally show diapause when they occur in the field have adapted to the greenhouse climate by no longer reacting to diapause inducing factors. As a result, rates of population growth are often much higher than in the field.

Green-house environment provides protection to pests from harsh environmental extremes. Specific chemicals suited for green-house conditions are lacking and there is no escape for toxic fumes of chemicals. Green-house grown crops require longer and frequent harvesting. It is difficult to use large sprayers due to limited space. Pesticide formulation is more phytotoxic within the protected environment of the greenhouse. Pesticide residues are higher in greenhouse vegetables compared to field vegetable produce due to the absence of environment eroding forces such as UV light, rain etc.

Nematode Problems in Green-houses

In the most widely practiced geoponic systems, the main sources of nematode infection are the infested soil and infected planting materials. Root-knot (*Meloidogyne* spp.), reniform (*Rotylenchulus reniformis*) and lesion nematodes (*Pratylenchus* sp.) have emerged as the major problems in green-house vegetables (tomato, cucumber, sweet pepper), ornamentals (gerbera, carnations, roses) and banana in many countries, including India (Tables 2, 3).

Ideal environment (soil temperature and moisture), availability of favourable host and intensive cultivation in green-house cultivation are most suited for unimpeded build-up and multiplication of nematodes (Fig. 1). Once introduced, the nematode populations shoot up exponentially and maintain damaging (pathogenic) levels continuously, resulting in huge crop losses. Further, unlike insect pest and other disease problems, nematode problems are relatively difficult to manage and perhaps impossible to eradicate.

Table 2. Nematode problems of vegetable crops in protected cultivation – literature review

Nematode	Crop	Results	Country	Reference
Meloidogyne arenaria *Meloidogyne incognita*	Tomato Cucumber	Infection rates in greenhouses with a sectional layout were lower (15%) than in those with a block layout (45.8%). Four 10-day old exposures of soil to sunrays (reaching 52-55ºC soil temperature) during summer break in cultivation, when repeated annually, reduced infections to acceptable levels.	Uzbek SSR	Rizaeva (1983)
Meloidogyne incognita	Tomato	Glasshouse tomato cv. Estafette gave the highest total yields when grown in soil, on rockwool or in peat bags, and was resistant to *M. incognita* and *Verticillium* wilt.	Nether-lands	Cools & Stolk (1984)
Meloidogyne incognita	Tomato Cucumber	The efficacy of Di-Trapex was compared with that of steam treatment. Di-Trapex at 500 litres/ha reduced the numbers of heavily infested plants to 18.3% from 35.7% and increased tomato yield by 3 kg/m2. The effect of Di-Trapex lasted into	USSR	Lashkova & Danilov (1982)

		the 2nd crop, increasing the cucumber yield by 4.2 kg/m2, and was more effective than steaming.		
Meloidogyne spp.	Tomato	The control of *Meloidogyne javanica, M. incognita* and *M. hapla* by chemical means or by use of resistant varieties is discussed. Resistance was detected in tomato varieties Restino, Boulba, Turquesa and Portanto, but not in Meltine.	Italy	Calabretta & Privitera (1985)
Meloidogyne spp.	Tomato	Covering greenhouses with polyethylene film and thereby raising soil temperature by some 7-14 C virtually eliminated natural populations of *M. incognita, M. acrita, M. javanica* and *M. arenaria* in tomato and increased yields comparably with aldicarb.	USSR	Khurramov (1990)
Meloidogyne javanica	Tomato	Tomato yield and growth in response to soil drench, single spray and multiple spray of 4, 8 and 12 mM of silver thiosulfate treatments were studied under plastic house conditions for two seasons in soil naturally infested with *M. javanica*. Gall formation and nematode reproduction were significantly reduced by all the treatments, with 4 mM multiple spray treatment showing the least phytotoxicity.	Jordan	Muhammad et al (1991)
Meloidogyne incognita	Tomato Cucumber	*M. incognita* causes widespread damage to tomato and cucumber and can regenerate rapidly after soil sterilization. The use of resistant tomato cv. Multiset and Tomosa and the new cucumber cv. TS 88122 reduced sterilant and nematicide costs. Initial experiments with the soil conditioner Clandosan as par of nematode control programme proved favourable.	Germany	Arndt (1994)
Meloidogyne spp. *Ditylenchus* spp.	Strawberry	Symptoms of damage and integrated control techniques for *Meloidogyne* spp. and *Ditylenchus* spp. are discus- sed, besides other pests & pathogens.	Italy	Manzali (1994)
Meloidogyne sp.	Tomato	Solarization together with dazomet or with calcium cyanamide gave best results in controlling nematodes - the infestation rates were 0 and 28%, respectively.	Italy	Fiume & Parisi (1995)

Meloidogyne hapla	Tomato	The chitin-urea amendments at 0.2 and 0.4% (v/v) were ineffective in reducing nematode populations. The final egg populations were significantly increased in amended soils.	Canada	Belair & Tremblay (1995)
Meloidogyne spp.	Aubergine	Trials using MBr at 20, 40 and 60 g/m2 soil and 2 types of covering (polyethylene and LMG Y181) were carried out in a plastic greenhouse. No interactions were observed between treatments and the type of plastic film used. Soil fumigation at 20 g MBr eradicated nematodes in the soil profile containing roots. Untreated plants gave a significantly lower yield.	Italy	Cartia et al (1996)
Meloidogyne spp.	Tomato	Experiments to find alternatives to MBr confirmed the possibility of using resistant cultivars and commercial cultivars grafted on resistant rootstocks in protected cultivation without the use of disinfectant.	Italy	Assenza et al (2000)
Meloidogyne spp.	Tomato	Studies were conducted on grafted and ungrafted plants for yield, nematode populations before and after tomato cultivation, root damage, and damage to lettuce (grown after tomatoes). Brigeor and Vogomax are recommended as rootstocks.	Germany	Graf et al (2001)
Meloidogyne spp.		Efficiency of solar heating with a soil warming system buried in a 60 cm deep radiation pipe was studied. Ethylene glycol was used as a heat conductor. Soil temperature was 10-12 C higher in solar heating with a warming system than in solar heating alone, and was more effective in controlling the nematode.	Japan	Tanaka et al (2000)
Meloidogyne incognita	Cucumber	Cucumber was treated with *Ambrosia martimala* dry matter at 1 kg/m2 or MBr 98% + 2% Chloropicrin at 70 g/m2 of soil. The nematodes in soil and surrounding cucumber roots decreased by 50 and 53%, for treatments with dry matter and MBr, respectively. Cucumber root galls were reduced to <1% & between 3.52 and 4.62% when treated with MBr and dry matter, respectively.	Egypt	El-Rab (2000)
Meloidogyne arenaria	Melon	Soil J2 population dynamics displayed a single peak on Oriental melon grown in a plastic tunnel.	Korea	Kim (2001)

		Nematode density increased slightly after planting but declined to 9/100 cc soil on 70th day, increased again on 100th day and this continued till 154 the day to reach 3817/100 cc. Thereafter the population declined till December. J2 were distributed relatively uniformly horizontally over the 180 cm wide row. The highest density of J2 was at 0-25 cm soil depth and only a few occurred at 40-50 cm.		
Meloidogyne spp.	Tomato Cucumber	*Trichoderma koningii* (strain 1/31), *Verticillium lecanii* (strain AFX-7) and *Paecilomyces lilacinus* (strain p-k1) resulted in good nematode control and increased yield.	Russia	Yarkulov (2000)
Meloidogyne sp.	Cucumber	A water solution with narcissus (registered as a plant growth regulator for increasing resistance to plant diseases) was used to control low & medium level infestations of root-knot nematodes. Yields also increased.	Russia	Dobrokhotov (2000)
Meloidogyne sp.	Cucumber	The reduction of *Meloidogyne* population and total yield was highest (100%) with emulsified solar oil treatment (drenching of seedlings with emulsified solar oil, at 20 ml/m2 dissolved in 5 litres of water, 15 days before transplanting), intercropping with Tagetes sp. & solarization.	Egypt	El-Haddad et al (2003)
Meloidogyne javanica	Tomato Lettuce	The effect of *Pochonia chlamydosporia* alone or in combination with oxamyl was evaluated in a double cropping system of lettuce & tomato, and compared with MBr. The fungus survived in the rhizosp-here at low densities. Final densities of *M. javanica* decreased after lettuce and increased after tomato. Fungus + oxamyl consistently reduced root gall ratings on tomato, but the numbers of eggs per g root varied. MBr treated plots remained free of nematodes at the end of 2-year study.	Spain	Verdejo-Lucas et al (2003)
Meloidogyne sp.	Cucumber	The nematode can cause 50-70% loss of cucumber in greenhouse. The effect of a fertilizer background (a mixture of peat+ sawdust at 1:0.5 + N60P60K90) + humus at 1, 2 and 3 kg/m2 + trace elements (crystalline	Russia	Bedin and Tokarev (2004)

		30%, wood ash 20%, a growth regulator 10%, copper sulphate 5%, manganese sulphate 5%, bone flour 15%, and a biological control agent Planriz 15% at 4 g /m2. A full pest control was achieved in variants with 2 and 3 kg/m2 humus + trace elements + background fertilizer.		
Nematodes	Cucumber	Grafting alone or in combination with fenamiphos was evaluated as an alternative to MBr. Grafted plants on nematicide treated soil yielded 85% more than non-grafted plants on non-treated soil. The use of fenamiphos increased production by 65% only.	Spain	Echevarria et al (2004)
Meloidogyne javanica	Tomato Cucumber	Studies conducted on effectiveness of soil treatment with heat and chemicals, use of biological control agents (*Arthrobotrys oligospora* as Nematofagin-BL and Trichodermin), use of resistant cultivars, use of plant growth regulators and growing of bait crops (a vetch-oat mixture) for protection of plants from nematode.	Russia	Trusevich (2004)
Nematodes	Tomato, Bell pepper, Cucumber, Tobacco	Studies conducted on combining PGPR with different modes of action with organic amendments. *Bacillus subtilis* (strain GB03) producing antibiotics, *B. amyloliquefaciens* (strain IN937a) and *B. subtilis* (strain (IN937b) both eliciting ISR were combined with chitosan (to stimulate microflora antagonistic to nematodes). The combination of 2 bacilli strains and chitosan resulted in significant growth promotion that was correlated with ISR in all the four crops. The preparation has been commercialized under the name BioYield.	USA	Kloepper et al (2004)
Meloidogyne spp.	Tomato	Tomatoes grown in a plastic tent were given biological treatments with or without plastic mulching or MBr. The biological agents Mycostop (a powder based on *Streptomyces griseoviridis* strain K6) at 5 g/m3 and Rootshield (granules based on *Trichoderma harzianum* strain T-22) at 4 g/m3 failed to protect the crop when applied alone or to enhance protection when used in conjunction with plastic mulching.	Italy	Polizzi et al (2004)

Meloidogyne javanica	Tomato	Plastic tunnel experiments were conducted to compare the ability of a native and introduced isolate of *Pochonia chlamydosporia*. Both isolates parasitized eggs of *M. javanica* but the native isolate increased parasitism after multiple applications. The native isolate was recovered more frequently from soil, and was a better colonizer of roots. Multiple applications of either isolate reduced the nematode gall rating, and the native isolate also reduced the final egg population in roots. Neither isolate reduced final nematode densities in soil or affected tomato yields compared to untreated plots.	Spain	Sorribas et al (2003)
Meloidogyne spp.	Tomato	Soil application of elemental sulfur was ineffective; applications of metam sodium gave satisfactory results, while a granular formulation of fenamiphos gave good results. Excellent results were obtained with manual soil injections of liquid fumigant formulation of 97% 1,3-dichloropropene. Doses of 200 litres/ ha when applied in July, effectively and economically protected the two subsequent tomato crops. Nematode population densities in soil and degree of root galling were considerably reduced by the treatment.	Italy	Lamberti et al (2003)
Meloidogyne incognita	Tomato	Meilisha – a new tomato hybrid having multiple resistance, including against *M. incognita,* is released. It is suitable for protected and open field cultivation.	China	Huang et al (2005)
Meloidogyne incognita *Meloidogyne javanica*	Tomato	A mixture of 1,3-dichloropropene 60.5% w/w and chloropicrin 33.3% w/w (Telone C35EC) suitable for soil drip fumigation, is compared with MBr. In sandy and sandy loam soils, no significant relationships were found between the rates of mixture applied (132.4, 268.4, 402.6 and 536.8 kg/ha) and the degree of nematode infestations.	Italy	Minuto et al (2006)
Meloidogyne incognita	Sweet pepper	Grafting on resistant rootstocks combined with biofumigation+ solarisation was evaluated in greenhouses. In biofumigated and solarised soils, resistance showed	Spain	Ros et al (2006)

		efficacy after 3 years of monoculture. In non-disinfected soil, nematode resistance was overcome in the 2nd year of monoculture. Combination of grafting and biofumigation (using sheep and chicken manure) plus solarisation was shown as a feasible alternative to MBr in sweet pepper greenhouses in South east of Spain.		
Meloidogyne incognita	Sweet pepper	The reiteration of biofumigation+ solarisation over 2 years recorded *M. incognita* control similar to MBr. Similar results were obtained to weed control. Plant growth was lower than MBr and yield was similar or higher than MBr and untreated.	Spain	Guerrero et al (2005)
Meloidogyne spp.	Tomato Cucumber	Root-knot nematodes cause losses from 14-44% in tomato and up to 100% in cucumber under plastic greenhouses. Ten tomato and 10 cucumber cultivars were evaluated against a mixed population of *M. incognita* race 1 and *M. javanica*. The tomato cvs. Debora and Akafuku and cucumber cvs. Meteoro, Sprint 440 and Dasher II were most productive under plastic greenhouse.	Brazil	Charchar and Aragao (2005)
Meloidogyne spp.	Tomato Cucumber	*Meloidogyne* spp. are considered damaging to tomato and cucumbers, besides other pests.	Bosnia-Herzegovina	Kohnic et al (2006)
Meloidogyne spp.	Cucumber	The application of *Paecilomyces lilacinus* controlled *Meloidogyne* spp. by 73.3%, resulting in a 19.7% increase in yield.	China	Xiao et al (2006)
Meloidogyne spp.	Tomato Cucumber	It is probably the most widespread and economically important of all nematode pests and in areas where root-knot nematode is not controlled, its damage has been responsible for losses up to 50% of tomato production. In some varieties of cucumbers, this pest is capable of destroying 100% of the crop.	Albania	Tedeschini and Jovani (2010)
Meloidogyne spp.	Vegetable crops	Average yield loss was estimated at about 30.8% of total vegetable production, and 38.1% of the farm advisors considered that root-knot nematode infections had increased over the last 5 years. Most farm advisors (78.3%) stated that the most	Spain	Talavera et al (2012)

commonly used control method was chemical soil fumigation. Non-fumigant nematicides were cited by 59.8% of them, grafting and resistant cultivars by 42.3%, soil solarisation 38.1%, plant extracts 14.4% and biofumigation 6.2%. The majority of farm advisors considered the application of fumigant nematicides combined with soil solarisation, and grafting onto resistant rootstocks as the most effective methods of root-knot nematode control.

Table 3. Nematode problems of ornamental crops in protected cultivation – Literature review

Nematode	Crop	Results	Country	Reference
Pratylenchus sp. *Paratylenchus* sp. *Longidorus* sp.	Roses	Symptoms include poor shoot and root growth, pale foliage, short inflorescences and vase life. The plants are starved and susceptibility to fungal diseases is increased.		Gabriel and Lucke (1979)
Nematodes	Roses	Hot water was circulated in 76.3 mm diameter pipes, 30 cm apart and 15 cm below the surface for 48-72 h from a boiler maintained at 90ºC. A temperature of 58-60ºC could be maintained for a depth of 30 cm and 43-50ºC to 40 cm. This proved adequate for the control of soil nematodes and the technique is cheaper than steam sterilization. Rose bushes planted after soil treatment grew well and no nematodes were found on the plants.	Japan	Hayashi (1979)
Pratylenchus sp.	Sweet pea	Roots turned black and flower stalks yellow. The bacterial nodules on the roots disintegrated. Partial nematode control was achieved with potatoes, beetroots and beans and by avoiding rotation with carrots, celery, peppermint and Primula. Cultivation of marigold (Tagetes) deterred nematodes. Soil steaming or chemical soil sterilization greatly reduced nematode populations but did not eliminate them.	Germany	Schreiner (1986)
Nematodes	Ornamentals	Pathogenicity and control of *Macroposthonia curvata, Xiphinema diversicaudatum, Ditylenchus dipsaci, Meloidogyne incognita,* and *Aphelenchoides ritzemabosi* are discussed	Bulgaria	Choleva (1982)

Heterodera daverti	Carnation	The rate of development of the nematode was favoured by a temperature over 24ºC. The total nematode number reached the highest level on the roots in Sept and maximum number of J2 in the soil and roots was found in Sept/Oct. The number of peaks resulting from the count of full eggs in the egg sacs indicates 7 generations per year.	Italy	Ambrogioni and D'Errico (1994)
Pratylenchus spp. *Meloidogyne* spp.	Rose Cucurbits	The monoculture of Tagetes in crop rotation reduced the population density of *Pratylenchus* spp. by 70-85% after 3 months. A cultivation of host plants over 3 years after *Tagetes patula* did not lead to an increase of the nematode population density. In the green-house, Tagetes reduced the population density of *Meloidogyne* spp. by 95% after a cultivation period of 2 months. In the green-house, Tagetes as cover crop in an intercropping system with row crops (rose and cucumber cultures) also reduced the population density but not as high as compared to the monoculture.	Germany	Lung et al (1997)
Meloidogyne incognita	Carnation Gerbera	Yield loss of 26.6% in carnation and 31.1% in gerbera was assessed due to nematode infestation by the application of phenamiphos @ 2.5 kg a.i./ha.	India	Nagesh and Reddy (2000)
Aphelenchoides ritzemabosi	Sweet basil	The nematode colonized and reproduced within the leaf, petiole and stem tissues. The epidermis and mesophyll were invaded by the nematode which caused interveinal discolouration and necrosis and collapse of the palisade and spongy parenchyma. All nematode developmental stages, including eggs were observed in the leaf tissues. Possible control measures, which are complicated by the short life cycle of the nematode, the broad host range and the short productive cycle of the basil, are also discussed.	Italy	Vovlas et al (2005)

Constant monitoring of pests and diseases in general, and nematodes in particular is the first and foremost requirement in protected cultivation. Lack of diagnostic above-ground symptoms and ignorance about nematode problems is chiefly responsible for their non-detection in the initial stages.

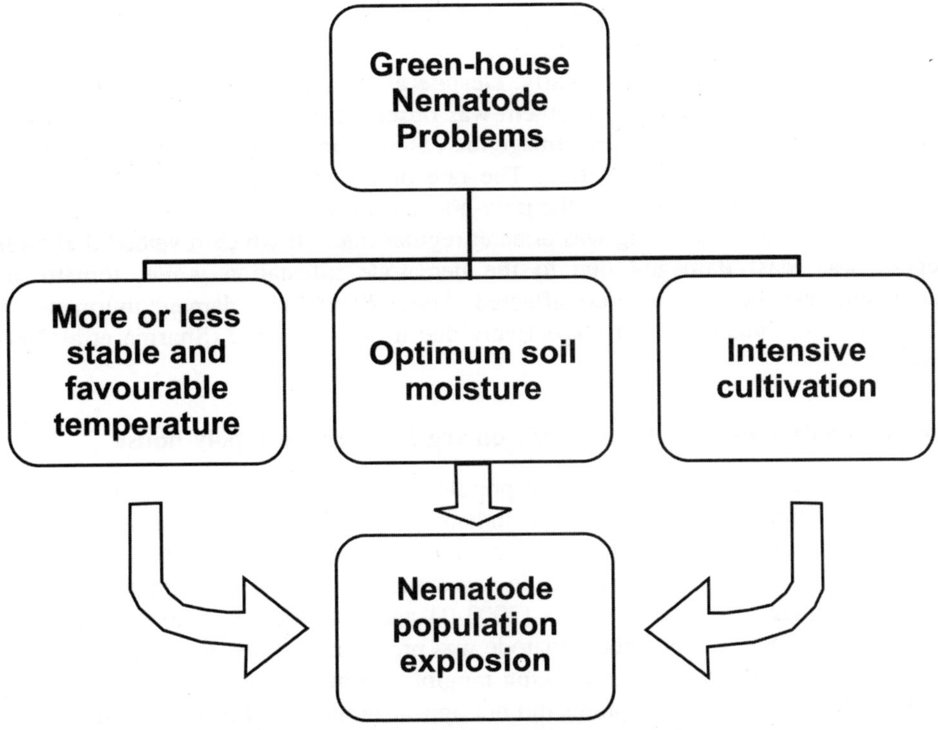

Fig. 1. Ecological factors favouring nematode population development in green houses

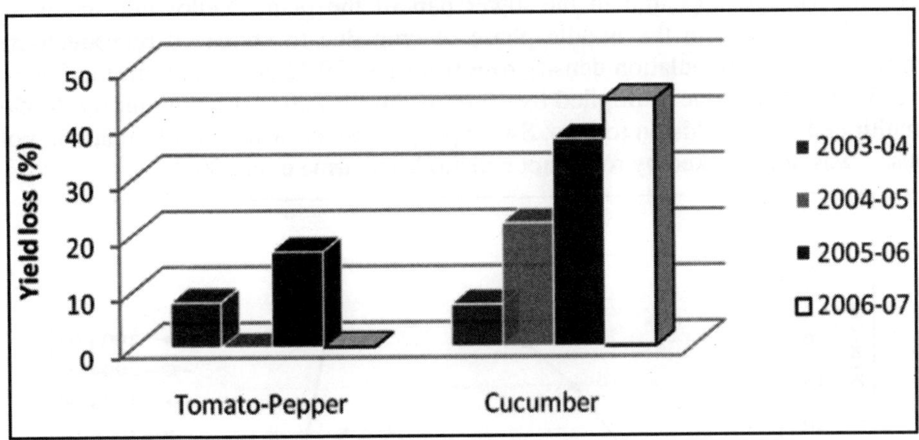

Fig. 2. Per cent yield loss in polyhouse I (Tomato-pepper) and Polyhouse II (Cucumber) caused by *Meloidogyne incognita* (Source: Sharma et al 2007)

Work done in India

In India, poly-house cultivation of agri-horticutural crops has recently been started. Root-knot nematodes, predominantly *Meloidogyne incognita* has emerged as the main problem. Nemic population dynamics was studied under Hi-tech cultivation of hybrid tomato, sweet pepper and hybrid cucumber under Indo-Israel project at Indian Agricultural Research Institute, New Delhi. Crop management practices such as fumigation (with formaldehyde @ 250 l/ha) of poly-house every year, removal/changing of top soil and regular supplement of pesticides (carbofuran 3G @ 1 kg a.i./ha) were followed regularly. Despite all these measures, intensity of root-knot nematode (*M. incognita*) problem was observed in tomato and cucumber. The intensity of infestation and crop damage increased exponentially even in plots treated with formaldehyde before planting. The rate of growth of population density and crop damage was much faster in the poly-house conditions than usually seen in open fields close by. Soil sampling was done at regular interval which revealed that sweet pepper was least damaged due to the nematode infestation while, tomato and cucumber were the most severely affected. About 20 and 45% damage in tomato and cucumber crop was observed respectively due to *M. incognita* (Sharma et al 2007) (Fig. 2).

Population dynamics of *M. incognita* on vegetable crops in poly-house

Studies conducted on nematode population dynamics in the Indo-Israel poly-house cultivation at IARI, New Delhi indicated *M. incognita* as a major problem in two poly-houses. The nematode problem was undetected in the first two years, but subsequently (2003 onwards), infection was severe on tomato. Tomato crop suffered root-knot damage in alternate years (2003-04 and 2005-06), whenever tomato was grown (Fig. 3). Since root-knot nematode was prevalent in poly-house, it was able to survive for one year without producing tangible symptoms on pepper during 2004-05. Root-knot nematode symptoms did not appear in the initial phase of tomato crop. Plant length was not affected as all the plants attained 5-7 m length. Symptom of root-knot damage appeared in the lower part of the plant. Yellowing and wilting appeared in tomato in the middle phase of crop due to enhanced reproduction of nematode and the population density rose from 1 to 30 J2 per cc soil during 2005-06 (Fig. 3). The nematode multiplied by 10 folds during 2003-04 while during 2005-06 it multiplied by 30 folds on tomato. Sweet pepper rotated with hybrid tomato in poly-house I was not attacked by root-knot nematode (Sharma et al 2007).

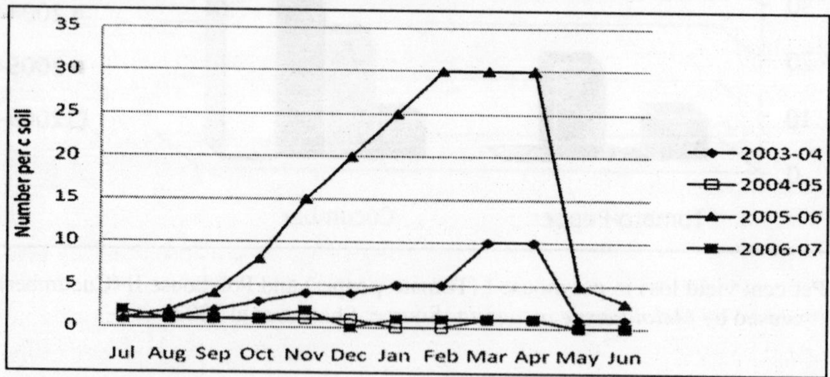

Fig. 3. Population dynamics of *Meloidogyne incognita* in polyhouse I (Totamo-Pepper) (Source: Sharma et al 2007)

In cucumber, however, two crops are taken in a year; therefore, the dynamics of *M. incognita* was similar to that in open cultivation (Fig. 4). Due to two crops of cucumber in poly-house, infestation of *M. incognita* spread very fast from few patches to whole area. There was seedling mortality in the initial phase of the crop. The progress of symptoms is fast as infected seedlings showed reduced number and size of fruits with time (Sharma et al 2007).

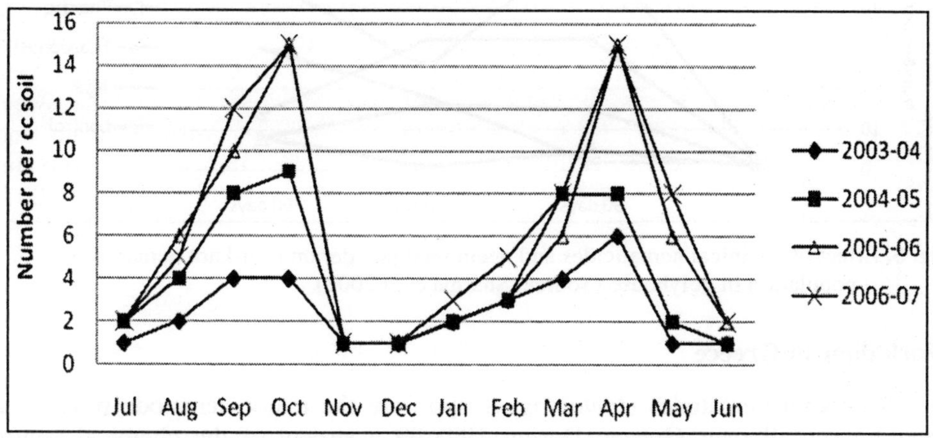

Fig. 4. Population dymaics of *Meloidogyne incognita* in polyhouse II (Cucumber) (Source: Sharma et al 2007)

Another major nematode, the reniform nematode, *Rotylenchulus reniformis* was observed in IARI poly-houses, however, build-up of this nematode was not recorded over the years. Perhaps, the fast multiplication of *M. incognita* resulted in the suppression of reniform nematode. Prevalence of *Tylenchorhynchus brevilineatus* was also recorded in all the poly-houses, before and after treatment of formaldehyde although population density was observed to be below 1 J2 per cc soil. The population of free-living nematodes was also meager (< 1 nematode per cc soil) than in open cultivation (Sharma et al 2007).

Effect of chemical nematicides (carbofuran 3G, phorate 10G, triazophos 40EC) and neem seed powder (NSP) was tested at IARI, New Delhi. The chemicals were applied @ 2 kg/2 l a.i./ha in two split doses one month apart; first dose was applied 30 days after transplanting. NSP was used @ 100 kg /ha. Root-knot population increased 8 fold in carbofuran treated plots; but two months after treatment it declined to 3 fold. In phorate treated plots, nematode population increased 15 times, but second application reduced it to 4 times, which increased again to 10 times after 2 months. Triazophos treated plots recorded minimum population. NSP also showed a gradual increase of 1-20 times (Fig. 5). The high temperature and humidity also had adverse effect on the efficacy of the applied chemicals. It was concluded that the management of root-knot nematodes on tomato under poly-house conditions could be achieved with triazophos when applied at 2 l a.i./ha in two equal split doses (Sharma et al 2008).

Other root-knot nematode management options in the poly-houses have been documented by Nagesh and Reddy (2005), who observed pre-plant application of dazomet (40 g/sq. m.) alone and in combination with neem cake 1 kg and *Paecilomyces lilacinus* significantly reduced root-knot count on gerbera.

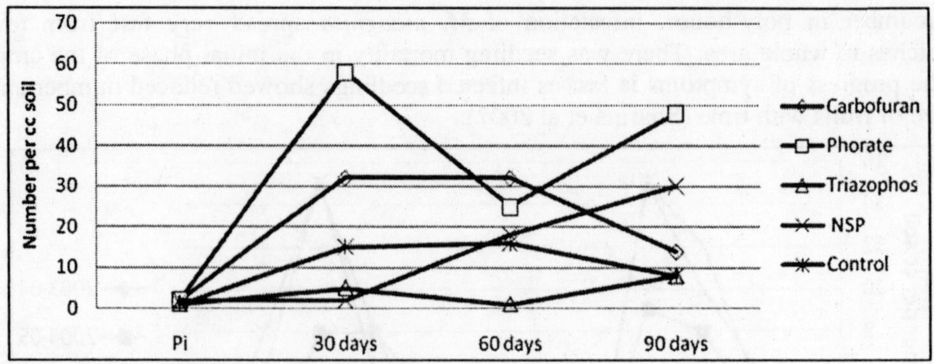

Fig. 5. Effect of chemical nematicides and neem seed powder on root-knot nematode population in polyhouse (Source: Sharma et al 2008)

Work done in Greece

Root-knot nematodes (*Meloidogyne* spp.) are the major nematode problem in green-houses all over Greece. During 1998-99 a survey of the chemical control methods used in green-houses for controlling nematodes was done. Farmers were asked to fill a questionnaire to monitor the specific type of chemical control they use as well as the crop and the variety they cultivated. All the green-houses were categorized in two groups according to the use of MBr or other nematicides. One group was consisted of green-houses where MBr was applied as a control method while the other of green-houses where non-fumigant nematicides were used. The results showed that 50% of 60 examined green-houses in North Greece were treated with MBr. The level of infestation was very low where MBr was applied while in some of the green-houses root-knot nematode was in non-detectable levels. However, in green-houses where non-fumigant nematicides were applied, the level of infestation was much higher. Numbers of nematodes were greater during summer of beginning of autumn than early in the spring. *M. javanica* was the predominant species (Prophetou-Athanasiadou et al 2002) (Table 4).

Table 4. Mean number of juveniles, galls, females and egg masses of *Meloidogyne* in 500 g soil after tomato plants were grown in pots with soil collected from 24 greenhouses from Greece during 1998-99

Greenhouse	J2 per 500 g soil		Number of galls per pot		Number of eggmasses per pot	
	Other nematicides	MBr	Other nematicides	MBr	Other nematicides	MBr
			1998			
1.	1300	0	4	0	3	0
2.	1833	0	6	0	3	0
3.	5067	0	9	0	4	0
4.	7433	0	11	0	6	0
5.	8233	0	16	0	5	0
6.	8500	0	20	0	8	0

7.	265	33	3	2	1	1
8.	1233	33	7	0	5	0
9.	1674	67	9	3	6	1
10.	4765	100	12	0	7	0
11.		133		4		3
12.		268		0		0
13.		433		0		0
14.		1167		7		5
			1999			
15.	335	0	12	0	5	0
16.	365	0	23	0	10	0
17.	300	0	28	0	14	0
18.	600	0	42	0	24	0
19.	965	0	49	0	20	0
20.	2065	0	58	0	26	0
21.	2565	168	70	19	29	10
22.	6100	450	171	49	141	22
23.	6800	135	140	11	121	9
24.	18535	900	770	103	503	53

Tzortzakakis et al (1999) compared two control strategies, chemical and non-chemical on the populations of *M. javanica* in green-house grown tomatoes. The chemical strategy consisted of an oxamyl application at the recommended dosages after transplanting a susceptible tomato in the autumn and again in the spring crop; and the non-chemical strategy was planting a resistant tomato in the autumn before a susceptible one in spring. An additional treatment was included where *Pasteuria penetrans* was applied in susceptible spring tomato following a resistant one. None of the treatments applied to the spring crop provided significant control of *M. javanica.* The growth of a resistant tomato, as the previous crop in autumn, reduced nematode population densities to undetectable levels but probably did not eliminate the nematode which subsequently increased on the following susceptible tomato crop. Eggs might have survived in the soil in a diapaused stage and juveniles hatched when susceptible crop was planted. An average of 65 ± 35% juveniles were found encumbered with a mean number of 3.4 ± 2.1 spores of *P. penetrans* in plots treated with the hyperparasite, indicating a degree of compatibility. However, infection in females was only 2.5-3%; thus the treatment did not have any significant effect. But earlier studies involving *P. penetrans* along with oxamyl, plant resistance or solarization showed its potential as a biological control agent in these crops under local conditions (Tzortzakakis and Gowen 1994).

This study shows that a single cropping cycle with a resistant tomato may not have a significant effect in root-knot nematodes population densities on the subsequent susceptible crop. In green-house conditions of vegetable production it is likely that more than one cropping cycle with resistant plants will be required, with subsequent nematicide applications in the following susceptible crop to reduce population densities below damaging thresholds.

Tzortzakakis (2000) tested the effect of the fungus *Verticillium chlamydosporium* (= *Pochonia chlamydosporia*) (applied @ 5000 chalmydospores per g soil) alone and in combination with oxamyl against *M. incognita* infecting tomato in a micro-plot experiment on sandy loam soil in plastic-house. The fungus survived in the soil and colonized the roots of tomato but did not reduce root galling, egg production and the remaining egg and juvenile inoculum in the soil after a six month period. A combined treatment with oxamyl did not have any additional effect. The apparent inability of oxamyl to reduce the nematode population could be attributed to the single application. Probably a second application would attain better control as is recommended commercially. The lack of nematode control by fungus could be attributed to the large nematode densities. On infected roots where two generations had developed galls were large with all egg masses remaining inside the gall. A partially resistant plant or a poor host supporting low nematode reproduction such that egg masses will be exposed to *V. chalmydosporium* in the rhizosphere will be a key factor in the development of integrated control strategies.

Work done in other countries

In Cyprus, *M. javanica* was successfully controlled on tomato under glass-house conditions using either dazomet with oxamyl or oxamyl alone. Significant yield increases ranging from 23.5 to 47.1% were achieved over the controls while the ratio of the incremental crop value over the nematicidal cost was significantly in favour of oxamyl being applied alone via drip irrigation (Philis 1994). This study also suggested that oxamyl should be applied early to protect the young root system of the seedlings from the invading juveniles. No residues of oxamyl were found in the fruit. Nematode resistant cultivars of tomatoes have been successfully tested to reduce crop losses (Philis 1990).

Nematode management options and their feasibility in PC

All the management options available for open-field cultivation may not find application in green-houses. The uniqueness of green-house cultivation permits cultivation of only selected vegetables crops; hence crop rotations involving non-hosts, antagonistic/trap crops become redundant. Similarly, practices like time of sowing, flooding, fallowing are also irrelevant for this system. Among the cultural practices, use of healthy planting material and general sanitation are of paramount importance. Soil solarization can also be very effective but warrants keeping the space without cultivation for a few weeks. Physical methods like hot water treatment offer excellent options in bulbous ornamentals. Replacement of top soil helps in the reduction of nematode inoculum but the endoparasitic nematodes are carried deep into the soil along with roots. Use of steam through network of underground pipes is practiced in certain high-tech green-houses but involves high energy consumptions.

Use of nematode resistant cultivars unequivocally is the best remedy for nematode management, particularly in green-house cultivation, but unfortunately, nematode resistant material is not available in most cases, like tomato. In cherry tomato which is commonly cultivated in green-houses, no resistant cultivars have been developed so far. However, resistant rootstocks can be exploited by grafting technique. Wild watermelon germplasm lines derived from *Citrullus lanatus* var.

citroides have been identified that may be useful as resistant rootstocks for managing root-knot nematodes in watermelon (Thies et al 2010).

Treatment of soil with a general biocide (formaldehyde) will alleviate most of the resident fauna and flora, including nematodes, at the start of cultivation. This is a general practice which is good for nematode management. Application of fumigants, in general, should be discouraged in consideration of problems involved in escape of fumes, besides other demerits mentioned earlier. The impact of nematicides is ephemeral and the residual populations shoot up due to favourable environmental conditions. For green-house applications, if at all required, it is best to use liquid formulations which can be applied with drip irrigations.

The potential of bio-control agents perhaps can best be harnessed in green-house cultivation. The favourable and stable environmental conditions both for nematode pests and their fungal (*Paecilomyces lilacinus*, *Pochonia chlamydosporia* etc.) and bacterial (*Pasteuria penetrans*) parasites, help in the establishment and amplification of introduced bio-agents. Initial supplements with compatible nematicides may help in the gradual build-up of suppressive soils; consequently, nematicides can be withdrawn from the system completely.

van Leteren (2000) has discussed specific advantages of biological pest control in greenhouses. Besides the general advantages, the specific reasons that make growers working in greenhouses prefer biological control are -

a. there are no phytotoxic effects on young plants, and premature abortion of flowers and fruit does not occur,

b. release of natural enemies takes less time and is more pleasant than applying chemicals in humid and warm greenhouses,

c. release of natural enemies usually occurs shortly after the planting period when the grower has sufficient time to check for successful development of natural enemies; thereafter the system is reliable for months with only occasional checks; chemical control requires continuous attention;

d. chemical control of some of the key pests is difficult or impossible because of pesticide resistance;

e. with biological control there is no safety period between application and harvesting fruit; with chemical control one has to wait several days before harvesting is allowed again;

f. biological control is permanent; once a good natural enemy – always a good natural enemy, and

g. biological control is appreciated by the general public.

The following preventive measures can be adopted as part of the Good Agricultural Practices (GAP) for protected cultivation (Sabir 2010).

- Solarization and Sanitation
- Augment bioagents vide fortified FYM (Store FYM under shade, mix bioagents such as *Trichoderma harzianum* culture (c.f.u. 2 x 10^9/g) @ 1 kg/500 kg of FYM and/or *Pseudomonas fluorescens* (c.f.u. 2 X 10^{12} cells/ml) 1 L/500 kg FYM and mix it with FYM one month before bed

preparation. Such enriched FYM may be used at the time of preparing planting beds.

- Implement GAP protocols especially
  - o Use of clean, uncontaminated water for irrigation from safe source
  - o Proper disposing of polythenes by burying
  - o Only need-based pesticide application
  - o Maintenance of worker hygiene
  - o Soil treatment/ drenching with carbofuran 3G @ 6 g/sq. m or captan 2 g/L of water
  - o Adoption of grafting technology (use of resistant root stock with suitable greenhouse varieties)
- Use of resistant varieties suitable for greenhouses
- Maintain weed free structures and remove all crop waste, especially diseased plant parts

Future thrust

Nematode management options such as selection of nematode resistant varieties; nematode-free and healthy seedlings; initial and periodic monitoring for nematode infestations during growth season, sanitation practices, and use of biological control can be judiciously framed in the IPM schedules designed for particular cropping and pest/disease green-house systems. In the light of little information available on the subject, the views expressed herein are only empirical and the nematode problems of PC systems need further detailed investigations to formulate tangible solutions.

References

Albajes R, Gullino ML, van Lenteren JC, and Elad Y (eds) 1999 *Integrated Pest and Disease Management in Greenhouse Crops*, Kluwer Publishers, Dordrecht

Ambrogioni L and D'Errico F P 1994 Seasonal population dynamics of a cyst forming nematode on carnation in Campania, *Italian Redia* 77 (1)11-22

Arndt M 1994 Which nematodes are injurious in vegetables grown under glass? *German Gartenbau Magazin* **3(5)** 14-15

Assenza M, Serges T, and Donzella G 2000 Rootstock: a possible alternative to methyl bromide, *Italian Informatore Agrario* **56 (28)** 34-36

Singh B 2005 Protected cultivation of vegetable crop, Kalyani Publishers

Bedin DP and Tokarev VS 2004 Against the gall nematode on cucumber in the greenhouse, *Russian Zashchita i Karantin Rastenii* **4** 28-29

Belair G and Tremblay N 1995 The influence of chitin-urea amendments applied to an organic soil on a *Meloidogyne hapla* population and on the growth of greenhouse tomato, Phytoprotection **76** (2) 75-80

Calabretta C and Privitera S 1985 Considerations on the present position in controlling root-knot nematodes of horticultural crops under protected cultivation and real prospects for other methods, Italian, *Tecnica Agricola* **37 (3-4)** 243-257

Cartia G, Causarano R and Primo P di 1996 Soil fumigation in protected environments: use of reduced doses of methyl bromide and coverings of different permeability, Third contribution, *Italian Difesa delle Piante* **19 (2/3)** 49-54

Charchar JM and Aragao FAS de 2005 Reproduction of *Meloidogyne* spp. on tomato and cucumber varieties under plastic greenhouse and field, *Portugues, Nematologia Brasiliera* **29 (2)** 243-249

Choleva B 1982 Parasitic nematodes of ornamental plants in greenhouses, *Bulgarian Rastitelna Zashchita* **30 (9)** 28-31

Cools MH and Stolk JH 1984 At last, root-knot nematode resistance and quality combined in one cultivar, *Dutch Groenten en Fruit* **39 (47)** 38-39

Dobrokhotov SA 2000 Narcissus against the root knot nematode in greenhouses, *Russian, Zashchita i Karantin Rastenii* **4** 21

Echevarria PH, Rodriguez-Castro A, Vivaracho SM, and Vallejo AD 2004 Influence of rootstocks and soil treatment on the yield and quality of greenhouse-grown cucumbers in Spain, *Acta Horticulturae* **633** 403-408

El-Haddad SA, Omar, MNA and El-Kattan MH 2003 Comparative studies on some components of integrated management on soil-borne pathogens affecting cucumber grown under protected agriculture, *Acta Horticulturae* **608** 219-226

El-Rab SMG 2000 Studies on the effect of *Damsis* sp. dry matter (*Ambrosia martimala* L.) as natural material to control nematodes of plastic-houses cucumber in Egypt, *Egyptian Journal of Horticulture* **27 (3)** 373-383

Fiume F and Parisi B 1995 Control of root-knot nematode and corky root on tomato grown in protected cultivation, *Italian, Informatore Fitopatologico* **45 (7/8)** 44-50

Gabriel G and Lucke 1979 Nematodes, danger for roses under glass, Gb + Gw: **79 (14)** 330-332, 334

Graf V, Augustin B, and Laun N 2001 Grafting of tomato plants-protection against root gall nematodes and cork root disease, *German Gemuse Munchen* **37 (3)** 8-12

Guerrero MM, Martinez MA, Martinez MC, Barcelo N, Lacasa A, Ros C, Guirao P, Bello A, and Lopez JA 2005 Biofumigation plus solarization efficacy for soil disinfestations in sweet pepper greenhouses in the southeast of Spain, *Acta Horticulturae* **698** 293-297

Hayashi I 1979 Sterilization, by buried hot-water pipes, against soil-borne pests and diseases, I, Adaptation of the technique for greenhouse-grown roses, *Japanese Bulletin of the Kanagawa Horticultural Experiment Station* **26** 60-72

Huang-TingTing, Liu-BingLu, Wang-ChangYi, and Liu-ShuQin 2005 A new tomato hybrid for protected cultivation - 'Meilisha', *Acta Horticultrae Sinica* **32 (1)** 179

Khurramov SK 1990 The use of solarization in control of gall-forming nematodes, *Russian Zashchita Rastenii Moskva* **4** 12-13

Kim DG 2001 Distribution and population dynamics of *Meloidogyne arenaria* on oriental melon (*Cucumis melo* L.) under greenhouse conditions in Korea, *Russian Journal of Nematology* **9 (2)** 61-68

Kloepper JW, Reddy MS, Rodriguez-Kabana R, Kenney DS, Kokalis-Burelle N, Martinez-Ochoa N, and Vavrina CS 2004 Application for rhizobacteria in transplant production and yield enhancement, *Acta Horticulturae* **631** 217-229

Kohnic A, Ostojic I, and Karic N 2006 Vegetable pests in greenhouses in territory of Herzegovina, Croatian Radovi Poljoprivrednog Fakulteta Univerziteta u Sarajevu Works of the Faculty of Agriculture University of Sarajevo **57(2)** 139-140

Lamberti F, Colombo A, D'Addabbo T, Caroppo S, D'Errico F P, Pelagatti O, Buonocore E, Ambrogioni L, Sasanelli N, Carella, A, Coiro, M I, Bacci L, D'Aloisio V and Vinci G

2003 Root knot nematodes on tomato: alternatives to methyl bromide, *Italian Colture Protette* **32 (12)** 69-78

Lashkova IA and Danilov LG 1982 Results of trials of different means of control against gall nematodes, Russian, in Sbornik Nauchnykh Trudov Vsesoyuznogo Nauchno Issledovatel skogo Instituta Zashchity Rasteneii Biologicheskie osnovy bor by s nematodami (eds) LA Guskova and AE Chumakova, 50-52 pages

Lung G, Fried A, and Schmidt U 1997 Biological control of nematodes with the enemy plant *Tagetes* spp., *German Gesunde Pflanzen* **49 (4)** 111-118

Manzali D 1994 Integrated control of strawberry *Italian Colture Protette* **23 (6)** 39-41

Minuto A, Gullino ML, Lamberti F, D'Addabbo T, Tescari E, Ajwa H, and Garibaldi A 2006 Application of an emulsifiable mixture of 1,3-dichloropropene and chloropicrin against root knot nematodes and soil-borne fungi for greenhouse tomatoes in Italy, *Crop Protection* **25 (12)** 1244-1252

Muhammad AG, Suwwan MA, and Abu-Gharbieh WI 1991 Silver thiosulfate effects on yield and growth of plastic house tomato in root-knot nematode infested soil, *Emirates Journal of Agricultural Sciences* **3** 41-65

Nagesh M and Reddy PP 2000 Crop loss estimation in carnation and gerbera due to the root-knot nematode, *Meloidogyne incognita* (Kofoid & White) Chitwood, *Pest Management in Horticultural Ecosystems* **6 (2)** 158-159

Nagesh M and Reddy PP 2005 Management of carnation and gerbera to control root-knot nematode, *Meloidogyne incognita* in commercial polyhouses, *Nematologia Mediterranea* **33** 157-162

Parrella MP, Stengard HL, and van Lenteren JC 1999 Glasshouse environments, in *Handbook of Biological Control* (eds) TS Bellows and TW Fisher, New York: Academic Press, 819-839 pages

Philis J 1990 The efficacy of nematode resistant tomato cultivars to *Meloidogyne javanica* under green-house conditions, *Nematologia Mediterranea* **18** 209-211

Philis J 1994 Use of dazomet and oxamyl for controlling the root-knot nematode, *Meloidogyne javanica* in glass-houses, *Nematologia Mediterranea* **22** 241-243

Polizzi G, Cascone G, D'Emilio A, and Castello I 2004 Physical and biological ways of containing *Pyrenochaeta lycopersici* and *Meloidogyne* spp. in protected tomato crops, *Italian Colture Protette* **33 (2)** 81-87

Prophetou-Athanasiadou DA, Giannakou I, and Sidiropoulos A 2002 The root-knot nematodes *Meloidogyne* (Tylenchida: Heteroderidae) in protected cultivation in Greece: Current control methods and their effect on nematode population, *Acta Horticulturae* **608**

Rizaeva SM 1983 Occurrence of root-knot nematodes in new greenhouse nurseries, *Russian Uzbekskii Biologicheskii Zhurnal* **6** 64

Ros C, Guerrero MM, Martinez MA, Lacasa A, and Bello A 2006 Integrated management of *Meloidogyne* resistance in sweet pepper in greenhouses, *Bulletin OILB/SROP* **29 (4)** 319-324

Sabir N, Singh B, Hasan M, Sumitha R, Deka S, Tanwar RK, Ahuja DB, Tomar BS, Bambawale OM, and Khah EM 2010 *Good Agricultural Practices (GAP) for IPM in Protected Cultivation,* Technical Bulletin No 23, National Centre for Integrated Pest Management, New Delhi, India, 16 pp

Schreiner U 1986 Nematode damage to sweet peas under glass. *German Deutscher Gartenbau* **40 (23)** 1076-1077

Sharma HK, Pankaj, and Jagan Lal 2008 Effect of chemical on root-knot nematode, *Meloidogyne incognita* infecting tomato under polyhouse cultivation, *Indian Journal of Nematology* **38** 124-125

Sharma HK, Pankaj, Gaur HS, and Singh B 2007 Nemic population dynamics on hybrid tomato, sweet pepper and hybrid cucumber under polyhouse cultivation, *Indian Journal of Nematology* **37** 161-164

Sorribas FJ, Ornat C, Galeano M, and Verdejo-Lucas S 2003 Evaluation of a native and introduced isolate of *Pochonia chlamydosporia* against *Meloidogyne javanica, Biocontrol Science and Technology* **13 (8)** 707-714

Talavera M, Sayadi S, Chirosa-Rios M, Salmeron T, Flor-Peregrin E, and Verdejo-Lucas S 2012 Perception of the impact of root knot nematode-induced diseases in horticultural protected crops of south-eastern Spain, *Nematology* **14** 517-527

Tanaka T, Ichikawa K, Matsusaki M, and Sakai K 2000 Effect of soil disinfection using solar heating with soil warming system to *Meloidogyne* sp. and *Ralstonia solanacearum, Japanese Research Bulletin of the Aichi ken Agricultural Research Center* **32** 105-110

Tedeschini J and Jovani V 2010 *Integrated pest management package on protected crops,* National catalogue of innovations in Albania, IDEASS - Innovation for Development and South Cooperation

Thies JA, Ariss JJ, Hassell RL, Olson S, Kousik CS, and Levi A 2010 Grafting for management of southern root-knot nematode, *Meloidogyne incognita,* in watermelon, *Plant Disease* **94** 1195-1199

Trusevich AV 2004 Control of the gall nematode, *Russian Zashchita i Karantin Rastenii* **2** 34-37

Tzortzakakis E A 2000 The effect of *Verticillium chlamydosporium* and oxamyl on the control of *Meloidogyne javanica* on tomatoes grown in a plastic house, *Nematologia Mediterranea* **28** 249-254

Tzortzakakis EA and Gowen SR 1994 The evaluation of *Pasteuria penetrans* alone and in combination with oxamyl, plant resistance and solarization for control of *Meloidogyne* spp. on vegetables grown in greenhouses of Crete, *Crop Protection* **13** 455-462

Tzortzakakis EA, Verdejo LS, Ornat C, Sorribos FJ, and Goumas DE 1999 Effect of previous resistant cultivar and *Pasteuria penetrans* on population densities of *Meloidogyne javanica* in green-house grown tomatoes, *Crop Protection* **18** 159-162

van Lenteren J C 2000 A greenhouse without pesticides: fact or fantasy? *Crop Protection* **19** 375-384

Verdejo-Lucas S, Sorribas FJ, Ornat C, and Galeano M 2003 Evaluating *Pochonia chlamydosporia* in a double-cropping system of lettuce and tomato in plastic houses infested with *Meloidogyne javanica, Plant Pathology* **52 (4)** 521-528

Vovlas N, Minuto A, Garibaldi A, Troccoli, A and Lamberti F 2005 Identification and histopathology of the foliar nematode, *Aphelenchoides ritzemabosi* (Nematoda: Aphelenchoididae) on basil in Italy, *Nematology* **7 (2)** 301-308

Xiao-Shaun, Zhang-ShaoSheng, and Liu-GuoKun 2006 Control effect of Paecilomyces lilacinus on Meloidogyne spp., Chinese Journal of Fujian Agriculture and Forestry University Natural Science Edition **35 (5)** 463-465

Yarkulov FY 2000 Control of root knot nematodes, *Russian Zashchita i Karantin Rastenii* **12** 20.

22

Climate Variables and their impact on Plant Diseases: Retrospect and Prospect

Suresh Pande and Mamta Sharma

International Crops Research Institute for the Semi-arid Tropics (ICRISAT), Patancheru - 502324, Andhra Pradesh, India

Climate variability and changing climate patterns are altering the equilibrium of host-pathogen interactions resulting in either increased epidemic outbreaks or emergence of new pathogens or less known pathogens causing severe yield losses. Climate variables (temperature, humidity, and greenhouse gases) are the key factors for these changes. Since agriculture plays a key role in overall economic and social well being of India, the country need to produce 50% more grain by 2020 to meet the demand for food. Meeting this difficult challenge will be even harder with the changing climate patterns. Further, plant pathogens inspite of sound management technologies still results in 10-16% of food yield losses due to the climate variability. This paper discusses the different climate variables and their effect on plant diseases in a range of crops and pathosystems and the mitigation strategies their management.

It is well established that temperature, moisture and greenhouse gases are the major elements of climate change. Current estimates of changes in climate indicate an increase in global mean annual temperatures of 1ºC by 2025 and 3ºC by the 2100. The carbon dioxide (CO_2) concentration is rising at the rate of 1.5 to 1.8 ppm per year and is likely to be doubled by the end of 21st century (IPCC 2007). Variability in rainfall pattern and intensity is expected to be high. Greenhouse gases (CO_2 and O_3) would result in increase in global precipitation of 2 ± 0.5ºC per 1ºC warming. Overall, changes in these elements will result in: i) warmer and more frequent hot days and nights, ii) erratic rainfall distribution pattern leading to drought or high precipitation and iii) drying of rainfed semi-arid tropics (SAT) in Asia and Africa. These changes will have major implications for food security, particularly in the developing countries, where the need to increase and sustain food production is most urgent. Even with minor deviations from the normal weather, the efficiency of applied inputs and food production is seriously impaired (Rotter et al 1999). Agriculture production of rainfed regions, which constitute about 65% of the area under cultivation and account for about 40-45% of the total production in India, is expected to suffer severe as a result of climate change (Agarwal 2003). It is estimated that the grain production has doubled over the last 40 years as a consequence of changes in plant protection and other agricultural technology, including a 15-20-fold increase in pesticide use worldwide. Despite this, the overall proportion of crop losses has increased during this period and excessive use of insecticides has increased pest outbreaks and losses in some crops and areas (Oerke, 2006).

Climate change may affect plant pathosystems at various levels viz. from genes to populations, from ecosystem to distributional ranges; from environmental conditions to host vigour/susceptibility; and from pathogen virulence to infection rates (Garret et al 2006). The three essential components in classic disease triangle; a virulent pathogen, a susceptible host and a favorable environment are required simultaneously for a disease to occur (Agrios 1997). Pathogens are dependent on the favourable environmental conditions for disease development/ and or epidemic to occur. The environment includes temperature, humidity, rainfall, sunlight, wind, nutrition and soil quality etc. However, Madden et al (1997) defined that epidemic develop when changes in disease intensity occur in host population over time and space referred as disease tetrahedran. These changes may show positive, negative and neutral impacts on host-pathogen interactions. In general, climatic changes could result in following changes in diseases: a) extension of geographical range; b) increased over-wintering and oversummering; c) changes in population growth rates; d) increased number of generations; (e) loss of resistance in cultivars containing temperature-sensitive genes (f) extension of crop development season; (g) changes in crop diseases synchrony; h) changes in inter-specific interactions; i) increased risk of invasion by migrant pathogens; and j) introduction of alternative hosts and 'green bridges' or over-wintering hosts. Plant diseases are a major impediment to the production and quality of important food stuffs, and diseases such as wheat Fusarium Head Blight affect quality and food safety. In addition to reducing yield, they are of particular concern because of their direct impacts on human and animal health. Mycotoxins and pesticide residues in food are among the top food safety concerns associated with a changing climate in Europe (Miraglia et al 2009). Changes in both physical climate and atmospheric composition influence severity of FHB. The most significant influences occur during the production phase, but impacts can affect the entire wheat value chain (Chakrabortya and Newton 2011). The chapter highlights the effect of climate variables on plant diseases in a range of crops and pathosystems with some specific examples on the effects of the climate variables on diseases and pathogens in semi-arid tropic (SAT) regions in India and the mitigation strategies for their management.

Climate variables and plant diseases

Effect of temperature

Changes in temperature particularly increase in temperature are predicted to lead to the geographic expansion of pathogen and vector distribution, bringing pathogens into contact with more potential hosts (Baker et al 2000; Olwoch et al 2003). Chakraborty et al (2000) reported that more aggressive strains of pathogen with broad host range, such as *Rhizoctonia, Sclerotinia* and other necrotrophic pathogens can migrate from agroecosystems to natural vegetation, and less aggressive pathogens from natural plant communities can start causing damage in monocultures of nearby regions. The geographic range of *Phytophthora cinnamoni* will increase in response to increased temperature that would allow the overwintering of this pathogen in new areas (Braiser and Scott 1994). Similarly, the introduction of new vectors and changes in overwintering and oversummering may have effects on pathogen survival, movement and reproduction. Higher minimum temperatures and reduced frequency or intensity of cold days will favor the survival of pathogens. Leaf rust epidemics (*Puccinia triticina*) in wheat were found to cause higher yield losses over nearly two decades in areas where the pathogen could overwinter (Eversmeyer and Kremer 2000). Also, change in temperature might lead to appearance of different races of the pathogens which are not active but might cause sudden epidemic.

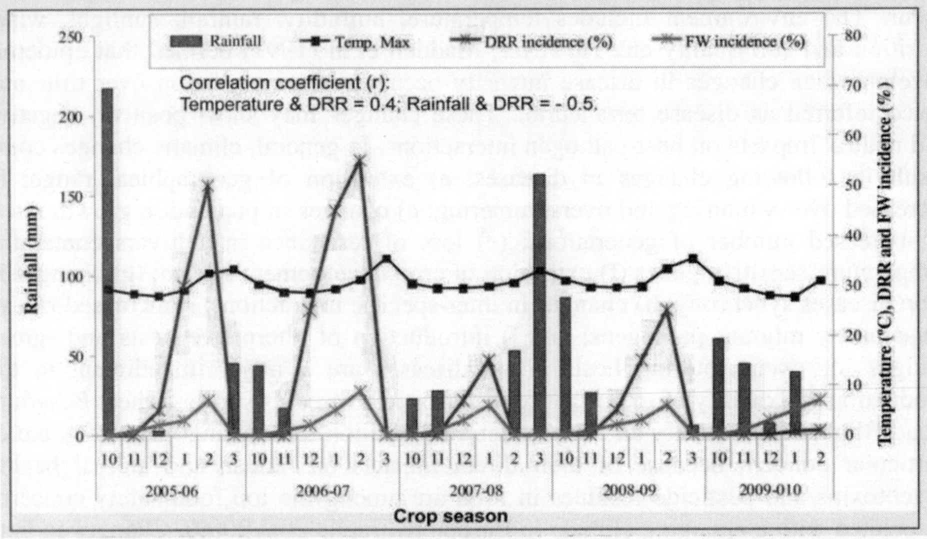

Fig. 1. Effect of weather variable (high temperature) on the incidence of dry root rot in cultivars resistant to Fusarium wilt in chickpea in semi arid tropic region in India.

Recent surveys conducted and reports from SAT regions indicated that soil borne diseases such as dry root rot (*Rhizoctonia bataticola*) in chickpea and charcoal rot (*Macrophomina phaseolina*) in sorghum increased many folds in last 2-3 years due to prolonged moisture stress. Studying data taken in India after the rainy season from 2005 to 2010, it was found that there is higher incidence of dry root rot in chickpea varieties that resist Fusarium wilt in years when temperatures exceed 33°C (Fig. 1; Pande et al 2010a; Sharma et al 2010). This is consistent with the greenhouse experiments where different soil moisture levels and temperatures were manipulated, showing that *R. bataticola* infected chickpea plants and caused dry root rot faster at 35 °C with soil moisture levels ≤ 60% (Sharma et al 2010) Temperature may have important repercussions on the effectiveness of host plant resistance. For some host-pathogen systems, resistance may get reduced or increased. Increase in temperature can modify host physiology and resistance by changing gene expression and activity. For example temperature above 20ºC deactivates temperature-sensitive resistance to stem rust in oat cultivars. Resistance mechanism in barley may be disrupted following drought stress as cells undergo expansion once an adequate water supply is restored (Newton and Young 1996). Lignifications of cell walls increased in forage species due to higher temperature resulting in enhanced resistance to fungal pathogens.

Effect of moisture (rainfall)

Moisture (rainfall) can affect both host and pathogens in various ways. Pathogens such as apple scab, late blight, and several other vegetable root pathogens are more likely to infect plants with increased moisture. Similarly prolonged moisture may create a new scenario of potential diseases in legume crops such as collar rot, wet root rot, anthracnose in chickpea and Alternaria blight, Phytophthora blight in pigeonpea and leaf spots and rusts in groundnut. Other pathogens such as powdery mildew tend to strive in conditions with lower moisture. In recent years, an erratic rainfall pattern is resulting in outbreak of Phytophthora blight of pigeonpea (*Phythophthora drechsleri* f. sp. *cajani*) in SAT regions (Sharma et al 2006; Pande and Sharma 2009). Preliminary analysis of weather data from 2000-2010 indicated that outbreak of Phytophthora blight of pigeonpea in SAT regions may be attributed to high intermittent rain (>350mm in 6-7 days) with in a short span of time leading to temporary flooding (Fig. 2). Similar environmental conditions manipulated in the greenhouse experiments indicated that Phytophthora blight infected the pigeonpea plants with in 48 hrs after inoculation at >90% humidity. The other pathogens favored by high rainfall include bacterial wilt in crop plants belonging to solanaceae and Zingiberaceae family.

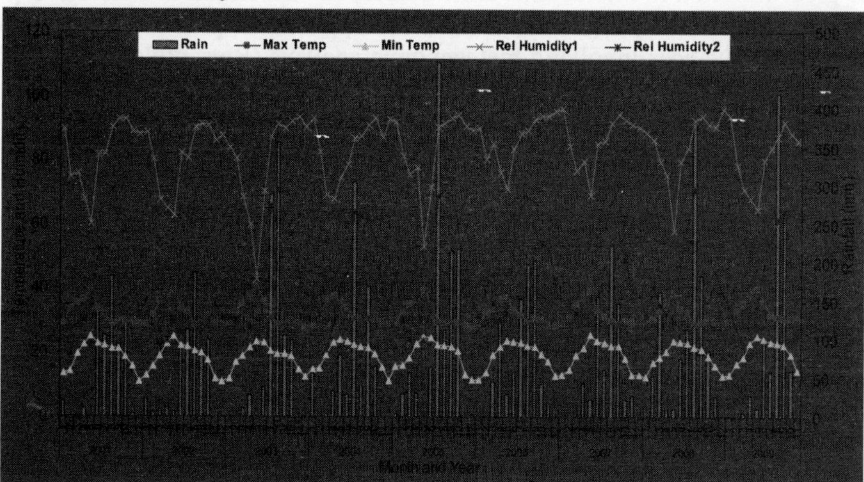

Fig. 2. Effect of weather variable (high rainfall) on the incidence of Phytophthora blight (PB) in pigeonpea during 2001-2009.

Effect of carbon dioxide (CO_2)

The impact of increased CO_2 can be observed in several stages of the host-pathogen interactions. Most of the available data suggests that atmospheric CO_2

enrichment asserts its greatest positive influence on infected as opposed to healthy plants. This influence in turn will modulate the balance of co-evolution between the host and the pathogen. Under elevated CO_2 the morpho-physiology of the crop plants is significantly influenced. Elevated CO_2 and associated climate change have the potential to accelerate plant pathogen evolution, which may, in turn, affect virulence. The increased biomass due to increase in CO_2 may become more conducive for rusts, mildews, leaf spots and blights etc. Chakraborty and Datta (2003) reported loss of aggressiveness of *Colletotrichum gloeosporioides* on *Stylosanthes scabra* over 25 infection cycles under elevated CO_2 conditions. On the contrary, foliar pathogens fecundity increased due to altered canopy environment due to the enhanced canopy growth that resulted in conducive microclimate for pathogen's multiplication. Exponential growth rates of *Phyllosticta minima* were 17% greater under elevated CO_2 (McElrone et al 2005). Lake and Wade (2009) found that *Erysiphe cichoracearum* aggressiveness increased under elevated CO_2 together with changes in the leaf epidermal characteristics of the model plant *Arabidopsis thaliana*.

Elevated CO_2 also have the potential to influence the effectiveness of host resistance (Plazek et al 2001; Plessl et al 2005). High levels of CO_2 may prevent induced resistance as plants grow more rapidly (Panagga et al 2004). However, at twice ambient CO_2, fecundity of the anthracnose pathogen *C. gloeosporioides* increased on both resistant and susceptible varieties of *Stylosanthes scabra* in a controlled environment (Chakraborthy and Datta 2003). The impacts of plant diseases under changing climate scenario may minimize or reverse any benefit from the CO_2 fertilization effect (Fernandes et al 2004; Butterworth et al 2010).

Elevated CO_2 will also affect the soil microbial communities. However because of the great variation in interactions among microbial species, it is difficult to predict the effect of climate change on the disease-suppressive qualities of soil. Recent technological advances involving metagenomic analysis (Riesenfeld et al 2004) will increase our understanding of microbial dynamics in soil.

Effect of greenhouse gases

The greenhouse gases such as ozone, nitrous oxide, carbon monoxide are secondary pollutants and current climate change scenarios predict an increase in these greenhouse gases which will interfere with photosynthesis and growth process of plants (Sandermann Jr 2000; Fuhrer 2003). Ozone can make plants sensitive for attack particularly for necrotrophic fungi, root rot fungi etc. Exposure of 5-10ppm ozone for a few hours can cause visible injury to sensitive crops like tomato, potato, onion, soybean, barley, wheat and tobacco (Plazek et al 2001). However, plants appear to be less sensitive to nitrous oxide, but higher concentration can cause water-soaked lesions, which soon turn brown. These symptoms may lead to problem in disease identification. The effects of ozone on leaf surface characteristics including wettability, led to increased rust incidence (Karonsky et al 2002). The increase in CO_2 can also alter the ozone concentration and thus the disease interactions (von Tiedemann et al 2000; Plessl et al 2007). Significant increase in disease incidence of powdery mildew, leaf spots, spot blotch and wheat flag leaves have been reported in plants exposed to ozone (von Tiedemann et al 1991).

Development of adaptation strategies

Climate change is a big challenge to the long-term sustainability of crop production. Disease management strategy will require adjustments under climate change, since regional impacts of climate change on plant diseases will be more. Some of the strategies for the management of diseases under climate change are discussed below:

Forecasting models

Forecasting/simulation models are used to predict the occurrence or change in the severity of plant diseases for a given crop production system and also shift in pathogen populations/fitness that may demand modifications in current production systems. Forecasting models which allow investigating multiple scenarios and interactions simultaneously will become most important for disease prediction, impact assessment and application of disease management measures. Two main types of simulation models can be particularly relevant: models simulating plant disease epidemics, and models simulating crop losses. These are useful tools for formulation of strategies, management decisions, guiding research, and future disease scenario predictions. These models are often driven by weather parameters and thus very suitable for studies of climate change effects on the crop- disease system. Coakley et al (1999) discussed several approaches that have been used for modeling the effect of climate change on disease. Several dynamic models have been developed in recent past but most of these are generally strong either in soils and crops, or in greenhouse gases (GHG) emissions. One approach is to use dynamic simulation models of pests and diseases coupled to a crop model (Kropff et al 1995; Agarwal et al 2005). Many weather-driven epidemiological models have also been developed, and can be mobilized to predict diseases epidemics under climate change (Pande et al 2005). Pande et al (2005) develop a disease prediction model based on 18 years data on weather variable and Botrytis gray mold severity in chickpea. Maximum temperature and afternoon relative humidity (referred as humid thermal ratio, Fig. 3) were identified as important component of the model and farmer friendly forecasting model was developed to predict the onset of botrytis gray mold and forms the basis for a predictive scheme to schedule fungicide sprays for managing BGM. CLIMEX exemplifies well these tools and has been used by plant pathologists to predict the likelihood of pathogen establishment under current climatic change scenarios (Sinha 2011). However, disease development, being sensitive to daily or hourly fluctuations in temperature and humidity, would need high resolution weather data for detailed simulation models. Such high resolution of weather data is in general not available. This problem can be overcome by the development of metamodels as shown by Sparks (Sparks, 2009). Global climate models such as General Circulation Models (GCM) was used by various researchers to simulate the potential impacts of temperature and rainfall on the expansion of *P. cinnamoni* in oak (Bergot et al 2004) by modeling phloem temperature of infected trees to evaluate overwintering probabilities; for *Plasmopara viticola* in grapes to evaluate disease pressure (Salinari et al 2006) and estimated the severity of Phoma stem canker epidemics on oilseed rape (Evans et al 2008). However, model simplification and testing will be required to address large scale future scenarios.

Also modeling the disease effects of varying production situation components in response to climate change elements such as fallow period duration, fertilizer input, crop rotation, or water stress has not been addresses adequately, and should be emphasized in future research. The research is underway where these forecasting models can become effective tools for technology targeting.

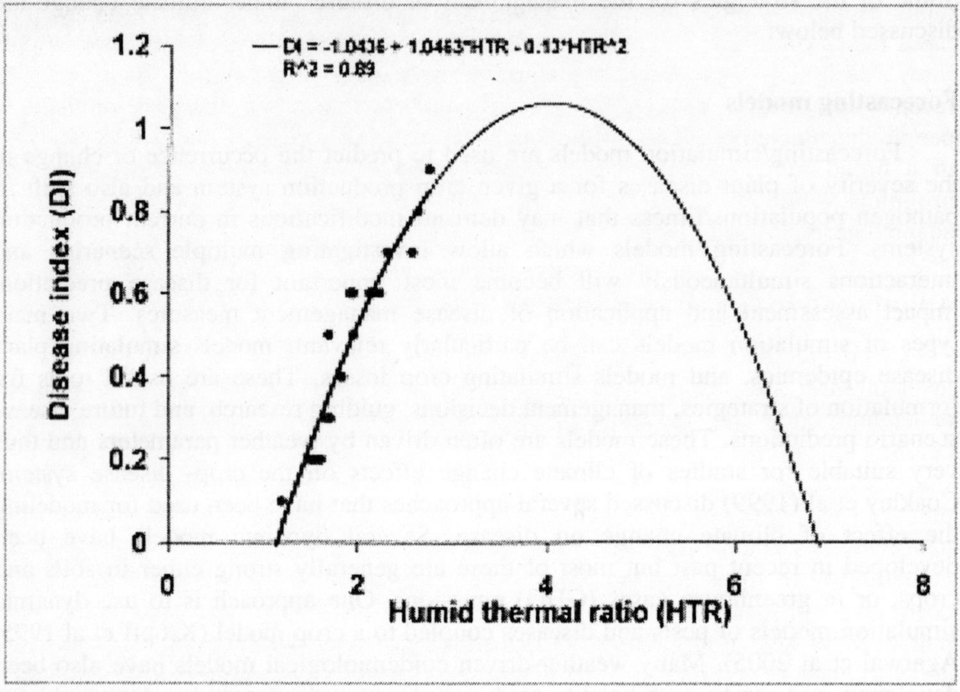

Fig. 3. Predictive weather based model for Botrytis gray mold infection in chickpea

Improving host plant resistance

Resistance to pathogens will become more important because of static and dynamic defenses from changes in physiology, nutritional status and water availability. Several studies provides evidences of these changes such as significant increases in photosynthetic rates, papillae productions, increased tannin content, higher carbohydrates accumulation in leaves, alteration in the production of resistance related enzymes (Hibberd et al 1996; Chakraborty et al 2000; Parsons et al 2003; Osswald et al 2006). Many changes in host physiology can potentially enhance plant resistance. Malmstrom and Field (1997) reported that CO_2 enrichment in oats may reduce losses of infected plants to drought and may enable yellow dwarf diseased plants to compete better with healthy neighbours. However, there will be greater risk to genetic resistance due to acceleration of pathogen cycles due to increased CO_2. Thus, identification of new sources of resistance to the emerging /threatening diseases is of utmost importance as the durability of resistance may be threatened and may lead to more rapid evolution of aggressive pathogen races. Progress in biotechnology, particularly marker assisted selection will contribute to making breeding for resistance against difficult traits more efficient. Tactics and methods might change, depending on the pathosystems. For example in chickpea, development and adoption of short duration cultivars will become more important to escape the terminal drought and thus minimizing losses due to emerging diseases such as dry root rot.

Gene pyramiding and the deployment of major genes could offer an option for a rapid response against a new threat. However, the effect may not last in the case of a rapid race evolution. An epidemic of stem rust on wheat caused by "new virulent form", designated as race Ug99 is currently spreading across Africa, Asia and most recently into Middle East and is causing major concern due to the large numbers of people dependent on wheat for sustenance (Vurro et al 2010). Since, wheat is grown in a broad range of environments; and pathogen is mutating rapidly, breeding programs would have extensive work remaining to get resistance into regionally adapted germplasm.

Adjustments in agronomic practices

To overcome the effects of climate change on plant diseases, adjustments at the level of farm are needed such as introduction of short or late maturing cultivars depending upon the pathogen outbreak, switching cropping sequence, changing planting dates, conserving soil moisture through appropriate tillage methods etc. The study of disease injury profiles (Savary et al 2006) under current and new agronomic practices will determine future needs in breeding and crop protection strategies. In many regions intensification has replaced diverse ecosystems and increased the vulnerability of disease attack. Monoculture or the use of megacultivars (varieties occupying millions of hectare) increases the chances of pathogen recombination or mutation by selection pressure. Changes in seasonal weather patterns will also contribute to the emergence of new diseases. Further, with the globalization of trade and international travel, quarantine measures and early intervention are essential to protect agro ecosystems from the introduction of exotic diseases and to prevent the spread of epidemics. This implies the development of adequate policies and the effective inspection and certification of seed and plant materials free from pathogens and pests.

Chemical control

Changes in the temperature and precipitation can alter the longevity and mode of action of fungicides due to changes in the plant morphology and physiology in plants grown under CO_2 enriched atmosphere (Coakley et al 1999). On the other hand, increase in disease infestation might lead to greater use of fungicides to control them. It has been estimated that the use of fungicides for controlling late blight in potato will increase by 15-20% in the coming decades (Fry and Goodwin 1997). Fungicide use could have detrimental effects on microorganism; however no studies have been done on the effect of increased temperature and CO_2 etc on the fungicide efficacy till date. In addition to this trans-boundary movement of toxic substances through natural and manmade pathways and their worldwide use in agriculture pose a threat to agriculture. Global wind currents and heavy rainfall are some of the elements which result in the spread of pesticides.

Future thrust

With the current trends in global warming and climate change, it is likely that most diseases will have a cosmopolitan range wherever the climate is favorable and the hosts are available. Rising temperatures, altered precipitation patterns, increase in atmospheric CO_2 concentration and erratic rainfall patterns will elicit complex change in plant pathosystems. Research has started only recently to understand the impacts of these climate change variables on agricultural systems and will be needed under realistic field conditions to clarify the situation. Prediction of such changes, and range and diversity in pathogens niches should help develop better control measures to minimize disease incidence. Regional impacts of climate change on plant disease management strategies need a relook using forecasting models and biotechnological approaches in understanding the emerging scenario of host pathogen interactions. Innovative methods may have to be adopted to develop adaptation strategies to overcome the impacts due to climate change and climate variability so that the food and livelihood security of rainfed farmers can be ensured. There is a need for a greater understanding of the effect of climate change on the efficacy of synthetic fungicides, their persistence in the environment, and development of resistance in pathogens populations to the fungicides. Recently, national and international net work is also actively anticipating and responding to biological complexity in the effects of climate change on agriculture and crop diseases (Garret et al 2009; Pande et al 2010b). The primary benefit of such studies to growers will be their ability to control the diseases that become severe as result of climate change, select varieties that are less vulnerable for diseases, and reduce fungicide application. The information will be useful to the crop growers, scientists and extension agencies, NGOs, research planners, and administrators. The outcomes of the research will have a major bearing on adopting IDM strategies in an environment friendly manner for sustainable crop production, and increasing food security in most vulnerable regions of the world.

References

Agarwal AA, Kotanen PM, Mitchell CE, Power AG, Godsoe W, and Klironomos J 2005 Enemy release? An experiment with congeneric plant pairs and diverse above- and below ground enemies, *Ecology* **86** 2979-2989

Aggarwal PK 2003 Impact of climate change on Indian agriculture, *Journal of Plant Biology* **30**(2) 189-198

Agrios G 1997 *Plant Pathology* fourth edition, New York: Academic Press

Baker RHA, Sansford CE, Jarvis CH, Canon RJC, MacLeod A, and Walters KFA 2000 The role of climatic mapping in predicting the potential geographical distribution of non-indigenous pests under current and future climates, *Agriculture, Ecosystems and Environment* **82** 57-71

Bergot ME, Cloppet V, Perarnaud M, Deque M, Marcais B, and Desprez-Loustou ML 2004 Simulation of potential range expansion of oak disease caused by *Phytophthora cinnamomi* under climate change, *Global Change Biology* **10** 1539-1552

Brasier CM and Scott JK 1994 European oak declines and global warming: a theoretical assessment with special reference to the activity of *Phytophthora cinnamomi, EPPO Bulletin* **24** 221-232

Butterworth MH, SemenovM, Barnes AP, Moran D, West JS, and Fitt BDL 2010 North-south divide; contrasting impacts of climate change on crop yields in Scotland and England, *Journal of the Royal Society Interface* **7** 123-130

Chakraborty S and Datta S 2003 How will plant pathogens adapt to host plant resistance at elevated CO_2 under a changing climate, *New Phytologist* **159** 733-742

Chakraborty S, Tiedemann AV, and Teng PS 2000 Climate change: Potential impact on plat diseases, *Environmental Pollution* **108** 317-326

Chakrabortya S and Newton AC 2011 Climate change, plant diseases and food security: an overview, *Plant Pathology* **60** 2-14

Coakley SM, Scherm H, and Chakraborty S 1999 Climate change and plant disease management, *Annual Review of Phytopathology* **37** 399-426

Evans N, Baierl A, Semenov MA, Gladders P, and Fitt BDL 2008 Range and severity of a plant disease increased by global warming, *Journal of the Royal Society Interface* **5** 525-531

Eversmeyer MG and Kramer CL 2000 Epidemiology of wheat leaf and stem rust in the central great plains of USA, *Annual Review of Phytopathology* **38** 491-513

Fernandes JM, Cunha GR, and Del Ponte E 2004 Modelling fusarium head blight in wheat under climate change using linked process-based models, in 2nd International Symposium on Fusarium HeadBlight Incorporating the 8th European Fusarium Seminar, Orlando, FL, USA (eds) SM Canty, T Boring, J Wardwell, and RW Ward RW 441-444

Fry W and Goodwin SB 1997 Resurgence of the Irish potato famine fungus, *Bioscience* **47** 363-371

Fuhrer J 2003 Agroecosystem responses to combinations of elevated CO_2, ozone, and global climate change, *Agriculture Ecosystems and Environment* **97** 1-20

Garrett KA, Dendy SP, Frank EE, Rouse MN, and Travers SE 2006 Climate change effects on plant disease: genomes to ecosystems, *Annual Review of Phytopathology* **44** 489-509

Garrett KA, Nita M, De Wolf ED, Gomez L, and Sparks AH 2009 Plant pathogens as indicators of climate change, in *Climate Change: Observed Impacts on Planet Earth,* (ed) T Letcher, Elsevier 425-437 pages

Hibberd JM, Whitbread R, and Farrar JF 1996 Effect of elevated concentrations of CO_2 on infection of barley by *Erysiphe graminis, Physiological and Molecular Plant Pathology* **48** 37-53

IPCC 2007 Climate Change 2007: The Physical Science Basis, in *Contribution of Working Group I to the Fourth Assessment Report of the Intergovernmental Panel on Climate Change*, (eds) S Solomon, D Qin, M Manning, Z Chen, M Marquis, KB Averyt, M Tignor, and HL Miller, Cambridge, United Kingdom and New York, NY, USA: Cambridge University Press

Karnosky DF, Percy KE, and Xiang BX 2002 Interacting elevated CO_2 and tropospheric O_3 predisposes aspen (*Populus tremuloides Michx.*) to infection by rust (*Melampsora medusae f. sp. tremuloidae*), *Global Change Biology* **8** 329-338

Kropff MJ, Haverkort AJ, Aggarwal PK, and Kooman PL 1995 Using systems approaches to design and evaluate ideotypes for specific environments, in *Ecoregional approaches for sustainable land use and food production*, (ed) J Bouma, 417-435 pages

Lake J and Wade 2009 Plant-pathogen interactions and elevated CO_2: morphological changes in favour of pathogens, *Journal of Experimental Botany* **60** 3123-3131

Madden LV, Hughes G, and van den BF 2007 The study of plant disease epidemics, St. Paul, Minnesota: American Phytopathological Society

Malmstrom CM and Field CB 1997 Virus-induced differences in the response of oat plants to elevated carbon dioxide, *Plant, Cell and Environment* **20** 178-188

McElrone AJ, Reid CD, Hoye KA, Hart E, and Jackson RB 2005 Elevated CO_2 reduces disease incidence and severity of a red maple fungal pathogen via changes in host physiology and leaf chemistry, *Global Change Biology* **11** 1828-1836

Miraglia M, Marvin HJP, and Kleter GA 2009 Climate change and food safety: an emerging issue with special focus on Europe, *Food and Chemical Toxicology* **47** 1009-1021

Newton AC and Young IM 1996 Temporary partial break down of Mlo-resistance in spring barley by the sudden relief of soil water stress, *Plant Pathology* **45** 973-977

Oerke EC 2006 Crop losses to pests, *Journal of Agricultural Science* **144** 31-43

Olwoch JM, de Rautenbach CJD, Erasmus BFN, Engelbrecht FA, and van Jaarsveld AS 2003 Simulating tick distributions over sub-Shaaran Africa: the use of observed and simulated climate surfaces, *Journal of Biogeography* **30** 1221-1232

Osswald WF, Fleischmann F, and Heiser I 2006 Investigations on the effect of ozone, elevated CO_2 and nitrogen fertilization on host parasite interactions, *Summa Phytopathologica* **32** 111-113

Pande S and Sharma M 2009 Effect of climate change on the diseases of food legumes in Semi-Arid Tropics: Retrospect and prospect, 11, *In* 5th International Conference on Plant Pathology in the Globalization Era, Indian Phytopathological Society, ICAR, New Delhi, India

Pande S and Sharma M 2010a Climate Change: Potential Impact on Chickpea and Pigeonpea Diseases in the Rainfed Semi-Arid Tropics (SAT), *In* 5th International Food Legumes Research Conference (IFLRC V) & 7th European Conference on Grain Legumes (AEP VII), Antalya, Turkey

Pande S, Desai S, and Sharma M 2010b Impacts of climate change on the rainfed crop diseases: current status and future research needs, *In* National Conference on Climate Change and Rainfed Agriculture, CRIDA, Hyderbad, Andhra Pradesh, India

Pande S, Kishore GK, and Rao JN 2005 Marigold: A Diagnostic tool for BGM forecasting and management in chickpea, *International Chickpea and Pigeonpea Newsletter* **12** 2-3

Pangga IB, Chakraborthy S, and Yates D 2004 Canopy size and induced resistance in *Stylosanthes scabra* determine anthracnose severity at high CO_2, *Phytopathology* **94** 221-227

Parsons WFJ, Kopper BJ, and Lindroth RL 2003 Altered growth and fine root chemistry of *Betula papyrifera* and *Acer saccharum* under elevated CO_2, *Canadian Journal of Forest Research* **33** 842-846

Plazek A, Hura K, Rapacz H, and Zur I 2001 The influence of ozone fumigation on metabolic efficiency and plant resistance to fungal pathogens, *Journal of Applied Botany* **75** 8-13

Plessl M, Elstner EF, Rennenberg H, Habermeyer J, and Heiser I 2007 Influence of elevated CO2 and ozone concentrations on late blight resistance and growth of potato plants, *Environmental and Experimental Botany* **60** 447-457

Riesenfeld CS, Schloss PD, and Handelsman J 2004 Matagenomics: genomic analysis of microbial communities, *Annual Review of Genetics* **38** 525-552

Rotter R and Van de Geijn SC 1999 Climate change effects on plant growth, crop yield and livestock, *Climate Change* **43**(4) 651–681

Salinary F, Giosue S, Tubiello FN, Rettori A, Rossi V, Spanna F, Rosenzweig C, and Gullino ML 2006 Downy mildew (*Plasmopara viticola*) Epidemics on grapevine under climate change, *Global Change Biology* **12** 1299-1307

Sandermann Jr H 2000 Ozone/biotic disease interactions:molecular biomarkers as a new experimental tool, *Environmental Pollution* **108** 327-332

Savary S, Teng PS, Willocquet L, and Nutter FWJr 2006 Quantification and modeling of crop losses: A review of purposes, *Annual Review of Phytopathology* **44** 89-112